RIGID-CHAIN POLYMERS

Hydrodynamic and Optical
Properties in Solution

MACROMOLECULAR COMPOUNDS

Series Editor: **M. M. Koton,** *Institute of Macromolecular Compounds*
Leningrad, USSR

CAPILLARY LIQUID CHROMATOGRAPHY
B. G. Belen'kii, E. S. Gankina, and V. G. Mal'tsev

ION-EXCHANGE SORPTION AND PREPARATIVE
CHROMATOGRAPHY OF BIOLOGICAL ACTIVE MOLECULES
G. V. Samsonov

MECHANISMS OF IONIC POLYMERIZATION: Current Problems
B. L. Erusalimskii

POLYIMIDES: Thermally Stable Polymers
M. I. Bessonov, M. M. Koton, V. V. Kudryavtsev, and L. A. Laius

RIGID-CHAIN POLYMERS: Hydrodynamic and
Optical Properties in Solution
V. N. Tsvetkov

RIGID-CHAIN POLYMERS

Hydrodynamic and Optical
Properties in Solution

V. N. Tsvetkov

Institute of Macromolecular Compounds
Academy of Sciences of the USSR
Leningrad, USSR

Translated from Russian by
E. A. Korolyova

CONSULTANTS BUREAU • NEW YORK AND LONDON

CHEMISTRY

0 335-2225

Library of Congress Cataloging in Publication Data

Tsvetkov, V. N. (Viktor Nikolaevich)
   Rigid-chain polymers.

   (Macromolecular compounds)
   Translation of: Zhestkofsepnye polimernye molekuly.
   Includes bibliographical references and index.
   1. Polymers and polymerization. I. Title. II. Series.
QD381.T7913   1989   547.7   88-34423
ISBN 0-306-11020-2

This translation is published under an agreement with the Copyright
Agency of the USSR (VAAP)

© 1989 Consultants Bureau, New York
A Division of Plenum Publishing Corporation
233 Spring Street, New York, N.Y. 10013

Printed in the United States of America

FOREWORD

A characteristic feature of polymer chain molecules is their flexibility, i.e., their ability to adopt various conformations and change them under the influence of thermal motion or deformational forces. This feature determines the behavior of any polymer material in bulk regardless of whether this material exhibits the properties of an elastic rubberlike body or is a high-modulus thermally stable plastic.

The most direct method of characterizing polymer chain flexibility under equilibrium conditions is through the investigation of molecular conformations in dilute solution. The history of the investigation of conformations and structural characteristics of polymer molecules in solutions is very extensive. It is sufficient to say that the fundamental concepts of the chain structure of high-molecular-weight substances formulated over half a century ago were based to a considerable extent on the behavior of polymer molecules in dilute solutions.

More recently the study of polymers in dilute solutions has been developed on a large scale and has become the principal source of information on molecular characteristics such as molecular weight and the dimensions and conformations of the macromolecules. The main experimental methods traditionally used for this purpose are viscometry, sedimentation–diffusion analysis, light scattering, flow birefringence, and others.

The extensive experimental data accumulated in this field have provided the basis for the development of statistical mechanics of chain molecules, a part of theoretical physics dealing with the calculation of conformational characteristics of polymer chains. The theory of the hydrodynamic properties of chain molecules – their translational and rotational friction, viscometry, and dynamo-optical characteristics – has been developed simultaneously. These theoretical investigations were also based on the experimental data obtained in the study of dilute polymer solutions.

The results of all these investigations have made it possible to develop a general concept (more or less complete) of the behavior of flexible chain molecules of synthetic polymers in dilute solutions. According to this concept, under the conditions of an ideal solvent ("θ-solvent") a flexible polymer chain has the shape of a random coil characterized by an array of conformations, the distribution of which corresponds to Gaussian statistics. Consequently, the average linear dimensions of the coil increase proportionally with the square root of the molecular weight M. In a nonideal solution molecular conformations are perturbed by "excluded volume" effects. As a result the distribution of conformations is no longer Gaussian, and with increasing M the linear dimensions of molecules increase faster than $M^{0.5}$. The main use of conformational statistics and thermodynamics of flexible polymer chains in nonideal solutions is in the theoretical evaluation of excluded volume effects and their influence on the conformational properties of molecules.

The equilibrium hydrodynamic properties of a flexible polymer molecule in solution, according to the generally accepted concept, correspond to the properties of a "nondraining" coil in the motion of which the solvent inside the coil is completely involved. When this simplifying concept is used, the hydrodynamic characteristics of the molecule (translational friction and intrinsic viscosity) readily allow the determination of its size and, if determined under θ conditions, the calculation of the equilibrium rigidity of the chain characterizing it by one of the accepted parameters, such as the length of the Kuhn segment. If the experimental data are not obtained in a θ solvent, then in order to characterize the "nonperturbed" dimensions it is necessary to extrapolate these data to conditions $M \rightarrow 0$ by using one of the semiempirical relationships proposed for this purpose.

The foregoing scheme for the interpretation of experimental data has been and is still being widely used in the investigation of dilute polymer solutions, and for all synthetic flexible-chain molecules investigated leads to reasonable values of their equilibrium rigidity parameters.

However, it has long been known that this scheme cannot be applied to the solutions of many natural compounds, such as viruses, native DNA, polypeptides in the helical conformation, and polysaccharide derivatives. Hence, the molecules of these polymers have traditionally been regarded as absolutely rigid "rodlike" molecules with the corresponding interpretation of their hydrodynamic properties. Attempts to use the nondraining coil model to describe the properties of cellulose ester and ether molecules have led to contradictory results.

In the last two decades the synthesis of new polymers with a wide variety of chemical structures has been greatly developed with the aim of preparation and practical application of materials used for special purposes. The molecules of these polymers often contain phenyl rings and complex aromatic heterocycles, conjugated and quasiconjugated (amide and ester) bonds, and many other specific fragments. Materials exhibiting good thermomechanical properties have been obtained from many of these polymers.

The experimental investigations of these polymers in dilute solutions have shown that although their molecules in solution can adopt coiled conformations, their hydrodynamic properties are not in agreement with the nondraining coil model, and to describe them quantitatively it is necessary to take into account the "draining" caused by a relatively weak intramolecular hydrodynamic interaction. The weakening of hydrodynamic interaction in these molecules is a direct result of the high equilibrium rigidity of their chains whose Kuhn segment length exceeds by one or more orders of magnitude that of typical flexible-chain polymers. The studies of flow birefringence in rigid-chain polymer solutions have suggested that the conformations of their molecules can differ greatly, varying from a rod to a Gaussian coil with increasing molecular weight. Hence, the conformational characteristics of rigid-chain molecules can be described most adequately by the wormlike chain model, which was proposed over thirty years ago but, for a long time, has not received due attention. This consideration also applies to natural compounds (DNA, helical polypeptides, and the derivatives of cellulose and other polysaccharides) which are typical rigid-chain polymers.

Recently the interest in rigid-chain polymers has particularly increased in connection with the possibility of using them for obtaining polymeric liquid crystals.

The high-equilibrium rigidity of molecules of many of these polymers favors the formation of the lyotropic mesomorphic (liquid-crystalline) structure in their concentrated solutions. This phenomenon has long been known for viruses and polypeptide solutions and is being widely used in spinning to obtain ultrahigh modulus fibers from such synthetic polymers as various aromatic polyamides and polyesters and other rigid-chain polymers. The potential of the polymer for lyotropic mesomorphism was found to be directly dependent on the equilibrium rigidity of its molecules characterized by the length of the Kuhn segment A.

Rigid-chain molecules are also of great importance in the synthesis of polymers capable of forming the thermotropic mesophase. These polymers are obtained by combining in the macromolecules being synthesized rigid-chain fragments with flexible chain sequences, the role of the mesogenic cores being played by the rigid-chain component.

Although the number of papers on the investigation of molecular characteristics of rigid-chain polymers in dilute solutions is relatively small (due to experimental difficulties involved in these investigations), the available data show that rigid-chain molecules in solution exhibit many specific properties, clearly distinguishing them from flexible-chain polymers. This refers not only to their hydrodynamic properties, as already mentioned, but even to a greater extent to their behavior in mechanical and electric fields, which is manifested in the phenomena of dynamic and electric birefringence of rigid-chain polymer solutions. In particular, the data obtained by these methods provide information not only on the equilibrium conformational properties of these molecules but also on the kinetics of their behavior in solution: the type of large-scale and local intramolecular motion characterizing the kinetic flexibility of the chain.

These problems form the main content of the present book, which deals with hydrodynamic, dynamo-optical, and electro-optical properties of rigid-chain polymers in dilute solutions. The experimental data obtained by various methods – hydrodynamic, dynamo-optical, and electro-optical – are interpreted from the standpoint of theoretical concepts in which a wormlike chain is used as the molecular model for a rigid-chain polymer. Moreover, in all the phenomena for which it is possible, the behavior of rigid-chain molecules is compared with that of flexible-chain molecules in order to show clearly both their similar features and their differences.

A considerable part of the experimental data is represented in the form of graphs, the number of which may seem excessive. The author has decided to use them instead of restricting himself to presenting tabular data because the comparison of experimental points with theoretical curves enables the reader to judge the degree of reliability of both the proposed relationships and the molecular characteristics based on them. With the same purpose in mind, molecular characteristics (for example, the length of the Kuhn segment) obtained by different methods, used in Chapters 6 and 7, and by various hydrodynamic methods (Chapter 4), such as translational friction and viscometry, are compared.

The experimental study of dilute solutions of many rigid-chain polymers requires the modification and development of present methods and instruments in order to satisfy the specific conditions of operation with aggressive solvents. These problems are considered in Chapter 3, which concerns the methodological aspects of transport processes.

The theories of hydrodynamic properties of macromolecules are considered in Chapter 2. They are based on the calculation of intramolecular hydrodynamic interactions, which are of major importance in the translational and rotational friction of rigid-chain molecules.

The existing concepts of molecular mechanisms of dynamic and electric birefringence in solutions of rigid-chain molecules are described in Chapters 5 and 7. Some of these concepts involve semiempirical relationships, but their validity is checked by experimental data.

Chapter 1 is essentially an introductory chapter which briefly considers the principles of conformational statistics of chain molecules required for discussing the subjects of the following chapters.

Much of the experimental data reported in this book were obtained in common publications with collaborators, as can be seen from the references in the text. Especially noted should be L. N. Andreeva, S. V. Bushin, I. P. Kolomietz, P. N. Lavrenko, N. V. Pogodina, E. I. Rjumtsev, and I. N. Shtennikova. These colleagues worked together with the author for a long period of time, and he thanks all of them.

                                                                        Victor Tsvetkov

CONTENTS

Chapter 1

CONFORMATIONAL CHARACTERISTICS
OF POLYMER CHAINS

Chapter 2

HYDRODYNAMIC PROPERTIES (THEORY)

Chapter 3

TRANSPORT METHODS

Chapter 4

HYDRODYNAMIC PROPERTIES.
EXPERIMENTAL DATA

Chapter 5

FLOW BIREFRINGENCE. THEORY

Chapter 6

FLOW BIREFRINGENCE.
EXPERIMENTAL DATA

Chapter 7

ELECTRIC BIREFRINGENCE

NOTATION

a	persistence length
a	spar twinning
$\Delta a = a_{\parallel} - a_{\perp}$	difference between the main polarizabilities of the monomer unit
a, b	exponents in Mark–Kuhn's equations
b_0	parameter of the shape asymmetry
c	weight concentration of solution
d	diameter of the molecular chain
e	parameter of the shape asymmetry
f	translational friction coefficient of the molecule
$f(p)$	function of the asymmetry of the shape of the particle or the molecule
g	velocity gradient (shear rate)
h	end-to-end distance of the chain
$\langle h^2 \rangle$	mean (overall conformations) square end-to-end distance of the chain
k	Boltzmann's constant
k_s	concentration parameter of the sedimentation coefficient
l	distance between neighboring beads in the chain
$m = M/N_A$	mass of the molecule
n	number of beads in the necklace model

n	mean value of the refractive index of the substance
n_k	refractive index of the polymer
n_s	refractive index of the solvent
$[n]$	intrinsic flow birefringence
Δn	birefringence of the solution
$\Delta n/\Delta\tau$	shear optical coefficient
p	shape asymmetry of the particle
r_{ik}	distance between ith and kth beads in the chain necklace model
S	sedimentation coefficient
S_0	sedimentation constant
t	time
u	velocity of the solvent
v	partial specific volume of the polymer
$x = 2L/A$	reduced (or relative) length of the wormlike chain
Z	excluded volume parameter
A	length of the statistical Kuhn segment
A_f	length of the statistical Kuhn segment with the assumption of free rotation about valence bonds
A_D, A_η	value of A obtained with translational friction and viscometric data, respectively
A_0	hydrodynamic parameter (invariant)
A_∞	value of A_0 at $x \to \infty$
B	optical coefficient
C	coefficient
D	translational diffusion coefficient
D_r	rotatory diffusion coefficient
E	power dissipated for friction per unit volume of the liquid
E	strength (intensity) of the electric field
F	force causing the motion of the particle
$F(p), F_\eta(p), F_r(p)$	functions of the particle shape asymmetry
G	coefficient
$\langle H_1 \rangle, \langle H_2 \rangle$	maximum and minimum dimensions of the molecular coil, respectively
I	ionic strength

K	Kerr constant
K_ν	Kerr constant at a frequency ν
mK	molar Kerr constant
K_∞	limiting value of Kerr constant when the chain reduced length $x \to \infty$
L	contour length of the chain
L	diffusion flow of the solute
L_A, L_B	lengths of the main and grafted chains, respectively, for a comb-shaped polymer
L_s	sedimentation flow of the solute
M	molecular weight of the macromolecule
M_L	average molecular weight per unit length of molecular chain
M_n, M_w, M_z	number-average, weight-average, and z-average molecular weight of the macromolecule
M_0	molecular weight of a monomer unit
M_s	molecular weight of a Kuhn segment
$M_{sD}, M_{D\eta}$	molecular weight of the polymer calculated from the experimental values of s, D and D, $[\eta]$, respectively
$N = L/A$	number of segments in a chain
N_A	Avogadro's number
P	function of hydrodynamic interactions in translational friction
P_∞	value of P at $x = 2L/A \to \infty$
P	degree of polymerization
$Q(\alpha), Q(\beta), Q(\gamma)$	orientation parameters in the electric field
R	universal gas constant
$\langle R^2 \rangle^{1/2}$	mean-square radius of gyration
S	number of monomer units in a statistical Kuhn segment
S_f	S value at free rotation about valence bonds
T	absolute temperature
$\mathbf{T}$	Oseen's hydrodynamic interaction tensor
$T_{ik}, T(r)_p$	hydrodynamic interaction tensors
U_{def}	deformation rate of the elastic necklace
$U = M_w/M_n$	parameter of polydispersity
V	volume of particle, molecular coil

W	rotational friction coefficient
X	hydrodynamic interaction parameter
Z	composition of the copolymer
$\mathscr{E}$	field acting on the molecule
$\mathscr{L}$	Langevin function
$\mathscr{F}(s)$	scattering intensity of x ray
$\alpha_1 - \alpha_2$	optical anisotropy of a segment
$\alpha_h,\ \alpha_R,\ \alpha_\eta$	expansion coefficients of the macromolecule
$\alpha,\ \beta_h$	force constants in Noda–Hearst theory
$\beta,\ \beta_M$	optical anisotropy of a chain per unit of its length and of its molecular weight, respectively
β_0	hydrodynamic parameter in shear flow
γ	coefficient in equations for translational and rotational friction of a rodlike particle
γ	degree of substitution in cellulose chain
γ	concentration parameter of the sedimentation coefficient
$\gamma_1,\ \gamma_2,\ \gamma_3$	main components of the tensor of optical polarizability of the molecule
$\langle \gamma_1 - \gamma_2 \rangle$	optical anisotropy of the molecule averaged on overall conformation
$\gamma_1 - \gamma_2$	optical anisotropy of the molecule
δ	temperature coefficient of viscosity
$\delta_1 - \delta_2$	anisotropy of the main dielectric polarizabilities of the molecule
$\delta_i,\ \Delta_i$	length of the virtual bond
ε	dielectric permittivity of the liquid
$\varepsilon_0,\ \varepsilon_\infty$	limiting values of ε at frequencies of electric field $\nu \to 0$ and $\nu \to \infty$, respectively
$\Delta\varepsilon/\Delta c$	dielectric increment of the solution
ζ	mean effective friction coefficient of a bead
η	viscosity of solution
η_0	viscosity of solvent
$\eta_r = \eta/\eta_0$	relative viscosity
$[\eta]$	intrinsic viscosity
θ	angle formed by the dipole of the molecule μ and the end-to end vector h

θ_i, θ_f, θ_{fs}	parameters of optical anisotropy of molecular coil – intrinsic anisotropy, macroform, and microform, respectively
ϑ	angle formed by the dipole moment of a monomer unit with the chain direction
ϑ	angle between the element of wormlike chain and end-to-end vector
λ	projection of the repeat unit on the direction of the extended chain
μ	dipole moment of the molecule
$\langle \mu^2 \rangle$	square of dipole moment averaged over all chain conformations
μ_0	dipole moment of a monomer unit
$\mu_{0\|}$, $\mu_{0\perp}$	parallel and perpendicular components of the dipole moment of the monomer unit with respect to the chain direction
μ_s	dipole moment of the segment
$\mu_{s\|}$, $\mu_{s\perp}$	parallel (longitudinal) and perpendicular (normal) components of the dipole moment of a segment
ν	frequency of the alternating electric field
$\nu(p)$	function of the asymmetry of the particle (in intrinsic viscosity equation)
ξ	hydrodynamic force per unit length of the cylindrical chain
ρ	polymer density; solution density
ρ_0	solvent density
$\rho(\varphi, \vartheta)$	orientation distribution function
ρ_u	depolarization of scattered light
σ	hindrance parameter in the chain
σ^2	dispersion of the diffusion and sedimentation curves
$\sigma = g/D_r$	molecular orientation parameter in the flow
τ	relaxation time of dipole orientation
τ_0	orientation relaxation time at free relaxation
τ_{or}, τ_{def}	relaxation times of orientation and deformation of the chain molecule, respectively
τ_i	relaxation time of ith motion mode of elastic necklace model
$\Delta\tau = g(\eta - \eta_0)$	shear stress in solution
φ_m	extinction angle of flow birefringence
$[\chi/g]$	intrinsic orientation in the flow

ψ   angle between the directions of two elements of the worm-like chain

ω   angular velocity

ω   cyclic frequency of the sinusoidal electric field

Φ   Flory coefficient (function of hydrodynamic interactions in viscosity phenomena)

Φ_∞ value of φ at $x = 2L/A \to \infty$

Ψ   interpenetration function

Ω   fraction of volume occupied by particles in the solution

Chapter 1

CONFORMATIONAL CHARACTERISTICS
OF POLYMER CHAINS

1. FREELY JOINTED CHAINS

The model for a freely jointed chain introduced and developed by Kuhn [1] and Mark [2] is of great importance for describing the conformational properties of polymer molecules.

In this model a real polymer chain is replaced by an *equivalent* chain consisting of N rectilinear segments of length A the spatial orientations of which are mutually independent. The total length of the equivalent chains L, where

$$L = NA, \tag{1.1}$$

is assumed to be equal to the length of a completely extended real chain without bond angle deformation (i.e., the contour length). The second indispensable condition that a model equivalent chain should satisfy is the coincidence of its beginning and end with those of the real chain at any conformation of the latter. Hence, the length of the vector **h**, the end-to-end distance of the chain, has the same value for a real chain and its model. This value is a very important characteristic of the conformational properties of the polymer chain.

Since the orientations of the segments of a freely jointed chain are mutually independent, statistical methods can be applied to them. The development of the statistics of polymer chains on the basis of the freely jointed chain model has led to the derivation of a very important relationship [2] – a Gaussian distribution of the end-to-end distance h in an assembly of long-chain molecules (each of length L) [1, 2]:

$$W(h)\,dh = \left(4/\pi^{1/2}\right)\left(3/2NA^2\right)^{3/2} e^{-3h^2/2NA^2} h^2 dh, \tag{1.2}$$

where $W(h)\,dh$ is the probability for the end-to-end distance of a chain, arbitrarily chosen from an assembly consisting of N segments of length A, to be in the range from h to $h + dh$. Chains satisfying such a distribution are usually called "Gaussian chains."

The maximum of the function $W(h)$ corresponds to the most probable value of h_m, for which $h_m = (2NA^2/3)^{1/2}$.

Equation (1.2) can be used to obtain any moment of the distribution $W(h)$ determined from the general equation

$$\langle h^k \rangle = \int_0^\infty W\,(h)\,h^k dh \Big/ \int_0^\infty W\,(h)\,dh, \tag{1.3}$$

where the brackets $\langle\ \rangle$ denote averaging over all possible chain conformations in an assembly.

In particular, for $k = 2$ the following value is obtained for the mean square h:

$$\langle h^2 \rangle = NA^2. \tag{1.4}$$

For $k = 4$, Eqs. (1.2) and (1.3) give for the fourth moment

$$\langle h^4 \rangle = 5\,(NA^2)^2/3 = 5\langle h^2 \rangle^2/3. \tag{1.5}$$

For $k = 6$, Eqs. (1.2) and (1.3) lead to the following for the sixth moment:

$$\langle h^6 \rangle = 35\langle h^2 \rangle^3/9. \tag{1.5'}$$

According to Eqs. (1.2) and (1.3), the mean "reciprocal distance" $\langle 1/h \rangle$ for a Gaussian chain is given by

$$\langle 1/h \rangle = (6/\pi)^{1/2}\,(NA^2)^{-1/2} = (6/\pi)^{1/2}\langle h^2 \rangle^{-1/2}. \tag{1.6}$$

The combination of Eqs. (1.1) and (1.4) eliminates the arbitrary choice of the number N of segments into which the freely jointed chain of length L is divided because both the real chain and its model must simultaneously satisfy conditions (1.1) and (1.4). It follows from these equations that the mean-square end-to-end distance of a Gaussian chain is given by

$$\langle h^2 \rangle = LA, \tag{1.7}$$

i.e., it is proportional to its contour length L. This is the most important property of Gaussian chains.

This property is also maintained when considering the distribution of chain segments with respect to the chain's center of mass. Thus, for the second moment of this distribution, which is the mean-square radius of gyration, the following relationship is valid:

$$\langle R^2 \rangle \equiv \langle \Sigma R_i^2 \rangle / n = \langle h^2 \rangle / 6 = (LA)/6. \tag{1.8}$$

In Eq. (1.8) summation is carried out over all n chain elements situated at distances R_i from the center of mass of the chain.

It follows from Eq. (1.7) that for a given contour length L the linear statistical dimensions of a Gaussian chain (characterized by the value $\langle h^2 \rangle^{1/2}$) are proportional to the square root of the length of the Kuhn segment A, and the "degree of coiling" of the Gaussian coil Q, where

$$Q = L/\langle h^2 \rangle^{1/2} = (L/A)^{1/2} \tag{1.9}$$

increases to the same extent with decreasing segment length. Hence, the length of the Kuhn segment A serves as a measure of the equilibrium (or static) rigidity of the polymer chain.

2. CHAINS WITH FIXED BOND ANGLES

The possibility of applying the statistics of freely jointed chains to real polymer molecules is based on the concept that if any real chain is sufficiently long, it acquires the properties of a Gaussian chain described by Eqs. (1.2)–(1.9).

This concept is confirmed by statistical theories of the conformational properties of polymer chains based on the consideration of their real chemical structure in which bond angles, the length of valence bonds, and the interactions hindering the rotation of atomic groups about these bonds are taken into account [3–5]. In these molecular models, which are more realistic than the freely jointed chain, the value of $\langle h^2 \rangle$ still serves as the main conformational parameter.

The initial and simplest of this type model for the polymer chain was first considered by Eyring [6]. It is a chain consisting of an arbitrary number n of linear units of length l jointed and freely rotating with respect to each other by bond angle $\pi - \vartheta$. The second moment calculated by the methods of vector algebra is then given by [6]

$$\langle h^2 \rangle_f = nl^2 [(1 + \alpha)(1 - \alpha)^{-1} - (2/n)\alpha(1 - \alpha^n)(1 - \alpha)^{-2}], \tag{1.10}$$

where $\alpha = \cos \vartheta$.

Similarly, for the radius of gyration of this chain, the following equation was obtained [7, 8]:

$$\langle R^2 \rangle_f = nl^2 [(n + 2)(1 + \alpha)/6(n + 1)(1 - \alpha) - \alpha/(n + 1)(1 - \alpha)^2$$
$$+ 2\alpha^2/(n + 1)^2(1 - \alpha)^3 - 2\alpha^3(1 - \alpha^n)/n(n + 1)^2(1 - \alpha)^4]. \tag{1.11}$$

Equations (1.10) and (1.11) are valid for any number of chain units n. In particular, for $n = 1$, i.e., for a rodlike molecule, they give

$$\langle h^2 \rangle = l^2, \ \langle R^2 \rangle = l^2/12. \tag{1.12}$$

For $n \to \infty$, i.e., when the chain is long, we have

$$\langle h^2 \rangle_f = nl^2 (1 + \cos \vartheta)/(1 - \cos \vartheta), \tag{1.13}$$

$$\langle R^2 \rangle_f = [(nl^2)/6](1 + \cos \vartheta)/(1 - \cos \vartheta). \tag{1.14}$$

Equations (1.13) and (1.14) show that a chain with free rotation at a fixed bond angle becomes a Gaussian chain if it is sufficiently long, because the second moments of distribution in h and R increase proportionally to the number of bonds n or to the contour length, and further, the ratio of $\langle h^2 \rangle$ to $\langle R^2 \rangle$ is 6 [compare with Eqs. (1.7) and (1.8)]. Hence, if Eqs. (1.7) and (1.13) are combined, it is possible to express the length of the Kuhn segment for the chain considered, having a fixed bond angle, in terms of its structural parameters l and ϑ:

$$A_f = (l^2/\lambda)(1 + \cos \vartheta)/(1 - \cos \vartheta), \tag{1.15}$$

where $\lambda = l \cos (\vartheta/2)$ is the projection of l on the direction of a trans-chain extended without bond-angle deformation.

A similar procedure can be used when considering any other model of a real chain of relatively large length. Thus, if each repeating structural unit in the chain consists of v bonds of lengths $\Delta_1, \Delta_2, ..., \Delta_v$ freely rotating about each other by the bond angle $\pi - \vartheta$, then the length of the Kuhn segment of this chain is determined by the equation [9, 10]

$$\lambda A_f = \left[(1 + \cos^v \vartheta) \sum_{i=1}^{v} \Delta_i^2 \right] \Big/ (1 - \cos^v \vartheta)$$
$$+ \left[2 \sum_{i=1}^{v-1} \sum_{k>i}^{v} \Delta_i \Delta_k (\cos^{k-i} \vartheta + \cos^{v-(k-i)} \vartheta) \right] \Big/ (1 - \cos^v \vartheta), \tag{1.16}$$

where λ is the projection of the repeat unit on the direction of the extended chain.

In real polymer chains the rotation of atomic groups about valence bonds is not free; it is hindered by interactions with neighboring groups. As a result, the experimental values of $\langle h^2 \rangle$, $\langle R^2 \rangle$, and A usually exceed the values calculated according to Eqs. (1.13)–(1.16). The value of σ characterizes the degree of hindrance and is determined from the equation

$$\sigma^2 \equiv \langle h^2 \rangle / \langle h^2 \rangle_f = \langle R^2 \rangle / \langle R^2 \rangle_f = A/A_f, \tag{1.17}$$

where $\langle h^2 \rangle$, $\langle R^2 \rangle$, and A are the experimental values, and $\langle h^2 \rangle_f$, $\langle R^2 \rangle_f$, and A_f are the values calculated from the theoretical equations [e.g., Eqs. (1.13)–(1.16)] assuming free rotation.

The theoretical expression of the second moment $\langle h^2 \rangle$ for long chains with rotation angle ϑ and bond length l is well known. It is obtained on making the simplifying assumption that for neighboring bonds the hindering potentials are symmetrical and mutually independent [11–14], i.e.,

$$\langle h^2 \rangle = nl^2 (1 + \cos \vartheta)(1 - \cos \vartheta)^{-1} (1 + \langle \cos \varphi \rangle)(1 - \langle \cos \varphi \rangle)^{-1}, \quad (1.18)$$

where φ is the azimuth of bond rotation measured from the position corresponding to the trans conformation of the chain, and the symmetry of the potential U means that $U(\varphi) = U(-\varphi)$.

In real chain molecules the interaction between side groups often increases the probability of conformations for which $\langle \cos \varphi \rangle > 0$; therefore, according to Eq. (1.18), the hindrance to rotation usually increases $\langle h^2 \rangle$ and, therefore, the equilibrium chain rigidity. In accordance with this, σ is usually greater than unity.

However, if the potential curve $U(\varphi)$ is so symmetrical that it can be expressed by the equation $U(\varphi) = U_0(1 - \cos 3\varphi)$, then $\langle \cos \varphi \rangle$ is equal to zero, Eq. (1.18) is transformed into Eq. (1.13), and σ is equal to unity. In this case the statistical chain size, the degree of its coiling, and hence its equilibrium rigidity, are the same as for unhindered (free) rotation. However, in this case, the internal rotation can be greatly hindered and actually has the character of twisting vibrations within potential holes (at $\varphi = 0, \pm 2\pi/3$) with jumps across the barrier U_0 at $\varphi = \pi$, $\pm \pi/3$.

In other words, transition from one chain conformation to another is accompanied by jumping the potential barrier and requires the time τ to be related to U_0 by the general relationship [15] $\tau \sim e^{U_0/kT}$. The greater is U_0, the longer is the time during which the chain conformation is unchanged and the higher is the kinetic rigidity of the chain. Hence, apart from the equilibrium flexibility of a polymer molecule characterizing the degree of its coiling in the equilibrium state, the concept of kinetic flexibility characterizing the rate at which the conformation and the degree of chain coiling can change is also of great importance.

For molecules for which Eq. (1.18) describes the chain conformation, the equilibrium rigidity is determined by the type of curve asymmetry of $U(\varphi)$, whereas the kinetic rigidity is determined by the height of the potential barrier. In other more general cases the relationship between the equilibrium and the kinetic rigidity is not so evident, and the determination of their correlation can prove to be a very difficult task.

For many real polymer molecules the simplifications on which Eq. (1.18) is based are not justified, and the calculation of hindrance to rotation requires the use of more rigorous theories involving a detailed analysis of energetic interactions in the chain taking into account the interdependence of bond rotations. Modern theories use, for the corresponding calculations, the scheme of rotational isomeric approach in which real rotation in the chain is replaced by an array of discrete rotational isomeric states differing in the energetic levels [3–5]. One theoretical achievement, very important in practice, has been the development of methods that make it possible to calculate the moments $\langle h^k \rangle$ for real chain molecules containing any number of valence bonds [5] without restricting oneself to the asymptotic limit for long (Gaussian) chains [3, 4].

However, it should be noted that these theories are based on the assumption of the absolute rigidity of valence bonds and bond angles in the chain (i.e., deformation-free chains). Hence, the only mechanism of chain flexibility considered in these theories is the rotation of atomic groups with bond angles and valence bonds maintained invariable. Although for many flexible-chain polymers for which the backbone is of the single-strain structure this assumption is quite justified, it cannot be valid for all rigid-chain polymers. Thus, for ladder polymers [16–18], polymers with conjugations in the chain [20–24], and polymers with a rigid secondary structure supported by hydrogen bonds (polypeptides [25, 26, 130], DNA [27–29], cellulose derivatives [30, 31], etc.) the mechanism of chain flexibility related to the deformation of bond angles and, sometimes, valence bonds (in particular, hydrogen bonds) should also exist. Without this mechanism it would be difficult to understand the fact that at relatively high molecular weights these molecules have a random-coil conformation and can even be described by the Kuhn model of an equivalent, freely jointed Gaussian chain. Moreover, for many rigid-chain polymers the length of the Kuhn segment A exceeds by more than one order of magnitude the corresponding value for flexible-chain molecules, for which the typical length A is $(15–30) \cdot 10^{-8}$ cm [32]. As a result, rigid-chain molecules, not only of small but even of moderately large length, which is not uncommon in practice, contain only a small number of Kuhn segments (or even sometimes fractions of a segment) and exhibit the conformational properties of a "weakly bending rod." Hence, they cannot be described in the framework of the statistics of freely jointed chains represented by Eqs. (1.2)–(1.9) or (1.13)–(1.18).

3. PERSISTENT (WORMLIKE) CHAINS

The model of "persistent," or wormlike, chains described by Kratky–Porod [33, 34] proved to be the most suitable for describing the conformational properties of rigid-chain molecules. In contrast to the Kuhn freely jointed chain, this model takes into account the orientational short-range interaction of chain elements. The Porod model (just as the Kuhn model) is based on a chain of length L consisting of n rectilinear elements of length ΔL so that $L = n \Delta L$. However, in this case, in contrast to the freely jointed chain, the spatial orientations of neighboring elements are not quite mutually independent: the direction of the first element is propagated ("remembered") to a certain extent along the chain. The correlation between the elements is expressed by the fact that the average value (for all conformations) of $\langle \cos \Delta\psi \rangle \equiv k$, where $\Delta\psi$ is the angle between the neighboring elements, is not equal to zero and is identical for all chain elements (evidently, for a freely jointed chain $k = 0$). Hence, the combination of the values of ΔL and k is a measure of orientational short-range interaction, i.e., of correlation or "persistence" in the chain. Mean cosines of the angles formed between the first chain element and the subsequent elements evidently correspond to the terms of the series $k, k^2, k^3, \ldots, k^n$, and, therefore, the angle ψ between the first and nth elements is determined by the equation

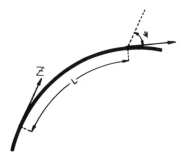

Fig. 1.1. A persistent chain.

$$\langle \cos \psi \rangle = \langle \cos \Delta\psi \rangle^{n-1} = \exp\left[(n-1)\ln\langle \cos \Delta\psi \rangle\right]$$
$$= \exp(-L/a)\exp(-\ln\langle \cos \Delta\psi \rangle), \qquad (1.19)$$

where

$$a = -\Delta L/\ln\langle \cos \Delta\psi \rangle. \qquad (1.20)$$

Here a is called the "persistence length." Equation (1.20) can also be written as

$$\langle \cos \Delta\psi \rangle = e^{-\Delta L/a}. \qquad (1.20')$$

If the values of L and a are left constant and one takes the limit $\Delta L \to 0$ [correspondingly, according to Eq. (1.20), $\langle \cos \Delta\psi \rangle \to 1$], the broken persistent chain is transformed into a continuous "wormlike" chain which from Eq. (1.19) is described by the equation

$$\langle \cos \psi \rangle = e^{-L/a}. \qquad (1.21)$$

Hence, the curvature of the wormlike chain is the same at all points and is determined by the value of a calculated according to Eq. (1.20), whereas the directions of bends at these points are chaotic. In other words, a wormlike chain can be characterized as a line in space having a constant curvature.

If the first chain element is directed along the Z axis, the origin of which coincides with the beginning of the chain (Fig. 1.1), evidently the averaged chain projection on the Z axis is given by

$$\langle Z \rangle = \int_0^L \langle \cos \psi \rangle \, dL = \int_0^L \exp(-L/a) \, dL = a(1 - e^{-L/a}). \qquad (1.22)$$

With increasing chain length ($L \to \infty$) the length of the projection tends to a limit

$$\langle Z \rangle_\infty = a. \qquad (1.23)$$

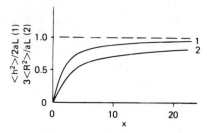

Fig. 1.2. Dependence of $\langle h^2 \rangle / 2aL$ and $3\langle R^2 \rangle aL$ on x for a wormlike chain (curves 1 and 2, respectively).

Besides Eq. (1.20), Eq. (1.23) also gives the persistence length as the length of projection of an infinitely long wormlike chain on the axis of its first element.

A chain with free rotation and constant bond angle $\pi - \vartheta$ [Eqs. (1.10) and (1.11)] is also a persistent chain since a constant value of $\cos \vartheta = \alpha$ corresponds to each of its elements l. For the limiting transition to the continuous wormlike chain it is also necessary to replace α in Eq. (1.10) according to Eq. (1.20'), i.e., $\alpha = \cos \vartheta = \exp(-l/a) \approx 1 - L/na$, after which one should neglect the term L/na and those of lower orders of magnitude. Hence, for the mean-square end-to-end distance of the wormlike chain, Eq. (1.10) yields the Porod equation [33]

$$\langle h^2 \rangle / 2aL = 1 - (1 - e^{-x})/x, \tag{1.24}$$

where $x = L/a$ is the reduced chain length. Similarly, the combined application of Eqs. (1.20) and (1.11) leads to the following expression for the radius of gyration $\langle R^2 \rangle$ of the wormlike chain [8]:

$$3\langle R^2 \rangle / aL = 1 - 3[1 - 2x^{-1} + 2x^{-2}(1 - e^{-x})]/x. \tag{1.25}$$

As $x \to 0$ (i.e., a short chain), the expansion

$$e^{-x} = 1 - x + x^2/2! - x^3/3! + \dots$$

can be used, and Eqs. (1.24) and (1.25) then give the values of $\langle h^2 \rangle = L^2$ and $\langle R^2 \rangle = (1/12)L^2$, which correspond to the conformation of a straight rod.

In the other limiting case as $x \to \infty$ (i.e., a long chain), it follows from Eqs. (1.24) and (1.25) that

$$\langle h^2 \rangle_\infty = 2aL, \quad \langle R^2 \rangle_\infty = aL/3. \tag{1.26}$$

Equations (1.26) are equivalent to Eqs. (1.7) and (1.8) under the condition that

$$a = A/2. \tag{1.27}$$

This means that at a large length L a wormlike chain becomes Gaussian with the length of the Kuhn segment equal to twice the persistence length.

Hence, when L increases from 0 to ∞, the conformation of the wormlike chain changes from a rod to a Gaussian coil. This change is described by Eqs. (1.24) and (1.25) and is illustrated in Fig. 1.2.

For a wormlike chain not only the second moment $\langle h^2 \rangle$ of distribution $W/(h)$ may be calculated, but also the fourth [35] $\langle h^4 \rangle$ and the sixth [36] $\langle h^6 \rangle$ moments, i.e.,

$$\langle h^4 \rangle /(LA)^2 = 5/3 - 52/9x - 2(1 - e^{-3x})/27x^2 + 8(1 - e^{-x})/x^2 - 2e^{-x}/x, \quad (1.28)$$

$$\langle h^6 \rangle /(LA)^3 = 35/9 - 259/9x + 904/9x^2 - 12286/81x^3$$
$$+ 18972e^{-x}/125x^3 + 1278e^{-x}/25x^2 + 21e^{-x}/5x$$
$$- 8e^{-3x}/81x^3 - 2e^{-3x}/27x^2 + 2e^{-6x}/1125x^3. \quad (1.29)$$

The limiting forms of expressions (1.28) and (1.29) give, as $x \to 0$, $\langle h^4 \rangle = L^4$ and $\langle h^6 \rangle = L^6$, i.e., a rodlike conformation, and as $x \to \infty$, $\langle h^4 \rangle = 5\langle h^2 \rangle^2/3$ and $\langle h^6 \rangle = 35\langle h^2 \rangle^3/9$, i.e., a Gaussian coil conformation in agreement with Eqs. (1.5) and (1.5').

It has been shown [35] that, in principle, it is possible to calculate any even moment of the distribution function $W(h)$. However, in contrast to the case for Gaussian chains, for wormlike chains the exact expression of the function $W(h)$ itself is unknown. Only approximate equations for $W(h)$ have been obtained that are valid at relatively high values of x. They are of a Gaussian distribution type containing a series of correcting terms in powers of x, h/L, and h/A [37–39], i.e.,

$$W(h)\,dh = (4/\pi^{1/2})(3/2LA)^{3/2} e^{-3h^2/2LA} h^2 dh [1 - 5/4x - 79/160x^2$$
$$+ 2(h/L)^2(1 - 329/240x) - (33/40)(h/L)^3(h/A) + 6799(h/L)^4/1600$$
$$- 3441(h/L)^5(h/A)/1400 + 1089(h/L)^6(h/A)^2/3200 + \ldots]. \quad (1.30)$$

The function $W(h)$ was found to be useful in the present theories of thermodynamic and hydrodynamic properties of solutions of chain molecules represented by a wormlike chain.

The dependence of the second moment on chain length, shown in Fig. 1.2, and a similar dependence for the higher moments, described by Eqs. (1.28) and (1.29), is a characteristic property of chain molecules if they are investigated over a wide range of x values.

Thus, for any real chain molecule, either for a rigid-chain (high A) or a flexible-chain (low A), according to Eqs. (1.24) and (1.26), respectively, and corresponding limits of x, the ratio $\langle h^2 \rangle/\langle h^2 \rangle_\infty$ increases with chain length and tends to unity in the Gaussian range.

As pointed out above, modern statistical mechanical methods developed by Flory and co-workers [5] for chain molecules allow the calculation of the dependence of $\langle h^2 \rangle$ on chain length for real chain molecules containing any number n of valence bonds from the known bond length l, bond angles $\pi - \vartheta$, and hindering

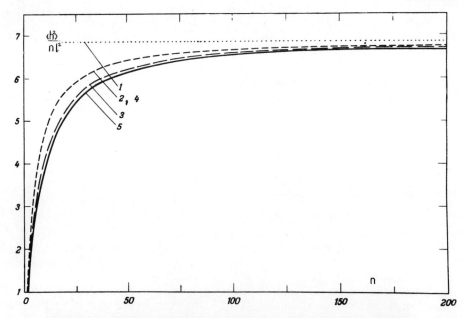

Fig. 1.3. Dependence of $\langle h^2 \rangle / nl^2$ on the number of bonds n for polymethylene obtained under different approximations [5]: 1) a freely jointed chain; 2) a chain with free rotation; 3) a chain with hindered rotation at independent potentials; 4) a wormlike chain; 5) a chain with hindered rotation at correlated potentials.

potentials in the chain. Thus, for a polymethylene chain the corresponding result is given by

$$\langle h^2 \rangle / nl^2 = (\langle h^2 \rangle / nl^2)_\infty \, f(n), \tag{1.31}$$

where the limiting value of $(\langle h^2 \rangle / nl^2)_\infty$, in the Gaussian range, is 6.87 and the theoretically calculated function $f(n)$ determines the shape of the curves, as shown in Fig. 1.3.

A similar dependence for a wormlike model can also be obtained by applying Eqs. (1.13), (1.15)–(1.17), and (1.29) in the form of Eq. (1.31), where

$$f(n) = 1 - (1 - e^{-x}) / x, \quad x = 2n \cos^2(\vartheta/2) / (\langle h^2 \rangle / nl^2)_\infty.$$

The corresponding curve is also shown in Fig. 1.3.

With the exception of curve 1 for a freely jointed chain, the curves in Fig. 1.3 are qualitatively similar and quantitatively close to each other. This implies that the wormlike chain model corresponds relatively closely to the conformational properties of real chain molecules both in the Gaussian range and in the range of molecular weights M (chain length L) in which the chain shape varies from a straight rod to a Gaussian coil.

It may be said that in this range the chain molecule behaves as a "semi-rigid" chain. This range exists for any real molecule. However, it is evident that the greater the equilibrium rigidity (A) of the polymer chain, the wider is this range. Experiments show that for typical flexible-chain polymers, with A on the order of $(15-30) \cdot 10^{-8}$ cm, considerable deviations from the Gaussian properties [according to Eq. (1.24) and Figs. 1.2 and 1.3] appear only in the oligomeric range, i.e., for the values of M less than $(5-10) \cdot 10^{-3}$. Moreover, the experimental determination of these deviations by direct measurement of $\langle h^2 \rangle$ is not possible, and more specific methods are required (e.g., flow birefringence [40–44]). In contrast, for chain molecules, whose rigidity corresponds to a value of A less than $100 \cdot 10^{-8}$ cm, deviations from the properties of a Gaussian coil can be distinctly observed, even at $M \leq 10^5$ g/mole, by using the direct measurement procedure for $\langle h^2 \rangle$ [31, 45]. This makes it possible to divide polymers into flexible-chain and rigid-chain polymer classes and to include in the latter those polymers which, for $M > 10^4$ g/mole, distinctly behave in solution as "semi-rigid" chains. The introduction of this criterion for equilibrium rigidity is justified experimentally since rigid-chain polymers classified in this manner reveal a number of characteristic properties markedly distinguishing them from flexible-chain polymers.

4. MOLECULAR STRUCTURE AND RIGIDITY OF A POLYMER CHAIN

The available experimental data reveal some peculiar features of the chemical structure of a polymer molecule contributing to the increasing rigidity of its chain.

The main mechanism determining the rigidity of typical flexible-chain polymers is the interaction between side groups which is taken into account in the conformational statistics of polymer chains [3–5]. However, this mechanism does not lead to the very high values of A which are characteristic of rigid-chain polymers. Systematic data on this problem were obtained in investigations of the conformational properties of "comb-shaped" polyalkyl methacrylates with different side-chain lengths [46]. When the alkyl side chain becomes longer, the rigidity of the main chain increases slightly, and for the highest homologs it is 2–3 times greater than that of polymethyl methacrylate. However, the interaction between side chains leads to a greater increase in their equilibrium rigidity and orientational order than for the main chain. This effect is much more noticeable in those molecules whose side groups include not only a chain structure but also a mesogenic group [47].

Further increase in the length of side chains up to several hundreds of carbon atoms in comb-shaped molecules can increase the rigidity of the main chain until the molecules begin to exhibit properties characteristic of rigid-chain polymers. This situation is observed for the molecules formed by the graft copolymerization of methyl methacrylate and styrene [47, 48]. However, a more effective method for increasing the rigidity of a chain is its cyclization and the corresponding decrease in the possibility of rotation about valence bonds. Cellulose derivatives whose chains are composed of glucoside rings [31, 46, 49, 50] are a classical example of such rigid-chain polymers. Moreover, hydrogen bonds can also stabilize a rigid structure, e.g., in polysaccharide molecules, because they lead to the formation of additional intramolecular rings [31, 46].

A still greater rigidity is exhibited by polypeptide and polynucleotide chains in a helical conformation completely "cyclized" by intramolecular hydrogen bonds [25–28, 48, 130]. The introduction of rings into the main chain of synthetic polymers can also increase their rigidity (i.e., polymaleinimides [51] and polyacenaphthylenes [52]) but only if the rings are not separated by many bonds which would allow rotation, such as with polycarbonate [40, 41, 53].

A drastic increase in the rigidity of a chain without hydrogen bonds requires virtually complete cyclization of the chemical bonds. As a result, the molecule acquires a "ladder" structure. Highly soluble ladder polymers were synthesized on the basis of double-strain siloxane chains (Fig. 1.4) with aliphatic and aromatic side groups [54, 55]. Investigations of the conformational properties of these polymers [16–18] have shown that their equilibrium rigidity (characterized by a length of the Kuhn segment A of several thousands of nanometers) can vary depending on the synthesis conditions. This implies that defects in the structure of the chains can play some part in their flexibility. However, the main flexibility mechanism is the deformation of bond angles and valence bonds in the double-strain "network" during its thermal vibrations. Other types of ladder polymers with a high equilibrium rigidity were obtained on the basis of chains containing naphthoylene–benzimidazole fragments [19] and polyquinoline chains [56].

Conjugation in the chain can lead to even stiffer polymer structures. High resonance energy in the amide group [57] leads to the quasiconjugation and coplanarity of its structure. Hence, the introduction of this group into the polymer chain decreases its flexibility [58, 59]. However, in common polyamides the amide groups are separated from one another along the chain by methylene or aromatic groups, ensuring some freedom of rotation, thus decreasing the chain rigidity to that of common flexible-chain polymers [5, 60]. The situation changes drastically when the amide groups, as in an aliphatic polyamide, are very close together in the chain. This is the case for polyalkyl isocyanates (nylon 1), whose molecules are composed entirely of amide groups in which the hydrogen at the nitrogen atom is replaced by an alkyl radical R [61]. Since each amide group has a coplanar (cis or trans) configuration, this structure should lead to the coplanarity of all the bonds of the main chain of polyalkyl isocyanates. Steric interactions between the radical R and the neighboring carbonyl oxygen can cause a departure from complete coplanarity, but the chain structure remains regular with a high degree of order: the amide groups are incorporated into the chain in a rigorous alternation of cis and trans configurations [21, 22, 62]. The replacement of a side aliphatic radical by an aromatic radical [poly(tolyl isocyanate)] destroys conjugation in the chain and leads to the appearance of a flexibility similar to that of common polymers [63].

Recently, the synthesis of polymer molecules containing both complex aromatic heterocycles and the amide groups in the chains has been widely developed. Many of these molecules exhibit a high equilibrium rigidity [23, 64] and can form a nematic mesophase in concentrated solutions [65]. These properties make it possible to use these polymers to obtain thermally stable materials with a high mechanical strength and high modulus [47]. The equilibrium rigidity parameter A

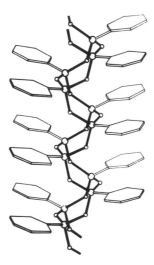

Fig. 1.4. Structure of a molecule of ladder polyphenylsiloxane.

of aromatic polymers can vary over a wide range, depending on their specific chain structure. This point will be considered in greater detail below.

In chain molecules of polyelectrolytes the repulsion of ionized groups in the chain leads to its uncoiling and an increase in the statistic size of the coil. This effect can be described as the increase in the equilibrium rigidity of the ionogenic chain with a corresponding increase in the persistence length a or the Kuhn segment length A [66–68]. This viewpoint corresponds to experimental data showing that when a flexible-chain polyion undergoes polyelectrolytic uncoiling, the values of a and A can increase by more than one order of magnitude [69–71]. Hence, polyelectrolytic "expansion" of a chain molecule is actually a conformational transition from a flexible chain structure to that of a "semi-rigid" chain.

The foregoing examples show that the structural features of molecules contributing to the appearance of properties characteristic of "semi-rigid" chains can be very diverse, and the equilibrium rigidity parameter A can vary from tens to thousands of angstroms, depending on the structure of the rigid-chain polymer. This property clearly distinguishes rigid-chain molecules from flexible-chain molecules. Although there are more flexible-chain polymers than rigid-chain polymers, for the synthetic polymers known at present, their segment lengths A vary over a relatively narrow range from 150 to 300 nm. Moreover, it should be noted that, in principle, present-day chemical synthesis allows polymer molecules to be prepared with a desired rigidity using rigid-chain polymers and a predetermined combination of flexible and rigid-chain fragments. Thus, a directed synthesis of materials with desired properties can be carried out.

However, regardless of structural details, the conformational properties of all these molecules can be described in the framework of a model in which short-range interactions in the chain are taken into account, i.e., the wormlike chain model. The

advantage of this model is its universal character because by definition it is possible
to determine chain flexibility without any *a priori* assumptions on the mechanism of
this flexibility. In contrast, according to experimental data on the molecular size
($\langle h^2 \rangle$ or $\langle R^2 \rangle$) and the degree of polymerization P, this theory makes it possible not
only to evaluate the rigidity of the chain but also to determine its conformation
since, in this case, the projection λ of the monomer unit on the chain axis is also
determined. For this purpose it is convenient to represent Eq. (1.24) in the
following form:

$$\langle h^2 \rangle / P = y \left[1 - z \left(1 - e^{-2P/z} \right) / 2P \right], \qquad (1.32)$$

where $y = A\gamma$ and $z = A/\lambda$. If the experimental values of $\langle h^2 \rangle / P$ are plotted as a
function of P, the initial slope of this curve is equal to λ^2 and the asymptotic limit is
equal to $A\lambda$. Similarly, Eq. (1.25) is equivalent to the equation

$$6\langle R^2 \rangle / P = y \left\{ 1 - 3z \left[1 - z/P + z^2 \left(1 - e^{-2P/z} \right) / 2P^2 \right] / 2P \right\}. \qquad (1.32')$$

The limit of the curve $6\langle R^2 \rangle / P = f(P)$ is still equal to $A\lambda$ and the initial slope is
$\lambda^2/2$. Hence, the dependences represented in Eqs. (1.32) or (1.32') allow the
determination of both λ and A from the experimental values of P and $\langle h^2 \rangle$, or P and
$\langle R^2 \rangle$, the initial slope and the limit of the curve.

5. INFLUENCE OF EXCLUDED–VOLUME EFFECTS

The conformational properties of chain molecules were considered above assuming that the interactions between the elements of the chain determining its
equilibrium rigidity are of the short-range type, i.e., they occur between the
elements situated close to each other in the chain sequence. This assumption is
based on the concept of the existence of "persistence" in the chain.

However, since any real chain molecule exhibits a certain degree of
flexibility, it is always possible that during its thermal motion the atoms and groups
situated at a considerable distance apart along the chain fortuitously approach one
another. This inevitably leads to an interaction between the approaching chain
elements, which is of the mutual repulsion type, and the greater the effective
volume occupied by the interacting pair of elements (i.e., the "excluded volume"),
the greater is this repulsion. These long-range interactions are usually called
excluded-volume effects since they are based on the impossibility of two chain
elements simultaneously occupying the same volume element in space. Excluded-
volume effects perturb the conformation of the random-coil molecule and lead (as a
result of the generated interactions) to an increase in the average spatial distances
between its elements, including an increase in $\langle h^2 \rangle$ and $\langle R^2 \rangle$. Quantitatively these
perturbations are characterized by the expansion coefficients α_h and α_R, i.e., by the
coefficients of a linear increase in the coil size determined by the equations

$$\langle h^2 \rangle = \alpha_h^2 \langle h^2 \rangle_0 \qquad (1.33)$$

Fig. 1.5. Density of a molecular coil of rigid (2) and flexible-chain (1) polymers at equal values of chain length L and diameter d.

and

$$\langle R^2 \rangle = \alpha_R^2 \langle R^2 \rangle_\theta. \tag{1.34}$$

where $\langle h^2 \rangle_\theta$ and $\langle R^2 \rangle_\theta$ are the mean-square end-to-end distance of the chain and its radius of gyration, respectively, in the absence of volume effects (i.e., unperturbed values) and $\langle h^2 \rangle$ and $\langle R^2 \rangle$ are the same values perturbed by volume effects.

Conformations of real polymer molecules are studied in dilute solutions in which volume effects greatly depend on the interaction between the molecules of the polymer and the solvent and can differ markedly in different solvents. It is possible to compensate the effect of the finite volume of the monomer unit by mutual attraction of chain units if a relatively "poor" solvent and the corresponding temperature ("θ-temperature") are selected. Under these conditions excluded-volume effects are absent, and in Eqs. (1.33) and (1.34) we have the coefficients $\alpha_h = \alpha_R = 1$ and, correspondingly, $\langle h^2 \rangle = \langle h^2 \rangle_\theta$ and $\langle R^2 \rangle = \langle R^2 \rangle_\theta$. In this case the osmotic pressure of the solution obeys the Van't Hoff law and the second virial coefficient in the osmotic equation $A_2 = 0$ ("θ-conditions"). When the thermodynamic strength of the solvent increases (with increasing A_2) and the polymer–solvent interactions increase correspondingly, attraction between chain elements cannot compensate their repulsion, the excluded-volume effect increases, and α becomes greater than unity.

The theories of volume effects in solutions of flexible-chain polymers based on Flory's ideas [72] have been widely developed in many papers and considered in detail in a number of monographs and collections [32, 73, 74]. Therefore, we are not going to dwell on them here. It needs only to be pointed out that these theories make it possible to represent the expansion parameter α and the second virial coefficient as functions of the excluded-volume parameter z. The parameter z characterizes the probability of single, double, and higher contacts between chain elements favoring their expansion. For each polymer–solvent system z depends on the temperature, and at $T = \theta$ (i.e., "θ-conditions") it is zero. With increasing chain length L, the parameter z increases proportionally to $L^{1/2}$.

For weak volume effects, if only a linear term is used in the dependence $\alpha(z)$ for a dilute solution of a flexible-chain polymer, the theory gives

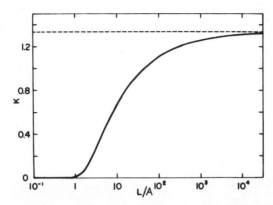

Fig. 1.6. Coefficient of z in the expansion of α_h^2 in powers of z vs. reduced chain length L/A {Eq. (1.37) [39]}.

$$\alpha_h^2 = 1 + 4z/3 + \dots \qquad (1.35)$$

Equation (1.35) together with Eqs. (1.33) and (1.34) means that, in a homologous series of chain molecules under "non-θ-conditions" (since $z \approx L^{1/2}$), the dependence of $\langle h^2 \rangle$ or $\langle R^2 \rangle$ on L, as in the form of Eqs. (1.7) and (1.8) which are characteristic of Gaussian coils (under θ-conditions), is not valid. The perturbed dimensions $\langle h^2 \rangle$ and $\langle R^2 \rangle$ of the molecules increase more rapidly than the direct proportionality to chain length L suggests and, to a rough approximation, the following empirical equation may be used [131] to express this dependence instead of Eq. (1.7):

$$\langle h^2 \rangle = A^{1-\varepsilon} L^{1+\varepsilon}, \qquad (1.36)$$

where $\varepsilon > 0$.

Both the theoretical considerations and the extensive experimental data available show that for the characterization of the conformational properties of flexible-chain polymer molecules in dilute solutions it is of great importance to take into account volume effects because the dimensions of these molecules in good solvents can be significantly greater than their unperturbed dimensions. Hence, quantitative determination of the parameters of the equilibrium chain rigidity is inevitably related to the exclusion of the influence of volume effects using θ-solvents or applying the extrapolation of experimental data to the low-molecular-weight range [32].

The validity of this assumption has been confirmed experimentally [16] and repeatedly stressed in a number of papers [46, 48]. The theory of excluded-volume effects for a rigid-chain polymer represented by a wormlike necklace consisting of spherical beads has been developed by Yamakawa and Stockmayer [39]. For this they used a distribution function $W(h)$ represented by Eq. (1.30) and the function of a bivariant distribution taking into account not only the distances h between the points of a wormlike chain, but also the directions of the tangent to the chain at these points [38]. The physical principles of this theory [39] are similar to those for flexible-chain polymers [72, 73]. However, apart from using the distribution equation (1.30) instead of Eq. (1.2), this theory takes into account the fact that in the unperturbed state the second moment $\langle h^2 \rangle$ for a chain molecule is determined by Eq. (1.24) (for a wormlike chain) rather than by Eq. (1.7) (for a Gaussian chain). This consideration leads to the following expression for the expansion parameter:

$$\alpha_h^2 = 1 + K\,(L/A)\,z + \dots . \tag{1.37}$$

This equation differs from Eq. (1.35) in that the coefficient K is a function of the reduced chain length $L/A = x/2$. Function $K(L/A)$ is tabulated in [39] and plotted in Fig. 1.6. With increasing chain length the coefficient K tends to a limit, 4/3, corresponding to a Gaussian coil. For a chain containing 10 Kuhn segments, K is close to half its limiting value, and for a chain with a length of one segment it becomes equal to zero. Hence, for molecules two to three segments in length, K is very small and the expansion parameter α is virtually unity.

However, even at relatively large values of L/A for which K is close to the asymptotic limit 4/3, the expansion parameter can also be close to unity since in this case the excluded-volume parameter may be small. This can be seen from an expression for z which has the form [39]

$$z = (3/2\pi)^{3/2}\,(\beta/l^2)\,(L^{1/2}/A^{3/2}). \tag{1.38}$$

where l is the distance between the neighboring beads of a wormlike necklace, and β characterizes the "effective excluded volume" of a pair of interacting chain units (beads). To a first approximation β is proportional to the cube of its diameter d. Hence, for a chain model with beads touching each other ($d = l$), the parameter z is approximately given by $z \approx (L/A)^{1/2}(d/A)$, and for relatively rigid thin chains (i.e., $d/A \ll 1$) it can be small even when the reduced chain length $L/A \geq 10$. Hence, a qualitative consideration of Eqs. (1.37) and (1.38) shows that the influence of excluded-volume effects on the chain conformation of typical rigid-chain polymers should be weak.

It is possible to evaluate this influence quantitatively using the theory of the second virial coefficient A_2 for a wormlike model. Within the approximation, taking into account double-contact long-range interactions, the following expression for a wormlike necklace has been obtained [39]:

$$A_2 = 4\pi^{3/2} N_A\,(\langle R^2 \rangle^{3/2}/M^2)\,\Psi, \tag{1.39}$$

where $\langle R^2 \rangle$ is the radius of gyration of a polymer molecule of molecular weight M whose conformation is perturbed by volume effects in dilute solution, N_A is Avogadro's number, and Ψ is the "interpenetration function."

Formally, Eq. (1.39) coincides with the corresponding dependence for flexible-chain polymers [73]. However, the function Ψ in Eq. (1.39) is of a more complex form in accordance with the properties of a wormlike chain:

$$\Psi = (LA/6 \langle R^2 \rangle_\theta)^{3/2} (2Q)^{-1} \ln (1 + 2Qz/\alpha_R^3), \qquad (1.40)$$

where $\langle R^2 \rangle_\theta$ is the radius of gyration of a chain unperturbed by volume effects [which is determined by Eq. (1.25)], and Q is a unique function of the reduced values of chain length L/A and diameter d/A which is expressed by the equation

$$Q(L/A; \ d/A) = Q(L/A; \ 0.1) + [0.4824A/d - 4.824$$
$$+ 2.497 \ln (10d/A)] (A/L)^{1/2} + [0.4824 \ln (10d/A)] (A/L)^{3/2}, \qquad (1.41)$$

where $Q(L/A; 0.1)$ is the function of $Q(L/A)$ at $d/A = 0.1$. This function is tabulated in Table II in [39].

Hence, the interpenetration function Ψ is related to the geometrical parameters L/A and d/A of an unperturbed wormlike chain and the parameters z and α_R determining the perturbation of its configuration by long-range interactions. For a chain of infinite length ($L/A \to \infty$), then, according to Eq. (1.25), $LA/6\langle P \rangle_\theta \to 1$ and, according to Eq. (1.41) and Table II in [39], $Q \to 2.865$. Hence

$$\Psi_{L \to \infty} = (5.73)^{-1} \ln (1 + 5.73z/\alpha_R^3), \qquad (1.42)$$

which corresponds to the well-known relationship for flexible-chain molecules [73]. By analogy with the expressions for flexible-chain polymers [73],

$$\alpha_h^5 - \alpha_h^3 = 4z/3, \ \alpha_R^5 - \alpha_R^3 = (134/105) z \qquad (1.43)$$

for rigid-chain molecules. Since the values of α_h and α_R are close to unity and applying Eq. (1.37), one can write

$$\alpha_h^2 - 1 = Kz/\alpha_h^3, \ \alpha_R^2 - 1 = (134/105)(3/4) Kz/\alpha_R^3 = (67K/70) z/\alpha_R^3. \qquad (1.44)$$

The exclusion of the parameter z from Eqs. (1.40) and (1.44) transforms Eq. (1.39) into the form

$$A_2 = 4\pi^{3/2}N_A (\langle R^2 \rangle^{3/2}/M^2)(LA/6 \langle R^2 \rangle_\theta)^{3/2} \ln [1 + 140Q (\alpha_R^2 - 1)/67K]/2Q. \qquad (1.45)$$

For $L/A \to \infty$, Eq. (1.45) transforms into Eq. (1.45'), which is well known for flexible-chain molecules [73]:

$$A_2 = 4\pi^{3/2}N_A (\langle R^2 \rangle^{3/2}/M^2)(5.73)^{-1} \ln [1 + 4.49 (\alpha_R^2 - 1)]. \qquad (1.45')$$

The expansion coefficient α_R in Eq. (1.45) is related to the experimental values of M, $\langle R^2 \rangle$, and A_2 together with the values Q, K, and $\langle R^2 \rangle_\theta/LA$ which are themselves unique functions of the reduced values of chain length L/A and diameter d/A. It follows from Eq. (1.45) that for a wormlike chain, as for any chain molecule, under the θ-conditions when $A_2 = 0$, the expansion parameter α_R is equal to unity.

In a thermodynamically good solvent for both flexible- and rigid-chain polymers A_2 has a relatively high value usual for these conditions. In accordance with this, the right-hand side of Eq. (1.45) is also sufficiently large. However, for a flexible-chain polymer, this corresponds to a much higher value of α_R than unity since for a flexible chain the factor $\langle R^2 \rangle^{3/2}/M^2$ is relatively low. In contrast, for a rigid-chain polymer the factor $\langle R^2 \rangle^{3/2}/M^2$ is high, and for a given A_2 the parameter α_R is close to unity. Equation (1.45) allows a quantitative evaluation of the expansion parameter α_R from the experimental values of A_2, M, and $\langle R^2 \rangle$.

6. SOME EXPERIMENTAL DATA

6.1. Small-Angle X-Ray Scattering

A direct method for experimentally determining the persistence length of a polymer chain by using small-angle X-ray scattering from a polymer solution has been proposed by Kratky and Porod simultaneously with the proposal for the wormlike chain model [34, 75]. This method is based on the principle that, according to the predictions of the theory [34], the dependence of the scattering intensity $\mathcal{I}$ on the scattering angle θ should have a specific shape for a long-chain molecule. The curve representing the dependence $\mathcal{I}(s)$ [where $s = (4\pi/\lambda) \sin(\theta/2)$, λ being the wavelength of the X rays] should contain regions in which $\mathcal{I}(s) \approx 1/s^2$ (at small θ angles) and in which $\mathcal{I}(s) \approx 1/s$ (at larger θ angles). The former range corresponds to the scattering from a Gaussian coil and the latter corresponds to rodlike scattering. The transition from coillike to rodlike scattering occurs for $s = s^*$. The value of s^* is related to the persistence length a by the equation $s^*a = \mu$, where μ is a constant derived in the theory and is equal to 1.5 according to [34].

In subsequent theories [76–81] the X-ray scattering curves were calculated for various chain-molecule conformations represented both in an analytical form and by generation using the Monte Carlo method. The general type of scattering curves calculated in these theories corresponds to the initial Kratky–Porod theory [34]. However, the values of μ obtained for various models range from 1.5 [34] to 2.87 [81].

The calculation of the persistence length a from the experimental X-ray scattering curve requires the determination of the value of s^* at which the shape of curve $\mathcal{I}(s)$ changes. This calculation is carried out with considerable error (usually by plotting the dependence of $\mathcal{I}s^2$ on s^4 [4]). There are few experimental studies in which the values of a for polymer molecules were determined by this method. Thus, for polyvinyl bromide $a = 11 \cdot 10^{-8}$ cm [34], for natural rubber $a = 5.9 \cdot 10^{-8}$ cm [82], for polystyrene $a = (8–10) \cdot 10^{-8}$ cm [83] and $(12.5–13) \cdot 10^{-8}$ cm [84], for

TABLE 1.1. Molecular Weights M_w, Second Virial Coefficients A_2, Radii of Gyration $\langle R^2 \rangle_w^{1/2}$, and Expansion Coefficients α_R for Cellulose Nitrate Molecules (N 13.9%) in Solution in Acetone Determined from Light Scattering Data and the Expansion Coefficients Calculated from These Data [103]

$M_w \cdot 10^{-3}$ g/mole	$A_2 \cdot 10^4$, mole·cm³/g²	$\langle R^2 \rangle_w^{1/2} \cdot 10^8$, cm	$\dfrac{\langle R^2 \rangle_w}{L_w} \cdot 10^8$, cm	α_R
0.81	10.8	290	59	1.029
1.63	10.3	415	60	1.041
3.5	10.7	615	62	1.058
7.5	10.1	920	64	1.079
16.6	8.4	1400	67	1.106
38.5	8.2	2220	73	1.144

TABLE 1.2. Molecular Weights M_z, Second Virial Coefficients A_2, Radii of Gyration $\langle R^2 \rangle_z^{1/2}$, and Expansion Coefficients α_R for Molecules of Ladder Polyphenylsiloxane in Benzene [120]

$M_z \cdot 10^{-5}$ g/mole	$A_2 \cdot 10^4$, mole·cm³/g²	$\langle R^2 \rangle_z^{1/2} \cdot 10^8$, cm	$\dfrac{6 \langle R^2 \rangle_z}{L_z} \cdot 10^8$, cm	α_R
11.3	1.9	490	132	1.10
5.20	2.4	350	147	1.07
3.78	1.6	270	121	1.05
1.98	2.3	200	126	1.05
1.68	2.2	190	133	1.04

TABLE 1.3. Molecular Weights M_w, Second Virial Coefficients A_2, Radii of Gyration $\langle R^2 \rangle_w$, and Expansion Coefficients α_R for Molecules of Polyamide Hydrazide in Dimethyl Sulfoxide Solutions [45] Determined from Light Scattering Data

$M_w \cdot 10^{-4}$ g/mole	$A_2 \cdot 10^3$, mole·cm³/g²	$\langle R^2 \rangle_w^{1/2} \cdot 10^8$, cm	$\dfrac{6 \langle R^2 \rangle_w}{L_w} \cdot 10^8$, cm	α_R	
				at $d/A = 0.1$, $A = 70 \cdot 10^{-8}$ cm	at $d/A = 0.02$, $A = 400 \cdot 10^{-8}$ cm
5.0	2.4	280	200	1.008	1.003
4.7	2.3	260	170	1.010	1.004
3.2	2.0	200	150	1.008	1.002
3.2	2.5	190	130	1.012	1.003
2.3	2.8	150	120	1.012	1.002
2.2	2.1	150	120	1.007	1.001

cellulose nitrate $a = 75 \cdot 10^{-8}$ cm [36, 85-87], whereas for cellulose carbonylate $a = 107 \cdot 10^{-8}$ cm [81], and for polyamide hydrazide $a = (90-100) \cdot 10^{-8}$ cm [88]. These limited data agree within an order of magnitude with the results obtained by other methods. It should also be noted that small-angle X-ray scattering allows a distinction to be made between the flexibility of rigid- and flexible-chain polymers.

The values of a obtained for cellulose derivatives and polyamide hydrazide are higher by an order of magnitude than those for the other polymers investigated.

In some papers [89, 90], the problem of the influence of excluded-volume effects on the X-ray scattering curves has been considered. Attempts to determine this influence experimentally have also been made [83, 91]. Theoretical considerations [89] suggest that volume effects do not influence the conformation of a chain of up to 30 persistence lengths, nor the shape of its scattering curve. However, experimental data on this problem are too contradictory to be discussed definitively.

As already mentioned, an advantage of the small-angle X-ray scattering technique is that it allows the most direct determination of the length of a chain section in which short-range interactions appear, i.e., the persistence length. However, this technique involves considerable experimental difficulties. The solvent investigated should ensure a relatively high electron-density increment in the polymer–solvent system, which is not always easy to achieve. To attain the required excess scattering of the polymer over that of the solvent, one should employ fairly high concentrations, and this leads to the complicating influence of intermolecular interactions and association effects. As a result, with increasing solution concentration the precision of determining the characteristic value s^* decreases [85, 88]. After s^* has been reliably determined, the difficulty in choosing a value for μ for the calculation of persistence length remains since the numerical values of μ suggested in different theories can differ almost by a factor of two.

Although small-angle X-ray scattering is a very attractive method, it has not yet been widely used and, at present, unfortunately, it cannot be considered to be a principal method for investigating the conformational properties of chain molecules.

6.2. Light Scattering

The light-scattering method in polymer solutions is much more widely used and is of incomparably greater importance for the characterization of macromolecular conformations. It was first proposed and developed by Debye [92] and then by Zimm [93], who proposed the "Zimm plot," so widely used in recent investigations. During the past four decades the light-scattering method has been widely developed in many papers and has become a principal method for the study of polymer molecules in solutions. Since both its theoretical and experimental aspects have been considered in detail in various monographs [74, 94–96], this method will not be discussed here.

It should be noted that, apart from the specific problems which can be solved by light scattering [94], this method offers three important advantages. First, light scattering is an absolute method for determining the molecular weight M of a polymer. Second, the study of the angular distribution of scattering makes it possible to determine directly the size (radius of gyration $\langle R^2 \rangle$) of the scattering molecules. Moreover, the osmotic second virial coefficient A_2 can be determined from the concentration dependence of light scattering, and thus the thermodynamic properties of the polymer–solvent system can be characterized.

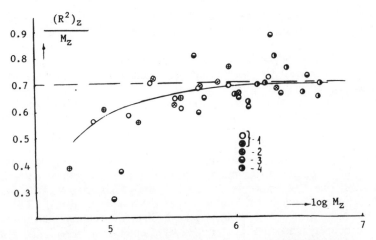

Fig. 1.7. Dependence of $\langle R^2 \rangle_z / M_z$ on M_z for cellulose nitrates in acetone and ethyl acetate according to data obtained by different authors: 1) [105]; 3) [97]; and 4) [106] – in acetone; 2) [98] – in ethyl acetate.

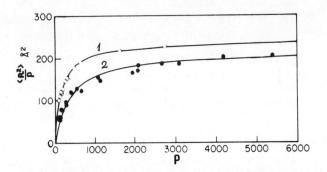

Fig. 1.8. Dependence of $\langle R^2 \rangle_w / P_w$ on P_w for fractions of 1) cellulose carbonylate and 2) amylose carbonylate in a θ-solvent.

6.2.1. Excluded-Volume Effect and Molecular Dimensions. The fact that the values of M, $\langle R^2 \rangle$, and A_2 can be obtained for the same polymer–solvent system is of fundamental importance because within a single experiment it is possible to solve the problem of the relative significance of the skeletal rigidity of the chain and the excluded-volume effects in the formation of conformational properties and the size of a polymer molecule in solution. This problem is of particular importance for rigid-chain polymers for which the part played by volume effects is not always correctly evaluated.

Cellulose derivatives can serve as an example of the foregoing considerations. The large molecular sizes and the deviation from Gaussian statistics experimentally observed for nitrates and other cellulose esters and ethers in solutions have for a long time been interpreted as manifestations of the high skeletal rigidity of their

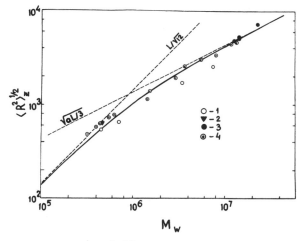

Fig. 1.9. Dependence of $\langle R^2 \rangle_z^{1/2}$ on M_w for samples and fractions of DNA according to data obtained by different authors: 1) [108]; 2) [109]; 3) [110]; 4) [111]. The solid curve is the theoretical dependence according to Eq. (1.32').

chains [97, 98]. However, the success of the theory of excluded-volume effects in solutions of flexible-chain polymers [72] suggested the application of the same concepts to solutions of rigid-chain molecules and, in the first place, to cellulose derivatives and other polysaccharides. As a result, the high values of the radii of gyration $\langle R^2 \rangle$ and the deviation from the Gaussian properties of the molecules of cellulose ethers and esters have been ascribed to long-range effects, and the concept of the high skeletal rigidity of these molecules has been disputed [99–101].

The existence of these two viewpoints has led to an animated discussion [102], and this controversy was resolved as a result of further experimental investigations of which Schulz and Penzel's work [103] was one of the most important. These authors have investigated the light scattering of fractions of cellulose nitrate in acetone in the molecular range $8 \cdot 10^4 - 4 \cdot 10^6$ g/mole, and have determined the corresponding values of M_w, $\langle R^2 \rangle_w$, and A_2 (Table 1.1). These experimental data were used to evaluate the expansion parameter α_R by applying equations similar to Eq. (1.45'). For molecular weights up to 10^6 g/mole the expansion parameter differs from unity at most by a few percent (Table 1.1). Hence, the $\langle R^2 \rangle_w / L_w$ ratio increases very slightly with increasing M [the exponent ε in Eq. (1.36) is 0.05]. Moreover, the values of the second virial coefficient A_2 are relatively high because acetone is a good solvent for the cellulose nitrates investigated. These results showed conclusively that the large size of cellulose nitrate molecules in both poor and good solvents is due to the high equilibrium rigidity of their chains rather than to excluded-volume effects. According to the data in [103], when the length of the monomer unit λ is $5.15 \cdot 10^{-8}$ cm, the length of the Kuhn segment A is $330 \cdot 10^{-8}$ cm.

Similar results were obtained for fractions of cellulose nitrate ($N \approx 13.5\%$) in ethyl acetate [104] and for ladder polyphenylsiloxane in benzene [120]. The data for the latter polymer are given in Table 1.2. As for the cellulose derivatives, the values of α_R differ from unity only by a few percent. The $6\langle R^2\rangle_z/L_z$ ratio is virtually constant in the series of molecular weights investigated and corresponds to a length of the Kuhn segment A of $130\cdot10^{-8}$ cm.

A quantitative evaluation of the significance of excluded-volume effects in determining the conformational properties of another rigid-chain polymer, aromatic polyamide hydrazide (PAH), in dimethyl sulfoxide has also been made [45]. The values of the expansion parameter α_R of the molecules were calculated from the experimental values of M_w, $\langle R^2\rangle_w$, and A_2 (obtained by light scattering) by using Eq. (1.45) for various possible values of the reduced chain diameter d/A. The results (Table 1.3) showed that although measurements were carried out in a good solvent (A_2 was relatively high), the expansion parameter α_R did not differ significantly from unity and, hence, volume effects have almost no influence on the size of the molecules in solution. Although the equilibrium rigidity of PAH molecules is of the same order of magnitude as that of cellulose nitrate (according to [45] A for PAH is $250\cdot10^{-8}$ cm), the α_R values for PAH are much closer to unity than those of cellulose nitrate (Tables 1.1 and 1.3). This is due to the fact that the molecular weight range of the cellulose nitrates investigated is an order of magnitude higher than for PAH. Hence, the increase in the $\langle R^2\rangle/L$ ratio with molecular weight M (Table 1.2) is much more dramatic for PAH than for cellulose nitrate. However, this increase is not due to long-range interactions but, rather, to the higher chain rigidity which is manifested in the deviation from Gaussian statistics at lower molecular weights in accordance with Eq. (1.25) or (1.32').

6.2.2. Dependence of Molecular Dimensions on Molecular Weight. The dependence of $\langle R^2\rangle/L$ on M for rigid-chain polymers has been studied experimentally by light scattering even in the early work by using fractions of cellulose ethers and esters. Figure 1.7 shows as an example the experimental data for cellulose nitrates obtained by several authors [105]. Although the scatter of the experimental points is rather wide, it is possible to follow the decrease in $\langle R^2\rangle_z/M_z$ with decreasing M_z, which qualitatively corresponds to dependence (1.25).

Burchard [31, 107] has carried out a quantitative comparison of the theoretical dependence (1.32') with the experimental data and used UV optics for the study of light scattering from fractions of cellulose and amylose carbonylates in dioxane and in a θ-mixture of dioxane with methanol. The results for solutions in the θ-solvent are shown in Fig. 1.8. The scatter of experimental points is relatively narrow and, therefore, their comparison with theoretical dependence (1.32') (solid curves) allows both the segment length A and the projection λ of the monomer unit on the chain direction to be evaluated. However, good agreement between the experimental points and the theoretical curves is attained for the values of $A = 350\cdot10^{-8}$ cm and $\lambda = 4.2\cdot10^{-8}$ cm for cellulose carbonylate and $A = 411\cdot10^{-8}$ cm and $\lambda = 3.08\cdot10^{-8}$ cm for amylose carbonylate. These low values of λ, instead of the $5.15\cdot10^{-8}$ cm calculated from the O_1–O_4 distance for a glucose ring, are interpreted by Burchard according to his proposal for a model of semiflexible helices [50]. If this molecular model is used, the length of the Kuhn segment A can be decreased to $280\cdot10^{-8}$ cm.

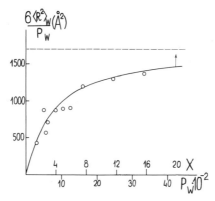

Fig. 1.10. Dependence of $6\langle R^2 \rangle_w / P_w$ on P_w for polyhexyl isocyanate in tetrahydrofuran. Points refer to experimental data [116]. The solid curve is the theoretical dependence according to Eq. (1.32') for $A = 850 \cdot 10^{-8}$ cm and $\lambda = 2 \cdot 10^{-8}$ cm.

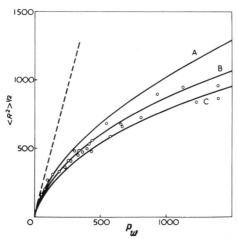

Fig. 1.11. Dependence of radius of gyration $\langle R^2 \rangle^{1/2}$ on the degree of polymerization P_w for PPhTPhA in H_2SO_4 [117]. Points refer to the experimental data. Solid curves A, B, and C are the theoretical dependences according to Eq. (1.32') for a value $\lambda = 12 \cdot 10^{-8}$ and the values of a persistence length given by $a = 300 \cdot 10^{-8}$ cm (A), $200 \cdot 10^{-8}$ cm (B), and $150 \cdot 10^{-8}$ cm (C).

Another polymer for which the dependence of $\langle R^2 \rangle_z^{1/2}$ on M_w was investigated for various fractions [108–111] was DNA. The results [111] are shown as the dependence of $\langle R^2 \rangle_z^{1/2}$ on M_w on a logarithmic scale in Fig. 1.9. The solid curve representing the theoretical dependence (1.25) is plotted for the values of $M_0/\lambda = 200 \cdot 10^8$ g/mole·cm (M_0 is the molecular weight of the monomer unit) and $A = 2200 \cdot 10^{-8}$ cm. The latter value seems overestimated [112]. However, it is important to note that the available experimental points make it virtually impossible to determine λ because of the considerable scattering and the absence of experimental data at relatively low M. This is clearly seen in Fig. 1.9: the asymptotic dashed straight line, corresponding to the equation $\langle R^2 \rangle_{x \to 0}^{1/2} = L/(12)^{1/2}$, is plotted in accordance with the theoretical curve rather than with the experimental points.

This difficulty is specific to the light-scattering method since, at relatively low molecular weights, the measurements of light-scattering asymmetry required for the determination of $\langle R^2 \rangle$ are subject to large experimental error and, therefore, the corresponding data are absent in most published papers.

The same is also true of investigations carried out with poly(alkyl isocyanate)s (nylon 1) [113–115]. The most reliable results on light scattering have probably been reported in [116, 117] for a limited series of fractions of poly(hexyl isocyanate) in tetrahydrofuran and hexane. The experimental data obtained according to [116], and shown as points in Fig. 1.10, represent the dependence of $6\langle R^2 \rangle_w / P_w$ on P_w. The solid curve represents the theoretical dependence plotted according to Eq. (1.32') for $A = 850 \cdot 10^{-8}$ cm and $\lambda = 2 \cdot 10^{-8}$ cm. The points are in relatively good agreement with the theoretical curve, and the values of A and λ are in good agreement with those obtained previously by other methods [20]. However, in this case the absence of points at low P also prevents the determination of λ from the initial slope of the experimental dependence of $\langle R^2 \rangle / P$ on P.

The light scattering method was also used to study the conformational characteristics of rigid-chain molecules of *para*-aromatic polyamides: poly-*p*-phenylene terephthalamide (PPhTPhA) and poly-*p*-benzamide (PPB) in concentrated sulfuric acid [118]. Although the measurement of light scattering in the aromatic polyamide–sulfuric acid system represents a difficult experimental problem, the authors have succeeded in obtaining reliable results. This was mainly due to thorough fractionation of samples. The high refractive index increment in the polymer–solvent system ($dn/dc \approx 0.3$ cm³/g) was also a favorable factor. The dependence of the molecular radius of gyration $\langle R^2 \rangle^{1/2}$ on the degree of polymerization P_w for PPhTPhA is shown by points in Fig. 1.11. Solid curves are plotted using Eq. (1.32') at various values of A and the accepted value of $\lambda = 12 \cdot 10^{-8}$ cm in accordance with which the initial slope of curves equal to $\lambda/(12)^{1/2}$ is shown by a dashed line. Although even in this case the value of λ cannot be reliable determined from experimental points at low P, the totality of points makes it possible to evaluate fairly reliably the length of segment $A = (400 \pm 100) \cdot 10^{-8}$ cm. Similar investigations of PPB in sulfuric acid lead to a value of A twice that for PPhTPhA in accordance with the results obtained by other methods [119, 132].

The light-scattering method has also been used to determine the molecular size of another rigid-chain heterocyclic polymer, polynaphthoylene benzimidazole (BBL) [19].

Experiments carried out with BBL fractions in methanesulfonic acid revealed a pronounced dependence of $\langle R^2 \rangle / P$ on P. Its comparison with Eq. (1.32') and the assumption that $M_0/\lambda = 26.5 \cdot 10^8$ g/mole·cm leads to a high value of the segment length, $A = 3000 \cdot 10^{-8}$ cm, for this ladder polymer.

6.3. Depolarization of Light Scattering and Optical Anisotropy of the Molecular Chain

The possibilities of using the angular asymmetry of light scattering for the determination of the size and rigidity of chain molecules are restricted to polymer samples of relatively high molecular weights for which this asymmetry is sufficiently pronounced. However, in reality this condition cannot always be fulfilled. Hence, recently attempts have been made to use the depolarization of scattered light for determining the conformational characteristics of rigid-chain polymers of low molecular weight.

It is known [121, 122] that the degree of depolarization of light ρ_u scattered by an assembly of molecules depends on their mean-square anisotropy δ^2, determined by the equation

$$\delta^2 = [(\gamma_1 - \gamma_2)^2 + (\gamma_1 - \gamma_3)^2 + (\gamma_2 - \gamma_3)^2]/(\gamma_1 + \gamma_2 + \gamma_3)^2, \qquad (1.46)$$

where γ_1, γ_2, and γ_3 are the principal polarizabilities of the molecule. When axial symmetry of the optical properties of the molecule exists ($\gamma_2 = \gamma_3$), Eq. (1.46) is transformed into the equation

$$\delta^2 = 2(\gamma_1 - \gamma_2)^2/9\bar{\gamma}^2, \qquad (1.47)$$

where $\bar{\gamma}$ is the average excess polarizability of the molecule over that of the solvent. The experimental value of δ^2 can be determined from the degree of depolarization ρ_u of excess scattering by using the equation [122]

$$\delta^2 = 10\rho_u/(6 - 7\rho_u). \qquad (1.48)$$

If the scattering system is a solution of chain molecules, each of them consisting of N structural elements, then by analogy to Eq. (1.47), the mean-square anisotropy of each chain element is given by

$$\delta_0^2 = 2(\alpha_1 - \alpha_2)_0^2/9\bar{\alpha}_0^2, \qquad (1.49)$$

and the anisotropy of the whole chain may be written as

$$\delta^2 = (\delta_0^2/N^2) \sum_i^N \sum_k^N \langle 3\cos^2\vartheta_{ik} - 1 \rangle/2, \qquad (1.50)$$

where $(\alpha_1 - \alpha_2)_0$ is the difference between the principal polarizabilities of a structural chain element (with uniaxial symmetry of the optical properties), ϑ_{ik} is the angle formed by the optical axes of the ith and the kth elements, and $\langle \; \rangle$ designates averaging over all the chain conformations [123].

If it is taken into account that the average excess anisotropy of a chain element is $\bar{\delta}_0 = \bar{\gamma}/N$, Eq. (1.50) can be written in the form

$$(\gamma_1 - \gamma_2)^2 = (\alpha_1 - \alpha_2)_0^2 \sum_i^N \sum_k^N \langle 3\cos^2\vartheta_{ik} - 1\rangle/2. \tag{1.51}$$

If a chain molecule is represented by the Kratky–Porod wormlike chain [34], the theory of its optical anisotropy leads to the equation [124, 125]

$$\delta^2 = (2\delta_0^2/x')[1 - (1 - e^{-x'})/x'], \tag{1.52}$$

which is equivalent to the expression

$$(\gamma_1 - \gamma_2)^2 = N^2(\alpha_1 - \alpha_2)_0^2 (2/x')[1 - (1 - e^{-x'})/x']$$
$$= (\beta L)^2 (2/x')[1 - (1 - e^{-x'})/x'], \tag{1.53}$$

where $x' = 3x = 3L/a = 3(2L/A)$, and β is the difference in the polarizabilities of a unit length of a wormlike chain. According to Eq. (1.53), the dependence of the ratio $(\gamma_1 - \gamma_2)^2/(\beta L)^2$ on the length L of the wormlike chain is analogous to the dependence of $\langle h^2 \rangle/L^2$ on L following Eq. (1.24). The only difference is that the "apparent" persistence length in Eq. (1.53) is shorter by a factor of three than Porod's persistence length a in Eq. (1.24). In particular, for a molecule in the conformation of a weakly bending rod, it follows from Eq. (1.53) that

$$(\gamma_1 - \gamma_2)_{x' \ll 1} = \beta L[1 - (2L/A) + (3/4)(2L/A)^2 - \cdots], \tag{1.54}$$

whereas in the conformation of a Gaussian coil

$$(\gamma_1 - \gamma_2)_{x' \to \infty} = \beta (AL/3)^{1/2}, \tag{1.55}$$

where A is the length of the Kuhn segment.

Hence, the mean-square difference between the polarizabilities of the Gaussian chain increases proportionally to the square root of its length L.

According to Eqs. (1.52) and (1.54), at infinitely low molecular weights the mean-square anisotropy of a chain molecule δ^2 is equal to that of its components δ_0^2, and with increasing chain length decreases asymptotically to zero. In accordance with this and Eq. (1.48) it is concluded that the degree of depolarization ρ_u of excess scattering of solution will also decrease with increasing molecular weight of the dissolved polymer M. In principle, an experimental study of the dependence of ρ_u on M and, correspondingly [by using Eq. (1.48)], of δ^2 on M and comparison of these dependences with the theoretical dependence from Eq. (1.52) allow the determination of δ_0 and A.

There are very few experimental papers in which the study of the depolarization of scattered light is used to determine the conformational characteristics of chain molecules. For both polystyrene [126–128] and the two rigid-chain aromatic polymers investigated [125, 19] the values of *a* obtained by this method are in qualitative agreement with those found using other methods.

However, the method of depolarization of light scattering has not yet found widespread use. This is due to the considerable experimental difficulties involved in this method and to the fact that experimental measurements of depolarization are virtually possible only in those specific cases in which the structural units of the polymer molecule under investigation exhibit very high optical anisotropy.

6.4. Conclusion

Hence, of all the methods considered above, at present the study of the intensity and angular asymmetry of light scattering in dilute solutions of macromolecular compounds is of major importance. As we have seen, this method has made it possible to obtain direct confirmation of the assumption that, even in thermodynamically good solvents, the size and conformation of rigid-chain polymer molecules are determined by the true skeletal rigidity of their chains rather than by the excluded-volume effects.

This reliably established fact is of great importance in many problems and especially in discussing the hydrodynamic properties of rigid-chain polymer molecules in dilute solutions.

REFERENCES

1. W. Kuhn, Kolloid-Z., **68**, 2 (1934); **76**, 258 (1936); **87**, 3 (1939).
2. E. Guth and H. Mark, Monatsh. Chem., **65**, 93 (1934).
3. M. V. Volkenstein, Configurational Statistics of Polymeric Chains, Wiley-Interscience, New York (1963).
4. T. M. Birshtein and O. B. Ptitsyn, Conformations of Macromolecules, Wiley-Interscience, New York (1966).
5. P. J. Flory, Statistical Mechanics of Chain Molecules, Wiley-Interscience, New York (1969).
6. H. Eyring, Phys. Rev., **39**, 746 (1932).
7. R. A. Sack, Nature, **171**, 310 (1953).
8. H. Benoit and P. M. Doty, J. Phys. Chem., **57**, 958 (1953).
9. T. M. Birstein, Vysokomol. Soedin., Ser. A, **19**, 54 (1977).
10. V. N. Tsvetkov, V. V. Korshak, I. N. Shtennikova, H. Raubach, E. S. Krongauz, G. M. Pavlov, and G. F. Kolbina, Macromolecules, **12**, 645 (1979).
11. S. Oka, Proc. Math. Soc. Jpn., **24**, 657 (1942).
12. W. J. Taylor, J. Chem. Phys., **15**, 412 (1947); **16**, 257 (1948).
13. H. Kuhn, J. Chem. Phys., **15**, 843 (1947).
14. H. Benoit, J. Chem. Phys., **44**, 18 (1947).
15. J. I. Frenkel, Kinetic Theory of Liquids, Izd. Akad. Nauk SSSR, Moscow (1945).
16. V. N. Tsvetkov, K. A. Andrianov, I. N. Shtennikova, E. L. Vinogradov, V. I. Pakhomov, and S. E. Yakushkina, J. Polym. Sci. C, No. 23, 385 (1968).

17. V. N. Tsvetkov, Makromol. Chem., **160**, 1 (1972).
18. V. N. Tsvetkov, K. A. Andrianov, N. N. Makarova, M. G. Vitovskaya, E. I. Rjumtsev, I. N. Shtennikova, and N. V. Pogodina, Eur. Polym. J., **9**, 27 (1973); **11**, 771 (1975).
19. J. C. Berry, J. Polym. Sci., Polym. Symp., No. 65, 143 (1978).
20. V. N. Tsvetkov, I. N. Shtennikova, E. I. Rjumtsev, and Y. P. Getmanchuk, Eur. Polym. J., **7**, 767 (1971); **10**, 55 (1974); **11**, 52 (1975).
21. U. Shmueli, W. Traub, and K. Rosenheck, J. Polym. Sci., **A2**, 7, 515 (1969).
22. V. N. Tsvetkov, E. I. Rjumtsev, and N. V. Pogodina, Dokl. Akad. Nauk SSSR, **224**, 112 (1975).
23. V. N. Tsvetkov, Eur. Polym. J., **12**, 867 (1976).
24. V. N. Tsvetkov, N. V. Pogodina, and L. V. Starchenko, Eur. Polym. J., **17**, 397 (1981).
25. P. Doty, J. Bradbury, and A. Holtzer, J. Am. Chem. Soc., **76**, 4492 (1954).
26. V. N. Tsvetkov, I. N. Shtennikova, V. S. Skazka, and E. I. Rjumtsev, J. Polym. Sci. C, No. 16, 3205 (1968).
27. P. Doty, Proc. Natl. Acad. Sci. USA, **42**, 791 (1956).
28. V. N. Tsvetkov, L. N. Andreeva, and L. N. Kvitchenko, Vysokomol. Soedin., **7**, 2001 (1965).
29. I. D. Watson and F. H. Crick, Nature, **171**, 737 (1953).
30. N. M. Bikels and L. Segal, Cellulose and Cellulose Derivatives, Wiley-Interscience, New York (1971).
31. W. Burchard, Makromol. Chem., **88**, 11 (1965).
32. V. N. Tsvetkov, V. E. Eskin, and S. J. Frenkel, Structure of Macromolecules in Solutions, National Lending Library for Science and Technology, Boston Spa, England (1971).
33. G. Porod, Monatsh. Chem., **2**, 251 (1949).
34. O. Kratky and G. Porod, Recl. Trav. Chim., **68**(12), 1106 (1949).
35. J. Hermans and R. Ullman, Physica, **18**, 951 (1952).
36. S. Heine, O. Kratky, G. Porod, and P. J. Schmitz, Makromol. Chem., **44**, 682 (1961).
37. H. Daniels, Proc. R. Soc. (Edinburgh), **A63**, 290 (1952).
38. W. Gobush, H. Yamakawa, W. H. Stockmayer, and W. S. Magee, J. Chem. Phys., **57**, 2839 (1972).
39. Y. Yamakawa and W. H. Stockmayer, J. Chem. Phys., **57**, 2843 (1972).
40. V. N. Tsvetkov, T. I. Garmonova, and R. P. Stankevich, Vysokomol. Soedin., **8**, 980 (1966).
41. V. N. Tsvetkov, T. I. Garmonova, M. G. Vitovskaya, P. N. Lavrenko, and E. V. Korovina, Vysokomol. Soedin., **A13**, 884 (1971).
42. E. V. Frisman and M. A. Sibileva, Vysokomol. Soedin., **7**, 674 (1965).
43. G. Thurston and J. Schrag, J. Polym. Sci., **6**, A-2, 1331 (1968).
44. J. V. Champion and A. Dandridge, Polymer, **19**, 632 (1978).
45. V. N. Tsvetkov and S. O. Tsepelevich, Eur. Polym. J., **19**, 267 (1983).
46. V. N. Tsvetkov, Usp. Khim., **38**, 1674 (1969).
47. A. Blumstein (ed.), Liquid Crystalline Order in Polymers, Chap. 2, Academic Press, New York (1978), p. 44.
48. V. N. Tsvetkov, Eur. Polym. J. Suppl. 237 (1969).
49. N. Yathindra and V. S. Rao, J. Polym. Sci., **8**, A-2, 2033 (1970); **9**, A-2, 1149 (1971); **10**, A-2, 1369 (1972).

50. W. Burchard, Br. Polym. J., **3**, 209 (1971).
51. V. N. Tsvetkov, M. G. Vitovskaya, P. N. Lavrenko, I. N. Shtennikova, N. N. Kuprianova, and V. S. Skaska, Vysokomol. Soedin., **A-9**, 1682 (1967); **A-10**, 903 (1968); **A-12**, 1974 (1970); **A-13**, 620 (1971).
52. V. N. Tsvetkov, M. G. Vitovskaya, P. N. Lavrenko, E. N. Sakharova, N. F. Gavrilenko, and N. N. Stefanovskaya, Vysokomol. Soedin., **A-13**, 2532 (1971).
53. G. Berry, H. Nomura, and K. G. Mayhan, J. Polym. Sci., **5**, A-2, 1 (1967).
54. J. F. Brown, J. Polym. Sci., **C1**, 83 (1963).
55. K. A. Andrianov, Vysokomol. Soedin., **7**, 1477 (1965); **A11**, 1362 (1969); **A13**, 253 (1971).
56. P. D. Sybert, W. H. Beever, and J. K. Stille, Macromolecules, **14**, 493 (1981).
57. L. Pauling, Nature of the Chemical Bond, Cornell Univ. Press, Ithaca (1960).
58. V. N. Tsvetkov and V. N. Bychkova, Vysokomol. Soedin., **6**, 600 (1964).
59. V. N. Tsvetkov, E. N. Zakharova, G. A. Fomin, and P. N. Lavrenko, Vysokomol. Soedin., **14**, 1956 (1972).
60. R. S. Saunders, J. Polym. Sci., **A2**, 3765 (1964); **A3**, 1221 (1965).
61. V. E. Shashoua, J. Am. Chem. Soc., **81**, 3156 (1959); **82**, 886 (1960).
62. N. S. Schneider, S. Furusaki, and R. W. Lenz, J. Polym. Sci., **A3**, 993 (1965).
63. M. G. Vitovskaya, I. N. Shtennikova, E. P. Astapenko, and T. V. Peker, Vysokomol. Soedin., **17**, 1161 (1975).
64. V. N. Tsvetkov and I. N. Shtennikova, Macromolecules, **11**, 306 (1978).
65. P. W. Morgan, Macromolecules, **10**, 1381 (1977).
66. A. Katchalsky, O. Kunzle, and W. Kuhn, J. Polym. Sci., **5**, 283 (1950).
67. A. Katchalsky and S. Lifson, J. Polym. Sci., **11**, 409 (1953).
68. O. B. Ptitsyn, Vysokomol. Soedin., **3**, 1084, 1252, 1401 (1961).
69. V. N. Tsvetkov and S. Y. Lyubina, Vysokomol. Soedin., **6**, 806 (1964); **8**, 846 (1966); **A10**, 74 (1968).
70. V. N. Tsvetkov, E. N. Sakharova, and M. M. Krunchjak, Vysokomol. Soedin., **A10**, 685 (1968).
71. E. N. Sakharova and V. N. Tsvetkov, Vestn. Leningr. Univ., No. 10, 41 (1967); No. 16, 55 (1970).
72. P. J. Flory, Principles of Polymer Chemistry, Cornell Univ. Press, Ithaca (1953).
73. H. Yamakawa, Modern Theory of Polymer Solutions, Harper and Row, New York (1971).
74. J. J. Hermans (ed.), Polymer Solution Properties, Parts 1 and 2, Dowden, Hutchinson, and Ross, Stroudsburg (1978).
75. G. Porod, Z. Naturforsch., **4a**, 401 (1949).
76. G. Porod, J. Polym. Sci., **10**, 157 (1953).
77. A. Guinier and G. Fournet, Small-Angle Scattering of X-Rays, Chap. 1, Wiley, New York (1955).
78. A. Peterlin, J. Polym. Sci., **47**, 403 (1960).
79. S. Heine, O. Kratky, and J. Poppert, Makromol. Chem., **56**, 150 (1962).
80. R. G. Kirste, Makromol. Chem., **101**, 91 (1967).
81. W. Burchard and K. Kajiwara, Proc. R. Soc. London, **A316**, 185 (1970).

82. O. Kratky and H. Sand, Kolloid-Z., **172**, 18 (1960).
83. E. Wada and K. Okano, Rep. Prog. Polym. Phys. Jpn., **6**, 1 (1963); **7**, 19 (1964); **8**, 15 (1965).
84. H. Durchschlag, O. Kratky, J. W. Breitenbach, and B. A. Wolf, Monatsh. Chem., **101**, 1462 (1976).
85. O. Kratky and N. Sembach, Makromol. Chem., **18–19**, 463 (1956).
86. O. Kratky, H. Leopold, and G. Puchwein, Kolloid-Z., **216–217**, 225 (1967).
87. P. Zipper, W. R. Krigbaum, and O. Kratky, Kolloid-Z., **235**, 1281 (1969).
88. W. R. Krigbaum and S. Sasaki, J. Polym. Sci., Polym. Phys. Ed., **19**, 1339 (1981).
89. S. Heine, Makromol. Chem., **71**, 86 (1964).
90. R. Koyama, J. Phys. Soc. Jpn., **34**, 1024 (1973); J. Phys. Soc. Jpn. Lett., **41**, 1077 (1976).
91. E. Wada, Y. Taru, S. Tatsumiya, H. Hiramatsu, K. Kurita, H. Tagama, A. Janosi, and G. Degovics, Rep. Prog. Polym. Phys. Jpn., **15**, 37, 41, 43 (1972), **19**, 34 (1976).
92. P. Debye, J. Appl. Phys., **15**, 338 (1944).
93. B. H. Zimm, J. Chem. Phys., **16**, 1093 (1948).
94. M. B. Huglin (ed.), Light Scattering from Polymer Solutions, Academic Press, New York (1972).
95. V. E. Eskin, Light Scattering from Polymer Solutions, Nauka, Moscow (1973).
95a. V. Degeorgio, M. Corti, and M. Giglio (eds.), Light Scattering in Liquids and Macromolecular Solutions, Plenum Press, New York (1980).
96. K. A. Stacey, Light Scattering in Physical Chemistry, Butterworths, London (1956).
97. A. M. Holzer, H. Benoit, and P. Doty, J. Phys. Chem., **58**, 624 (1954).
98. M. L. Hunt, S. Newman, H. A. Scheraga, and P. J. Flory, J. Phys. Chem., **60**, 1278 (1956).
99. M. Kurata and W. H. Stockmayer, Fortschr. Hochpolym. Forsch., **3**, 196 (1963).
100. W. Brown, D. Henley, and J. Ohman, Makromol. Chem., **62**, 164 (1963); **64**, 49 (1963).
101. W. Brown and D. Henley, Makromol. Chem., **75**, 179 (1964); **79**, 68 (1964).
102. P. J. Flory, Makromol. Chem., **98**, 128 (1966).
103. G. V. Schulz and E. Penzel, Makromol. Chem., **112**, 260 (1968).
104. S. O. Tsepelevich, G. N. Marchenko, and V. N. Tsvetkov, Vysokomol. Soedin., **B23**, 773 (1981).
105. A. Munster and H. Diener, Symposium uber Makromolekule in Wiesbaden, October 1959, Kurzmitteilungen, Sektion II.A2, Verlag Chemie, Julius Beltz, Weinheim (1959).
106. M. M. Huque, D. A. I. Goring, and S. G. Mason, Can. J. Chem., **36**, 952 (1958).
107. W. Burchard, Br. Polym. J., **3**, 214 (1971).
108. P. Doty, Proc. Natl. Acad. Sci. USA, **44**, 432 (1958).
109. G. Cohen and H. Eisenberg, Biopolymers, **4**, 429 (1966).
110. J. A. Harpst, A. I. Krasna, and B. H. Zimm, Biopolymers, **6**, 595 (1968).
111. R. G. Kirste, Discuss. Faraday Soc., No. 49, 51 (1970).

112. H. Yamakawa and M. Fujii, Macromolecules, **7**, 649 (1974).
113. L. J. Fetters and H. Yu, Macromolecules, **4**, 385 (1971).
114. D. N. Rubingh and H. Yu, Macromolecules, **9**, 681 (1976).
115. M. R. Ambler, D. McIntyr, and L. J. Fetters, Macromolecules, **11**, 300 (1978).
116. M. N. Berger and B. M. Tidswell, J. Polym. Sci., Polym. Symp., No. 42, 1063 (1973).
117. H. Murakami, T. Norisuye, and H. Fujita, Macromolecules, **13**, 345 (1980).
118. M. Arpin and C. Strazielle, Polymer, **18**, 591 (1977).
119. V. N. Tsvetkov and I. N. Shtennikova, Dokl. Akad. Nauk SSSR, **224**, No. 5, 1126 (1975); **231**, No. 6, 1373 (1976).
120. V. E. Eskin, O. S. Korothina, P. N. Lavrenko, and E. V. Korneeva, Vysokomol. Soedin., **A15**, 2110 (1973).
121. R. Gans, Lichtzerstreuung, in: Handbuch der Experimental Physik, W. Wien and F. Harms (eds.), Vol. 19, Akadem. Verlegsges. Leipzig (1928).
122. S. Bhagavantam, Scattering of Light and Raman Effect, Chem. Publ. Co., New York (1942).
123. H. Benoit, C. R. Acad. Sci., **236**, 687 (1953).
124. N. Saito, K. Takahashi, and Y. Yunoki, J. Phys. Soc. Jpn., **22**, 217 (1967).
125. M. Arpin, C. Strazielle, G. Weill, and H. Benoit, Polymer, **18**, 262 (1977).
126. G. Weill, C. R. Acad. Sci. Paris, **246**, 272 (1958).
127. E. P. Piskareva, E. G. Erenburg, and I. Ya. Poddubnyi, Dokl. Akad. Nauk SSSR, **180**, 1395 (1968).
128. H. Utiyama and Y. Tsunashima, J. Chem. Phys., **56**, 1626 (1972).
129. J. Stejskal and P. Kratochvil, Polymer, **22**, 1301 (1981).
130. A. Teramoto and H. Fujita, Adv. Polym. Sci., **18**, 65–150 (1975).
131. O. B. Ptitsyn and Yu. Eisner, Zh. Fiz. Khim., **32**, 2464 (1958).
132. N. V. Pogodina, V. N. Tsvetkov, and I. N. Bogatova, Vysokomol. Soedin., **A27**, 1405 (1985).

Chapter 2

HYDRODYNAMIC PROPERTIES
(THEORY)

When a polymer molecule moves in a solvent, the resistance to this motion depends on the size and shape of the molecule. Hence, a study of this resistance can provide information regarding the conformational characteristics of a polymer chain. Generally, the principal phenomena used for this purpose are the translational and rotational friction of the macromolecules which become apparent in the diffusion, sedimentation, and viscometry of polymer solutions. These experimental data are quantitatively interpreted on a molecular level by using the theories relating the conformational characteristics of a molecule to its hydrodynamic properties investigated in solution. In each theory the polymer molecule under investigation is represented by a body of a certain configuration, and the translational and rotational motion of this body in a solvent are described in terms of the laws of the hydrodynamics of macroscopic bodies in a viscous medium.

A rigorous solution of hydrodynamic problems involving translational and rotational motion has been obtained only for a model of a compact sphere and an ellipsoid of revolution (spheroid). This model can represent fairly adequately the shape of rigid globular molecules and was therefore widely used in the study of biological molecules and, primarily, globular proteins [1–4]. Although the model of a compact ellipsoid is unlike the conformation of a real flexible-chain molecule, for the solution of some specific (e.g., dynamo-optical) problems it may sometimes prove useful if applied to molecular coils of flexible-chain polymers [5–7]. As to rigid-chain polymers represented by a wormlike chain, whose asymptotic limit as $L/A \to 0$ is a straight rod, under these limiting conditions an elongated ellipsoid of revolution may serve as a fairly good model for the description of the hydrodynamic properties of rigid-chain molecules. Hence, the principal expressions for the theory of translational and rotational friction of ellipsoids of revolution are directly related to the hydrodynamic characteristics of rigid-chain molecules and, therefore, should be briefly considered.

1. HYDRODYNAMICS OF ELLIPSOIDAL PARTICLES

1.1.   Translational Friction

The translational friction coefficient f is the quantitative characteristic of resistance to the translational motion of a body (particle) in the surrounding liquid. It is determined by the equation

$$F = fu, \tag{2.1}$$

where u is the velocity of motion of the body caused by force F. The direction of u coincides with that of F.

A rigorous solution of the hydrodynamic Navier–Stokes equations [8] for the simplest case of the motion of a sphere in a viscous liquid leads to the well-known Stokes equation [9]

$$f = 3\pi\eta_0 a, \tag{2.2}$$

where η_0 is the viscosity coefficient of the surrounding liquid, and d is the sphere diameter.

Equation (2.2) was obtained by making certain assumptions. The most important of them include the absence of slippage on the interface between the spherical particle and the surrounding liquid, a sufficiently large particle size so that the surrounding liquid (solvent) might be regarded as a continuous medium, and the absence of interaction between the particles.

The same assumptions are made in calculating the friction coefficient of a spherical body using the Navier–Stokes equations.

For an aspherical body (including an ellipsoidal body) the translational friction coefficient f is a tensor value; i.e., it depends on the direction of motion with respect to a system of coordinates rigidly related to the particle.

Calculations of f carried out in some papers [10–12] yield the values of f for an elongated ellipsoid of revolution (spheroid) ($p > 1$) when it moves in a direction parallel (f_1) or normal (f_2) to the axis of symmetry, respectively:

$$f_1 = 8\pi\eta_0 L \left(1 - \frac{1}{p^2}\right) \Big/ \left(-2 + \frac{2p^2 - 1}{p\sqrt{p^2 - 1}} \ln \frac{p + \sqrt{p^2 - 1}}{p - \sqrt{p^2 - 1}}\right), \tag{2.3}$$

$$f_2 = 8\pi\eta_0 L \left(1 - \frac{1}{p^2}\right) \Big/ \left(1 + \frac{2p^2 - 3}{2p\sqrt{p^2 - 1}} \ln \frac{p + \sqrt{p^2 - 1}}{p - \sqrt{p^2 - 1}}\right). \tag{2.4}$$

where L and a are the major and minor axes of the ellipsoid, respectively, and $p = L/a$ characterizes the asymmetry of its shape.

If the orientational distribution in an assembly of ellipsoidal particles is random, the average value of f is evidently given by

$$f = 3f_1 f_2/(2f_1 + f_2) = 6\pi\eta_0 L \sqrt{p^2 - 1} \Big/ \left(p \ln \frac{p + \sqrt{p^2 - 1}}{p - \sqrt{p^2 - 1}} \right). \qquad (2.5)$$

For a sphere $(p = 1)$, Eq. (2.5) is equivalent to Eq. (2.2).

For a very elongated ellipsoid of revolution $(p \gg 1)$, Eq. (2.5) becomes

$$3\pi\eta_0 L/f = \ln 2p = \ln (L/a) + 0.693. \qquad (2.6)$$

It can easily be shown using Eqs. (2.2) and (2.5) that, when the volumes of spherical and ellipsoidal particles are equal, the friction coefficient of the latter is higher than that of the former, and this difference increases with increasing value of p [13].

If the model of a compact rigid ellipsoid is used for the characterization of the hydrodynamic properties of polymer molecules, it follows from Eq. (2.5) that the molecular-weight (M) dependence of the friction coefficient is of a definite type. Thus, if M increases in a homologous series of molecules and the similarity of their shape is retained $(p = \text{const})$, then, according to Eq. (2.5), the friction coefficient f increases proportionally to the linear dimensions of the ellipsoid L and a, i.e., proportionally to the cube root of its volume. Hence, in this case we have

$$f = CM^b, \qquad (2.7)$$

where $b = 1/3$ and C is a constant.

Of particular interest is the dependence of f on M for "rodlike" molecules, for which M increases proportionally to the length L, and a remains constant. It follows from Eq. (2.6) that when these molecules are sufficiently elongated, their f/L ratio changes with L only because, in this case, $\ln M$ changes. This implies that for rodlike molecules the dependence of f on M cannot be described by Eq. (2.7) with constant values of b and C. Analysis of Eq. (2.5) shows that in this case the exponent b in Eq. (2.7) increases with M (and p proportional to it) approaching unity [3]. Thus, for p values ranging from 20 to 1000, b varies from 0.7 to 0.9.

1.2. Rotational Friction

The rotational friction of a spherical body in a viscous medium has been studied from the standpoint of classical hydrodynamics by Stokes, who showed that when a compact sphere rotates about the central axis at an angular velocity ω in a viscous liquid, it is subjected to forces of rotational friction whose moment $\mathscr{L}$ is proportional to ω:

$$\mathscr{L} = W\omega. \qquad (2.8)$$

The proportionality factor W is the rotational friction coefficient for a sphere. It is proportional to the volume of the sphere V and the viscosity of the solvent η_0 according to the equation

$$W = \pi \eta_0 a^3 = 6 \eta_0 V, \tag{2.9}$$

where a is the diameter of the sphere.

The equations expressing the rotational friction of ellipsoids of revolution (spheroids) have also been obtained by rigorous solutions of hydrodynamic equations [10, 13]. The rotational friction coefficient W, just as that for translational motion, is a tensor whose components have values corresponding to different axes of rotation. The value of W corresponding to rotation about the central axis normal to the axis of symmetry of the spheroid is of primary importance. For an elongated spheroid ($p > 1$) this corresponds to rotation about the transverse (minor) axis. In this case we have

$$\dot{W} = \eta_0 V f_0(p), \tag{2.10}$$

where $f_0(p)$, the function of the shape asymmetry p, is determined by the equation

$$f_0(p) = \frac{4(p^4 - 1)}{p^2} \left(\frac{2p^2 - 1}{2p\sqrt{p^2 - 1}} \ln \frac{p + \sqrt{p^2 - 1}}{p - \sqrt{p^2 - 1}} - 1 \right)^{-1}. \tag{2.11}$$

It follows from Eqs. (2.9) and (2.10) that $f_0(p)/6$ is the ratio of the rotational friction coefficients for an ellipsoid and a sphere of equal volumes in the same liquid. The function $f_0(p)$ for $p = 1$ is equal to 6 and increases drastically with increasing p [5]. Hence, it can be used for determining the degree of asymmetry of particles according to experimental data on their rotational friction.

For a very elongated ellipsoid of revolution ($p \gg 1$) Eqs. (2.10) and (2.11) give [14]

$$W = (\pi \eta_0 L^3 / 3) / (\ln 2p - 0.5). \tag{2.12}$$

Equations (2.10) and (2.11) determine the type of dependence of the rotational friction coefficient W on the molecular weight M of polymer molecules if they can be represented by rigid, compact spheroids. Thus, if the shape asymmetry p does not change in a series of molecular weights, the dependence of W on M corresponds to Eq. (2.7), where the exponent $b = 1$. For a series of "rodlike," i.e., spheroidal, molecules with a constant value of diameter a, the value of W evidently varies proportionally to $M \cdot f_0(p)$, where $p = \text{const} \cdot M$. In this case the dependence of W on M cannot be described by Eq. (2.7) with constant values of b and C. According to Eqs. (2.10) and (2.12), for rodlike molecules the exponent b increases with M and has the values $b = 2.55$ for $p = 10$, $b = 2.75$ for $p = 100$, $b = 2.82$ for $p = 500$, and $b = 2.86$ for $p = 1000$. This type of dependence, in which the exponent b approaches 3 with increasing length of the molecule L, follows directly from Eq. (2.12).

The translational and rotational friction of ellipsoidal particles (just as that of macromolecules of any other shape) in solutions are directly reflected in the viscous properties of these solutions.

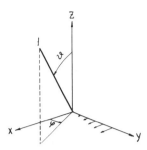

Fig. 2.1. Axially symmetric molecule in a laminar flow. X) Flow direction; Y) direction of the velocity gradient; 1) axis of symmetry of the molecule.

1.3. Viscosity of Solutions of Spheroidal Particles

The viscosity of a liquid can be observed in its flow if a velocity gradient g exists in this flow. In the simplest case of a flow with a constant velocity gradient g (i.e., laminar flow) in a spatial system of X, Y, Z coordinates (Fig. 2.1) the velocity components of the liquid are given by

$$v = v_x = gy, \; v_y = v_z = 0. \tag{2.13}$$

Under these conditions the shear stress $\Delta \tau$ in the liquid is proportional to the velocity gradient according to Newton's law

$$\Delta \tau = \Delta \tau_x = \eta \frac{dv_x}{dy} = \eta g. \tag{2.14}$$

The viscosity coefficient η is the proportionality factor. When a velocity gradient exists, liquid flow is accompanied by the dissipation of energy expended on friction, and the viscosity coefficient may serve as a measure of this dissipation. Applying Eq. (2.14), it is possible to show that the power E consumed for friction per unit volume of the liquid is given by

$$E = \eta g^2. \tag{2.15}$$

If the particles or macromolecules are much larger than the molecules of the liquid, they perturb the flow, and this leads to additional losses in energy for friction and to a corresponding increase in the viscosity of a suspension or a solution. Hence, the difference in the viscosities of the solution η and of the solvent η_0 is determined by the general equation

$$\eta - \eta_0 = (E - E_0)/g^2, \tag{2.16}$$

where E and E_0 are the powers consumed for friction per unit volume of the solution and the pure solvent, respectively.

The problem of the viscosity of a solution of spherical particles has been solved by Einstein [15] on the basis of the Navier–Stokes equations under the limiting conditions of equal velocities of the liquid and a particle on their interface and equal velocities of the liquid and the unperturbed flow at points situated far from the particle.

The calculation of the energy dissipated in a dilute solution under these conditions and its comparison with the energy dissipated in the absence of particles leads to Einstein's equation:

$$\eta = \eta_0 (1 + 2.5\Omega), \tag{2.17}$$

where Ω is the fraction of volume occupied by particles in solution. Equation (2.17) is equivalent to the equation for intrinsic viscosity $[\eta]$:

$$[\eta] \equiv \lim \left(\frac{\eta - \eta_0}{\eta_0 c} \right)_{c \to 0} = 2.5 \bar{v}, \tag{2.18}$$

where c is the weight concentration of the solution, and $\bar{v}$ is the partial specific volume of the solute.

Equation (2.18) shows that the intrinsic viscosity of a solution of spherical compact particles (macromolecules) is independent of their size, i.e., their molecular weight M, and is determined by their density ρ alone, where $\rho \approx 1/v$.

Einstein's method has been applied by Simha [16] to the calculation of the viscosity of solutions containing ellipsoidal particles (macromolecules) under the conditions of weak hydrodynamic flow and the prevailing effect of Brownian rotational motion of particles.

In the general case, when the hydrodynamic flow cannot be regarded as weak, it is necessary to consider in greater detail the motion of asymmetrical dissolved macromolecules. This investigation has been carried out by Jeffery [17], who showed that in a laminar flow [described by Eqs. (2.13) and shown in Fig. 2.1] a spheroidal particle moves nonuniformly. Here the motion is described by the equations

$$\omega_\varphi = d\varphi/dt = -(g/2)(1 - b_0 \cos 2\varphi), \tag{2.19a}$$

$$\omega_\vartheta = d\vartheta/dt = (gb_0/4) \sin 2\vartheta \sin 2\varphi, \tag{2.19b}$$

where

$$b_0 = (p^2 - 1)/(p^2 + 1) \tag{2.20}$$

characterizes the shape asymmetry of the spheroid. According to Eq. (2.19a), the rotation of particles in the XY plane, i.e., about the Z axis (Fig. 2.1), is nonuniform, which results in the appearance of a "kinematic orientation" of particles with a maximum of the distribution function $\rho(\varphi, \vartheta)$ of their longitudinal axes in the flow direction X. In order to take into account the effect of Brownian rotational motion on the distribution function $\rho(\varphi, \vartheta)$ under stationary conditions, it is necessary to solve the differential equation of rotational diffusion:

$$\frac{1}{\sin \vartheta} \frac{\partial}{\partial \vartheta}\left(\sin \vartheta \frac{\partial \rho}{\partial \vartheta}\right)+\left(\frac{1}{\sin^2 \vartheta}\right) \frac{\partial^2 \rho}{\partial \varphi^2}-\frac{3 g b_0 W}{2kT} \sin^2 \vartheta \sin 2\varphi = 0. \qquad (2.21)$$

This solution has been obtained by Peterlin [18] in the form of a power series of the parameter $\sigma = gW/kT$:

$$4\pi\rho(\varphi, \vartheta) = 1 + (\sigma b_0/4) \sin 2\varphi \sin^2 \vartheta$$
$$- (\sigma^2 b_0/8)[(\cos 2\varphi \sin^2 \vartheta)/3 - b_0(\cos 4\varphi \sin^4 \vartheta - \sin^4 \vartheta + 8/15)/8] + \cdots \qquad (2.22)$$

It follows from Eq. (2.22) that in a weak flow, i.e., under the conditions of the prevailing Brownian motion ($\sigma \ll 1$), the maximum of the distribution function $\rho(\varphi, \vartheta)$ in the XY plane corresponds to the direction at an angle $\varphi_m = p/4$ to the flow (Fig. 2.1). By using Eq. (2.22), Kuhn [19] calculated the intrinsic viscosity [h] of solutions of spheroidal particles and explained the dependence of [η] on the shear rate g. For weak flows ($\sigma \ll 1$) Kuhn's result for [η] virtually coincides with Simha's equations [16] and can be expressed by the general equation

$$[\eta] = N_A V \nu(p)/M, \qquad (2.23)$$

where V and M are the volume and molecular weight of the particle, respectively, and $\nu(p)$ is the function of shape asymmetry, where $p = L/a$.

Over various ranges of p values the function $\nu(p)$ may be expressed [19, 7] by the equations

$$\nu(p) = 2.5 + 0.4075(p - 1)^{1.508}, \qquad (2.24)$$

for $1 \leq p \leq 15$ and

$$\nu(p) = 1.6 + \frac{p^2}{5}\left[\frac{1}{3(\ln 2p - 1.5)} + \frac{1}{\ln 2p - 0.5}\right], \qquad (2.25)$$

for $p > 15$. For $p = 1$, $\nu(p) = 2.5$ and Eq. (2.23) coincides with Eq. (2.18).

For a very high degree of elongation ($p \gg 1$), Eq. (2.25) becomes

$$\nu(p) = \frac{4p^2}{15} \frac{1}{\ln 2p - 0.75}. \qquad (2.26)$$

It follows from Eqs. (2.23)–(2.26) that the intrinsic viscosity of a solution of rodlike molecules whose molecular weight M increases proportionally to L (for constant a) cannot be expressed by a simple power dependence on M [of the type of Eq. (2.7)] with a constant value of the exponent b. According to Eqs. (2.23)–(2.26), for these molecules for $p \approx 20$ the value of b is 1.55, for $p \approx 100$ it is 1.8, and as p increases (and, correspondingly, M), it approaches 2.

A universal relationship exists between the rotational friction coefficient W of spheroidal molecules and the intrisic viscosity $[\eta]$ of their solutions. This relationship results from a comparison of Eqs. (2.10) and (2.23):

$$M\,[\eta]\,\eta_0/WN_A = \nu\,(p)/f_0\,(p) = F\,(p). \tag{2.27}$$

The function of shape asymmetry of a spheroid $F(p)$ changes with p over the range determined directly from a comparison of the limiting expressions for $[\eta]$ and W. Thus, for a sphere ($p = 1$), according to Eqs. (2.9) and (2.18), we have $F(p) = 5/12$, and for a rod with the maximum extension ($p \to \infty$), according to Eqs. (2.12) and (2.15), $F(p) \to 2/15$. For $p = 10$, $F(p)$ is equal to $1/6$.

1.4. "Equivalent" Spheroid

The above expressions describing the transport properties of spheroidal particles are the most reliable because they are obtained by rigorous solutions of the equations of classical hydrodynamics. However, for the study of the hydrodynamic properties of long rodlike molecules a straight thin cylinder appears to be a model that corresponds better to the true shape of the molecule than an ellipsoid of revolution. Hence, when experimental data are compared with the above theoretical equations, the latter are sometimes regarded as relationships describing the hydrodynamics of an "equivalent" spheroid, i.e., a spheroid whose volume and major axis are equal to the volume and length L of a cylinder taken as the model for a rodlike molecule. This assumption implies that the transverse (i.e., minor) axis a of the equivalent ellipsoid should be longer than the diameter d of the cylindrical model: $a = (3/2)1/2d$. In accordance with this condition, if all the above equations are applied to equivalent spheroids, the degree of asymmetry should be taken to be $p = (2/3)1/2L/d$, where L is the contour length of the macromolecule, and d is its diameter. Proceeding from this assumption, for an "equivalent" spheroid Eqs. (2.6), (2.12), and the combination of Eqs. (2.23) and (2.26) are transformed into Eqs. (2.6'), (2.12'), and (2.26'), respectively:

$$f = 3\pi\eta_0 L/[\ln\,(L/d) + \gamma], \tag{2.6'}$$

where $\gamma = 0.49$;

$$W = (\pi\eta_0 L^3/3)/[\ln\,(L/d) + \gamma], \tag{2.12'}$$

where $\gamma = -0.01$; and

$$[\eta] = (2\pi N_A L^3/45M)/[\ln\,(L/d) + \gamma], \tag{2.26'}$$

where $\gamma = -0.26$.

Although these equations probably improve the situation and the properties of this model approach the behavior of real rodlike macromolecules, the question about the extent to which a spheroid is hydrodynamically equivalent to a rigid rod of the same length and volume still remains unanswered. Consequently, the next step of the theory bringing it closer to the description of real rodlike molecules was the calculation of the rotational and translational friction of long cylindrical particles.

2. TRANSLATIONAL AND ROTATIONAL FRICTION OF CYLINDRICAL PARTICLES

2.1. The Oseen–Burgers Method. Hydrodynamic Interaction

Of great importance to these investigations was the possibility of using the method developed by Oseen [20], who represented hydrodynamic equations as relationships describing the perturbation in a liquid due to the effects of external forces applied directly to volume elements (points) of the liquid. Burgers has shown [21] that Oseen's equations can be simplified to

$$v'_x = \frac{F_x}{8\pi\eta_0}\left(\frac{1}{r} + \frac{x^2}{r^3}\right),$$

$$v'_y = \frac{F_x}{8\pi\eta_0}\frac{xy}{r^3}, \tag{2.28}$$

$$v'_z = \frac{F_x}{8\pi\eta_0}\frac{xz}{r^3},$$

where F_x is the x component of the point force developed in a liquid (with viscosity η_0) at the origin of a rectangular system of the XYZ coordinates, and v_x', v_y', and v_z' are the velocity components of the perturbed motion of the liquid caused by the force F_x at the point xyz located at a distance r from the origin, i.e., from the point of application of force F_x. Expressions analogous to Eq. (2.28) can be written for the velocity components of the perturbed motion caused by the force components F_y and F_z.

In vector form, Eq. (2.28) becomes

$$V' = TF, \tag{2.29}$$

where T is Oseen's hydrodynamic interaction tensor, which is determined by the equation

$$T = \frac{1}{8\pi\eta_0 r}\left(I + \frac{rr}{r^2}\right), \tag{2.30}$$

where I is the unit tensor, and r is the vector of the distance between the point of application of force F and the point of perturbation. According to Eqs. (2.27) and (2.30), the perturbation due to a point force in a liquid (hydrodynamic interaction) is inversely proportional to distance.

Equation (2.28) has been used by Burgers to calculate the translational and rotational resistance experienced by a long straight cylinder moving in a viscous liquid. Here the cylinder is mentally removed from the liquid and replaced with a system of forces acting on the liquid and distributed along the axis of the cylinder.

According to Eq. (2.29), these forces perturb the velocities v_x', v_y', and v_z' at those points in the liquid whose geometrical locus was the surface of the cylinder when it was present in the liquid. The totality of forces distributed along the axis of the cylinder is equal in value and opposite in sign to the force of viscous resistance $-F$ experienced by the cylinder in moving translationally at a velocity u in the liquid. For rotational motion the sum of moments of forces distributed along the axis of the cylinder is equal in value and opposite in sign to the moment of forces of viscous resistance of the surrounding liquid. To find the totality of these forces, one should adopt the usual hydrodynamic condition that the velocities of the liquid and the moving body at its boundary are equal. According to Burgers and Oseen, this principal condition is formulated as the requirement that, for any section of the cylinder cut by a plane normal to its axis, the average (over the contour of the section) velocity v' of the perturbed motion of the liquid should be equal to the velocity of the motion of the cylinder u in this section, i.e., $v' - u = 0$.

2.2. Translational Friction of a Cylinder

If a cylinder moves translationally in the direction of the X axis coinciding with the axis of the cylinder (Fig. 2.2) and the origin is located at the center of the cylinder, then the force $\xi(\zeta)\,d\zeta$ applied to an element of length $d\zeta$ of the axial line causes perturbation dv_x' at point P on the surface of the cylinder. According to Eq. (2.28), this perturbation is given by

$$dv_x' = \frac{\xi(\zeta)\,d\zeta}{8\pi\eta_0}\left[\frac{1}{r}+\frac{(x-\zeta)^2}{r^3}\right],\tag{2.31}$$

where x is the coordinate of the normal plane section to which point P belongs, and r, the distance between the point of application of force and the point P, is evidently given by

$$r = [(x-\zeta)^2+d^2/4]^{1/2},\tag{2.32}$$

where d is the cylinder diameter.

Since the cylinder is symmetrical with respect to the X axis (Fig. 2.2), the value of dv_x' is identical for all points of the chosen section x. The average value of components dv_y and dv_z taken at the same section x is equal to zero because Eq. (2.28) is symmetrical for these components.

The total value of perturbation v_x' at any point of section x on the surface of the cylinder is obtained if Eq. (2.31) is integrated over all values of ζ from $-L/2$ to $+L/2$, where L is the cylinder length. The application of the principal condition $v' - u = 0$ to Eqs. (2.31) and (2.32) gives

$$\frac{1}{8\pi\eta_0}\int_{-L/2}^{+L/2}\xi(\zeta)\left\{\frac{1}{[(x-\zeta)^2+d^2/4]^{1/2}}+\frac{(x-\zeta)^2}{[(x-\zeta)^2+d^2/4]^{3/2}}\right\}d\zeta - u = 0.\tag{2.33}$$

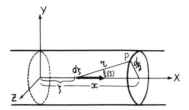

Fig. 2.2. Cylindrical particle moving in a viscous liquid in the direction parallel to its axis.

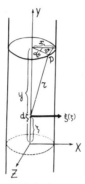

Fig. 2.3. Cylindrical particle in a viscous liquid moving translationally in the direction X normal to its long axis Y or rotating about the Z axis.

Equation (2.33) is an integral equation for the unknown function on $\xi(\zeta)$ related by the equation

$$F = \int_{-L/2}^{+L/2} \xi(\zeta)\,d\zeta \tag{2.34}$$

to the total force F moving the cylinder in the liquid at a velocity u.

Equation (2.33) is solved approximately with some simplifying assumption, one of which is the condition $L/d \gg 1$. The expression for $\xi(\zeta)$ obtained in this manner is substituted into Eq. (2.39) whose integration leads to a dependence of F on u that is analogous to Eq. (2.1). In this case, the translational friction coefficient of the cylinder in its motion along the long axis is given by

$$f_1 = 2\pi\eta_0 L/[\ln(L/d) - 0.03]. \tag{2.35}$$

A similar treatment is also carried out for the motion of the cylinder along the X axis normal to its longitudinal axis Y (Fig. 2.3). In this case, forces $\xi(\zeta)$ moving the cylinder are directed along the X axis but are distributed along the $d\zeta$ elements of the Y axis. Accordingly, the distance between the point of application of force and the point P of perturbation is $r = [(y - \zeta)^2 + d^2/4]^{1/2}$, and the x coordinates of points P of the normal section are different for different points and should be averaged over the contour of the section: $\bar{x}^2 = (d/2)^2 \overline{\cos^2 \vartheta} = d^2/8$. If the foregoing considerations and the principal condition $v' - u = 0$ are taken into account, Eqs. (2.28) yield the integral equation

$$\frac{1}{8\pi\eta_0} \int_{-L/2}^{+L/2} \xi(\zeta) \left\{ \frac{1}{[(y-\zeta)^2 + d^2/4]^{1/2}} + \frac{d^2}{8[(y-\zeta)^2 + d^2/4]^{3/2}} \right\} d\zeta - u = 0. \qquad (2.36)$$

The solution of Eq. (2.36) under the condition that $L/d \gg 1$ and with the application of Eq. (2.34) gives the following expression for the translational friction coefficient of the cylinder in the direction normal to its axis:

$$f_2 = 4\pi\eta_0 L/[\ln(L/d) + 1.19]. \qquad (2.37)$$

According to Eqs. (2.36) and (2.37), the average value for the random distribution of axial orientations is given by

$$f = 3\pi\eta_0 L/[\ln(L/d) + 0.58]. \qquad (2.38)$$

Hence, the form of the expression for the friction coefficient f of a long cylinder is identical to that for a long spheroid, and the value of $\gamma = 0.58$ obtained by Burgers lies between those for a long spheroid [according to Eq. (2.6)] and a long "equivalent" spheroid [according to Eq. (2.6')].

2.3. Rotational Friction of a Cylinder

Equations (2.28) can also be used to calculate the rotational friction of a cylinder. The rotational motion of a cylindrical particle is also illustrated in Fig. 2.3. In this case the local forces $\xi(\zeta)$ are distributed along the Y axis of the cylinder and are parallel to the X axis. However, in contrast to what happens in translational motion, these forces are directed toward positive x values at $y > 0$ and toward negative x values at $y < 0$. Hence, these forces induce a torque with respect to the Z axis. The value of this moment $\mathscr{L}$ is evidently given by

$$\mathscr{L} = \int_{-L/2}^{+L/2} \xi(\zeta)\zeta\,d\zeta = W\omega, \qquad (2.39)$$

where ω is the angular velocity of rotation of the cylinder, and W is the rotational friction coefficient of the cylinder corresponding to its rotation about the central transverse axis.

The integral equation for the determination of the function $\xi(\zeta)$ is found using Eq. (2.28) from the principal condition that the perturbed velocity v_x' of a liquid in the y section of the cylinder is equal to the linear velocity of motion of the cylinder ωy at this section. In accordance with this condition, by analogy with Eq. (2.36) we have

$$(8\pi\eta_0)^{-1} \int_{-L/2}^{+L/2} \xi(\zeta) \left\{ [(y-\zeta)^2 + d^2/4]^{-1/2} + d^2/8 [(y-\zeta)^2 + d^2/4]^{3/2} \right\} d\zeta = \omega y. \qquad (2.40)$$

In the approximation $L/d \gg 1$ the solution of Eqs. (2.40) and (2.39) yields the expression for the rotational friction coefficient

$$W = \pi\eta_0 L^3/3 \, [\ln(2L/d) - 0.8] = \pi\eta_0 L^3/3 \, [\ln(L/d) + \gamma], \qquad (2.41)$$

where $\gamma = -0.11$.

Equation (2.41) differs from the corresponding Eqs. (2.12) and (2.12') for a spheroid in a slightly higher negative value of the coefficient γ.

2.4. Refinement of Burgers' Equations

When Burgers' method was subsequently used in the study of translational and rotational friction, attempts were made to take into account the finite length of the cylinder and the effect of the top and bottom of the cylinder on its hydrodynamic properties. Thus, Broersma [22] proposed to take into account the top and bottom effect in the rotation of a cylinder bounded by plane ends expressing γ in Eq. (2.41) as a function $\ln(L/d)$. According to [22], for values of $\ln(2L/d) > 2$, the coefficient γ can be approximately written in the form

$$\gamma = -0.877 + 7 \{[1/\ln(2L/d) - 0.28]^2\}. \qquad (2.42)$$

The asymptotic limit of γ as $L/d \to \infty$ is -0.447. The absolute magnitude of this value exceeds the corresponding values for both a spheroid [Eqs. (2.12) and (2.12')] and a long cylinder according to Burgers [Eq. (2.41)].

Analogous calculations in which terms of the order of magnitude of d/L are taken into account have also been carried out for the determination of the translational friction of cylindrical particles of finite length bounded at their ends by planes [23]. For values of $\ln(2L/d) > 2$, equations expressing the translational friction coefficients f_1 and f_2 for the motion of a cylinder along its axis and transverse to it, respectively, have been proposed. These equations are similar to Burgers' equations (2.35) and (2.37), except that the constant coefficients -0.03 and 1.19 are replaced by the functions γ_1 and γ_2, respectively:

$$\begin{aligned} \gamma_1 &= -0.607 + 8 \{[1/\ln(2L/d) - 0.30]^2\}, \\ \gamma_2 &= 0.343 + 4 \{[1/\ln(2L/d) - 0.43]^2\}. \end{aligned} \qquad (2.43)$$

The values of $\gamma_1 = -0.12$ and $\gamma_2 = 0.88$ are asymptotic limits as $L/d \to \infty$. Hence, according to Broersma, the average value of the translational friction coefficient f for a long cylinder is determined by Eq. (2.38), in which $\gamma = (1/2)(\gamma_1 + \gamma_2) = 0.38$, instead of the value 0.58 found by Burgers.

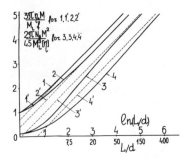

Fig. 2.4. Translational friction and intrinsic viscosity vs. particle length for spheroids and spherocylinders. Curves 1 and 2: $3\pi\eta_0 M/fM_L$ vs. $\ln{(L/d)}$ according to Eqs. (2.5) and (2.44), respectively. Curves 3 and 4: $2\pi N_A M^2/45 M_L^3[\eta]$ vs. $\ln{(L/d)}$ according to Eqs. (2.49) and (2.45), respectively. Broken straight lines 1'–4' are asymptotic limits (at $L/d \to \infty$) of curves 1-4, respectively.

Rotational friction of cylinders of finite length (for $L/d \geq 62$) has been estimated [24] by computer, taking into account terms up to $(d/L)^3$ but neglecting the top and bottom effects. The result is described by Eq. (2.41), in which the coefficient γ is expressed as a power series in d/L. As $L/d \to \infty$, the value of γ is -0.697 and its absolute value is higher than those for a spheroid [Eq. (2.26)] and an "equivalent" spheroid [Eq. (2.26')]. Comparison of expressions for the rotational friction coefficient and the intrinsic viscosity [η] yields an equation that is analogous to Eq. (2.27). In this case the function $F(p)$ varies from $\bar{F} = 2/15 = 0.133$ to $F \approx 0.126$, when the shape asymmetry of the cylinder $p = L/d$ changes from ∞ to 10^2 similarly to the relationship that applies for spheroids.

2.5. Hydrodynamic Characteristics of Spherocylinders

Great progress in the calculation of the hydrodynamic characteristics of cylindrical particles has been made in papers which considered cylinders bounded at their tops and bottoms by spherical surfaces (spherocylinders) [25] or spheroidal surfaces (spheroidocylinders) [26]. The use of these models, whose surfaces are continuous at all points, made it possible to obtain expressions for f, W, and [η] for straight cylindrical particles over the entire range $1 \leq p \leq \infty$.

2.5.1. Translational Friction. Using the Oseen–Burgers method, an expression relating the translational friction coefficient f of spherocylindrical particles to their length L and diameter d has been obtained [25]:

$$3\pi\eta_0 L/f = C_1 \ln{(L/d)} + C_2, \tag{2.44}$$

where $C_1 = 1$, $C_2 = 0.3863 + 0.6863(d/L) - 0.06250(d/L)^2 - 0.01042(d/L)^3 - 0.000651(d/L)^4 + 0.0005859(d/L)^5 + \dots$. At its limiting values, Eq. (2.44) is equivalent to Eq. (2.5) for spheroids. Thus, for $L/d = 1$ we have $f = 3\pi\eta_0\,d$, which corresponds to the f value for a sphere [see Eq. (2.2)] whose diameter is equal to that of the cylinder. As $L/d \to \infty$, Eq. (2.44) assumes the form of Eq. (2.6) or

(2.6') with the value of $\gamma = 0.3863$, which corresponds to that obtained by Broersma [23]. The overall trend of the dependence of $3\pi\eta_0 L/f = 3\pi\eta_0 M/(M_L f)$ on $\ln (L/d)$ is shown in Fig. 2.4 for an ellipsoid of revolution (spheroid) according to Eq. (2.5) (curve 1) and for a spherocylinder according to Eq. (2.44) (curve 2). Here $M_L \equiv M/L$ is the average molecular weight per unit length of the particle. Curves 1 and 2 coincide for $L/d = 1$, whereas for high L/d ratios they degenerate to straight lines with equal slopes separated from each other along both the abscissa and the ordinate by the value $0.693 - 0.386 = 0.307$. Dashed lines 1 and 2 show the asymptotic dependences according to Eqs. (2.6) and (2.44) for $C_2 = 0.386$, respectively.

2.5.2. Intrinsic Viscosity.

Yamakawa [26] has calculated not only f but also the coefficients of rotational friction W and intrinsic viscosity $[\eta]$ for straight cylinders of arbitrary length bounded at the top and bottom by the surfaces of spheres or elongated and flattened spheroids. The results of calculating f for spherocylinders are in complete agreement with the data in [25], i.e., with Eq. (2.44). Analysis of the equations obtained for spheroidocylinders has shown that the shape of cylinder ends (sphere, elongated or flattened spheroid) affects only very slightly its translational and rotational friction. For viscosity this effect is somewhat more important but is pronounced only for L/d values close to unity. Hence, in the following discussion only the data for spherocylinders with a degree of asymmetry in the range $1 < p \leq \infty$, i.e., for particles representing rodlike molecules, will be reported.

The equations for rotational friction and viscosity of solutions of spherocylinders for high L/d ratios have been obtained [24, 25] by using the Oseen–Burgers method. For short spherocylinders, friction forces were determined by the methods of classical hydrodynamics. Empirical interpolation formulas adequately satisfying the limiting precise equations for $L/d \approx 1$ and $L/d \to \infty$ were constructed for the entire range of L/d ratios.

Thus, the interpolation formula for the intrinsic viscosity is given by

$$2\pi N_A M^2/45 M_L^3 [\eta] = F_\eta(p), \tag{2.45}$$

where, as before, $p = L/d$.

The function $F_\eta(p)$ is determined by the expression

$$F_\eta(p) = \ln p + 2\ln 2 - 25/12 + [\ln 2/\ln (1 + p)]\left(8/75 - 2\ln 2 + 25/12 - \sum_{i=1}^{5} b_i\right)$$

$$+ b_1/p^{1/4} + b_2/p^{1/2} + b_3/p^{3/4} + b_4/p + b_5/p^{5/4}, \tag{2.46}$$

where the coefficients b_i have the values $b_1 = 3.60517$, $b_2 = -15.35634$, $b_3 = 28.0617$, $b_4 = -14.8934$, and $b_5 = 3.553442$ and, correspondingly, $\Sigma_{i=1}^{5} b_i = 4.970572$.

For $L/d = 1$, it follows from Eq. (2.45) that $F(1) = 8/75$ and, hence,

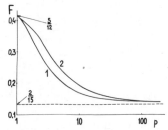

Fig. 2.5. Coefficient $F = \eta_0 M[\eta]/N_A W$ vs. degree of asymmetry p for spheroids (curve 1) and spherocylinders (curve 2).

$$[\eta] = 5\pi N_A d^3/12M = 2.5 N_A v/M, \tag{2.47}$$

which corresponds to Eq. (2.18), i.e., to the intrinsic viscosity of a solution of spheres whose diameter is equal to that of the cylinder.

In another limiting case of very long cylinders (i.e., $L/d \rightarrow \infty$), Eq. (2.46) gives

$$F_\eta(p) = \ln p + 2\ln 2 - 25/12 = \ln(L/d) - 0.697. \tag{2.48}$$

This corresponds to the results reported in [24] and to Eq. (2.26') with the value of $\gamma = -0.697$.

Curve 4 in Fig. 2.4 shows the dependence of $2\pi N_A M^2/45 M_L^3[\eta]$ on $\ln p$ corresponding to Eq. (2.45).

Dashed line 4' represents the asymptotic limit of curve 4 according to Eq. (2.48).

Curve 3 in Fig. 2.4 represents a similar dependence for spheroids according to Eq. (2.23), which is transformed into the equivalent equation

$$2\pi N_A M^2/45 M_L^3[\eta] = 4p^2/15v(p), \tag{2.49}$$

where $v(p)$ is determined by Eqs. (2.24) and (2.25).

Dashed line 3' represents the asymptotic limit of curve 3, in which $v(p)$ is expressed according to Eq. (2.26), which corresponds to Eqs. (2.26') or (2.48) with the coefficient $\gamma = -0.057$.

Curves 3 and 4, representing the viscous properties of spheroids and spherocylinders, just as curves 1 and 2 do for translational friction, start at one point with the ordinate 8/75 and, as the particle length increases, approach their asymptotic limits, i.e., straight lines 3' and 4' displaced with respect to each other along the axes of the coordinates by the value $0.697 - 0.057 = 0.64$. However, the asymptote is approached much more slowly (for higher L/d ratios) for the viscosity curves than for the translational friction curves. Hence, it might be expected that in a homologous series of long straight molecules, when they decrease in length, their "rodlike" properties will be retained much longer in the phenomena of translational friction than in viscometry.

2.5.3. Rotational Friction. Similarly to Eq. (2.45), the rotational friction coefficient W of a spherocylinder with respect to the central axis normal to the axis of symmetry of a particle can be expressed [26] by the equation interpolating the hydrodynamic properties of the particle from a sphere to an infinitely long cylinder:

$$W = (\pi \eta_0 L^3/3) F_r(p) = (\pi \eta_0 d^3/3) p^3 F_r(p), \tag{2.50}$$

where $F(p)$ is determined by the equation

$$1/F_r(p) = \ln p + 2 \ln 2 - 11/6 + [\ln 2/\ln(1 + p)]$$
$$\times \left(1/3 - 2 \ln 2 + 11/6 - \sum_{i=1}^{6} a_i \right) + \sum_{i=1}^{6} a_i p^{-i/4}. \tag{2.51}$$

The numerical coefficients a_i have the following values: $a_1 = 13.04468$, $a_2 = -62.6084$, $a_3 = 174.0921$, $a_4 = -218.8356$, $a_5 = 140.26992$, and $a_6 = -33.27076$.

For $p = 1$, according to Eq. (2.50), we have $F_r(p) = 3$ and, correspondingly, $W = \pi \eta_0 d^3$; i.e., it is equal to the rotational friction coefficient of a sphere whose diameter is equal to that of a spherocylinder.

As $p \to \infty$,

$$1/F_r(p) = \ln p + 2 \ln 2 - 11/6 = \ln p - 0.447, \tag{2.52}$$

which corresponds to the results reported in [22, 24] and to Eq. (2.12') with the value $\gamma = -0.447$.

Comparison of Eqs. (2.45) and (2.50) shows that for spherocylindrical particles, just as for spheroids, the values of W and $[\eta]$ are related to each other by a general equation (2.27) in which the function $F(p)$ is given by

$$F(p) = 2/(15 F_\eta F_r). \tag{2.53}$$

The dependence of $F(p)$ on p is shown in Fig. 2.5 by curve 1 for spheroids according to Eqs. (2.27), (2.11), (2.23), and (2.24), and by curve 2 for spherocylinders according to Eqs. (2.27), (2.53), (2.46), and (2.51). Curves 1 and 2 coincide at low and high p values. In the intermediate range they are similar in shape but the curve for spherocylinders is characterized by a slower approach to the asymptotic limit than for spheroids.

2.6. Possibility of Application to Real Extended Chain Molecules

Hence, when the hydrodynamic properties of rodlike molecules are described by using ellipsoids of revolution or cylinders as models, the quantitative relationships obtained for these two models are similar.

Consequently, in many cases calculations of the geometrical characteristics of rodlike macromolecules based on the data of their hydrodynamic properties in solutions can lead to identical results regardless of whether ellipsoidal or cylindrical particles are used as models.

However, a necessary condition for the validity of these results is that the real macromolecules have to be actually rodlike over the molecular-weight range investigated. A plot based on the relationships shown in Fig. 2.4 may serve as a practical method for the verification of this point. In fact, if the experimental values of M/f or $M^2/[\eta]$ for a homologous series of rodlike molecules with a constant diameter d are represented as functions of $\ln M$, the experimental points should fit the asymptotic straight lines satisfying the equations

$$\frac{M}{f} = \frac{M_L}{3\pi\eta_0} [\ln M - \ln(M_L d) + \gamma_f], \tag{2.54}$$

$$\frac{M^2}{[\eta]} = \frac{45 M_L^3}{2\pi N_A} [\ln M - \ln(M_L d) + \gamma_\eta]. \tag{2.55}$$

The presence of a distinct straight region on the experimental curve supports the fact that over the corresponding range of molecular weights the macromolecules may be regarded as rodlike.

In this case, it is possible to determine an important structural parameter, the molecular weight per unit length of the molecule $M_L = M_0/\lambda$, from the slopes of the straight lines regardless of whether the molecule is represented by an ellipsoid or a cylinder.

In principle it is also possible to determine the diameter d of the rodlike molecule from the intercepts of the asymptotic straight lines with the axes of the coordinates according to Eqs. (2.53) and (2.55). However, the determination of d is much less reliable than that of M_L. This is due both to a pronounced effect of experimental errors and to the dependence of the result on the values of γ_f and γ_η which differ in different theories and for different models.

If any real chain molecule is sufficiently short (i.e., if M is small), its hydrodynamic properties deviate from those of a thin rod because the chain

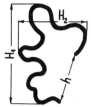

Fig. 2.6. Parameters characterizing the size of a chain molecule according to Kuhn. h is the end-to-end distance of the chain, H_1 is the distance between the most distant points of the chain (longitudinal chain dimensions), and H_2 is the distance between the most distant points of the chain in the direction normal to H_1 (maximum transverse chain dimensions).

diameter is finite. These deviations are taken into account in the theories of spheroids and spherocylinders considered above.

However, there is another reason for the deviation of the hydrodynamic properties of any real molecule from those of a rod: its chain flexibility is increasingly manifested with increasing chain length. This characteristic is of paramount importance for most known polymer molecules. Hence, the existence of the linear region of the curves in Fig. 2.4 cannot be a general rule for real chain molecules. It may be expected only in relatively rare cases of polymers exhibiting high-equilibrium chain rigidity. However, the models used above, i.e., spheroids and straight spherocylinders, are not intended for the discussion of these problems.

3. CHAIN MOLECULES

The hydrodynamic properties of chain molecules in dilute solutions have long been the object of comprehensive theoretical and experimental studies yielding extensive results widely considered in the literature [5, 7, 27–31]. Hence, in this book only the aspects of the problem important for the hydrodynamics of rigid-chain polymers will be considered. The hydrodynamic characteristics of flexible-chain polymer molecules are profoundly affected by the thermodynamic strength of the solvent and the excluded-volume effects since these factors determine considerably the size of a flexible molecular coil in solution. Hence, the modern theory of hydrodynamic behavior of real flexible-chain molecules devotes much attention to the influence of volume effects on the friction and viscometric properties of chains in dilute solutions [30].

For rigid-chain polymers the excluded volume effects have little influence on chain conformation (see Chapter 1) and, hence, to the first approximation may be neglected in discussing the hydrodynamic properties of the molecule. On the other hand, the extended chain conformation of a rigid-chain polymers and relatively "loose" structure of its molecular coil (Fig. 1.5) should weaken the average hydro-dynamic interaction between chain segments as compared with that for a flexible-chain polymer. However, in contrast to the conformation of a real flexible-chain molecule, that of different molecules of a rigid-chain polymer can differ over a wide range, from a rodlike conformation to a Gaussian coil, accompanied by changes in the average value of hydrodynamic interaction between chain segments and, correspondingly, by hydrodynamic changes. Hence, the problems of intramolecular hydrodynamic interaction and its dependence on chain conformation are of major importance in the theory of hydrodynamic properties of rigid-chain polymers.

3.1. Kuhn's Model Experiments

Among early works dealing with these problems, the papers by Kuhn and his co-workers should be mentioned. They have investigated the translational and rotational friction of chain molecules in experiments with wire models [7, 32–36]. A cylindrical wire of length L and diameter d served as a model for a chain molecule. It curved continuously such that it consisted of flat round segments of equal length but different radii of curvature and spatial orientations of segment planes. Moreover, the condition that the spatial orientations of tangents to a segment at its beginning and end be mutually independent was satisfied. Evidently, the contour of this model was a continuous line in space exhibiting a curvature changing from one segment to the other.

The translational friction of wire models was determined from the velocity of their fall in a viscous liquid at different possible orientations in space. In order to determine the rotational friction coefficient, the model, under the influence of a constant moment, was rotated in a viscous medium about the axis to the direction H_1 corresponding to the maximum size of the model (Fig. 2.6). The direction of the axis of rotation was varied (remaining normal to H_1) and the values of rotational friction coefficients were averaged.

To apply the results obtained from models to real chain molecules, a principle of hydrodynamic similarity was used according to which, if the similarity of shape of the macromolecule is retained, its translational friction coefficient varies proportionally to its linear dimensions while the rotational friction coefficient varies proportionally to the cube of linear dimensions.

Although hydrodynamic experiments were carried out on models with circular segments, in the final equations friction parameters are expressed by the length A of linear segments of a freely jointed chain equivalent to a real chain molecule. Hence, the expressions for the translational (f) and rotational (W) friction of a chain molecule have been obtained. Using the symbols employed in the foregoing discussion, these expressions are given by

$$\eta_0 M/f = 0.136\,(M_L/A)^{1/2}\,M^{1/2} + 0.069 M_L\,[\ln{(A/d)} - 0.43], \qquad (2.56)$$

$$\eta_0 M^2/W = 0.47\,(M_L/A)^{3/2}\,M^{1/2} + 0.47\,(M_L)^2\,A^{-1}\,[\ln{(A/d)} - 1.54]. \qquad (2.57)$$

Although the numerical coefficients in these empirical equations can be regarded only as approximate values, the general type of equations (2.56) and (2.57) enabled Kuhn to correctly characterize the change in the hydrodynamic properties of chain molecules occurring with a variation in their length (i.e., their molecular weight M).

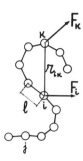

Fig. 2.7. Intramolecular hydrodynamic interaction in a chain molecule according to Kirkwood (pearl necklace model). i and k are beadlike chain elements: sources of forces F_i and F_k acting upon the liquid at points i and k, l is the bond length between neighboring beads, and r_{ik} is the distance between the interacting elements.

Of the two terms on the right-hand side of Eqs. (2.56) and (2.57), the second term is independent of molecular weight, whereas the first term is proportional to $M^{1/2}$. Hence, at relatively high M, the second term may be neglected, and for a Gaussian chain it follows from Eqs. (2.56) and (2.57) that

$$f = (1/0.136)\,\eta_0\,(LA)^{1/2} = 7.3\eta_0\,\langle h^2\rangle^{1/2}, \qquad (2.58)$$

$$W = (\eta_0/0.47)\,(LA)^{3/2} = 2.13\eta_0\,\langle h^2\rangle^{3/2}. \qquad (2.59)$$

Equations (2.58) and (2.59) indicate that at high L values translational friction is proportional to the linear dimensions of a Gaussian coil, whereas the rotational friction is proportional to the cube of its linear dimensions just as for compact particles with a constant degree of asymmetry [Eqs. (2.2) and (2.9)]. This fact implies that, under these conditions, the molecular coil in its motion completely retains the solvent situated in the interior of the coil, i.e., the molecule behaves as a "nondraining" particle.

According to Eqs. (2.56) and (2.57), with decreasing M the role played by the first term on the right-hand side of these equations decreases and at the limit where $M \to 0$ the left-hand side becomes independent of M. Under these conditions, f varies proportionally to M, i.e., to chain length L, and W varies proportionally to L^2. Hence, the resistance experienced by the chain in its translational or rotational motion is simply equal to the sum of resistances which would be experienced by all its elements if each of them moved in the absence of the others (i.e., if the motion of the solvent with respect to some element were not perturbed by the motion of other elements). Kuhn called this molecular coil a "freely draining coil."

According to Eqs. (2.58) and (2.59), the hydrodynamic behavior of a "nondraining" molecular coil is equivalent to that of a compact body, with linear dimensions proportional to $\langle h^2\rangle^{1/2}$ and, correspondingly, a volume proportional to $\langle h^2\rangle^{3/2}$. The transport behavior of flexible-chain molecules is generally considered from this viewpoint [30] and has been widely used by Flory [37], who also took into account the effect of the thermodynamic strength of the solvent on the value of $\langle h^2 \rangle$ [38].

3.2. Hydrodynamic Interaction. Kirkwood's Method

The development of the theory of hydrodynamic properties of polymer molecules is in many respects due to the fundamental papers by Kirkwood et al., who have developed methods for the calculation of hydrodynamic interactions in polymer chains of various configurations [39–43]. Kirkwood's method, just as that of Burgers, is based on the use of Eqs. (2.28)–(2.30). However, Kirkwood uses as a hydrodynamic model for a chain molecule a system of spherical beads joined by frictionless bonds l (the pearl necklace model) (Fig. 2.7) rather than a compact body (cylinder). This model has already been used by Kuhn [43] but he did not take into account hydrodynamic interaction between the beads.

Each ith chain element (bead) in Kirkwood's theory is characterized by a friction coefficient ζ determined according to Eqs. (2.1) and (2.2) from the condition that when an element moves in a fluid at a velocity $\mathbf{u}_i$, it exerts upon it a force $\mathbf{F}_i$ equal to

$$\mathbf{F}_i = \zeta(\mathbf{u}_i - \mathbf{v}_i), \tag{2.60}$$

where $\mathbf{v}_i$ is the velocity that the fluid would have at the point of location of the ith element in the absence of this element. Hence, chain elements are regarded as sources of point forces exerted upon the fluid and producing in it perturbations described by Eqs. (2.28)–(2.30).

Hence, the velocity of the fluid $\mathbf{v}_i$ at a point i is the sum of the velocities of the unperturbed flow $\mathbf{v}_i^0$ (i.e., the velocity of the fluid at this point in the absence of the whole macromolecule) and the perturbation $\mathbf{v}_i'$ caused by the overall action of point forces of all n chain elements other than the ith element (Fig. 2.7):

$$\mathbf{v}_i = \mathbf{v}_i^0 + \mathbf{v}_i'. \tag{2.61}$$

According to Eq. (2.29), $\mathbf{v}_i'$ is given by

$$\mathbf{v}_i' = \sum_{\substack{k=1 \\ k \neq i}}^{n} \mathbf{F}_k \mathbf{T}_{ik}, \tag{2.62}$$

where $\mathbf{T}_{ik}$ is the hydrodynamic interaction tensor of the ith and the kth elements which, in accordance with Eq. (2.30), is given by

$$T_{ik} = \frac{1}{8\pi\eta_0 r_{ik}} \left(I + \frac{r_{ik} \cdot r_{ik}}{r_{ik}^2} \right),$$ (2.63)

where r_{ik} is the vector of the distance between the k and i chain elements (i.e., between the centers of the kth and ith beads).

Substitution of Eqs. (2.61) and (2.62) into Eq. (2.60) gives

$$F_i = \zeta(u_i - v_i^0) - \zeta \sum_{\substack{k=1 \\ k \neq i}}^{n} F_k T_{ik},$$ (2.64)

where T_{ik} is determined using Eq. (2.63).

Equation (2.64) is of fundamental importance because it allows the calculation of the force exerted by the ith chain element on the surrounding solvent. The forces F_i summed in an appropriate manner over all chain elements can characterize the resistance experienced by the molecule in its translational and rotational motion in a solvent and the related losses for friction which contribute to the viscosity of a polymer solution.

The first term on the right-hand side of Eq. (2.64) characterizes the force exerted upon the solvent by the ith chain element in the absence of all other elements, i.e., in the absence of hydrodynamic interaction. The second term characterizes the contribution to this force by all other chain elements, i.e., by hydrodynamic interaction. According to Eq. (2.63), this second term is dependent upon the reciprocal distances $1/r_{ik}$ between chain elements and, hence, on its configuration. However, the calculations carried out for chains exhibiting different conformations and for different types of motion (translational or rotational) show that the expression on the right-hand side of Eq. (2.64) defined by the sign of the sum Σ is always positive. Consequently, since the sum is contained in Eq. (2.64) with the negative sign, intramolecular hydrodynamic interaction always leads to a decrease in the friction forces between the macromolecule and the solvent.

The value of the Oseen tensor averaged over all conformations and spatial orientations of the molecule is often used for the solution of hydrodynamic problems. In this case, $\langle r_{ik} r_{ik}/r_{ik}^2 \rangle = 1/3$, and Eq. (2.63) becomes

$$\langle T_{ik} \rangle = \frac{1}{6\pi\eta_0} \left\langle \frac{1}{r_{ik}} \right\rangle.$$ (2.65)

Using Eqs. (2.64) and (2.65), Kirkwood [43] has obtained a general expression for the friction coefficient f of a molecule represented by a chain necklace of n beads for the translational motion of this molecule as a whole in a solvent with viscosity η_0

$$\frac{1}{f} = \frac{1}{n\zeta} \left(1 + \frac{\zeta}{6\pi\eta_0 n} \sum_{\substack{i \\ i \neq k}}^{n} \sum_{k}^{n} \left\langle \frac{1}{r_{ik}} \right\rangle \right).$$ (2.66)

The first term on the right-hand side of Eq. (2.66) corresponds to a completely "draining" molecule when f increases proportionally to the number n of chain units. The second term characterizes hydrodynamic interaction. Since Eq. (2.66) was obtained with the preaveraged Oseen tensor according to Eq. (2.65), it corresponds to the value of f averaged over all chain orientations and conformations. However, Eq. (2.66) is valid for a chain necklace in any conformation and, therefore, can be used in the calculation of f both for a rodlike molecule and a Gaussian chain.

3.3.   Friction Characteristics of a Rodlike Necklace

If a chain molecule exhibits a rigid rodlike conformation represented by a straight necklace of length $L = nl$, then $\langle 1/r_{ik} \rangle = 1/(|i - k|l)$. In this case, the summation in Eq. (2.66) leads to the equation [40, 41]

$$\frac{1}{f} = \frac{1}{n\zeta}\left[1 + \frac{\zeta}{3\pi\eta_0 l}(\ln n - 1)\right], \qquad (2.67)$$

which for a long chain (i.e., at large n values) takes an approximate form

$$f = 3\pi\eta_0 L/\ln (L/l). \qquad (2.68)$$

If a molecule is represented by a chain of touching beads, then $l = d$, where d is the bead diameter, and it follows from Eq. (2.68) or directly from Eq. (2.67) that

$$f = 3\pi\eta_0 L/\ln (L/d), \qquad (2.69)$$

where d may be regarded as the chain diameter.

The form of Eqs. (2.67)–(2.69) is identical with that of the equations for the translational friction coefficient of long ellipsoids of revolution and cylinders [see Eqs. (2.6), (2.38), and (2.43)] differing from it only in the value of $\gamma = 0$. Hence, according to Eq. (2.67), the factor $\ln (L/d)$ contained in all these equations characterizes the hydrodynamic interaction in a long rodlike molecule, increasing proportionally with the logarithm of its length.

The problem of the rotational friction of a rodlike necklace has been solved by Kirkwood et al. [39–41] using Eq. (2.64) and the general theory of irreversible processes [41]. The following expression has been obtained for the rotational friction coefficient with respect to the central axis normal to that of the rod:

$$W = \pi\eta_0 L^3/3 \ln (L/l). \qquad (2.70)$$

For a straight necklace with the touching beads ($l = d$) one obtains, instead of Eq. (2.70),

$$W = \pi\eta_0 L^3/3 \ln (L/d), \qquad (2.71)$$

Equation (2.71) is equivalent to Eqs. (2.12), (2.41), and (2.52) for long ellipsoids and cylinders with the value $\gamma = 0$.

The following method is used to calculate the viscosity of molecules represented by a chain necklace. In a laminar flow the unperturbed velocity v^0 of which is defined by Eq. (2.13), a chain element (bead) i exerts a force F_i upon the solvent. Hence, the friction work done by the bead per unit time is given by

$$F_i \cdot v_i^0 = F_i v_i^0 \cos (F_i v_i^0) = F_{ix} v_i^0 = F_{ix} g y_i,$$

since, according to Eq. (2.13), $v_i^0 = v_{ix}^0 = g y_i$. Hence, the mean energy loss per unit time caused by the presence in solution of one molecule (n beads) is evidently equal to $g \Sigma_i^n F_{ix} y_i$, where the averaging is carried out over all chain conformations. Consequently, the power loss for friction in unit volume caused by the presence in it of $N_0 = c N_A / M$ molecules is given by

$$E - E_0 = (c N_A g / M) \sum_i^n \langle F_{ix} y_i \rangle. \tag{2.72}$$

Comparison of Eq. (2.72) with Eqs. (2.16) and (2.18) gives for the intrinsic viscosity of the solution

$$[\eta] = (N_A / M g \eta_0) \sum_i^n \langle F_{ix} y_i \rangle, \tag{2.73}$$

where F_{ix} is determined from Eq. (2.64).

In the case of steady flow with the shear rate g constant in time for a long rodlike necklace, the problem of viscosity is solved [40], just as for ellipsoids, by taking into account the effect of Brownian motion and using Eq. (2.21) and the corresponding distribution function (2.22) in which b_0 is taken to be unity since the chain length L is large compared with the distance l between the beads. Under these conditions, the expression for $[\eta]$ is obtained using Eq. (2.73):

$$[\eta] = 2\pi N_A L^3 / 45 M \ln (L/l). \tag{2.74}$$

For a necklace with touching beads ($l = d$) this expression corresponds to the equations expressing the viscosity of elongated ellipsoids (2.26') and cylinders (2.48) with the value $\gamma = 0$.

Comparison of Eqs. (2.70) and (2.74) shows that the rotational friction coefficient and the intrinsic viscosity for the straight necklace model are related to each other by the equation

$$M [\eta] \eta_0 / W N_A = 2/15, \tag{2.75}$$

which is independent of hydrodynamic interaction in the chain [the factor $\ln (L/l)$ is absent] and coincides with the corresponding expressions for long spheroids [Eq. (2.27)] and cylinders [Eq. (2.53)].

Hence, both in the application to the translational and rotational friction of rodlike molecules and in the calculation of viscosity properties of their solutions, the chain necklace model leads to results which virtually coincide with those obtained for models of long spheroids and cylinders. This point confirms the universal character of the chain necklace model and justifies its application to chain molecules exhibiting a more complex configuration.

3.4. Gaussian Necklace

A Gaussian necklace is the main model used for the description of the hydro-dynamic properties of a flexible chain molecule in solution. For this model, according to Eq. (1.4), the mean-square distance $\langle r_{ik}^2 \rangle$ between chain elements i and k (Fig. 2.7) is proportional to the difference $i - k$ and the square of the effective bond length l_0:

$$\langle r_{ik}^2 \rangle = l_0^2 |i - k|. \tag{2.76}$$

The effective bond length l_0 is dependent upon the equilibrium rigidity of the chain and, generally speaking, is not equal to the hydrodynamic bond length l (Fig. 2.7) depending on friction characteristics of the chain. The value of l is related to the contour length L_{ik} of the chain region between the ith and kth elements by the equation

$$l = L_{ik}/|i - k|, \tag{2.77}$$

whereas

$$l_0 = (L_{ik}A/|i - k|)^{1/2}, \tag{2.78}$$

where A is the length of the Kuhn segment.

According to Eq. (2.65), for the calculation of hydrodynamic interaction in a chain molecule the value of $\langle 1/r_{ik} \rangle$ averaged over all chain orientations and conformations is necessary. According to Eq. (1.6), for a Gaussian chain this value is given by

$$\langle 1/r_{ik} \rangle = (6/\pi)^{1/2} \langle r_{ik}^2 \rangle^{-1/2} = (6/\pi)^{1/2}/l_0 |i - k|^{1/2}. \tag{2.79}$$

3.4.1. Translational Friction Coefficient. The average translational friction coefficient f of a Gaussian chain is determined by the substitution of Eq. (2.79) into Eq. (2.66),

$$\frac{1}{f} = \frac{1}{n\zeta}\left(1 + \frac{\zeta}{\sqrt{6\pi^3}\,\eta_0 l_0 n}\sum_i^n\sum_k^n\frac{1}{|i - k|^{1/2}}\right) = \frac{1}{n\zeta}\left(1 + \frac{8}{3}X\right), \tag{2.80}$$

where

$$X = \zeta n^{1/2}/\sqrt{6\pi^3}\,\eta_0 l_0. \tag{2.81}$$

Equation (2.80) shows that the hydrodynamic interaction in translational friction of the Gaussian chain may be taken into account if the effective value of the friction coefficient of a chain element $\bar{\zeta}$ is introduced. It is defined by the equation

$$\bar{\zeta} = \zeta \Big/ \Big(1 + \tfrac{8}{3} X \Big), \qquad (2.82)$$

where X, the parameter of hydrodynamic interaction in the chain, according to Eq. (2.81) is proportional to the square root of the number of its elements n or its length L. In this case, $f = n\bar{\zeta}$.

If Eq. (2.81) is substituted into Eq. (2.80) and if we recall that $l_0 n^{1/2} = \langle h^2 \rangle^{1/2} = (LA)^{1/2}$ and $M/n = M_L l$, then Eq. (2.80) is transformed into the equation

$$\eta_0 M / f = 8 \big(3 \sqrt{6\pi^3}\big)^{-1} (M_L/A)^{1/2} M^{1/2} + M_L (l\eta_0/\zeta). \qquad (2.83)$$

Equation (2.83) is similar to the Kuhn equation (2.55) and predicts the same dependence of f on M upon variation of the hydrodynamic interaction (or "draining") in the chain.

If the friction coefficient ζ of a chain element is represented in the Stokes expression (2.2) for a sphere $\zeta = 3\pi\eta_0 d$, then for the model of touching beads ($l = d$) the second term on the right-hand side of Eq. (2.83) is equal to $M_L/3\pi \approx 0.1 M_L$.

Hence, the numerical factor 0.1 in the "draining" term of Eq. (2.83) differs by 30% from that in Eq. (2.56) (if the factor containing a logarithmic term is not taken into account).

The first term on the right-hand side of Eq. (2.83) characterizes hydrodynamic interaction, and for a relatively long "nondraining" chain we have

$$f = P_\infty \eta_0 (LA)^{1/2}, \qquad (2.84)$$

where $P_\infty = (3/8)(6\pi^3)^{1/2} = 5.11$, which is 30% lower than the numerical coefficient in the Kuhn equation (2.58).

In subsequent calculations [44] the value of P_∞ equal to 5.2 has been obtained; it virtually coincides with the Kirkwood–Risemann value mentioned above.

3.4.2.   Intrinsic Viscosity.   Equation (2.73), in which the x-components of point forces, F_{ix}, are determined from Eq. (2.64), is the initial equation for the calculation of the intrinsic viscosity of a solution of molecules represented by a Gaussian necklace. Moreover, since for rotational motion in a laminar flow the spatial distribution of elements (beads) of the Gaussian necklace is, on the average, spherically symmetric, the averaged value of the tensor T_{ik} expressed by Eq. (2.65) is used in Eq. (2.64) from the outset. The average values of reciprocal distance between chain elements $\langle 1/r_{ik} \rangle$ calculated in Eq. (2.65) are defined by Eq. (2.76), as before.

TABLE 2.1. Limiting Values of the Parameter Φ_∞
Obtained in Different Theories

$\dfrac{\Phi_\infty \cdot 10^{-23}}{\text{mole}}$	Reference	$\dfrac{\Phi_\infty \cdot 10^{-23}}{\text{mole}}$	Reference
3.62	[39]	2.68	[50]
3.37	[47]	2.76	[51]
2.862	[48]	2.19	[52]
2.84	[49]	2.51	[53]

The use of these expressions, and the solution of the corresponding integral equations, yields [39]

$$[\eta] = (N_A \zeta n^2 l_0^2 / 36 \eta_0 M)\, F(X). \tag{2.85}$$

where the hydrodynamic interaction parameter X is defined by Eq. (2.81), and the function $F(X)$ is defined by Eq. (2.86):

$$F(X) = \frac{6}{\pi^2} \sum_{k=1}^{\infty} [k^2 (1 + X/k^{1/2})]^{-1}. \tag{2.86}$$

In the absence of hydrodynamic interaction ($X \to 0$) the function $F(X)$ is equal to unity and, according to Eq. (2.85), we have

$$[\eta] = N_A \zeta n^2 l_0^2 / 36 \eta_0 M = (\zeta / 36 \eta_0 m_0)\,(L/A), \tag{2.87}$$

where m_0 is the mass of the friction element (bead) of the Gaussian chain. Equation (2.87) has previously been obtained by Debye [46]. It shows that for a draining Gaussian chain $[\eta]$ increases proportionally to its contour length L (or molecular weight).

If hydrodynamic interaction exists, its effect on viscosity may be taken into account, just as for translational friction, by introducing the effective value of the friction coefficient of a chain element $\bar{\zeta}$ defined by the equation

$$\bar{\zeta} = \zeta F(X). \tag{2.88}$$

In this case, $[\eta]$ is determined from Eq. (2.87), where ζ is replaced by $\bar{\zeta}$.

Applying Eqs. (2.81) and (2.78), the equation for viscosity (2.85) becomes

$$[\eta] = \Phi (L/A)^{3/2} / M, \tag{2.89}$$

where the function

$$\Phi = (\pi/6)^{3/2} N_A X F(X) \tag{2.90}$$

is usually called the Flory coefficient.

Under the conditions of a limiting strong hydrodynamic interaction ($X \to \infty$), Eq. (2.86) yields [39] $[F(X)]_\infty = C/X_\infty$, where $C = 1.588$. Correspondingly, for a nondraining Gaussian chain the limiting value of the Flory coefficient is $\Phi_\infty = (\pi/6)^{3/2} N_A C = 3.62 \cdot 10^{23}$ mole^{-1}. Under these conditions, according to Eq. (2.89), for a nondraining Gaussian coil the value of $[\eta]$ increases with chain length proportionally to the square root of molecular weight M.

In subsequent papers, the function $F(X)$, the coefficient C, and the corresponding value of Φ_∞ were calculated using various mathematical approaches with various approximations. Depending on the character of the latter, different authors obtained different results, some of which are listed in Table 2.1. If these data are considered in chronological order, a decrease in the calculated value of Φ_∞ is observed. This decrease is to a certain extent related to the fact that in later calculations the authors desired to avoid the preaveraging to T_{ik} which always yields somewhat lower values of Φ_∞. The lowest value of $\Phi_\infty = 2.19 \cdot 10^{-23}$ mole^{-1} has been obtained by Hearst and Tagami [52] who, in contrast to other workers, used as a molecular model a wormlike necklace instead of a Gaussian necklace. Zimm [53], in his calculations of $[\eta]$, has fundamentally revised the problem concerning the effect of preaveraging of hydrodynamic interaction. By using the Monte Carlo method and generating in a computer simulation a set of random configurations of a polymer chain, the author solved the problem of hydrodynamic interactions in translational friction and viscosity for each conformation and, only after this, averaged the results. Table 2.1 gives the value of $\Phi_\infty = 2.51 \cdot 10^{23}$, according to Zimm, extrapolated to a chain with an infinite number of segments. The value of $P_\infty = 5.99$ in Eq. (2.84) was obtained by the same method. Thus, the value of Φ_∞ was found to be lower, and that of P_∞ higher, than the usually applied values. The value of $P_\infty = 5.99$ seems particularly unusual, but it should be noted that it is nevertheless much lower than the value of 7.3 obtained by Kuhn in experiments with wire models [Eq. (2.58)].

The following relation may serve as a fairly good approximation to Eq. (2.86):

$$F(X) = 1/(1 + X/C), \tag{2.91}$$

where $C = [XF(X)]_\infty$. If Eq. (2.91) is applied, Eq. (2.85) can be written in the form

$$M/[\eta] = (1/\Phi_\infty)(M_L/A)^{3/2} M^{1/2} + (36 \eta_0 m_0 / \zeta)(M_L/A). \tag{2.92}$$

Just as in Eq. (2.83) for translational friction, the first term on the right-hand side of Eq. (2.92) for viscosity characterizes hydrodynamic interaction, while the second term describes the properties of the draining Gaussian coil.

Substitution of Eq. (2.91) into Eq. (2.88), which describes the effective value of the friction coefficient of a chain element ζ, gives

$$\bar{\zeta}=\zeta/(1+aX). \tag{2.93}$$

Equation (2.93) coincides in form with Eq. (2.82) expressing the effective coefficient of translational friction. However, the numerical coefficient $1/C$ [Eq. (2.93)] according to various theories (Table 2.1) ranges between 0.6 and 1, i.e., it is three or four times lower than the coefficient 8/3 of X in Eq. (2.82). Hence, in a Gaussian chain with a given value of X, the hydrodynamic interaction, although it decreases friction in both the translation motion and the viscosity phenomenon, is less pronounced in the latter case. The reason for this can be easily understood if it is taken into account that during the translational motion of a molecule all its elements move in one direction entraining the solvent in this direction and, therefore, they interact "positively," decreasing the overall friction in the coil. The main contribution to the viscosity behavior of a solution in the field of shear rates is provided by the rotational motion of molecules with respect to the center of their hydrodynamic resistance. During this rotation of the coil some of its elements move in the opposite directions (e.g., elements k and j in Fig. 2.7) and this causes their "negative" interaction and an increase in the overall friction in the coil.

A close relationship between the intrinsic viscosity of the solution and the rotational friction of chain molecules can also be clearly seen in the following discussion.

3.4.3. Rotational Friction. The rotational friction coefficient W of the Gaussian necklace defined by Eq. (2.8), just as that for any other body, is a tensor value depending on the position and direction of the selected axis of rotation. However, usually the Gaussian chain is characterized by an average rotational friction coefficient corresponding to the rotation of the molecule as a whole about its center of mass (or the center of hydrodynamic resistance); i.e., the chain elements exhibit a spherically symmetric space–time distribution with respect to this center [40].

It can be easily shown that for any body represented by a stiff distribution of n elements (beads) the rotational friction coefficient determined in this manner in the absence of hydrodynamic interaction between the elements is given by

$$W=(2/3)\,n\zeta\langle R^2\rangle, \tag{2.94}$$

where $\langle R^2\rangle$ is the radius of gyration of the body. In accordance with this, for a freely drained Gaussian necklace, if Eq. (1.8) is applied, we have

$$W=(1/9)\,n\zeta\langle h^2\rangle=(1/9)\,n\zeta LA. \tag{2.95}$$

The evaluation of the effect of hydrodynamic interaction for the rotational friction of the Gaussian necklace [40] has shown that Eq. (2.95) can be applied to this case also if the friction coefficient of a chain element ζ in this equation is replaced by its effective value $\bar{\zeta}$, defined by Eq. (2.88). Moreover, it is important that the function $F(X)$ be identical in the problems of rotational friction and viscosity.

If ζ in Eq. (2.95) is replaced by $\bar{\zeta}$, then applying Eq. (2.88) and comparing Eqs. (2.85) and (2.95), one obtains

$$M\,[\eta]\,\eta_0/WN_A = 1/4, \qquad (2.96)$$

i.e., Eq. (2.27) in which $f(p) = 1/4$.

Equation (2.96) is independent of the hydrodynamic interaction in the chain and can be derived for a freely drained chain by direct comparison of Eqs. (2.87) and (2.95) without the replacement of ζ by $\bar{\zeta}$.

The numerical value of the coefficient, 1/4, on the right-hand side of Eq. (2.96) is not a specific feature of the Gaussian chain. Actually this value corresponds to a rigid-chain necklace in any conformation if the value W in Eq. (2.96) designates the average coefficient of rotational friction of the chain with respect to its center of mass.

If $\bar{\zeta}$ is replaced by ζ in Eq. (2.95) according to Eq. (2.88), and Eqs. (2.91), (2.90), and (2.81) are used, then Eq. (2.95) becomes

$$\eta_0 M^2/W = (N_A/4\Phi_\infty)(M_L/A)^{3/2}\,M^{1/2} + (9\eta_0 l/\zeta)(M_L)^2\,A^{-1}, \qquad (2.97)$$

The form of Eq. (2.97) is similar to that of the Kuhn equation (2.57). This resemblance can be enhanced if ζ is expressed by $\zeta = 3\pi\eta_0 d$ and l is taken to be equal to d (necklace with touching beads). Then Eq. (2.97) becomes

$$\eta_0 M^2/W = (N_A/4\Phi_\infty)(M_L/A)^{3/2}\,M^{1/2} + 0.95 M_L^2 A^{-1}. \qquad (2.98)$$

The numerical coefficient in the "draining" term is twice that in the corresponding term of Eq. (2.57) [if the factor containing $\ln(Ad)$ is not taken into account in Eq. (2.57)]. For a nondraining Gaussian coil according to Eq. (2.98), we have

$$W = (4\Phi_\infty/N_A)\,\eta_0\,(LA)^{3/2}. \qquad (2.99)$$

Depending on the selected value of Φ_∞ (Table 2.1), the numerical coefficient in Eq. (2.99) ranges between 1.5 and 2.4. In order of magnitude this value corresponds to the value 2.13 in the Kuhn equation (2.59).

3.4.4. Significance of the Draining Effect. It can be seen in the preceding section that in the description of the hydrodynamic properties of chain molecules the Gaussian necklace model, which takes hydrodynamic interaction into account according to Eq. (2.65), yields results qualitatively similar to those obtained in model experiments [Eqs. (2.56) and (2.57)]. Thus, at high molecular weights when the draining effect is negligible, it can be expected that for a Gaussian chain f and $[\eta]$ increase proportionally to $M^{1/2}$ whereas W increases proportionally to $M^{3/2}$. According to the theory the draining effect should increase with decreasing molecular weight and, hence, one should expect a power dependence of f, $[\eta]$, and W on M with higher exponents. Moreover, for real flexible-chain molecules it is naturally assumed that the excluded volume effects are absent, i.e., that θ-solvents are used (see Chapter 1).

However, there is some difference between Eqs. (2.83) and (2.97) on the one hand, and Eqs. (2.56) and (2.57) on the other hand. In Eq. (2.56) at $\ln (A/d) = 0.432$ and in Eq. (2.57) at $\ln (A/d) = 1.540$ the draining term becomes zero and, hence, at these values of A/d the hydrodynamic properties of the molecule in any range of molecular weights investigated should correspond to those of the nondraining Gaussian chain. In contrast, the term in Eqs. (2.83), (2.92), and (2.98) characterizing draining is positive and does not become zero at any values of the structural parameters of the model which are contained in this term. In accordance with this, in all cases of a decrease in molecular weight one should expect the manifestation of the draining effect in the hydrodynamic and viscosity behavior of chain molecules. Nevertheless, experimental data for solutions of flexible-chain polymers do not support this prediction of the theory and this resulted in a discussion of the adequacy of the Gaussian necklace model and the limits of applicability of Kirkwood's method.

Yamakawa [54] has somewhat clarified this problem. He modified to some extent the tensor of hydrodynamic interaction T_{ik} [Eq. (2.63)], taking into account the finite character of the size of interacting beads in the necklace. In this consideration it is assumed that the point force F_i exerted by the ith bead on the surrounding liquid is not applied to the bead center, but is distributed on the surface of a sphere of diameter d. In the modified form the Oseen tensor contains the additional term proportional to the square of the d/r_{ik} ratio

$$ \mathbf{T}_{ik} = \frac{1}{8\pi\eta_0 r_{ik}}\left[\left(1 + \frac{\mathbf{r}_{ik}\cdot\mathbf{r}_{ik}}{r_{ik}^2}\right) + \left(\frac{d}{2r_{ik}}\right)^2\left(\frac{1}{3}I - \frac{\mathbf{r}_{ik}\cdot\mathbf{r}_{ik}}{r_{ik}^2}\right)\right]. \qquad (2.100) $$

The application of the modified expression for T_{ik} according to Eq. (2.100) to the problems of viscosity and translational friction showed that for polymer chains of relatively large length the result remains the same since the term characterizing hydrodynamic interaction in Eqs. (2.83) and (2.82) does not vary. In contrast, the significance of the draining term decreases. This can be seen most clearly in the properties of the friction coefficient for which the second term on the right-hand side of Eq. (2.83) becomes $M_L(l\eta_0/\zeta)[1 - 2(6/\pi)^{1/2}(d/l)]$. In accordance with this, at $l/d = 2(6/\pi)^{1/2} = 2.76$ the draining term in Eq. (2.83) becomes zero and the friction behavior of the model should correspond to the properties of the nondrained Gaussian chain.

This result improves the situation to some extent because it shows that with a certain choice of model parameters the theory can explain the absence of the draining effect in flexible-chain molecules in accordance with experimental data in θ-solvents.

However, the problem of the elucidation of structural characteristics of a real Gaussian chain, which correspond to the Gaussian necklace representing this chain with the distance between the neighboring beads greater than the bead diameter by a factor of 2.76, remains unsolved.

This problem is better clarified if the hydrodynamic properties of chain molecules are studied using a wormlike chain model.

4. WORMLIKE PEARL NECKLACE

The chain pearl necklace model of spherical beads (Fig. 2.7) may also be used to describe the hydrodynamic properties of a wormlike chain. For this purpose it is only necessary to take into consideration the fact that the conformational characteristics of the necklace should correspond to those of the wormlike chain. This implies that in calculating the distance r_{ik} between the ith and kth beads of the necklace (Fig. 2.7) one should use, instead of Eq. (2.76), the equation

$$\langle r_{ik}^2 \rangle = A L_{ik} \{1 - (A/2L_{ik})[1 - \exp(-2L_{ik}/A)]\}, \qquad (2.101)$$

which is equivalent to Porod's equation (1.24).

In this case, as before, the length of the chain region L_{ik} between the beads i and k is related to that between the neighboring beads l by Eq. (2.77).

However, to calculate the friction characteristics of the molecule taking into account hydrodynamic interaction according to Eq. (2.63) or (2.65), it is necessary to use the values of mean reciprocal distances $\langle 1/r_{ik} \rangle$.

4.1. Mean Reciprocal Distances between Chain Elements

These values can be calculated using the distribution function $W(r_{ik})$, which is known in the complete form only for a Gaussian chain [Eq. (1.2)], whereas for a wormlike chain it was obtained only in the range of relatively high $x = 2L/A$ in the form of the correction factor for the Gaussian distribution [Eq. (1.30)]. Hence, to calculate $\langle 1/r_{ik} \rangle$ or $\langle 1/h \rangle$ over the entire range of the x values for a wormlike chain, approximate methods were used. Thus, Ptitsyn and Eizner [55] have approximated the distribution function $W(h)$ for a wormlike chain by a combination of the Gaussian distribution and the delta-function. Hearst and Stockmayer [56] have used for the same purpose a combination of a part of function (1.30) at high x and a "cubic approximation" at low x. The results of the calculation of translational friction in a wormlike chain obtained in these two papers are qualitatively close to each other. However, it is more convenient to use them in discussion in the form presented by Hearst and Stockmayer, and they will be considered here in this form.

Using in Eq. (1.30) the correction terms containing $1/x$, $(h/L)^2$, and $(h/L)^3$ (Daniels' approximation), it is easy to calculate

$$\langle 1/h \rangle = (6/\pi)^{1/2} (LA)^{-1/2} [1 - (1/40)(A/L)] \qquad (2.102)$$

and, correspondingly, for beads relatively distant from each other along the chain when $L_{ik}/A > \sigma$, to apply the equation

$$L_{ik}\langle 1/r_{ik}\rangle = (6/\pi)^{1/2}(L_{ik}/A)^{1/2}[1-(1/40)(A/L_{ik})]. \qquad (2.103)$$

For beads located close to each other along the chain ($L_{ik}/A \to 0$) the part of the wormlike chain located between them is straight and, therefore, for them we have $L_{ik}\langle 1/r_{ik}\rangle = 1$. Expanding Porod's equation (1.24) in a power series in x, it is easy to see that at $L_{ik}/A \ll 1$ the following equation holds:

$$L_{ik}\langle 1/r_{ik}\rangle = 1 + (1/3)(L_{ik}/A) + \cdots \qquad (2.104)$$

To calculate $\langle 1/r_{ik}\rangle$ in the range of intermediate values of L_{ik}/A, Hearst and Stockmayer have proposed an interpolation equation in the form of a series (cubic approximation):

$$\langle 1/r_{ik}\rangle L_{ik} = 1 + (1/3)(L_{ik}/A) + \alpha(L_{ik}/A)^2 + \beta(L_{ik}/A)^3. \qquad (2.105)$$

The coefficients α and β are found from the condition that the two curves describing the dependence of $\langle 1/r_{ik}\rangle L_{ik}$ on L_{ik}/A plotted according to Eqs. (2.103) and (2.105), respectively, converge (i.e., they intersect and have equal slopes and curvatures) at the point $L_{ik}/A = \sigma$.

These conditions yield the following values:

$$\sigma = 2.2, \quad \alpha = 0.118, \quad \beta = -0.026. \qquad (2.106)$$

Hence, to calculate the reciprocal distances between the beads of a wormlike necklace according to Hearst and Stockmayer, Eqs. (2.105) and (2.106) should be used at $L_{ik}/A \le 2.2$, and Eq. (2.103) should be applied at $L_{ik}/A > 2.2$.

4.2. Translational Friction Coefficient [56]

The translational friction coefficient f of the molecule is calculated according to Kirkwood's equation (2.66) using, for $\langle 1/r_{ik}\rangle$, Eqs. (2.103) and (2.105). The final result is represented in the form of analytical expressions of f for the two limiting cases: at a contour length of the chain $L/A \le 2.2$ (weakly bending rod) and at $L/A > 2.2$ (wormlike coil). At $L/A \le 2.2$, by applying Eqs. (2.66) and (2.105) and neglecting terms of the order of magnitude $(L/A)^2$ and higher and the term proportional to l/L, Hearst and Stockmayer obtain

$$1/f = (1/n\zeta)\{1 + (\zeta/3\pi\eta_0 l)[\ln(L/l) - 1 + 0.166L/A + \ldots]\}. \qquad (2.107)$$

This expression differs from Eq. (2.67) for a rodlike necklace by the presence of the term $0.166L/A$ characterizing the finite flexibility of a wormlike chain leading to a lower value of f as compared to a rodlike chain at the same values of L, n, and ζ.

If the friction coefficient of a bead is expressed according to Stokes, $\zeta = 3\pi\eta_0 d$, and a model of touching beads is used, then instead of Eq. (2.107), one obtains

$$3\pi\eta_0 L/f = \ln(L/d) + 0.166 L/A. \tag{2.108}$$

In the other limiting case, for a long chain ($L/A > 2.2$), using Eqs. (2.66), (2.105), and (2.103) these authors obtain

$$1/f = (1/n\zeta)\{1 + (\zeta/3\pi\eta_0 l)[\ln(A/l) - 2.431 + (4/3)(6/\pi)^{1/2}(L/A)^{1/2}$$
$$+ 0.138(L/A)^{-1/2} - 0.305(L/A)^{-1}]\}. \tag{2.109}$$

If the terms $(L/A)^{-1/2}$ and $(L/A)^{-1}$ are neglected, (2.109) becomes

$$\eta_0 M/f = P_\infty^{-1}(M_L/A)^{1/2}M^{1/2} + (\eta_0 l/\zeta)M_L + (M_L/3\pi)[\ln(A/l) - 2.431], \tag{2.110}$$

where P_∞ is determined from Eq. (2.84).

Equation (2.110) for a wormlike chain differs from Eq. (2.83) for a Gaussian coil in the presence of a term containing $\ln(A/l)$ on its right-hand side.

If ζ is expressed according to Stokes: $\zeta = 3\pi\eta_0 d$, and a model of a pearl necklace with touching beads is used ($l = d$), then instead of Eq. (2.110) one obtains

$$\eta_0 M/f = P_\infty^{-1}(M_L/A)^{1/2}M^{1/2} + (M_L/3\pi)[\ln(A/d) - 1.431]. \tag{2.111}$$

Equation (2.111) coincides with Kuhn's equation (2.56) (only the numeric coefficients are different) and differs from Eq. (2.83) by the factor $\ln(A/d) - 1.431$ contained in the term that takes into account the draining effect. It follows from Eq. (2.111) that for molecules for which $\ln(A/d) = 1.431$, i.e., $A \approx 4d$, the draining effect is absent over the entire range of values of $L/A > 2.2$. If it is considered that for many flexible-chain polymers the ratio of the length of the Kuhn segment to the chain diameter is actually not far from four, it is easy to understand the well-known fact [5] that under θ-conditions these molecules behave as nondrained Gaussian coils over the entire range of molecular weights available experimentally (including the oligomer range).

4.3. Rotational Friction

Rotational friction of a wormlike chain has been considered by Hearst [57], who used and further developed the formalism developed by Kirkwood [43] in his theory of a rodlike pearl necklace.

In Hearst's model the positions of the centers of hydrodynamic resistance (beads) with the friction coefficient ζ are determined in a molecular system of XYZ coordinates the origin of which coincides with the middle point of the chain and the direction of the Z axis coincides with the chain direction at this point (Fig. 2.8). A cylindrical symmetric distribution of chain elements with the axis of symmetry Z is assumed. Using the methods developed for a wormlike chain [58], Hearst calculated $\langle x_i^2 \rangle$, $\langle y_i^2 \rangle$, and $\langle z_i^2 \rangle$, the mean-square coordinates of the ith chain element located along the chain contour at a distance L_i from the origin:

$$\langle z_i^2 \rangle = AL_i/3 - (A^2/18)(1 - e^{-6L_i/A}), \tag{2.112}$$

$$\langle x_i^2 \rangle = \langle y_i^2 \rangle = AL_i/3 - (2/9) A^2 + (A^2/4) e^{-2L_i/A} - (A^2/36) e^{-6L_i/A}. \tag{2.113}$$

At low L_i values Eqs. (2.112) and (2.113) become

$$\langle z_i^2 \rangle/L_i^2 = 1 - 2L_i/A + (3/4)(2L_i/A)^2 - \ldots, \tag{2.114}$$

$$\langle x_i^2 \rangle = \langle y_i^2 \rangle = (1/12)(2L_i/A)^3 - (2/3)(2L_i/A)^4 + (13/240)(2L_i/A)^5 - \ldots \tag{2.115}$$

At $L_i \to 0$ it follows from Eqs. (2.114) and (2.115) that $\langle Z_i^2 \rangle \to L_i^2$ and $\langle x_i^2 \rangle = \langle y_i^2 \rangle \to 0$. This corresponds to the conformation of a thin straight rod. At $L_i \to \infty$ it follows from Eqs. (2.112) and (2.113) that $\langle z_i^2 \rangle = \langle x_i^2 \rangle = \langle y_i^2 \rangle = A/L_i/3$, which corresponds to a Gaussian coil with the mean segment distribution spherically symmetric with respect to the middle chain point.

Equations (2.103) and (2.105) are used, as before, to calculate hydrodynamic interaction between the elements of the chain performing rotational motion.

Hearst obtained the final expressions for the rotational friction coefficient W of a wormlike chain in its rotation about the axis X or Y (Fig. 2.8) for the two limiting cases.

For short chains, when $L/A \ll 1$ (weakly bending rod) we have

$$\pi \eta_0 L^3/W = 3 \ln (L/l) - 7.0 + 12\pi\eta_0 l/\zeta + (L/A)$$
$$\times [2.25 \ln (L/l) - 6.66 + 6\pi\eta_0 l/\zeta]. \tag{2.116}$$

For long chains, when $L/A \gg 1$ (wormlike coil) we have

$$\eta_0 M^2/W = 0.72 (M_L/A)^{3/2} M^{1/2} + 0.64 M_L^2 A^{-1} [\ln (A/l) - 2.43 + 3\pi\eta_0 l/\zeta]. \tag{2.117}$$

If ζ is expressed as the diameter d of an equivalent Stokes sphere, $\zeta = 3\pi\eta_0 d$, and a wormlike necklace with touching beads is taken as a model ($l = d$), then Eqs. (2.116) and (2.117) are transformed into Eqs. (2.118) and (2.119), respectively:

$$\pi \eta_0 L^3/3W = \ln (L/d) - 1 + 0.75 (L/A) [\ln (L/d) - 2.07], \tag{2.118}$$

$$\eta_0 M^2/W = 0.72 (M_L/A)^{3/2} M^{1/2} + 0.64 M_L^2 A^{-1} [\ln (A/d) - 1.43]. \tag{2.119}$$

Equations (2.116) and (2.118) differ from Kirkwood's equations (2.70) and (2.71) for a rodlike necklace in the presence of a term proportional to L/A. This

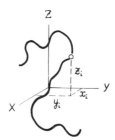

Fig. 2.8. Coordinate system of the wormlike chain. X, Y, and Z are orthogonal axes of the system, the origin of which is at the middle point of the chain. The Z axis is tangent to the chain contour at the origin. x_i, y_i, and z_i are the coordinates of the ith chain element.

term characterizes a decrease in the rotational friction W of a weakly bending rod as compared to that of a straight rod due to the flexibility of the former.

Equation (2.119) coincides with Kuhn's equation (2.57) (only the numerical coefficients are different) and predicts the disappearance of the draining effect for molecules for which $A \approx 4d$, just as Eq. (2.111) does.

4.4. Intrinsic Viscosity

The molecular model shown in Fig. 2.8 has also been used to calculate the intrinsic viscosity of a wormlike chain [52, 59]. Moreover, since this model is aspheric, the function ρ of orientational distribution of molecules in a flow shear field should be determined. For this purpose the equation of rotational diffusion [Eq. (2.21)] is solved. As expected for molecules with a cylindrical symmetry, its solution coincides with Peterlin's result (2.22). Furthermore, it was found that the parameter b_0 contained in Eq. (2.22) for a wormlike chain, just as that for a spheroid, can be expressed by Eq. (2.20), in which p^2 for a wormlike chain is given by

$$p^2 = \sum_{i=1}^{n} \langle z_i^2 \rangle \bigg/ \sum_{i=1}^{n} \langle y_i^2 \rangle . \tag{2.120}$$

Hence, Eq. (2.120) is a hydrodynamic characteristic of the mean asymmetry of shape of a wormlike chain in the molecular system of coordinates shown in Fig. 2.8.

Reciprocal distances $\langle 1/r_{ik} \rangle$ required for the calculation of hydrodynamic interaction in the molecule are determined, as before, according to Eqs. (2.103) and (2.105).

According to Hearst and Tagami, the expressions for the intrinsic viscosity in the limiting cases of a short and a long wormlike chain are given by [52], at $L/A \ll 1$:

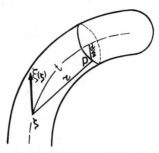

Fig. 2.9. Wormlike spherocylinder. ζ is the point of the application of the force $\xi(\zeta)$ at the axial line of the cylinder. P is the point on
the contour of the normal section of the cylinder at which the perturb-
ing action of the force $\xi(\zeta)$ is observed, t is the distance between ζ and
the plane of the normal section of the cylinder along the contour of the
axial line, and d is the diameter of the cylinder.

$$[\eta] = \frac{\pi N_A L^3}{90M} \left[\frac{1}{\ln(L/l) - 2.72 + 0.66(3\pi\eta_0 l/\zeta)} + \frac{3 - L/4A}{\ln(L/l) - 2.72 + 1.33(3\pi\eta_0 l/\zeta)} \right] \qquad (2.121)$$

and at $L/A \gg 1$:

$$M/[\eta] = \Phi_\infty^{-1} (M_L/A)^{3/2} M^{1/2} + 0.89\Phi_\infty^{-1} (M_L^2/A) [\ln(A/l) - 2.431 + 3\pi\eta_0 l/\zeta], \qquad (2.122)$$

where $\Phi_\infty = 2.19 \cdot 10^{23}$ mole^{-1}.

If $\zeta = 3\pi\eta_0 d$ is substituted into Eqs. (2.121) and (2.122) and the touching
bead model is used ($l = d$), they become, respectively,

$$[\eta] = \frac{2\pi N_A M^2}{45 M_L^3} \frac{1}{\ln(L/d) - 1.39} \left[\frac{3}{4} + \frac{1}{4} \frac{\ln(L/d) - 1.39}{\ln(L/d) - 2.06} - \frac{L}{16A} \right]$$

$$\approx (2\pi N_A M^2/45 M_L^3) [\ln(L/d) - 1.39]^{-1} [1 - (1/16)(L/A)], \quad \frac{L}{A} \ll 1, \frac{L}{d} \gg 1, \qquad (2.123)$$

$$M/[\eta] = \Phi_\infty^{-1} (M_L/A)^{3/2} M^{1/2} + 0.89\Phi_\infty^{-1} (M_L^2/A) [\ln(A/d) - 1.43], \quad \frac{L}{A} \gg 1. \qquad (2.124)$$

At very high values of L/d (thin wormlike chain) and at $L/A \to 0$, Eq. (2.123)
becomes Kirkwood's equation (2.74) for a rodlike necklace. An important difference between Eqs. (2.123) and (2.74) is the presence in the former of the term
$L/16A$ describing the decrease in the viscosity of a weakly bending rod as compared
to a straight rod. It should be noted, however, that if the theory is compared to
experimental data for real rigid-chain molecules [60], the theoretical value of the
coefficient 1/16 in this term proves to be too low.

Equation (2.124) for a wormlike coil differs from the Kirkwood–Riseman
equation (2.92) for a Gaussian chain by the numerical value of the coefficient $\Phi_\infty =
2.19 \cdot 10^{23}$ mole^{-1}. Moreover, the presence of the factor $\ln(A/d) - 1.43$ in the
draining term in Eq. (2.124) predicts the disappearance of this term for chain
molecules for which $A \approx 4d$.

Just as for other molecular models, a general expression exists for a wormlike chain relating the intrinsic viscosity [η] to the rotational friction coefficient of the molecule. This can be clearly seen if the above expressions for [η] and W are compared.

At ultimate small lengths L when $L/A \to 0$, Eqs. (2.118) for the rotational friction and (2.123) for the viscosity of a wormlike chain are transformed into Kirkwood's equations (2.71) and (2.74), respectively, for a rod. In accordance with this, Eq. (2.75), characteristic of thin rodlike molecules, holds for a wormlike chain in the range $x = 2L/A \to 0$.

In the range $x = 2L/A \gg 1$ for a wormlike coil with a very long chain, if Eqs. (2.119) and (2.124) are compared, one obtains Eq. (2.96) characteristic of a necklace with a spherically symmetric spatial bead distribution. This result is due to the specificity of the molecular system of coordinates (Fig. 2.8) fixing the axes with respect to which the rotational friction coefficient is calculated. The mean asymmetry p of the shape of a wormlike chain leading to different friction coefficients for the rotation about the X (or Y) axis and about the Z axis is defined by Eq. (2.120). Substitution of the expressions for $\langle y_i^2 \rangle$ and $\langle z_i^2 \rangle$ from Eqs. (2.112) and (2.113) into Eq. (2.120) and summation over all chain elements gives

$$p^2(x) = \frac{x^2 - (2/3)\,x + (2/9) - (2/9)\,e^{-3x}}{x^2 - (5/3)\,x + (26/9) - 3e^{-x} + (1/9)\,e^{-3x}}, \qquad (2.125)$$

where $x = 2L/A$.

It follows from Eq. (2.125) that at $x \to 0$ we have $p^2(x) \to \infty$, which corresponds to an infinitely thin straight rod. In the Gaussian range when $x \to \infty$, $p^2(x)$ approaches unity, which corresponds to a model with spherically symmetric distribution of chain elements.

5. POSSIBILITY OF APPLYING THEORETICAL CONSIDERATIONS TO REAL CHAIN MOLECULES

The material considered in Sections 3 and 4 shows that the pearl necklace model introduced by Kirkwood for the description of the hydrodynamic interaction and hydrodynamic properties of chain molecules can also be applied to macromolecules in conformations varying from a straight rod to a Gaussian coil. Moreover, it is found that when the chain length increases, the hydrodynamic interaction of rodlike molecules increases proportionally to the logarithm of chain length and that of Gaussian coils increases proportionally to the square root of chain length.

If a chain is relatively long and coil-shaped in the theories of both the Gaussian and the wormlike necklace, expressions containing $1/f$, $1/W$, and $1/[\eta]$ can be represented as a sum of two terms, the first of which characterizes hydrodynamic interaction and the second, the "draining" term of the molecule.

The first term contains only the parameters M_L, M, and A characterizing the structure, the length, and equilibrium rigidity of a real polymer chain. Hence, the term representing hydrodynamic interaction may be used to characterize these parameters. Thus, if the experimental dependence of $\eta_0 M/f$, $\eta_0 M^2/W$, or $M/[\eta]$ on $M^{1/2}$ is plotted, the slope of the resulting straight line may be used to determine the M_L/A ratio according to Eqs. (2.83) and (2.110) or (2.97) and (2.117), or (2.92) and (2.122). Furthermore, if M_L is known from other data (see Section 2.6), it is possible to determine the value of A.

The draining term appearing in the above equations contains the parameters ζ and l characterizing the properties of the model. As we have seen, the model parameters can be excluded if ζ is replaced by $3\pi\eta_0 d$ and the touching bead model ($l = d$) is used. In the Gaussian necklace model this leads to equations in which the draining term is determined in translational friction by the value of M_L [see Eq. (2.83)] and in rotational friction and viscosity by the value of $M_L^2 A^{-1}$ [see Eqs. (2.98) and (2.92)]. However, as already mentioned, the presence of the draining term differing from zero contradicts experimental data for flexible-chain polymers in θ-solvents. Hence, in the Gaussian pearl necklace model the draining term does not reflect the structural characteristics of real polymer molecules and its comparison with experimental data obtained for these molecules is meaningless.

The application of the condition of touching beads to the model of a long wormlike necklace transforms the draining term in Eqs. (2.111), (2.119), and (2.124) into a form agreeing with experimental data for flexible-chain polymers in θ-solvents in which the draining effect disappears. Hence, it is reasonable to use the draining term in these equations to determine the hydrodynamic diameter d of a wormlike chain. It can be determined from the intercept with the ordinate of one of the straight lines in the plots of $\eta_0 M/f$, $\eta_0 M^2/W$, or $M/[\eta]$ vs. $M^{1/2}$. This procedure is widely used and yields reasonable results (see Chapter 3).

However, it should be noted that the application of the touching beads model to short chain parts is not essentially rigorous since the calculation of hydrodynamic interaction between neighboring beads in Kirkwood's method involves arbitrary, artificial assumptions. This refers both to the draining term in the equations of a wormlike necklace at large L/A and to the equations for a straight (or a weakly bending) rod [e.g., Eq. (2.68)] in which the distance l between neighboring beads is arbitrarily replaced by their diameter d.

For this reason, in modern theories of hydrodynamic properties of chain molecules the necklace model with an inhomogeneous structure is often replaced by the model of a compact homogeneous body, a Gaussian cylinder [61], or a wormlike cylinder [62, 63].

6. WORMLIKE CYLINDER

A wormlike cylinder is a cylinder curved in such a way that the shape of its axial line is described by the equation of the wormlike chain [Eq. (1.21)] and, correspondingly, the distance between any two points on this line is determined by Eq. (2.101). The hydrodynamic resistance experienced by this body in its motion in a viscous liquid is calculated by the Oseen–Burgers method described in Sections 2.1–2.3.

Just as for a straight cylinder, for a wormlike cylinder (Fig. 2.9) the forces of hydrodynamic resistance $\xi(\zeta)$ are assumed to be distributed along its axial line and the perturbations in the liquid caused by these forces are considered at points P on its surface [64]. In accordance with Burgers' method described above, e.g., for the translational motion of the cylinder, the following equation analogous to Eq. (2.33) is valid:

$$\int_0^L T(r)_P \, \xi(\zeta) \, d\zeta = u, \qquad (2.126)$$

where r is the distance between the point ζ of the application of the force $\xi(\zeta)$ and the point P. The value of $T(r)_p$ is the hydrodynamic interaction tensor [according to Eq. (2.30)] averaged over the entire contour of the normal section in which the point P is located, u is the velocity of the cylinder at the point P, and L is the contour length of the cylinder.

6.1. Translational Friction

The use of Oseen's tensor in the preaveraged form [Eq. (2.65)] and the solution of the integral in Eq. (2.126) according to a well-known method [39] yield a general expression for the translational friction coefficient [64]

$$3\pi\eta_0 L^2 / f = \int_0^L (L - t) \langle 1/r \rangle \, dt, \qquad (2.127)$$

where t is the distance along the contour of the axial line of the cylinder between the point of application of the force and the plane of the normal section of the cylinder.

Further, for calculating f according to Eq. (2.127) the mean reciprocal distance $\langle 1/r \rangle$ should be expressed as a function of t. For this purpose the Hearst-Stockmayer method is used [56].

At large t the distribution function [Eq. (1.30)] with all the correction terms is used, and instead of Eqs. (2.102) and (2.103) the following equation is obtained [64]:

$$\langle 1/r \rangle t = (6/\pi)^{1/2} (t/A)^{1/2} \{1 - (1/40) (A/t) [1 + 5 (d/A)^2]$$
$$- (73/4480) (A/L)^2 [1 - (294/73) (d/A)^2 - (63/73) (d/A)^4 + \ldots]\} \qquad (2.128)$$

for $t/A > \sigma$.

At small t, an approximation is used, analogously to Eq. (2.105), in the form of the series

$$\langle 1/r \rangle [t^2 + (d^2/4)]^{1/2} = 1 + \gamma (t/A) + \alpha (t/A)^2 + \beta (t/A)^3 + \delta (t/A)^4 + \ldots \qquad (2.129)$$

for $t/A \leq \sigma$.

In contrast to Eq. (2.105), Eq. (2.129) takes into account the fact that for a straight cylinder (at $t/A \to 0$), analogously to Eq. (2.32), we have $r^2 = t^2 + (d/2)^2$.

If calculations are carried out for the wormlike cylinder model without taking into account the influence of its ends [64], then it is possible to use in Eq. (2.129) only the terms corresponding to the cubic approximation. Moreover, the values of the coefficients α, β, and γ virtually do not differ from those in Eq. (2.105), and the value of σ is found to be $\sigma = 2.278$.

A refined theory [25] that considers the effect of the form of cylinder ends, and uses for this purpose the spherocylinder model (Fig. 2.9), takes into account the fact that near cylinder ends the value of d in Eq. (2.129) is different for different normal sections. Accordingly, the coefficients γ, α, β, δ, ... in Eq. (2.129) are functions of d/t.

Substitution of Eqs. (2.128) and (2.129) into Eq. (2.127) permits the calculation of the translational friction coefficient f for a wormlike spherocylinder over the entire range of L/A and L/d values, which is of practical interest.

The results are given by the following: At $L/A \leq 2.278$ [25],

$$3\pi\eta_0 L/f = C_1 \ln(L/d) + C_2 + C_3(L/A) + C_4(L/A)^2$$
$$+ C_5(L/A)^3 + C_6(L/A)^4 + C_7(L/A)^5 + \cdots \qquad (2.130)$$

where C_1 and C_2 have the same values as in Eq. (2.44) and the coefficients C_3-C_7 are defined by Eqs. (2.131):

$$C_3 = 0.1667 - 0.06838(d/L)^2 + 0.02083(d/L)^3 - 0.01693(d/L)^4$$
$$- 0.008594(d/L)^5 + \cdots,$$

$$C_4 = 0.01111 + 0.07917(d/L)^2 - 0.1799(d/L)^3 + 0.1055(d/L)^4$$
$$+ 0.02461(d/L)^5 + \cdots,$$

$$C_5 = 0.001058 - 0.004960(d/L)^2 + 0.001653(d/L)^3 \qquad (2.131)$$
$$- 0.07348(d/L)^4 - 0.03281(d/L)^5 + \cdots,$$

$$C_6 = 0.0001587 - 0.0007275(d/L)^2 + 0.0003638(d/L)^3$$
$$- 0.08630(d/L)^4 + 0.4000(d/L)^5 + \cdots,$$

$$C_7 = 0.00003848 - 0.0001714(d/L)^2 + 0.0001142(d/L)^3$$
$$+ 0.006183(d/L)^4 - 0.002897(d/L)^5 + \cdots$$

At $L \ll A$, Eq. (2.130) is transformed into Eq. (2.44) for a straight sphero-cylinder, and when L decreases further to the value $L = d$, it is transformed into the equation for a spherical particle with the diameter d.

At $L/A \geq 2.278$ [64],

$$3\pi\eta_0 L/f = B_1(L/A)^{1/2} + B_2 + B_3(A/L)^{1/2} + B_4(A/L) + B_5(A/L)^{3/2}. \qquad (2.132)$$

where

$$B_1 = (4/3)(6\pi)^{1/2} \equiv (3\pi/P_\infty) = 1.843,$$
$$B_2 = [1 - 0.01412\,(d/A)^2 + 0.00592\,(d/A)^4]\ln(A/d) - 1.0561$$
$$\qquad - 0.1667\,(d/A) - 0.1900\,(d/A)^2 - 0.0224\,(d/A)^3 + 0.0190\,(d/A)^4,$$
$$B_3 = 0.1382 + 0.6910\,(d/A)^2, \qquad\qquad\qquad\qquad\qquad (2.133)$$
$$B_4 = [0.04167\,(d/A)^2 + 0.00567\,(d/A)^4]\ln(A/d) - 0.3301$$
$$\qquad + 0.5\,(d/A) - 0.5854\,(d/A)^2 - 0.0094\,(d/A)^3 - 0.0421\,(d/A)^4,$$
$$B_5 = -0.0300 + 0.1209\,(d/A)^2 + 0.0259\,(d/A)^4.$$

In contrast to Eqs. (2.107), (2.109), and (2.110) for a wormlike pearl neck-lace, these equations for a wormlike cylinder do not contain model parameters and express the friction coefficient of the chain only through the values of L, d, and A characterizing its structure and conformation. This is an indisputable advantage of the solid cylinder model. Moreover, the possibilities of using these equations are much wider because they describe the friction characteristics of a rigid-chain molecule not only in the range of variations from a random coil to a thin straight rod (small d/L ratio) but also in those cases when the decreasing chain length L be-comes comparable to the chain diameter.

In order to carry out a quantitative comparison of the equations for a wormlike cylinder and a wormlike pearl necklace, the latter should be used in the form containing no model parameters, i.e., corresponding to the touching bead model with the Stokes bead diameter.

The part of Eq. (2.130) containing only the terms with the coefficients C_1, C_2, and C_3 at $d/L \to 0$ corresponds to the conformation of a weakly bending rod. Com-parison of this part with Eq. (2.108) shows that the values of C_1 and C_3 in both equations coincide, whereas in Eq. (2.130), $C_2 = 0.3863$ and in Eq. (2.108), $C_2 = 0$. This difference is due to the mathematical approximations made in the derivations of Eqs. (2.108) and (2.130) and, therefore, is not of fundamental significance. It is noteworthy that the term $C_3(L/A)$ characterizing the effect of chain flexibility on the value of f is identical in both equations and, hence, if they are used for the determination of A, the results should coincide.

For a thin and relatively long chain molecule in the wormlike coil conforma-tion, if the terms A/L and d/A in Eq. (2.132) are neglected, one obtains the equation

$$\eta_0 M/f = P_\infty^{-1}(M_L/A)^{1/2}M^{1/2} + (M_L/3\pi)[\ln(A/d) - 1.0561], \qquad (2.134)$$

differing from Eq. (2.111) only in the term 1.0561 (instead of 1.431). Hence, in the determination of the chain rigidity parameter A from the slope of the straight line representing the dependence of $\eta_0 M/f$ on $M^{1/2}$, the wormlike pearl necklace and the wormlike cylinder models should give coinciding results. However, the value of the chain diameter d determined from Eq. (2.134) for a wormlike cylinder will exceed that determined from Eq. (2.111) for a wormlike necklace by a factor of $e^{0.375} = 1.455$.

6.2. Intrinsic Viscosity

The theory of intrinsic viscosity [η] of a solution of rigid-chain molecules based on the wormlike cylinder model has been developed by Yamakawa et al., who employed the Oseen–Burgers method and a procedure similar to that used in the calculation of translational friction.

In the initial theory they calculated [η] without taking into account the effects at the top and bottom of the wormlike cylinder [65]. Subsequent calculations were supplemented by taking into account the effect of the shape of surfaces limiting the top and bottom of the wormlike cylinder [66], which is important at low values of L/A when the molecule has the shape of a weakly bending rod of finite thickness. Consequently, for this range of L/A values the friction properties of molecules may be described by a combination of the theory of a straight spherocylinder [26] taking into account the finite value of d/L and the theory of a wormlike cylinder [65] in which chain flexibility (finite value of L/A) is taken into account.

As a result of this combination, an expression for the intrinsic viscosity [η] of a solution of wormlike spherocylindrical particles has been obtained for the range $L/A \leq 2.278$ [66]:

$$2\pi N_A M^2/45 M_L^3 [\eta] = F_\eta (p)/f (L/A). \tag{2.135}$$

The function $F_\eta(p) = F_\eta(L/d)$ in Eq. (2.135) describing the viscous properties of a straight spherocylinder over the entire range of values of $1 \leq p \leq \infty$ is determined by Eq. (2.46). The function $f(L/A)$ taking into account the flexibility (bending) of a wormlike chain is determined by the power series

$$f (L/A) = 1 - 0.321593 (L/A) - 0.0466384 (L/A)^2 + 0.106466 (L/A)^3 \\ - 0.0379317 (L/A)^4 + 0.00399576 (L/A)^5 \ldots \tag{2.136}$$

In the range of values $L/A > 2.278$ the expression for viscosity can be represented in the form [66]

$$M/[\eta] = \Phi_\infty^{-1} (M_L/A)^{3/2} M^{1/2} \varphi (L, d, A), \tag{2.137}$$

where $\Phi_\infty = 2.87 \cdot 10^{23}$ and the function φ is given by

$$\varphi = 1 - [C_1 (A/L)^{1/2} + C_2 (A/L) + C_3 (A/L)^{3/2} + C_4 (A/L)^2]. \tag{2.138}$$

The coefficients C_1-C_4 in Eq. (2.138) depend on L/A, and in different ranges of d/A values they are represented by the following series:

For $d/A \leq 0.1$,

$$C_1 = 3.230981 - 143.7458 (d/A) - 1906.263 (d/A)^2 \\ + \ln (d/A) [2.463404 - 1422.067 (d/A)^2],$$
$$C_2 = -22.46149 + 1347.079 (d/A) + 19387.400 (d/A)^2 \\ + \ln (d/A) [-5.318869 + 13868.57 (d/A)^2],$$
$$C_3 = 54.81690 - 3235.401 (d/A) - 49357.06 (d/A)^2 + \tag{2.139}$$

$$+ \ln(d/A)[15.41744 - 34447.63\,(d/A)^2],$$
$$C_4 = -32.91952 + 2306.793\,(d/A) + 36732.64\,(d/A)^2$$
$$+ \ln(d/A)[-8.516339 + 25198.11\,(d/A)^2];$$

and for $0.1 < d/A < 1.0$,

$$C_1 = 6.407860 - 25.43785\,(d/A) + 23.33518\,(d/A)^2$$
$$+ \ln(d/A)[3.651970 - 25.73698\,(d/A)^2],$$
$$C_2 = -115.0086 + 561.0286\,(d/A) - 462.8501\,(d/A)^2$$
$$+ \ln(d/A)[-33.69143 + 523.6108\,(d/A)^2],$$
$$C_3 = 318.0792 - 1625.451\,(d/A) + 1451.374\,(d/A)^2 \qquad (2.140)$$
$$+ \ln(d/A)[92.13427 - 1508.112\,(d/A)^2],$$
$$C_4 = -144.5268 + 661.6760\,(d/A) - 1057.731\,(d/A)^2$$
$$+ \ln(d/A)[-42.41552 + 211.6622\,(d/A)^2].$$

The combination of Eqs. (2.137)–(2.140) ensures a smooth transition from the equations for viscosity at $L/A \leq 2.278$ to Eqs. (2.135) and (2.136) at $L/A > 2.278$.

The dependence of intrinsic viscosity on the molecular parameters $M, L, d,$ and A for a wormlike chain can also be represented in the form of the well-known Flory equation

$$[\eta] = \Phi\,(LA)^{3/2}/M, \qquad (2.141)$$

where Φ is a function of these parameters.

The form of this function for the wormlike spherocylinder model follows directly from the above expressions.

Thus, comparison of Eqs. (2.141) and (2.135) gives, for $L/A \leq 2.278$,

$$\Phi = (2\pi N_A/45)\,(L/A)^{3/2}\,[f\,(L/A)/F_\eta\,(p)]. \qquad (2.142)$$

Comparison of Eqs. (2.141) and (2.137) yields, for the range $L/A > 2.278$,

$$\Phi = \Phi_\infty/\varphi\,(L, \; d, \; A). \qquad (2.143)$$

The combination of Eqs. (2.141)–(2.143) permits the calculation of the intrinsic viscosity for the wormlike cylinder model over the entire range of possible molecular shapes from a sphere with diameter d (when $L = d$) to the nondraining Gaussian coil for which $L \gg d$ and $L \gg A$.

The values of Φ obtained for various values of d/A and $L/A > 2.278$ are given in Table 1 of [65].

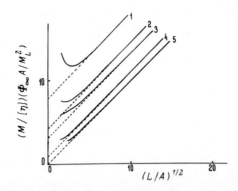

Fig. 2.10. Dependence of $(M/[\eta])(\Phi_\infty A/M_L^2)$ on reduced chain length L/A for a wormlike spherocylinder according to Eq. (2.137'). The normalization factor $\Phi_\infty A/M_L^2$ on the ordinate is independent of chain length. Curves 1-3 are plotted according to Eqs. (2.138) and (2.139) and curves 4 and 5 are plotted using Eqs. (2.138) and (2.140). The numbers at the curves correspond to the following d/A values: 1) 0.01; 2) 0.06; 3) 0.10; 4) 0.30; 5) 0.47. Broken straight lines represent asymptotic equation (2.145).

The equations for the intrinsic viscosity of wormlike spherocylinders may be compared to those based on the model of a wormlike pearl necklace with touching beads.

For this purpose, at $L/A \ll 1$ it should be assumed in Eq. (2.135) that $p = L/d \gg 1$ and, correspondingly, $F_\eta(p)$ should be expressed according to Eq. (2.48), and the function $f(L/A)$ in Eq. (2.136) should be limited to the linear term in the expansion in powers of L/A. Then Eq. (2.135) reduces to

$$[\eta] = (2\pi N_A M^2/45 M_L^3)\,[\ln(L/d) - 0.697]^{-1}\,[1 - 0.32\,(L/A)], \qquad (2.144)$$
$$L/A \ll 1, \ L/d \gg 1.$$

Equation (2.144) coincides in form with Eq. (2.123) for a wormlike necklace but differs from it in the much higher value of the numerical coefficient of the term L/A, taking into account chain flexibility (0.32 rather than 1/16). This difference is important in the determination of A from viscometric measurements in solutions of rigid-chain polymers [60].

To compare the theories of viscous properties of a wormlike spherocylinder and a wormlike necklace at high L/A one should analyze the dependence of $M/[\eta]$ on $M^{1/2}$ for spherocylinders, since for a wormlike necklace this dependence is known and expressed by an inclined straight line described by Eq. (2.124).

For this purpose it is convenient to transform Eq. (2.137) into the equivalent expression

$$(M/[\eta])\,(\Phi_\infty A/M_L^2) = (L/A)^{1/2}\,\varphi \qquad (2.137')$$

and to plot its left-hand side as a function of $(L/A)^{1/2}$ using the function φ according to Eq. (2.138). This plot is shown in Fig. 2.10. As the reduced chain length, L/A, increases, all the curves (corresponding to various values of d/A) degenerate into asymptotic straight lines with slope equal to unity.

With decreasing chain length L/A the asymptotic straight line becomes a curve with upward curvature, which corresponds to a change in chain conformation from a random coil to a bending rod. The lower the d/A value corresponding to the curves in Fig. 2.10 (i.e., the higher the equilibrium chain rigidity), the more pronounced is their curvature.

With increasing d/A (i.e., with decreasing equilibrium rigidity) the intercepts of the asymptotes of curves 1, 2, 3, ... with the ordinate decrease, and at $d/A = 0.47$ or $\ln A/d = 0.755$, the asymptote passes through the origin.

The type of curve in Fig. 2.10 in the region of $L/A \gg 1$ corresponds qualitatively to the dependence of $M/[\eta]$ on $M^{1/2}$ for the wormlike necklace model described by Eq. (2.124). Hence, for the asymptotic limit of Eq. (2.137') (at $L/A \gg 1$) the following equation may be used with a good approximation for a wormlike spherocylinder [67]:

$$M/[\eta] = \Phi_\infty^{-1} (M_L/A)^{3/2} M^{1/2} + 2.2\Phi_\infty^{-1} (M_L^2/A) \lfloor \ln (A/d) - 0.755 \rfloor. \qquad (2.145)$$

The broken lines in Fig. 2.10 are plotted according to Eq. (2.145) at the values of the parameter d/A corresponding to those for curve 1-5. It can be seen that at $L/A > 20$ the asymptotic straight lines virtually coincide with the corresponding curves and, hence, in this range of L/A values Eq. (2.145) is identical to the general equation (2.137).

7. EXCLUDED-VOLUME EFFECT

7.1. Flexible-Chain Polymers

The thermodynamic nonideality of the solution and the related excluded volume effect (Chapter 1, Section 5) profoundly affect the hydrodynamic behavior of flexible chain polymer molecules, thus complicating the interpretation of experimental data [5, 30, 31].

However, if the aim of the investigations is to determine hydrodynamic and conformational characteristics of molecules unperturbed by volume effects, one can use for this purpose fairly simple procedures of extrapolation of experimental data to the range of low molecular weights. These procedures are based on the fact that with decreasing molecular weight M the excluded volume parameter z [Eq. (1.38)] for a flexible-chain polymer decreases proportionally to $M^{1/2}$ according to the equation

$$z = (3/2\pi)^{1/2} B (M/LA)^{3/2} M^{1/2}, \qquad (2.146)$$

where B is dependent on solvent quality (under θ-conditions $B = 0$) but independent of the molecular weight of the polymer.

The hydrodynamic properties of molecules of flexible-chain polymers correspond to those of nondraining coils and, hence, the increase in their mean-square length $\langle h^2 \rangle^{1/2}$ upon expansion in a good solvent should be accompanied by an increase in the translational friction coefficient f proportional in the first approximation to an increase in $\langle h^2 \rangle$ or, to be precise [68] [compare to Eqs. (1.34) and (1.35)],

$$f = f_\theta (1 + 0.609z), \qquad (2.147)$$

where f_θ is determined according to Eq. (2.84).

A comparison of Eqs. (2.147), (2.89), and (2.146) gives [69]

$$f/M^{1/2} = \eta_0 P_\infty (LA/M)^{1/2} [1 + 0.2B (M/LA)^{3/2} M^{1/2}]. \qquad (2.148)$$

Equation (2.148) predicts a linear dependence of experimental values of $f/M^{1/2}$ on $M^{1/2}$; the extrapolation of this dependence to $M \to 0$ yields the value of f_θ corresponding to Eq. (2.84).

Similar considerations lead to Eq. (2.149) for the viscosity $[\eta]$ of a flexible-chain polymer under non-θ-conditions [70]

$$[\eta]/M^{1/2} = (LA/M)^{3/2}\Phi_\infty + 0.51\Phi_\infty BM^{1/2}. \qquad (2.149)$$

The extrapolation of the dependence of experimental values of $[\eta]/M^{1/2}$ on $M^{1/2}$ to the conditions $M \to 0$ makes it possible to obtain the value of $(a/M_L)^{3/2}$ and to determine the parameter A of the equilibrium chain rigidity unperturbed by the excluded volume effect.

7.2. Rigid-Chain Polymers

As has been shown in Chapter 1, for typical rigid-chain polymers over the range of molecular weights accessible in practice, even in thermodynamically good solvents the excluded-volume effects have little influence on the size of the molecules and, hence, should not influence their hydrodynamic properties either. However, in principle, in the study of rigid-chain polymers of very high molecular weight the possibility of this effect cannot be ruled out. For example, this effect occurs in DNA [71].

The translational friction of a rigid-chain polymer represented by a wormlike pearl necklace has been calculated by Hearst et al. [72] taking into account the excluded-volume effects. Calculations were carried out using the parameter ε characterizing the deviation of the statistical chain dimensions from Gaussian properties caused by volume effects. According to Eqs. (1.36) and (1.7) the second moment, $\langle h^2 \rangle$, of the end-to-end distance distribution $W(h)$ in Gaussian chains perturbed by volume effects is related to the unperturbed value $\langle h^2 \rangle_\theta = LA$ (under θ-conditions) by the equation

$$\langle h^2 \rangle = \langle h^2 \rangle_\theta (L/A)^\varepsilon. \tag{2.150}$$

Similarly, according to Eqs. (1.6) and (2.150) we have

$$\langle 1/h \rangle = \langle 1/h \rangle_\theta (L/A)^{-\varepsilon/2}. \tag{2.151}$$

In accordance with this, when hydrodynamic interaction in a wormlike pearl necklace was calculated, the reciprocal distances $\langle 1/r_{ik} \rangle$ between chain elements were determined by using the following relationships. At $r_{ik}/A > \sigma$ it was assumed that $L_{ik}\langle 1/r_{ik} \rangle = L_{ik}\langle 1/r_{ik} \rangle_0 (L_{ik}/A)^{-\varepsilon/2}$, where $\langle 1/r_{ik} \rangle_0$ is determined from Eq. (2.103). At $r_{ik}/A < \sigma$, the product $L_{ik}\langle 1/r_{ik} \rangle$ was calculated according to Eq. (2.105) but the coefficients α and β were selected as functions of ε to achieve a smooth joining of one curve to another at $r_{ik}/A = \sigma$ and at the corresponding value of ε.

As a result, the following expression was obtained for the translational friction coefficient at high L/A ratios:

$$\eta_0 M/f = P_\infty^{-1} [(1-\varepsilon)(1-\varepsilon/3)]^{-1} M_L^{\frac{1+\varepsilon}{2}} A^{\frac{\varepsilon-1}{2}} M^{\frac{1-\varepsilon}{2}}$$
$$+ (M_L/3\pi)[\ln(A/d) + 1 + \Psi(\varepsilon)], \tag{2.152}$$

where the function $\Psi(\varepsilon)$ has been tabulated in [72]. At $\varepsilon = 0$ we have $\Psi(\varepsilon) = -2.431$ and Eq. (2.152) becomes Eq. (2.111).

Hence, when excluded-volume effects exist, at high L/A the M/f ratio is linear with respect to $M^{(1-\varepsilon)/2}$. The same method was used to show that under these conditions $M/[\eta]$ is linear with respect to $M^{(1-3\varepsilon)/2}$ [73].

In accordance with this theory, if the experimental values of M/f are plotted as a function of $M^{1/2}$, when volume effects exist, one obtains, instead of a straight line, a curve concave toward the abscissa. By plotting M/f as a function of M^k and selecting the value of $k = k_0$ in such a manner that the experimental points fit a straight line, the experimenter can determine ε from the condition $(1-\varepsilon)/2 = k_0$.

REFERENCES

1. J. H. Edsall, Fortschr. Chem. Forsch., **1**, 119 (1949).
2. J. T. Yang, Adv. Protein Chem., **16**, 323 (1961).
3. H. Benoit, L. Freund, and G. Spach, in: Poly-α-amino Acids, G. D. Fasman (ed.), Marcel Dekker, New York (1967).
4. A. Teramoto and H. Fujita, Adv. Polym. Sci., **18**, 65 (1975).
5. V. N. Tsvetkov, V. E. Eskin, and S. Ya. Frenkel, Structure of Macromolecules in Solutions, Nauka, Moscow (1964); National Lending Library for Science and Technology, Boston Spa, England (1971). Vol. 3.
6. V. N. Tsvetkov, in: Newer Methods of Polymer Characterization, B. Ke (ed.), Wiley, New York (1964).

7.   W. Kuhn, H. Kuhn, and P. Buchner, Ergeb. Exakten. Naturwiss., **25**, 1 (1951).

8.   H. Lamb, Hydrodynamics, Cambridge Univ. Press (1932).

9.   G. Stokes, Trans. Cambridge Philos. Soc., **9**, 8 (1856).

10.   R. Gans, Ann. Phys., **86**, 628 (1928).

11.   R. Herzog, R. Illig, and H. Kudar, Z. Phys. Chem., **A167**, 329 (1933).

12.   F. Perrin, J. Phys. Rad., **7**, 1 (1936).

13.   D. Edwards, Q. J. Math., **26**, 70 (1893).

14.   F. Perrin, J. Phys. Rad., **5**, 497 (1934).

15.   A. Einstein, Ann. Phys., **19**, 289 (1906); **34**, 598 (1911).

16.   R. Simha, J. Phys. Chem., **44**, 25 (1940).

17.   G. B. Jeffery, Proc. R. Soc. London, **A102**, 161 (1922).

18.   A. Peterlin, Z. Phys., **111**, 232 (1938).

19.   W. Kuhn and H. Kuhn, Helv. Chim. Acta, **28**, 97, 1533 (1945); **29**, 72 (1946).

20.   C. W. Oseen, Hydrodynamik, Akademische Verlagsgesselscheft, Leipzig (1927).

21.   J. M. Burgers, Second Report on Viscosity and Plasticity of the Amsterdam Academy of Science, Nordemann, New York (1938).

22.   S. Broersma, J. Chem. Phys., **32**, 1626 (1960).

23.   S. Broersma, J. Chem. Phys., **32**, 1632 (1960).

24.   H. Yamakawa, Macromolecules, **8**, 339 (1975).

25.   T. Norisuye, M. Motowoka, and H. Fujita, Macromolecules, **12**, 320 (1979).

26.   T. Yoshizaki and H. Yamakawa, J. Chem. Phys., **72**, 57 (1980).

27.   C. Tanford, Physical Chemistry of Macromolecules, New York (1961).

28.   H. Morawetz, Macromolecules in Solution, Wiley-Interscience, New York (1965).

29.   G. Champetier and L. Monnerie, Introduction á la Chimie Macromoleculare, Masson et Editeurs, Paris (1969).

30.   H. Yamakawa, Modern Theory of Polymer Solutions, Harper and Row, New York (1971).

31.   J. J. Hermans (ed.), Polymer Solution Properties, Parts I and II, Dowden, Hutchinson and Ross (1978).

32.   H. Kuhn, Habilitationsschrift, Basel (1946).

33.   H. Kuhn, J. Colloid Sci., **5**, 331 (1950).

34.   H. Kuhn and W. Kuhn, J. Polym. Sci., **5**, 519 (1950); **9**, 1 (1952).

35.   H. Kuhn, F. Moning, and W. Kuhn, Helv. Chim. Acta, **36**, 731 (1953).

36.   H. Kuhn, W. Kuhn, and A. Silberberg, J. Polym. Sci., **14**, 193 (1954).

37.   P. J. Flory, Principles of Polymer Chemistry, Cornell Univ. Press, Ithaca (1953).

38.   P. J. Flory and T. G. Fox, J. Am. Chem. Soc., **73**, 1904 (1951).

39.   J. G. Kirkwood and J. Riseman, J. Chem. Phys., **16**, 565 (1948).

40.   J. Riseman and J. G. Kirkwood, J. Chem. Phys., **17**, 442 (1949).

41.   J. Riseman and J. G. Kirkwood, J. Chem. Phys., **18**, 512 (1950).

42.   J. G. Kirkwood and P. L. Auer, J. Chem. Phys., **19**, 281 (1951).

43.   J. G. Kirkwood, J. Polym. Sci., **12**, 1 (1954).

44.   W. Kuhn, Z. Phys. Chem., **A161**, 1 (1932).

45.   M. Kurata and H. Yamakawa, J. Chem. Phys., **29**, 311 (1958).

46.   P. Debye, J. Chem. Phys., **14**, 636 (1946).

47. J. G. Kirkwood, R. W. Zwanzig, and R. J. Plock, J. Chem. Phys., **23**, 213 (1955).
48. P. L. Auer and C. S. Gardner, J. Chem. Phys., **23**, 1545 (1955).
49. B. H. Zimm, J. Chem. Phys., **24**, 269 (1956).
50. C. W. Pyun and M. Fixman, J. Chem. Phys., **42**, 3838 (1965).
51. M. Bixon and R. W. Zwanzig, J. Chem. Phys., **68**, 1890 (1978).
52. J. E. Hearst and Y. Tagami, J. Chem. Phys., **42**, 4149 (1965).
53. B. H. Zimm, Macromolecules, **13**, 592 (1980).
54. H. Yamakawa, J. Chem. Phys., **53**, 436 (1970).
55. O. B. Ptitsyn and Yu. E. Eizner, Vysokomol. Soedin., **3**, 1863 (1961).
56. J. E. Hearst and W. H. Stockmayer, J. Chem. Phys., **37**, 1425 (1962).
57. J. E. Hearst, J. Chem. Phys., **38**, 1062 (1963).
58. J. J. Hermans and R. Ullman, Physica, **18**, 951 (1952).
59. J. E. Hearst, J. Chem. Phys., **40**, 1506 (1964).
60. M. G. Vitovskaya and V. N. Tsvetkov, Eur. Polym. J., **12**, 251 (1976).
61. S. F. Edwards and M. A. Oliver, J. Phys. A, Gen. Phys., **4**, 1 (1971).
62. R. Ullman, J. Chem. Phys., **49**, 5486 (1968).
63. R. Ullman, J. Chem. Phys., **53**, 1734 (1970).
64. H. Yamakawa and M. Fujii, Macromolecules, **6**, 407 (1973).
65. H. Yamakawa and M. Fujii, Macromolecules, **7**, 128 (1974).
66. H. Yamakawa and T. Yoshizaki, Macromolecules, **13**, 633 (1980).
67. A. V. Lesov and V. N. Tsvetkov, Vysokomol. Soedin., **A26**, 494 (1984).
68. W. H. Stockmayer and A. C. Albrecht, J. Polym. Sci., **32**, 215 (1958).
69. J. M. G. Cowie and S. Bywater, Polymer, **6**, 197 (1965).
70. W. H. Stockmayer and M. Fixman, J. Polym. Sci., **C1**, 137 (1963).
71. D. M. Crothers and B. H. Zimm, J. Mol. Biol., **12**, 525 (1965).
72. H. B. Gray, V. A. Bloomfield, and J. B. Hearst, J. Chem. Phys., **46**, 1493 (1967).
73. P. Sharp and V. A. Bloomfield, J. Chem. Phys., **48**, 2149 (1968).

Chapter 3

TRANSPORT METHODS

As already mentioned, the main phenomena that provide information about the translational friction of macromolecules are transport processes: diffusion and sedimentation of polymer molecules in solution.

The theory behind these phenomena and their experimental application to polymers has been extensively developed and considered in detail in many excellent monographs [1–11]. Therefore, here we will deal with these problems only insofar as they are important to the specific polymer systems considered below.

The diffusion method is of particular importance to many current rigid-chain polymers and will be considered first.

1. DIFFUSION

1.1. Principles Governing the Phenomenon

Translational diffusion in solutions is a process directly related to the thermal motion of solute molecules and may serve as the most direct method for the determination of their translational friction coefficient.

Macroscopic-directed diffusion in solution takes place when a concentration gradient of the solute, dc/dx, exists in the direction x. It is described by two Fick's equations (3.1) and (3.2), the first of which also serves for the determination of the diffusion coefficient D:

$$L = -D \, (dc/dx). \tag{3.1}$$

where L is the flow of the substance diffusing in the direction x (mass of the solute passing in unit time through unit area of a section normal to x).

The second Fick's equation characterizing the nonstationary diffusion flow expresses the change in solution concentration with time, dc/dt, occurring as a result of this diffusion. For the one-dimensional case (diffusion in one direction) this equation follows from Eq. (3.1) if we recall that $\partial c/\partial t = -\partial L/\partial x$. Then we obtain

$$\partial c/\partial t = \partial/\partial x\,[D\,(\partial c/\partial x)]. \tag{3.2}$$

If the diffusion coefficient D in this system is independent of x, then Eq. (3.2) becomes

$$\partial c/\partial t = D\,(\partial^2 c/\partial x^2). \tag{3.3}$$

Of great practical importance in the investigation of diffusion in polymer solutions is the case where two solutions separated by a planar interface (at $x = 0$) normal to x at the initial concentrations c_1 and c_2, respectively, come into contact (at the moment $t = 0$).

If these liquids extend along the x axis for relatively large distances to both sides of the plane where $x = 0$, the solution of Eq. (3.3) yields the following concentration distribution at the moment t (integral distribution):

$$c = c_1 + \frac{\Delta c}{2}\left(1 + \frac{2}{\sqrt{\pi}}\int_0^y e^{-y^2}dy\right), \tag{3.4}$$

where $y = x/2\sqrt{Dt}$ and $\Delta c = c_2 - c_1 \equiv c_0$.

The dependence $c(x)$ corresponding to Eq. (3.4) is shown in Fig. 3.1a at various times t.

Evidently, it is possible to determine D from the experimental curve $c(x)$.

In many modern methods the experimentally determined value is the concentration gradient dc/dx rather than the concentration of the solution. In these cases it is more convenient to use Eq. (3.5) describing the differential concentration distribution

$$dc/dx = \left(\Delta c/\sqrt{2\pi}\sigma\right)e^{-x^2/2\sigma^2}, \tag{3.5}$$

where

$$\sigma^2 = 2Dt. \tag{3.6}$$

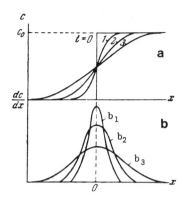

Fig. 3.1. a) Integral concentration distribution in diffusion at time $t = 0$ and $t_1 < t_2 < t_3$.; c_0 is the initial difference in concentrations. b) Differential concentration distribution in diffusion. The times are as in Fig. 3.1a.

The curves dc/dx plotted according to Eq. (3.5) are shown in Fig. 3.1b for various values of σ ($\sigma_1 < \sigma_2 < \sigma_3$).

Hence, in the simplest case considered here, corresponding to Eq. (3.3) ($D =$ const), the diffusion distribution dc/dx has the shape of a Gaussian curve with the dispersion σ^2 determined according to Eq. (3.6).

The experimental curves in Fig. 3.1b make it possible to determine the dispersion σ^2 and, hence, the value of D.

To find σ^2, the method of moments of the distribution curve can be used, according to which

$$\sigma^2 = m_2/m_1 - (m_1/m_0)^2, \tag{3.7}$$

where m_0, m_1, and m_2 are the distribution moments of the zeroth, first, and second orders with respect to an arbitrary point on the abscissa (x) lying beyond the distribution curve dc/dx:

$$m_k = \int_0^\infty x^k \, (dc/dx) \, dx. \tag{3.8}$$

Evidently, m_0 is equal to the area under the distribution curve dc/dx, whereas m_1/m_0 gives the abscissa value of its center (centroid).

If the distribution dc/dx in a diffusion experiment is a Gaussian distribution, i.e., if it corresponds to Eq. (3.5), then it follows from Eqs. (3.5) and (3.6) that

$$\sigma^2 = (Q/H)^2/2\pi, \tag{3.9}$$

where Q is the area enclosed by the diffusion curve and H is its maximum ordinate. This permits the determination of the diffusion coefficient $D = D_A$ from the values of Q and H using Eqs. (3.9) and (3.6) ("height–area" method).

As has been shown by Einstein [12], the diffusion coefficient in solution macroscopically determined by Eqs. (3.1) and (3.2) is directly related to the thermal mobility of the solute molecules. For example, Eq. (3.6) is an expression of this relationship. In fact, the dispersion σ^2 of the Gaussian distribution curve (3.5) is the mean-square value x^2 of the displacement of dissolved molecules from their initial position $x = 0$ in the time t characterizing their thermal mobility.

Another basic expression relating the macroscopic diffusion process to the properties of diffusing molecules is, according to Einstein, the dependence of the coefficient D on the translational friction coefficient of the molecule, f (Chapter 2).

This dependence is expressed by the equation

$$D = \frac{1}{f} \frac{M}{N_A} \frac{dp}{dc}, \qquad (3.10)$$

where p is the osmotic pressure of the solution.

For an ideal solution of noninteracting particles the osmotic pressure is determined by the Van't Hoff law

$$p = cRT/M. \qquad (3.11)$$

Equations (3.10) and (3.11) yield the Einstein equation

$$D = kT/f. \qquad (3.12)$$

Hence, if the diffusion rate (i.e., D) of a polymer in solution is measured experimentally, it is possible to determine, according to Eq. (3.12), the friction coefficient of its molecule, f, which is directly related to the geometrical characteristics of this molecule (Chapter 2).

1.2. Effect of Concentration and Polydispersity on the Shape of Diffusion Curves

The distribution curves $c(x)$ for Eq. (3.4) or dc/dx for Eq. (3.5) considered above were obtained by solving Fick's equation (3.3) for the case where the solution is characterized by a constant value of the diffusion coefficient D.

This situation is possible if two conditions are met: 1) when the concentration dependence $D(c)$ is absent and 2) when the dissolved polymer is a very narrow fraction that can be regarded as a monodisperse system.

If one of these conditions (or both) is not fulfilled, the type of distribution curves changes. In particular, the distribution dc/dx is no longer Gaussian, and the shape of the dc/dx curve can differ greatly from that corresponding to Eq. (3.5).

1.2.1. Concentration Effect. Since in most cases polymer solutions are not ideal, the concentration dependence of their osmotic pressure does not obey the Van't Hoff law (3.11) and is expressed as a virial power series of c with the coefficients $A_2, A_3, \ldots$ [13]. If this series is substituted into Eq. (3.10), the following dependence is obtained:

$$D = (kT/f)(1 + 2A_2Mc + 3A_3Mc^2 + \cdots),\tag{3.13}$$

which becomes Einstein's equation (3.12) for a solution at infinite dilution ($c \rightarrow 0$).

Since the sum of the terms depending on c on the right-hand side of Eq. (3.13) is positive, the nonideality of the solution should lead to an increase in D with concentration. The higher the molecular weight M of the dissolved polymer and the greater the second virial coefficient A_2 in the polymer–solvent system, the greater is this increase. These relationships are confirmed by experimental data [7].

The experimental investigation of the concentration dependence of diffusion under θ-conditions (when $A_2 = A_3 = 0$) shows that even for high-molecular-weight polymer samples in a range of concentrations up to $c = 0.01$ g/cm³ the diffusion coefficient virtually does not vary with changing c [7]. This fact implies that f in Eq. (3.13) also shows no marked concentration dependence.

If the concentration dependence $D(c)$ in the polymer–solvent system is relatively pronounced and the experiment is carried out with a relatively great concentration difference $c_1 - c_2$, then the dc/dx curve is asymmetric. In fact, since D increases with concentration (positive dependence), the diffusion process occurs more slowly in the regions of solution exhibiting a lower value of c. Hence, the dc/dx curve descends more steeply at lower concentrations (or pure solvent) and less steeply at higher concentrations of the solution.

A particularly strong positive concentration dependence of diffusion is observed in polyelectrolyte solutions of low ionic strength and, hence, these solutions exhibit very asymmetric differential dc/dx curves.

It is impossible to calculate the diffusion coefficient from these curves using Eqs. (3.4) and (3.5), and rather complex analytical and graphical procedures leading to considerable errors have been employed for this purpose [14–16].

More reliable results can be obtained only for very dilute solutions. In this case, even at a strong concentration dependence of D, the difference in diffusion coefficients on both sides of the diffusion interface can, in principle, be made small. This will ensure symmetric diffusion curves and will allow Eqs. (3.4) and (3.5) to be used.

1.2.2. Polydispersity. If concentration effects are ruled out (the solution is thermodynamically ideal and contains no polyelectrolyte, or a small difference between concentrations $c_1 - c_2$ is used in the experiment), then for a substance monodisperse with respect to molecular weight the diffusion curves dc/dx correspond to the Gaussian distribution [Eq. (3.5)].

However, if the polymer is polydisperse, then, even if concentration effects are absent, the dc/dx curves prove to be non-Gaussian.

In fact, during diffusion under these conditions each monodisperse component forms its own Gaussian curve at the time t determined by the dispersion $\sigma_i^2 = 2D_i t$. The superposition of these Gaussian curves for all components gives the distribution dc/dx which, although still symmetric, is, generally speaking, not a Gaussian distribution (Fig. 3.2).

These curves exhibit a characteristic sharp and high maximum with broad "wings" at the base. This is due to the presence of high- and low-molecular-weight components, respectively.

The overall distribution dc/dx is characterized by the dispersion σ^2,

$$\sigma^2 = \sum_i w_i \sigma_i^2 = 2t \sum_i w_i D_i, \tag{3.14}$$

where w_i is the weight fraction of the ith component ($\Sigma_i w_i = 1$). The value of

$$\sum_i w_i D_i = D_w = \int_0^c D\,dc \bigg/ \int_0^c dc \tag{3.15}$$

may be called the weight-average diffusion coefficient for a polydisperse system. It is determined from the dispersion σ^2 of the experimental curve of dc/dx according to Eq. (3.14)

$$D_w = \sigma^2/2t. \tag{3.16}$$

For a curve of any shape, σ^2 is determined by the method of moments according to Eq. (3.7).

Hence, the diffusion coefficient D_m calculated by the method of moments [Eqs. (3.7) and (3.16)] is always equal to the weight-average coefficient $D_m = D_w$ regardless of the shape of the diffusion curve.

For an ideal Gaussian curve obtained for a monodisperse substance in the absence of concentration effects, the dispersion σ^2 is defined by Eq. (3.9). Hence, Eq. (3.16) becomes

$$D_A = (1/4\pi t)(Q/H)^2. \tag{3.17}$$

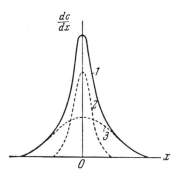

!Fig. 3.2. Theoretical diffusion curve 1 of a polydisperse
substance consisting of two monodisperse components
represented by Gaussian curves 2 and 3.

For a polydisperse substance the value of D_A calculated from the area Q under the curve, and the height H of the curve, has no such direct physical meaning as $D_m = D_w$. However, it can be shown [16] that, at any distribution dc/dx,

$$D_A = \left(\int_0^c dc \Bigg/ \int_0^c \frac{dc}{D^{1/2}} \right)^2 = \left(\int_0^\infty q_w(M)\, dM \Bigg/ \int_0^\infty \frac{q_w(M)}{D^{1/2}}\, dM \right)^2 , \qquad (3.18)$$

where $q_w(M)$ is the weight function of the molecular-weight distribution. In this case the following inequality is always valid:

$$D_A < D_m = D_w. \qquad (3.19)$$

Hence, the D_m/D_A ratio calculated according to Eqs. (3.16) and (3.17) using the experimental values of σ^2, Q, and H may serve as a measure of polydispersity of the sample.

Information concerning the polydispersity of the polymer can also be obtained by investigating the changes occurring in its diffusion characteristics with time during diffusion [17–19]. For the diffusion of a polydisperse sample the leveling off of concentrations, i.e., the decrease in their gradients, is due mainly to the low-molecular-weight components which diffuse more rapidly. After diffusion they can no longer take part in the overall diffusion process, and as a result the shape of the diffusion curve approaches a Gaussian shape and the area under it, $Q \approx \Sigma_i \Delta c = \int_0^\infty (dc/dx)dx$, seems to decrease.

Evidently the apparent decrease in this area during diffusion can characterize the polydispersity of the sample.

Moreover, the removal of low-molecular-weight components leads to a decrease in diffusion rate with time because the remaining high-molecular-weight components diffuse more slowly. As a result, the diffusion coefficient calculated at different times t according to Eqs. (3.17) or (3.16) decreases with increasing t.

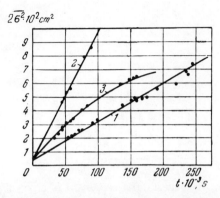

Fig. 3.3. Dependence of $2\sigma^2$ on time t for polystyrene fractions in dichloroethane. 1) $M = 1.32 \cdot 10^6$; 2) $M = 0.25 \cdot 10^6$; 3) 50% mixture of these fractions.

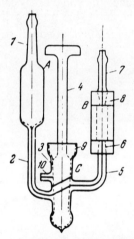

Fig. 3.4. Cell for common organic solvents.

This decrease is clearly seen in the curvature toward the abscissa of the experimental curve of the dependence of σ^2 on t. This can be illustrated in Fig. 3.3, which shows the dependence obtained for two polystyrene fractions in dichloro-ethane and for their 50% mixture in the same solvent [17]. For each fraction the plot $\sigma^2(t)$ is linear whereas for their polydisperse mixture this dependence is a curve concave toward the abscissa.

These data not only illustrate the polydispersity of the system qualitatively [17], but also, when the experiment is carried out carefully, they can be used to obtain quantitative information about the type of molecular-weight distribution existing in this system [18, 19].

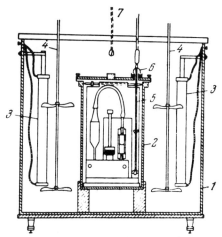

Fig. 3.5. Thermostat of a polarizing diffusiometer. 1) Outer thermostat; 2) inner thermostat with the cell; 3) heating elements; 4) stirrers; 5) stirrer of the inner tank; 6) mercury seal; 7) key for turning the cell valve.

1.3. Experimental Devices

1.3.1. Diffusion Cells. In order to observe diffusion in a cell, a sharp interface between two layers of liquid (solution and solvent) should be established. In diffusiometers the initial sharp interface is achieved by layering the lighter of the two liquids on top of the heavier liquid or by the introduction of the heavier liquid under the lighter liquid.

Many types of cells have been proposed and used by various authors (see, e.g., [7, 11]). Here only those types that were used in the experiments discussed below will be described.

The cell construction shown in Fig. 3.4 was found to be suitable for the investigation of solutions in common organic and aqueous solvents. A liquid of higher density is poured into a cylindrical vessel, A, ending in a filling funnel 1 at the top and passing at the bottom into a capillary 2 through which the solution is fed into the diffusion cell B.

The layering rate can be regulated with a silver or platinum hair placed inside the capillary 2. The cell B is fixed on the ground end 6 of a capillary 5 which is a continuation of the capillary 2 and is connected to it by a three-way valve 4. The cell has the shape of a rectangular parallelepiped. Its walls are made of optically homogeneous glass and cemented together with special resins (at 250–260°C) or fusible glass. Operating surfaces are polished to within 0.1 band. The cell thickness along the light path is usually a few centimeters.

First, the cell B is half filled with the liquid of lower density. A ground capillary tube 7 is connected by a rubber hose with the filling funnel A for equalizing the pressure in the sealed system. To avoid water leaking from the thermostat into the cell, the valve has a mercury seal 9 and the cell B has ground ends 6 and 8. The outlet of the valve 10 and the connections with the rubber hose are in some cases covered with paraffin. The cell prepared for the experiment is fixed in a holder rigidly attached with screws to the inner tank of the thermostat (Fig. 3.5).

After thermal equilibrium is established, the valve 4 is placed in the position shown in Fig. 3.4, and the layering starts. A good interface is obtained if the layering is continued for at least 40–50 min.

Figure 3.5 shows the thermostat of the diffusiometer [20, 21] consisting of two tanks 1 and 2, and ensuring constant temperature to within a few thousandths of a degree K. The diffusion cell is fixed in the inner tank.

For operations with aromatic rigid-chain polymers soluble only in very corrosive solvents, it is of great importance to employ a diffusion cell [11, 22] stable to these solvents. In this cell the liquid being investigated comes into contact only with plane-parallel glasses and a fluoroplastic insert, and the chambers are sealed by mechanical pressing of the optical glasses to their flat end surfaces. The thickness of the insert along the path of the light beam is 1–5 cm. It has two rectangular chambers connected at the bottom by a narrow capillary (0.01 to 0.1 mm²) (Fig. 3.6).

The liquids are introduced into the chamber through the upper openings (Fig. 3.6). The lighter liquid is introduced into the right-hand viewing chamber, and the heavier liquid is poured into the left chamber. The openings are sealed with couplings connected by a hose and the cell is thermostated in the diffusiometer tank (Fig. 3.6, position I). Compressed air is introduced into the left chamber and, as a result, the heavier liquid is pressed through the capillary into the right chamber, forming a layer under the lighter liquid. For nonvolatile liquids the introduction of compressed air into the left chamber may be replaced by degassing the right chambers. After the concentration interface reaches the side opening, this interface can be made sharper by drawing off the liquid through a side coupling (position II). After the layering is finished, the air flow is stopped and the upper openings are connected (position III). The cell volume is small, and it is possible to use as little as 1 cm³ of the solution.

All experimental data on the diffusion of molecules of rigid-chain aromatic polymers soluble in concentrated sulfuric acid were obtained with the aid of this cell.

1.3.2. Optical Scheme of Diffusiometer. As has been emphasized in Section 1.2.1 of this chapter, to obtain reliable experimental data on the type of diffusion curves it is very important to minimize the concentration effects distorting their shape. For this purpose the concentration c of the solutions to be investigated should be as low as possible and, naturally, this increases the requirements imposed on the sensitivity of the instrument, i.e., its ability to record small changes in concentration and in the concentration gradient, dc/dx.

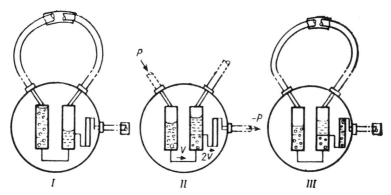

Fig. 3.6. Scheme of the cell for corrosive media and sequence of operations during experiment.

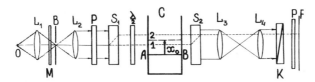

Fig. 3.7. Scheme of the optical part of the polarizing diffusiometer. AB is the initial interface of the two liquids coming into contact. x_0 is the coordinate of the point located at equal distances from the two interfering beams 1 and 2.

The local changes in c and dc/dx are usually recorded by refractometric methods in which the dependence of the refractive index of the solution n on its concentration is utilized:

$$dn/dx = \nu\, dc/dx, \qquad (3.20)$$

where $\nu = dn/dc$ is the refractive index increment in the polymer–solvent system.

The sensitivity of refractometric methods may be increased using various interferometric schemes [7, 11]. Among these schemes, a polarizing interferometer [23] has certain advantages and on its basis a polarizing diffusiometer has been developed [20]. Here a scheme of this instrument will be considered because all experimental data on diffusion discussed in subsequent sections were obtained with its use. The optical scheme of the diffusiometer is shown in Fig. 3.7. The light from the source O passes through a monochromatizing device M, and is focused by lens L_1 on the slit diaphragm B. With the aid of lens L_2 it travels further as a parallel beam. The beam is polarized with a polaroid P_1 and separated into two beams 1 and 2 by an Iceland spar plate S_1. After passing a semiwave plate $\lambda/2$ (rotating the planes of polarization of beams 1 and 2 by 90°) and a diffusion cell C, the light beams are combined into one beam again by the second spar plate S_2 equivalent to S_1. A quartz wedge (Babinet compensator) K and a polaroid P_2 crossed with P_1 make it possible to observe the interference pattern of the two polarized beams.

With the aid of a telescopic system of lenses L_3 and L_4, this interference pattern is projected on a photographic film F. The back surface of the cell C is imaged on the film F also (this surface is situated in the focal plane of lens L_3). The optical axis, the edge of the wedge K, and the axes of spar plates S_1 and S_2 are mutually parallel and vertical. Consequently, the interfering beams 1 and 2 in the cell are mutually displaced by the distance a (spar twinning) in the direction normal to the plane of the solution–solvent interface.

In the absence of optical inhomogeneities in the liquid filling the cell, the interference pattern on the screen has the type of alternating vertical light and dark bands whose intervening distance is dependent on the wedge angle of the compensator.

If there are optical inhomogeneities in the cell due to changes in solution concentration c in the direction X, then, in accordance with Eq. (3.20), the optical retardation δ of the two interfering beams 1 and 2, expressed in wavelengths, is given by

$$\delta = \beta \int_{x_0 - a/2}^{x_0 + a/2} (dc/dx)\, dx, \tag{3.21}$$

$$\beta = (l/\lambda)(dn/dc), \tag{3.22}$$

where x_0 is the x coordinate of the point located at equal distances from the two interfering beams, l is the cell thickness, λ is the wavelength of light, and dc/dx is the concentration gradient at the point x.

The optical retardation δ is revealed in the nonuniform shift of interference bands in the direction parallel to the liquid interface (normal to the x axis) and in their corresponding curvature. The shape of interference bands makes it possible to determine the character of the spatial distribution of the solute concentration and to calculate the diffusion coefficient by taking into account the change in this distribution with time. The procedure of this calculation is profoundly affected by the character of the diffusion investigated, and the most unequivocal results are obtained in the simplest case when the distribution dc/dx may be considered to be Gaussian.

1.3.3. Interference Pattern and Calculation of D for the Gaussian Distribution of the Concentration Gradient. If the distribution of the concentration gradient during diffusion corresponds to Eq. (3.5), then the substitution of Eq. (3.5) into Eq. (3.21) and the introduction of the variable $y = x/(\sigma\sqrt{2})$ readily transforms Eq. (3.21) into the form

$$\delta = (\beta \Delta c/2)\, \varphi = (\delta_m^0/2)\, \varphi, \tag{3.23}$$

where

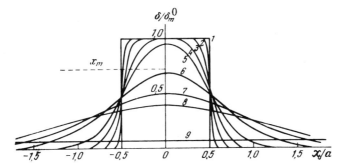

Fig. 3.8. Dependence of the shape of the interference band δ/δ_m^0 on x_0/a at various values of $2\sigma/a$. The values of $2\sigma/a$ are: 1) 0; 2) 0.05; 3) 0.15; 4) 0.3; 5) 0.5; 6) 1.0; 7) 1.5; 8) 2.0; 9) 10.

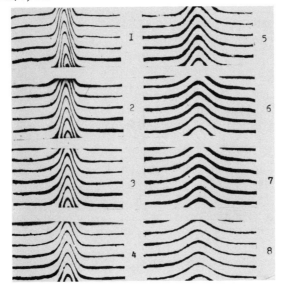

Fig. 3.9. Interference bands obtained in a polarizing diffusiometer in the diffusion of poly-(hydroxyphenylbenzoxazole-terephthalamide) in sulfuric acid, $M = 42 \cdot 10^3$ g/mole. Diffusion time in hours: 1) 9; 2) 15; 3) 27; 4) 41; 5) 99; 6) 190; 7) 235; 8) 305. Concentration $c = 0.06 \cdot 10^{-2}$ g/cm³. Spar twinning, $a = 1$ mm.

$$\varphi = \frac{2}{\sqrt{\pi}} \left[\int_0^{(x_0+a/2)/\sigma\sqrt{2}} e^{-y^2} dy - \int_0^{(x_0-a/2)/\sigma\sqrt{2}} e^{-y^2} dy \right] = \psi\left(\frac{x + a/2}{\sigma\sqrt{2}}\right) - \psi\left(\frac{x_0 - a/2}{\sigma\sqrt{2}}\right). \quad (3.24)$$

where ψ is the probability integral, a is the spar twinning, σ^2 is the dispersion of diffusion curve (3.5), and x_0 is the x coordinate of points on the interference curve measured with the origin at the interface of the liquids brought into contact.

Figure 3.8 shows the curves of the dependence of δ/δ_m^0 on x_0/a representing the contour of the interference band and calculated from Eqs. (3.23) and (3.24) at various dispersions σ^2 (expressed in relative units $2\sigma/a$). At the initial moment ($t = 0, \sigma = 0$) the function $\delta(x_0)$ and, correspondingly, the band contour have the form of a rectangle with base equal to a, height equal to

$$\delta_m^0 = \Delta n \,(l/a) = \beta \Delta c \qquad (3.25)$$

and area equal to

$$Q = \delta_m^0 a = \beta a \Delta c. \qquad (3.26)$$

Moments later, diffusion leads to a change in the band shape, which approaches a Gaussian curve, whereas the area Q remains unchanged.

In all the equations the area Q under the curve is expressed in unit length multiplied by the number of interference orders. To express the value of Q in the usual units of area, it should be multiplied by the distance between the interference bands.

The photographs of the corresponding experimental interference bands taken in a polarizing diffusiometer at different moments are shown in Fig. 3.9. They may be used to calculate dispersion σ^2 (and, hence, D) at the corresponding moments [7, 20].

If spar twinning a is sufficiently great and the initial boundary in layering is relatively sharp, in principle it is possible to carry out measurements in relatively short experimental time when $\sigma \leq a/2$. Under these conditions, we may use the total area Q under the curve and its "external" part ΔQ lying in the range of abscissa x_0 which satisfy the conditions $1 \leq 2x_0/a \leq \infty$ and $-1 \geq 2x_0/a \geq -\infty$.

It follows from Eqs. (3.23) and (3.24) at $\sigma \leq a/2$ to within 2% that

$$\sigma = 1.25\Delta Q/Q.$$

This expression may serve to determine σ under the above conditions. However, only the measurements at longer experimental times, i.e., at relatively high σ values, are usually of practical importance.

From the moment when $2\sigma/a > 0.5$, according to Eqs. (3.23)–(3.26) the following equation can be used:

$$\delta_m/Q = [\psi\,(a/\sqrt{8}\,\sigma)]/a, \qquad (3.27)$$

where δ_m and Q are the maximum ordinate (height) and the area under the interference curve, respectively. Hence, σ is calculated from the experimental values of δ_m and Q.

The results obtained for some samples of poly(hydroxyphenylbenzoxazole-terephthalamide) of various molecular weights in sulfuric acid are shown in Fig. 3.10 as an example. The values of σ^2 are plotted vs. time t. For each sample the points fit a straight line in accordance with a relatively narrow molecular-weight distribution of the polymer. The slopes of the straight lines allow the determination of the diffusion coefficients D of the samples according to Eq. (3.6).

The intercepts with the ordinate of these lines extrapolated to $t \to 0$ correspond to the "zero dispersion" σ_0^2 or to the initial width of the concentration interface. The values of σ_0^2 characterize the quality of the formation of the interface between the solvent and the solution when they are layered in the cell. For a reliable determination of the slope of the straight line and, hence, of the value of D, the experimental time should be sufficiently long in order that the value of σ^2 at the end of the experiment might exceed several times the value of σ_0^2. This condition is fulfilled in the experiments shown in Fig. 3.10.

With a cell length of 3–5 cm, it is possible to measure D with the aid of a polarizing diffusiometer with sufficient reliability in polymer solutions with concentration not exceeding a few hundredths of a percent if the refractive index increment in the polymer–solvent system is approximately 0.1 or higher. It can be clearly seen [see Eq. (3.13)] that at these concentrations even in thermodynamically good solvents ($A_2 \approx 10^{-3}$ cm^3·mole/g^2) for a polymer with a molecular weight of about $M \approx 5 \cdot 10^4$ g/mole, the measured diffusion coefficient D differs from the D value at $c \to 0$ by as little as a few percent. This conclusion is confirmed by experimental data [7].

Actually, the situation for rigid-chain aromatic polymers is even more favorable, since the molecular weight $5 \cdot 10^4$ g/mole exceeds the upper limit possible in practice for most of them, and their refractive index increment in the solvents used is generally higher than 0.2. Consequently, for this class of polymers, quite reliable values of D may be obtained from measurements at a single sufficiently low concentration and thus it is possible to avoid long procedures of studying the concentration dependence of D with subsequent extrapolation to $c \to 0$.

These considerations (supplemented by others that will be mentioned later) make the diffusion method an indispensable method for investigating the mobility and the friction characteristics of molecule of rigid-chain aromatic polymers.

1.3.4. Interference Pattern for the Non-Gaussian Distribution dc/dx. If the distribution dc/dx is not a Gaussian distribution [Eq. (3.5)], then the shape of the interference band, $\delta(x)$, does not correspond to Eq. (3.23) and Eqs. (3.23) and (3.24) cannot be used to calculate the dispersion σ^2 of the distribution dc/dx. In this case, the differential (dc/dx) and the integral [$c(x)$] concentration distributions in a cell may be obtained from the contour of the experimental interference band by a procedure described below [24].

Figure 3.11 shows a scheme for a part of the interference band B observed in the diffusiometer cell. The polymer concentration increases in the direction of the x

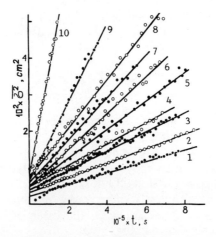

Fig. 3.10. Dispersion σ^2 of diffusion boundary vs. time t for a series of samples of poly-(hydroxyphenylbenzoxazole-terephthalamide) of various molecular weights M in sulfuric acid. $M \cdot 10^{-4}$ is equal to: 1) 42.0; 2) 26.0; 3) 15.5; 4) 10.3; 5) 14.0; 6) 8.0; 7) 7.0; 8) 6.2; 9) 2.6; 10) 0.84. Concentration $c = 0.05 \cdot 10^{-2}$ g/cm³.

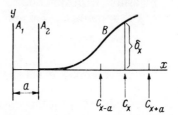

Fig. 3.11. Part of interference band B observed in a cell of a polarizing diffusiometer.

axis, i.e., this axis is opposite to the direction of polymer diffusion. A_1 and A_2 are the two images of the meniscus (or the cell base, depending upon the direction of polymer diffusion) corresponding to the twinning of the spar, a. The x axis coincides with the unshifted part of the band (determining the baseline) and the y axis coincides with the first (A_1) image of the meniscus (base). According to Eq. (3.21), the shift of the interference band (expressed in the number of bands) δ_x at a point with the coordinate x (measured from the origin at A_1) is given by

$$\delta_x = \beta\,(c_x - c_{x-a}),\tag{3.28}$$

where c_x and c_{x-a} are the solution concentrations at two points of the cell with the coordinates x and $x - a$, respectively.

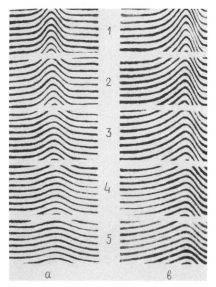

Fig. 3.12. Interference curves obtained during the diffusion of a polyacrylic acid fraction. $M = 7 \cdot 10^4$ g/mole in a salt solution at the ionic strength $I = 0.006 \cdot 10^{-3}$ mole/cm³ at two different concentrations c [25]. a) $c = 0.02 \cdot 10^{-2}$ g/cm³, cell length $L = 5$ cm, experimental time in minutes: 1) 60; 2) 105; 3) 195; 4) 285; 5) 345. b) $c = 0.2 \cdot 10^{-2}$ g/cm³, $L = 1$ cm, experimental time in minutes: 1) 60; 2) 105; 3) 150; 4) 195; 5) 285.

If the y axis coincides with the second (A_2) image of the meniscus (base), then

$$\delta_x = \beta \, (c_{x+a} - c_x). \tag{3.29}$$

If the axis y coincides with A_1 (Fig. 3.11), and two points of the cell with the abscissas x and $x - ma$ (where m is the natural number) are considered, then according to Eq. (3.28) the relationship between concentrations c_x and c_{x-ma} at these points is given by

$$c_x = c_{x-ma} + \frac{1}{\beta} \sum_{i=0}^{m-1} \delta_{x-ia}, \quad m = 1, \ 2, \ 3, \ \ldots, \tag{3.30}$$

where δ_{x-ia} are the ordinates of the points on the interference curve with the absicssa $x, x - a, \ldots, x - (m - 1)a$, respectively.

In a similar manner, if the origin of the abscissas is A_2 in Fig. 3.11, then, in accordance with Eq. (3.29), we have

$$c_x = c_{x+ma} - \frac{1}{\beta} \sum_{i=0}^{m-1} \delta_{x+ia}, \quad m = 1, \ 2, \ 3, \ \ldots, \tag{3.31}$$

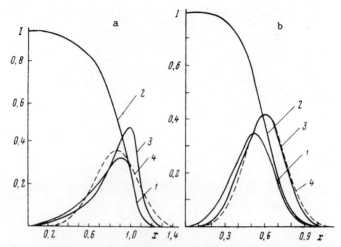

Fig. 3.13. Example of the treatment of interference curve obtained during the diffusion of a polyacrylic acid fraction $M = 4.2 \cdot 10^5$ g/mole in a salt solution at spar twinning of 1 mm. 1) Contour of interference band; 2) integral distribution $c(x)$; 3) differential distribution dc/dx; 4) Gaussian distribution dc/dx. a) $c = 0.11 \cdot 10^{-2}$ g/cm^2, cell length 3 cm; b) $c = 0.01 \cdot 10^{-2}$ g/cm^2, cell length 5 cm.

where δ_{x+ia} are the ordinates of the points on the interference curves with the abscissas $x, x + a, \ldots, x + (m - 1)a$, respectively.

It is possible to calculate the integral distribution curve $c = c(x)$ with the aid of Eqs. (3.30) or (3.31) using the contour of the interference band $\delta = \delta(x)$.

In this case, it is convenient to use Eq. (3.30), beginning the measurements in the region of the pure solvent. For this purpose the base point with the abscissa $x - ma$ is chosen in the region of the solvent and, correspondingly, c_{x-ma} is assumed to be zero. The value of m is found by measuring the distance along the x axis between the point considered and the base point. The values of δ_{x+ia} are determined from the contour of the interference curve.

It is more convenient to apply Eq. (3.31) when the concentrations in the region of the interface are compared to the initial solution concentration c_0. In this case, the base point with the abscissa $x + ma$ is chosen in the region of the initial solution (relatively far from the interface), where $c_{x+ma} = c_0$.

After plotting of the integral curve $c = c(x)$, the concentration gradient distribution curve dc/dx is found by graphic differentiation of the curve $c(x)$.

In order to demonstrate the application of this method, experimental data on the diffusion of aqueous salt solutions of polyacrylic acid (PAA) at low ionic strength, $I = 0.006 \cdot 10^{-3}$ mole/cm^3 [25], are reported. The interference curves in Fig. 3.12b were obtained at the initial solution concentration $c = 0.21$ g/100 cm^3. In contrast to the curves shown in Fig. 3.9, they are very asymmetric and are steeper toward the solvent side (compare Section 1.2.1). This fact indicates that the concentration dependence of the diffusion rate is higher due to charge effects [1]. Figure 3.12a shows the interference curves obtained for the same system at the same ionic strength but at a concentration lower by one order of magnitude ($c = 0.02$ g/100 cm³). These curves are virtually symmetric and may be used to calculate the diffusion coefficient according to Eqs. (3.23)–(3.27).

General equations (3.30) and (3.31) make it possible to treat both symmetric (Fig. 3.12a) and asymmetric (Fig. 3.12b) interference curves. This is shown in Fig. 13a, b. Curve 1 in Fig. 3.13a represents the contour of the interference band $\delta(x)$, obtained in the diffusion of another fraction of the same PAA sample ($M = 4.2 \cdot 10^5$ g/mole) into a salt solvent with ionic strength $I = 0.006 \cdot 10^{-3}$ mole/cm³ at a polymer concentration $c = 0.11$ g/100 cm³. Curve 2 is plotted according to curve 1, using Eq. (3.30), and represents the integral concentration distribution in the cell, $c(x)$. Curve 3, the differential concentration distribution dc/dx, was obtained by differentiation of curve 2. The dispersion σ^2 of the distribution dc/dx is calculated by the method of moments according to Eq. (3.7). The broken line 4 in Fig. 3.13a is a Gaussian curve [Eq. (3.5)], the area under which, Q, and dispersion, σ^2, are equal to the area and dispersion of curve 3, respectively.

A peculiar feature of curve 3, distinguishing it from the Gaussian curve, is its marked asymmetry due to the strong concentration dependence of polyelectrolyte diffusion. It should be emphasized that if Eq. (3.6) is used, the dispersion σ^2 of curve 3 allows only the calculation of the "effective" diffusion coefficient characterizing the overall mobility of macro-ions, including both their displacement during the true diffusion process and their migration caused by charge effects. This effective diffusion coefficient is not related to the friction coefficient f of the macro-ion by a simple expression [Eq. (3.12)] and, hence, cannot be used for the determination of f. This is equally true for the macro-ions of flexible- and rigid-chain polymers, particularly because when a flexible-chain polyion extends as a result of the interaction of its ionogenic groups, its conformational and hydrodynamic properties approach those of a rigid-chain macromolecule.

Similar plots are constructed (Fig. 3.13b) for the diffusion of the same PAA fraction ($M = 4.2 \cdot 10^5$ g/mole) at the same ionic strength but at a PAA concentration lower by one order of magnitude ($c = 0.01$ g/100 cm³). Under these conditions the distribution dc/dx is virtually symmetric and close to the Gaussian distribution, and its dispersion σ^2 is directly related to the true diffusion coefficient D of PAA molecules by Eq. (3.6).

Hence, even at a low ionic strength of the solution ($I = 0.006$) and, correspondingly, at strong intramolecular electrostatic interactions, polarized interferometry makes it possible to carry out measurements at low polyelectrolyte concentrations at which the distorting influence of charge effects is virtually excluded, and allows the determination of the true value of the translational diffusion coefficient (and, correspondingly, the friction coefficient f) of an expanded macro-ion.

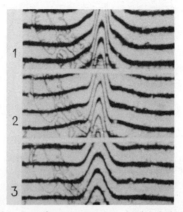

Fig. 3.14. Interference curves obtained during the diffusion of a graft copolymer: poly(methyl methacrylate)–polystyrene at various times t after the beginning of diffusion: 1) 0.5; 2) 2; 3) 15 h.

1.3.5.   Interference Curves in the Diffusion of Polydisperse Polymers and Copolymers with Inhomogeneous Composition. If N components differing in both diffusion coefficients D_i and refractive indices n_i are mixed in a solution and the solution concentration is relatively low, the Gaussian curves dc_i/dx and dn_i/dx correspond to each component and, according to Eqs. (3.23) and (3.24), the overall effect is characterized by the interference curve

$$\delta = \sum_i^N \delta_i = (l/\lambda) \sum_i^N \left(\sqrt{2\pi}\,\sigma_i\right)^{-1} \int\limits_{x_0 - a/2}^{x_0 + a/2} \exp\left(-x^2/2\sigma_i^2\right) dx \qquad (3.32)$$

with the total area under the curve equal to

$$Q = \sum_i^N Q_i = (la/\lambda) \sum_i^N \nu_i c_i, \qquad (3.33)$$

where $\nu_i = (dn/dc)_i$ and c_i are the refractive index increment and the concentration of the ith component, respectively.

If the polymer exhibits molecular weight polydispersity but a homogeneous chemical structure, then the ν_i values for all components are identical, at least in sign. In this case, the interference curves are of the shape shown in Fig. 3.14 [26]. This shape corresponds to that of schematic curves shown in Fig. 3.2 because, in this case, the graft copolymer consists virtually of two components with widely different molecular weights. The percentage of these components in the copolymer and their diffusion coefficients can be determined according to Eqs. (3.32) and (3.33).

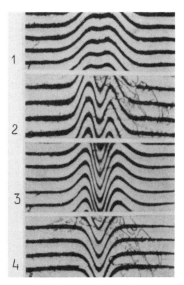

Fig. 3.15. Interference curves obtained during the diffusion of an unfractionated PMMA–PS block copolymer in solvents exhibiting different refractive indices n equal to: 1) 1.55; 2) 1.56; 3) 1.57; 4) 1.6.

This method has been developed further and used to characterize the polymer polydispersity by studying the rate at which the redistribution of the area Q under the interference curve occurs between its various parts during diffusion [27, 28].

If the substance investigated is a copolymer (random, block, or graft copolymer) whose components exhibit different refractive indices, in some cases the application of a polarizing diffusiometer makes it possible to obtain information on the compositional inhomogeneity of the copolymer if it is related to a considerable inhomogeneity in molecular weights [26, 29]. For this purpose, diffusion curves should be studied, using several solvents with different refractive indices n_k. Figure 3.15 shows, as an example, the interference curves for the diffusion of a block copolymer consisting of blocks of polystyrene (PS, $n_i = 1.59$) and poly(methyl methacrylate) (PMMA, $n_i = 1.50$) in solvents with different n_k values. One can see that the block copolymer consists of two fractions with different styrene content and different M. The fraction of lower molecular weight (broader diffusion curve) in all solvents (except bromoform, $n_k = 1.60$) exhibits a positive refractive index increment: the curve has an upward curvature. The second fraction of higher molecular weight in all solvents exhibits a negative refractive index increment: the curve is directed downward. Quantitative investigations of the interference curves in Fig. 3.15 and the application of Eqs. (3.32) and (3.33) allow the determination of diffusion coefficients (molecular weights) and contents of the copolymerized components in both fractions.

2. SEDIMENTATION
IN AN ULTRACENTRIFUGE

2.1. Main Concepts

In diffusion phenomenon the translational motion of molecules is due to the forces caused by the osmotic pressure gradient [Eq. (3.10)], whereas in sedimentation the driving force of the molecule is the centrifugal force F_c caused by the rotation of the ultracentrifuge rotor

$$F_c = (M/N_A)(1 - \bar{v}\rho_0)\,\omega^2 x. \tag{3.34}$$

where $M/N_A = m$ is the mass of the molecule, x is its distance from the rotation axis, ω is the angular velocity of the rotor, and $1 - v\rho_0$ is the buoyancy factor dependent upon the density of the solvent ρ_0 and the partial specific volume of the polymer $\bar{v}$.

In sufficiently strong centrifugal fields (high rotation velocities) capable of overcoming the Brownian motion of macromolecules, a directed flow of these macromolecules develops and, as a result, the interface between the pure solvent and the solution is formed. This interface moves in the radial direction, spreading with time as a result of the diffusion and polydispersity of the polymer (transport or velocity sedimentation).

Under the conditions of velocity sedimentation the centrifugal force F_c is equilibrated by the friction force F acting on the molecule from the side of the solvent, and defined by Eq. (2.1). Recalling that in Eq. (2.1) $u = dx/dt$ and equating $F = F_c$, one obtains

$$(1 - \bar{v}\rho_0)\, M/N_A f = (dx/dt)/(\omega^2 x) \equiv s. \tag{3.35}$$

Equation (3.35) shows that the sedimentation rate dx/dt at point x is proportional to the centrifugal acceleration $\omega^2 x$ at this point, and the proportionality coefficient s is called the sedimentation coefficient.

The sedimentation coefficient of macromolecules is determined experimentally in high-speed centrifugation by fixing the position (x coordinate) of the sedimentation boundary at the corresponding moments t. In this case the following expression is used:

$$s \equiv (1/\omega^2 x)(dx/dt) = (1/\omega^2)[d(\ln x)/dt] = (1/0.434\omega^2)(d\lg x/dt). \tag{3.36}$$

which makes it possible to determine s from the slope of the curve representing the experimental dependence of $\log x$ on t at a given value of ω. Other forms of Eq. (3.36) are

$$s = \ln(x/x_0)/\omega^2 t, \tag{3.36a}$$

$$x = x_0 e^{s\omega^2 t}, \tag{3.36b}$$

where x_0 is the coordinate of the sedimentation boundary at the initial moment.

The position and width of the sedimentation boundary are fixed experimentally by the methods used in the study of diffusion. The value of $x = x_m$ corresponding to the maximum of the differential distribution curve dc/dx is usually taken as the coordinate determining the position of the boundary "as a whole" [3, 4, 7, 11].

The dependence of the sedimentation coefficient s on the concentration c of the polymer solution is usually much more pronounced than the concentration dependence of diffusion. The value of s decreases with increasing c. Hence, to determine the sedimentation constant $s_0 = \lim_{c \to 0} s$, one must extrapolate the experimental values of s to the conditions of infinite dilution. This extrapolation is carried out using the empirical linear relationship

$$1/s = (1/s_0)(1 + K_s c). \qquad (3.37)$$

It was found for many flexible-chain polymers in thermodynamically good solvents that the coefficient K_s is equal to $\gamma[\eta]$, where $[\eta]$ is the intrinsic velocity of the solution and $\gamma \approx 1.7$. This empirical rule has been interpreted on the basis of some semiquantitative considerations [30, 31]. However, it is not a general rule. Thus, for flexible-chain polymers under θ-conditions, γ reduces to a few tenths and is not a constant [7, 32].

It follows from Eq. (3.35) that the sedimentation constant s_0 for a monodisperse polymer is proportional to the M/f ratio

$$s_0 = (M/f)(1 - \bar{v}\rho_0)/N_A. \qquad (3.38)$$

By definition, the translational friction coefficient of the macromolecule in an infinitely dilute solution should have the same value in the phenomena of sedimentation and diffusion. Hence, comparison of Eqs. (3.38) and (3.12) yields

$$M_{sD} = (s_0/D) RT/(1 - \bar{v}\rho_0). \qquad (3.39)$$

Equation (3.39) is the well-known Svedberg equation which allows the determination of the molecular weight of the polymer M_{sD} from the experimental values of D and s in infinitely dilute solutions.

It should be emphasized that the application of Eq. (3.39) to solutions of finite concentrations without extrapolation to $c \to 0$ cannot give correct values of M because the concentration dependences of D and s differ greatly [7].

If the centrifugal fields are relatively weak (low ω) or the polymer is of low molecular weight, the flow of the solute $L_s = c(dx/dt) = cs\omega^2 x$ due to sedimentation can be equilibrated by a diffusion flow L having the opposite direction. Under these conditions the sedimentation equilibrium is attained in the centrifuge cell and

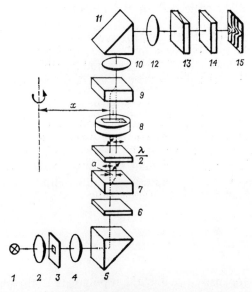

Fig. 3.16. Scheme of a polarizing interferometric attachment for an analytical ultracentrifuge. 1) Light source (mercury lamp); 2, 4) lenses; 3) slit; 5, 11) prisms with total internal reflection; 6, 14) polaroids; 7, 9) spars, $\lambda/2$ is a half-wave plate; 8) cell; 10, 12) telescopic lens system projecting the image of the middle of the cell middle on the photographic film 15; 13) wedge.

the concentration does not vary with time at all the points in the solution. Putting L_s $= -L$ and using Eq. (3.1), we obtain

$$cS\omega^2 x = D\,(dc/dx). \tag{3.40}$$

The substitution of s/D according to Eq. (3.39) yields the second Svedberg equation

$$M = [RT/(1 - \bar{v}\rho_0)]\,(dc/dx)\,(1/cx)/\omega^2, \tag{3.41}$$

allowing the determination of the molecular weight of the polymer from the experimental values of c and dc/dx in any section x of the solution. When Eq. (3.41) is applied, the data should be extrapolated to $c \to 0$ to find the true value of M. For a polydisperse polymer the molecular weight determined by this method is the weight-average molecular weight M_w.

The possibility of using the state when the sedimentation equilibrium is not yet attained is of great practical importance. In this state the solution in the middle parts of the cell has the same concentration c_p (plateau region) and the value of $(dc/dx) \neq 0$ is observed in the regions near the meniscus (where $c < c_p$) and near the base (where $c > c_p$). Under these conditions, Eq. (3.41) remains valid [33, 2] only for the x values corresponding to solution regions near the meniscus and the cell

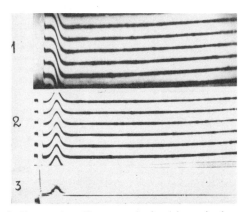

Fig. 3.17. Sedimentation diagrams obtained in a single experiment with the aid of: 1, 2) the polarizing interferometric method; and 3) Philpot–Svensson optics for poly(acetyl acrylate) in heptane (dn/dc = 0.1 cm^3/g). 1) Interference at spar twinning of 1.1 mm; 2) interference at spar twinning of 0.25 mm; 3) Philpot–Svensson's method. Solution concentration $0.05 \cdot 10^{-2}$ g/cm^3. Experimental time 20 min.

bottom. The measurement of c and dc/dx in these regions makes it possible to determine M_w according to Eq. (3.41) in a short time, thus avoiding a very long operation of the ultracentrifuge before the sedimentation equilibrium is attained.

An ultracentrifuge can also have other applications apart from the determination of the molecular weight of the polymer. It is one of the main instruments which permits the analysis of the molecular-weight distribution (MWD) of high-molecular-weight substances. This analysis is based on the fact that the concentration distribution $c(x)$ and the concentration gradient distribution dc/dx in the region of the sedimentation boundary mainly reflect the molecular-weight heterogeneity of the polymer and, hence, may be used for the characterization of its MWD.

However, to establish a quantitative relationship between the distribution curves $c(x)$ and dc/dx on the one hand, and the MWD of the polymer on the other, one must take into account the role played by the diffusion process in broadening the sedimentation boundary and exclude the influence of concentration effects complicating the interpretation of experimental data. Hence, to obtain reliable results the sedimentation analysis of polymers should be carried out with instruments which make it possible to use solutions at maximum dilution. Furthermore, sedimentation investigations should be supplemented with independent measurements of the diffusion coefficient D of the polymer, not only for the determination of its molecular weight M_{sD} according to Eq. (3.39), but also for the correct analysis of its MWD using the concentration distribution curves in the sedimentation boundary regions.

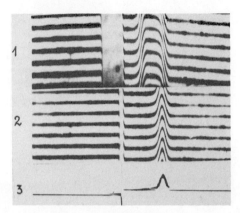

Fig. 3.18. The same system as in Fig. 3.17. Solution
concentration $0.1 \cdot 10^{-2}$ g/cm. Experimental time 40 min.

2.2. Polarizing Interferometry
in Sedimentation Analysis [24, 34–36]

Optical methods used in diffusion measurements can also be used for the
analysis of distribution curves $c(x)$ and dc/dx in sedimentation. The scheme for a
polarizing interferometer described in Section 1.3.2 proved efficient for this purpose. A schematic diagram of the polarizing–interferometric attachment to the
ultracentrifuge is shown in Fig. 3.16. It differs from the optical scheme for a polarizing diffusiometer (Fig. 3.7) only in that the interfering beams pass the cell of
the ultracentrifuge in the vertical (and not horizontal) direction. These beams are
mutually displaced by a distance a in the radial direction x of the rotor normal to the
sedimentation boundary. In the absence of optical inhomogeneities in the cell [Fig.
3.16(8)] the interference bands obtained using a wedge and recorded on photographic film are parallel to the radial direction x of the rotor, and during sedimentation they are curved in the direction normal to x (parallel to the sedimentation
boundary).

The polarizing–interferometric attachments were designed, constructed, and
used in combination with ultracentrifuges of the MOM company (Hungary) of various series (constructions): G–100, G–110, G–120, 3170, and 3180.

Just as for diffusion, the contour of interferometric curves obtained in sedimentation experiments is described by Eqs. (3.21), (3.22), and (3.28)–(3.31), depending on whether the measurement of the x coordinate is carried out from the left-
or the right-hand image of the meniscus (A_1 and A_2 in Fig. 3.11). Moreover, the
polymer concentration in Fig. 3.11 increases in the direction of the x axis; i.e., the
direction of this axis coincides with that of the sedimentation or flotation of the
polymer. In the case of flotation, A_1 and A_2 are the images of the cell base.

In the same experiment the appearance of interference bands is dependent
upon the twinning a of the spars used.

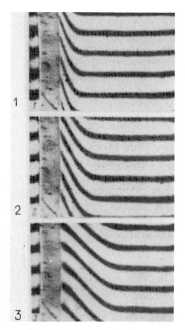

Fig. 3.19. Interference bands obtained in the region of the meniscus for a solution of poly(1,3-dimethylene–4,6-dimethylbenzene) in toluene, $c = 0.327 \cdot 10^{-2}$ g/cm^3. Rotor speed is 50,000 rpm. Experimental time (min): 1) 23; 2) 41; 3) 80.

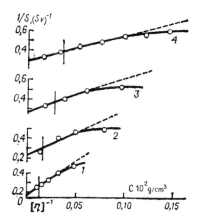

Fig. 3.20. Dependence of $1/s$ on solution concentration c for fractions of poly(butyl isocyanate) in tetrachloromethane. M equal to: 1) $13.8 \cdot 10^5$; 2) $7.67 \cdot 10^5$; 3) $3.66 \cdot 10^5$; 4) $2.66 \cdot 10^5$.

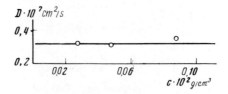

Fig. 3.21. Diffusion coefficient D vs. concentration c for the solution of a fraction ($M = 7.67 \cdot 10^5$) of poly(butyl isocyanate) in tetrachloromethane.

TABLE 3.1. Parameter γ Characterizing the Dependence of the Sedimentation Coefficient s on the Concentration c for Rigid-Chain Polymers

Polymer–solvent	$M \cdot 10^{-5}$ g/mole	γ
Poly(butyl isocyanate) in tetrachloromethane	2.5—13.8	0.39 ± 0.04
Poly(chlorohexyl isocyanate) in tetrachloromethane	0.5—3.1	0.24 ± 0.07
Ladder poly(phenyl siloxane) in benzene	0.88—8.1	0.85 ± 0.20
Ladder poly(chlorophenyl siloxane) in benzene	1.24—16.0	0.72 ± 0.19
Ladder poly(phenylisohexyl siloxane) in butyl acetate	2.9—5.13	0.41 ± 0.10
Ladder poly(methylbutene siloxane) in butyl acetate	6—7.4	0.42 ± 0.11
Cellulose carbonilate in ethyl acetate (DS 2.2)	1.5—25 [25]	0.85 ± 0.05
Cellulose monophenyl acetate in benzene (DS 2.6)	2.3—54	1.25 ± 0.20
Ethyl cellulose in ethyl acetate (DS 2.5)	0.19—0.30	1.0 ± 0.2
Cellulose benzoate in dioxane (DS 2.2)	1.9—6.6	0.80 ± 0.10
Cellulose diphenylphosphonocarbamate in dioxane (DS 2.6)	16.7—45	0.6 ± 0.3
Cellulose nitrate in ethyl acetate [41, 44] (DS 2.3–2.4)	0.25—2.8	0.6 ± 0.1
Cellulose nitrate in dimethylacetamide [42] (DS 1.14)	0.67—5.66	1.3 ± 0.2

Note: DS is the degree of substitution.

If the visible width of the sedimentation boundary (the region in which $dc/dx \neq 0$) is less than a, then the contour of the interference band corresponds to the integral concentration distribution $c(x)$ in the cell. This can be seen in Figs. 3.17(1) and 3.18(1), which show the interference patterns obtained for velocity sedimentation with spar twinning $a = 1.1$ mm. According to Eq. (3.28), the contour of the left wing of the interference curves is described by the expression $\delta_x = \beta c_x$ because the measurements are carried out from the region of the pure solvent in which $c_{x-a} = 0$. According to Eq. (3.29), the right branch of the curve corresponds to the expression $\delta_x = \beta(c_p - c_x)$ because the measurements are carried out from the plateau region in which concentration is equal to c_p.

If the sedimentation boundary is much broader than a, then the contour of the interference band qualitatively resembles the differential distribution curve dc/dx but does not coincide with it quantitatively (because the value of a is finite). This can be seen in Figs. 3.17(2) and 3.18(2), which show the interference patterns for the same solutions as in Figs. 3.17(1) and 3.18(1), but at the spar twinning $a = 0.25$ mm.

The method described in Section 1.3.4 and in [24, 36] is used for the quantitative treatment of experimental interference curves at any spar twinning and Eqs. (3.30) and (3.31) are applied. As a result of this treatment the curves of integral distribution $c(x)$ at the sedimentation boundary are obtained, and their graphical differentiation gives the differential curve dc/dx.

Polarizing interferometry is a useful tool for the study of not only velocity sedimentation but also sedimentation equilibrium in all its stages. This can be seen in Fig. 3.19, which shows the interference curves obtained in the region of the meniscus for the solution of a low-molecular-weight polymer ($M = 3400$ g/mole) in toluene at different moments t after the beginning of the experiment. The photograph distinctly shows the plateau zones, which justifies the use of Archibald's method [33] and the application of Eq. (3.41). The intersection of the curve $\delta(x)$ with the meniscus ($x = 0$) allows the determination of c in the region of the meniscus, whereas the value of dc/dx is determined directly from the slope of the curves $\delta(x)$ in this region.

The main advantage of the polarizing interferometer is its high sensitivity as compared to the well-known Philpot–Svensson optics supplied with almost all commercial ultracentrifuges. This can be seen in Figs. 3.17 and 3.18, which show the sedimentation curves obtained simultaneously using a MOM G–120 ultracentrifuge (Hungary) with the aid of two optical systems: the interferometric system and the Philpot–Svennson system. At concentrations of about 0.05 g/100 cm³ the Philpot–Svennson optics system virtually does not function any longer, whereas the polarizing interferometer makes it possible not only to reliably measure the position of the sedimentation boundary, but also to investigate the shape of the interference contour characterizing the MWD of the polymer.

2.3. Influence of Concentration on the Experimental Values of Sedimentation Coefficients

The linear concentration dependence of the sedimentation constant defined by Eq. (3.37) is valid only at relatively low concentrations. For typical flexible-chain polymers it is in the range of c values between zero and 1 or 2 g/100 cm³ [37], whereas for rigid-chain polymers its upper limit is much lower. This can be seen in Fig. 3.20, which shows the dependence of $1/s$ on c for several poly(butyl isocyanate) fractions in tetrachloromethane [38, 11]. For the lowest-molecular-weight fraction the deviation from linearity appears at $c > 0.1$ g/100 cm³, whereas for the highest-molecular-weight fraction it is already noticeable at $c = 0.05$ g/100 cm³.

Similar dependences have also been obtained in the investigations of a number of other rigid-chain polymers [39, 40].

It is noteworthy that, for the same rigid-chain polymers, the concentration dependence of the diffusion coefficient D is much less pronounced than that for sedimentation. This can be seen in Fig. 3.21, which shows the values of D at various concentrations for one fraction ($M = 7.67 \cdot 10^5$) of poly(butyl isocyanate) in tetrachloromethane [38]. At concentrations of up to 0.1 g/100 cm^3 the value of D is virtually independent of c, whereas the value of $1/s$ for the same polymer (Fig. 3.20, curve 2) in the same concentration range increases almost twice with increasing c. This fact is of great importance because it shows the advantages of the diffusion method over sedimentation in determining friction characteristics of rigid-chain polymer molecules in dilute solutions.

Since polarizing interferometry permits measurements over the concentration range in which the dependence of $1/s$ on c is linear (Fig. 3.20), it seems possible to determine the coefficients K_s in Eq. (3.37) for rigid-chain polymers. It is shown experimentally that K_s usually increases with the molecular weight of the polymer. If this dependence is determined in the same form as is generally done for flexible-chain molecules, by assuming $K_s = \gamma[\eta]$, then the systematic changes in the parameter γ with the variation in the molecular weight of the same polymer are usually unnoticed. However, the average value of γ for rigid-chain polymers is much lower than that for flexible-chain polymers. The experimental data summarized in [11, 39] according to published experimental evidence are listed in Table 3.1, which also includes data reported in some other papers [41, 42].

Lower values of γ for rigid-chain polymers than for flexible-chain polymers are often accounted for by the draining effect. Although a considerable scattering of the experimental data given in Table 3.1 is observed, it is evident that molecules with the highest rigidity (polyisocyanates) display the lowest values of γ. It is also characteristic that for cellulose nitrate molecules the value of γ increases drastically with a decreasing degree of substitution. Moreover, this is accompanied by the coiling of the chain in an amide solvent and the corresponding decrease in its equilibrium rigidity. The latter is manifested in a decrease in the intrinsic viscosity of the solution and the friction coefficient of the molecule (Chapter 4, Table 4.9).

Although the values of γ listed in Table 3.1 are lower by a factor of 2–4 than the value of $\gamma \approx 1.7$ characteristic of flexible-chain polymers in good solvents, the coefficients K_s in Eq. (3.37) for rigid-chain polymers exceed those for flexible-chain polymers of the same molecular weight. This is due to the fact that, at equal values of M, the value of $[\eta]$ is much higher for a rigid-chain polymer. Therefore, the concentration effects on the experimental values of s for rigid-chain polymers are much more pronounced. This implies that if the sedimentation method is used, then in order to obtain reliable quantitative data on the friction characteristics of rigid-chain polymer molecules, detailed investigations of the concentration dependence of the sedimentation coefficient are required even at high dilution.

2.4. Concentration Effects on Polymer Polydispersity

As already indicated, the concentration distribution $c(x)$, and its gradient distribution dc/dx at the sedimentation boundary, reflect the molecular-weight distribution (MWD) of the polymer dw/dM, because the x coordinates of the molecules undergoing sedimentation are dependent upon the corresponding sedimentation coefficients which, in turn, are dependent upon molecular weight M.

For a solution at infinite dilution the relationship between s_0 and M is determined by Eq. (3.42) obtained by comparing Eqs. (3.38) and (2.7):

$$s_0 = K_s M^{1-b}, \tag{3.42}$$

where K_s is a constant in a homologous series of polymer molecules, and the exponent b determines the dependence of the friction coefficient of the molecule on its molecular weight. Applying Eqs. (3.42) and (3.36a) one obtains

$$dw/dM = (dc/dx)\, x\, (x^2/x_0^2)\, [\ln (x/x_0)]\, (1-b)/M, \tag{3.43}$$

where the factor $(x/x_0)^2$ is introduced to take into account the sector dilution of the solution in sedimentation.

In principle, Eq. (3.43) makes it possible to plot the curve of dw/dM, the MWD of the polymer, from the experimental distribution curve dc/dx. However, Eq. (3.43) does not account for the broadening of the sedimentation boundary caused by diffusion and the influence of concentration effects leading to a very complex shape in the experimental curves dc/dx [3–7].

Hence, the most reliable method for studying the polydispersity of a polymer with a broad MWD is its fractionation and the subsequent determination of percentage and molecular weights of fractions using the sedimentation–diffusion or the chromatographic procedure [43–45].

In the study of hydrodynamic properties of polymer molecules the investigations of the dependence of their friction characteristics on molecular weight is of great importance. In these investigations samples (or fractions) of a polymer with a relatively narrow unimodal MWD are generally used. In these cases it is sufficient to apply the characteristic of sample polydispersity to the Gaussian approximation, considering the experimental curves of dc/dx obtained in velocity sedimentation to be Gaussian distributions with the width determined by the dispersion $\Delta^2 \equiv \overline{x^2} - \bar{x}^2$.

With this approximation it can be assumed [46] that the value of Δ^2 is the sum of the dispersion Δ_s^2 caused by the polydispersity of the polymer and the dispersion Δ_D^2 caused by the broadening of the sedimentation boundary due to polymer diffusion toward pure solvent:

$$\Delta^2 = \Delta_s^2 + \Delta_D^2. \tag{3.44}$$

The value of Δ_D^2 can be determined by independent measurements of diffusion rate in a diffusiometer. When polarizing interferometry (Section 1.3.3) is used, it is not very difficult to obtain the value

$$\left(\Delta_D^2\right)_{c\to 0} = \sigma_D^2 = 2Dt, \tag{3.45}$$

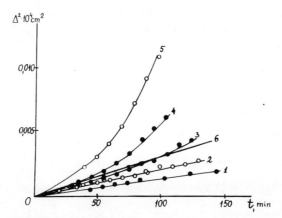

Fig. 3.22. Dispersion of sedimentation boundary Δ^2 vs. sedimentation time t (taking into account rotor acceleration time) for a cellulose nitrate fraction with $M_{sD} = 192 \cdot 10^3$ g/mole (degree of substitution 2.4) at various concentrations c of the initial solution in ethyl acetate. 1) $c = 0.18 \cdot 10^{-2}$ g/cm^3; 2) $c = 0.14 \cdot 10^{-2}$ g/cm^3; 3) $c = 0.10 \cdot 10^{-2}$ g/cm^3; 4) $c = 0.06 \cdot 10^{-2}$ g/cm^3; 5) $c = 0$; 6) $\sigma_D^2 = 2Dt$ [42].

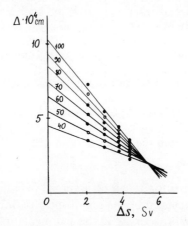

Fig. 3.23. Standard deviations Δ of sedimentation curves vs. the parameter $\Delta s = s_0 - s$ for the cellulose nitrate fraction shown in Fig. 3.22. Each straight line corresponds to a definite sedimentation time t (in minutes, as indicated by the numbers along the curves). The points on each line correspond to four concentrations at which the measurements were carried out (see Fig. 3.22). The intercepts of these straight lines with the abscissa give the values of Δ_0 corresponding to $c = 0$.

characterizing the diffusion coefficient D and the dispersion σ_D^2 of the diffusion boundary spreading in the absence of concentration effects.

The dispersion Δ^2 of the dc/dx curve measured in an ultracentrifuge strongly depends on solution concentration. This can be seen in Fig. 3.22, which shows the dependence of Δ^2 on sedimentation time t at various initial solution concentrations c for a cellulose nitrate fraction with the polydispersity parameter $M_z/M_w = 1.15$ [42]. The increase in concentration (from curve 4 to curve 1) drastically decreases the observed dispersion Δ^2 and at relatively high c it appears even lower than the dispersion σ_D^2 of the diffusion boundary spreading measured in a diffusiometer (curve 6). This phenomenon is an effect of "autocontraction" of the sedimentation boundary [3, 4]. The reason for this effect is that, for macromolecules which have equal molecular weight but are located in different concentration regions (which can occur as a result of their diffusive motion), sedimentation occurs at different rates: it is higher for macromolecules situated in regions of lower concentration (because s decreases with c) than for those in higher concentration regions.

In a similar manner, the molecules with lower M and, hence, lower s, which lag behind during sedimentation, are situated in the region of a lower local concentration and, hence, their sedimentation rate is higher than that for heavier molecules at the same concentration which left them behind. As a result, the sedimentation boundary, and correspondingly the sedimentation curve, become more narrow and Δ decreases.

Naturally, the autocontraction effect, as with any other concentration effect, decreases with decreasing concentration. Therefore, it can be neglected if a reliable extrapolation of Δ to the conditions $c \to 0$ is carried out. A convenient method of this extrapolation has recently been suggested [47]. Experimental values of Δ are plotted as a function of $\Delta s = s_0 - s$, where s is the value of the sedimentation coefficient measured at the same concentration as Δ, and s_0 is the sedimentation constant obtained by the usual extrapolation [according to Eq. (3.37)] of the value of s to the conditions $c \to 0$. Figure 3.23 shows an example of the Δ vs. Δs plot according to [47] obtained using the data presented in Fig. 3.22. For each moment of sedimentation t the points corresponding to four different concentrations (shown in Fig. 3.22) fit a straight line, the extrapolation of which to $\Delta s \to 0$ gives the value of $\Delta_0 \equiv \lim_{c \to 0} \Delta$ corresponding to the time t. The dependence of Δ_0^2 on t is represented by curve 5 in Fig. 3.22.

After the function $\Delta_0^2(t)$ has been determined, the dispersion of the distribution dc/dx due to polydispersity is extrapolated to zero concentration $(\Delta_s^2)_0 \equiv \lim_{c \to 0}(\Delta_s^2)$ and determined at any moment t according to Eqs. (3.44) and (3.45) by the equation

$$(\Delta_s^2)_0 = \Delta_0^2(t) - 2Dt. \tag{3.46}$$

The transition from the dispersion $(\Delta_s^2)_0$ characterizing the width of the distribution dc/dx to σ_s^2 characterizing that of dc/ds may be carried out to a good approximation [7] by using the equation

$$\sigma_s^2 = (\Delta_s^2)_0/(\omega^2 t x_m)^2, \tag{3.47}$$

where x_m is the coordinate of the maximum of the sedimentation curve.

It follows approximately from Eq. (3.42) that

$$\sigma_s/\sigma_w = s_0 (1 - b)/M_w, \tag{3.48}$$

where σ_w^2 is the dispersion of the molecular-weight distribution of the polymer and M_w is its weight-average molecular weight. It follows from Eq. (3.48) that the polydispersity parameter M_z/M_w is given by

$$M_z/M_w = 1 + \sigma_u^2/M_w^2 = 1 + (1 - b)^{-2}(\sigma_s/s_0)^2. \tag{3.49}$$

Hence, M_z/M_w is determined from the experimental values of the sedimentation constant of the polymer s_0 and the dispersion σ_s^2 of its distribution according to sedimentation constants. This result may be obtained only if the influence of concentration effects is completely excluded by using the above procedure [47] and the data of both sedimentation and diffusion measurements. The latter point is significant in the study of rigid-chain polymers with sedimentation coefficients strongly dependent upon concentration.

Experiments show that the use of interference optics in diffusion and sedimentation investigations makes it possible to reliably determine the polydispersity of samples and fractions of rigid-chain polymers whose equilibrium rigidity is close to that of cellulose derivatives and ladder polysiloxanes, and whose molecular weights attain tens of thousands and even higher.

2.5. Resolution of the Method

The resolution of molecular-weight characteristics and polydispersity for polymers with different chain rigidities using sedimentation–diffusion methods is a very important problem. It follows from Eq. (3.42) that the possibility of distinguishing different-molecular-weight polymer samples according to the difference in their sedimentation constants s_0 (even when concentration effects are entirely excluded) is dependent upon the value of the exponent b. The closer the value of b to unity, the lower is this possibility, because the sensitivity of the sedimentation constant to molecular weight decreases as b approaches unity. The possibility of determining the polydispersity of the polymer (i.e., σ_s) using the sedimentation method [Eqs. (3.48) and (3.49)] decreases correspondingly.

According to Eq. (2.7), the exponent b determining the dependence of the friction coefficient of the macromolecule on its molecular weight for real existing macromolecules (see Chapter 2) falls within the range with an upper limit of unity, which corresponds to the hypothetical case of an absolutely draining chain molecule (without hydrodynamic interaction). The lower limit of b is dependent upon the model used to describe the hydrodynamic properties of the molecule. Thus, compact spherical particles or a homologous series of spheroids with a constant value for the shape asymmetry p are characterized by the value $b = 1/3$. For nondraining Gaussian chains $b = 1/2$, which corresponds to the hydrodynamic properties of a flexible-chain molecule under θ-conditions (see Chapter 2). In thermodynamically good solvents for relatively high-molecular-weight flexible-chain molecules we have $0.5 < b < 1$. Hence, the maximum resolution of an ultracentrifuge as an instrument for the determination of molecular weight from the values of sedimentation constants is attained for a flexible-chain polymer if a θ-solvent is used ($b = 0.5$).

For rigid-chain polymers the situation may be quite different. As already indicated, for cellulose derivatives and ladder polysiloxanes the resolution of a centrifuge allows a reliable determination of molecular weights over any range of M values and is also sufficient for the characterization of the polydispersity of samples if their molecular weight is not too low. This is due to the fact that for these polymers the exponent b in Eq. (3.42) is usually close to 0.6–0.7, i.e., it differs markedly from unity.

For polymers with higher equilibrium rigidity, such as poly(alkyl isocyanate)s, the conditions under which sedimentation analysis can be used are less favorable. For these molecules draining is higher and, therefore, the exponent b is closer to unity, particularly for samples of low molecular weight. Hence, for poly(alkyl isocyanate)s the real possibilities of using the sedimentation method are limited in practice to the determination of average molecular weights M_{sD} by velocity sedimentation, and those of M_w by equilibrium centrifugation. Although in principle the characterization of polydispersity is possible in this case also (at least for relatively high-molecular-weight samples), the required procedures are too complex and laborious to be widely used.

The situation is very unfavorable for sedimentation analysis for many aromatic rigid-chain polymers widely employed in present-day industrial chemistry. High equilibrium rigidity and relatively low molecular weights (usually not exceeding $M \approx 5 \cdot 10^4$ g/mole) favor a weakening of hydrodynamic interaction in the molecules of these polymers, increasing the exponent b to a value of about 0.8. As a result, the resolution of an ultracentrifuge for these polymers is very low and the possibilities of using it are limited to the determination of average molecular weights M_{sD} for samples of the highest molecular weights provided these samples are soluble in appropriate solvents. However, the difficulties are usually increased by the fact that the majority of the most important rigid-chain aromatic polymers are soluble only in corrosive high-viscosity liquids (sulfuric acid), the use of which in sedimentation analysis is virtually impossible.

Equation (2.7) should be applied to evaluate the sensitivity of the diffusion method to molecular weight variations. Its combination with Eq. (3.12) gives

$$D = K_D M^{-b}. \tag{3.50}$$

It follows from Eq. (3.50) that the dependence of D on molecular weight increases with the exponent b and, hence, in contrast to the sedimentation constant, the higher the draining and the equilibrium rigidity of the chain molecules investigated, the stronger this dependence.

This result is of great practical importance because it means that the diffusion method can also be successfully used for the study of friction characteristics of molecules of rigid-chain polymers in those cases when the sedimentation method cannot be employed. A favorable factor in this respect is the relatively weak concentration dependence of D mentioned above (Section 1.3.3, Fig. 3.21).

However, in this case, the problem of the molecular-weight determination of rigid-chain aromatic polymers soluble in concentrated inorganic acids should be solved because the primary absolute method, the Svedberg method, based on the combination of measurements of D and s and on Eq. (3.39), is inapplicable to these polymers. This difficulty can be avoided if diffusion and viscometric measurements are combined, making the sedimentation method unnecessary.

3. VISCOMETRY

The procedure used in viscometric measurements of liquids is relatively simple and more easily available to research workers than sedimentation and diffusion methods. Hence, the viscometry of dilute polymer solutions carried out with the aim of characterizing hydrodynamic and conformational properties is much more frequently used in practice than other transport methods and has been described in the literature in sufficient detail [7, 10]. Therefore it is not necessary to dwell in detail on various instruments and experimental techniques used in viscometry. It is sufficient to consider some methodological problems that seem of importance in the study of rigid-chain polymers. The most important of these problems is the evaluation of the dependence of the viscosity of a polymer solution on shear stress used in viscometric measurements.

3.1. Effect of Shear Rate

It is known that over a wide concentration range polymer solutions exhibit the properties of non-Newtonian fluids. For these fluids the Newtonian equation (2.14) is formally valid, but the shear stress $\Delta\tau$ is not proportional to the rate gradient g and, correspondingly, the viscosity coefficient η varies with g and $\Delta\tau$.

These anomalies in the rheological properties of more or less concentrated polymer solutions [49, 50] are due to intermolecular interactions between macromolecules and are often produced by the formation of their entanglements, associates, and crosslinked structures. Consequently, these anomalies are sometimes called structural viscosity. The larger the size of the macromolecules, the lower the concentrations at which intermolecular interaction appears and deviations from Newtonian flow begin. However, when the concentration of the polymer solution decreases, the phenomena of structural viscosity for any polymer are less pronounced and at sufficient dilution (e.g., in the determination of intrinsic viscosity $[\eta]$) can always be eliminated.

At the same time, experimental data show that the intrinsic viscosity $[\eta]$ itself [determined by Eq. (2.18)] depends on g or on $\Delta\tau$, and the larger the size of the macromolecule, the greater is this dependence. This effect is determined by the structure and hydrodynamic behavior of individual macromolecules in flow and, hence, will interest us the most. Moreover, the primary reason for non-Newtonian behavior at high concentrations is the structurization of the solution, whereas at very low concentrations the effect of the dependence of [η] on g prevails since, under these conditions, intermolecular interaction can be minimized.

Hence, the experimental determination of the dependence of [η] on g includes the following procedures: 1) The measurement of the relative viscosity of polymer solutions $\eta_r = \eta/\eta_0$ at various concentrations c and rate gradients g, and the plotting of the dependence of η_r on g at various c with the extrapolation of the curves to the conditions $g \to 0$; 2) the use of the curves of the dependence $\eta_r = \eta_r(g)$ for plotting the curves of the dependence of $(\eta_r - 1)/c \equiv \eta_{sp}/c$ on c at various values of g (including $g = 0$); and, 3) the extrapolation of the curves of the dependence of η_{sp}/c on c to the conditions $c \to 0$, which yields the values of [η] at various values of g, including $[\eta]_0 \equiv \lim_{g \to 0}[\eta]$.

3.2. Some Methodological Information

The primary types of viscometers used in the measurements of the viscosity of polymer solutions are Couette-type rotary instruments and Ostwald-type capillary instruments [49, 51]. The main important advantage of rotary viscometers over capillary viscometers is that when the gap between the rotor and the stator of a cylindrical instrument is narrow, the rate gradient g may be considered to be constant over the entire volume occupied by the solution, and g may be determined from the simple equation

$$g = \pi v \frac{R_1 + R_2}{R_1 - R_2},\qquad(3.51)$$

where v is the number of rotor revolutions per unit time, and R_1 and R_2 are the radii of the outer and inner cylinders, respectively.

It should be noted that Eq. (3.51) holds both for Newtonian and non-Newtonian fluids and in all cases allows the determination of the true value of g from the geometric data (R_1 and R_2) of the instrument and the rotation speed of the rotor.

Coaxial rotary viscometers with both an inner and an outer rotor may be constructed as precision instruments with high sensitivity permitting the viscosity measurements of dilute solutions at very low rate gradients, e.g., as the Eisenberg instrument [52]. However, a defect of many rotary viscometers is their complex design, making them difficult to construct and operate, mainly because the coaxial cylinders must be very precisely aligned.

For this reason the Zimm coaxial viscometer [53] is the most widely used instrument among viscometers of this type; it is convenient for operation with aqueous solutions of high-molecular-weight polymers. A modified construction of this viscometer [54] is shown schematically in Fig. 3.24. It consists of two main parts: the stator–rotor system and a rotating magnet. The solution being studied (in the measurements of η) or the solvent (in the measurements of η_0) is poured into the glass stator S. The rotor r has the shape of a glass tube and its weight can be regu-

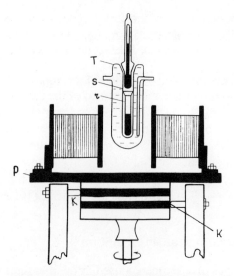

Fig. 3.24. Scheme of a modified Zimm viscometer.

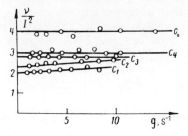

Fig. 3.25. Dependence of v/I^2 on g for the solvent (aqueous-salt solution, 0.2 N NaCl, curve c_0) and DNA solutions at concentrations $c_1 = 1\cdot10^{-4}$ g/cm³, $c_2 = 0.732\cdot10^{-4}$ g/cm³, $c_3 = 0.517\cdot10^{-4}$ g/cm³, and $c_4 = 0.369\cdot10^{-4}$ g/cm³.

lated. It floats freely in the liquid contained in the stator and is centered by forces of surface tension. The stator is surrounded with the thermostatting jacket T. A tablet of soft iron is fixed on the bottom of the rotor. A rotating force applied to the rotor is provided by a regularly rotating magnetic field acting on the iron tablet. A field of direct-current solenoids fixed on the rotating plate P is used as a source of a uniform magnetic field. The current is fed to the solenoids through sliding contacts KK. The rotor operates as an asynchronizing motor with a very high slip coefficient with the number of its revolutions per second v virtually independent of the speed of rotation of the field. However, if the viscosity η of the liquid filling the gap between the stator and the rotor is constant, v is proportional to the square of the intensity of the current feeding the solenoids, I^2. Hence, the viscosity of the liquid is determined from the simple equation

$$\eta = cI^2/v, \qquad (3.52)$$

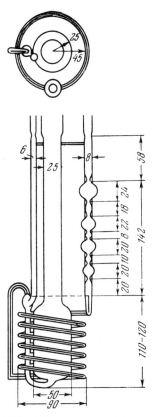

Fig. 3.26. Viscometer with a spiral capillary. All the dimensions are given in millimeters.

where c is the instrument constant determined by calibration.

Figure 3.25, illustrating the instrument operation, shows the dependence of v/I^2 (in arbitrary units) on g for solutions of a high-molecular-weight ($[\eta]_0 = 80 \cdot 10^2$ cm^3/g) DNA sample at various concentrations c and for a pure solvent (aqueous salt solution, 0.2N NaCl, curve c_0). Straight lines (c_0, c_3, and c_4) parallel to the abscissa indicate the Newtonian type of flow of the solvent and of the DNA solutions at relatively low concentrations. At higher DNA concentrations (c_1 and c_2) the curves exhibit a slope, which corresponds to a decrease in η with increasing g. The treatment of these data using the procedures described before (Section 3.2) allows the determination of $[\eta]_0$ and shows that for this DNA sample the intrinsic viscosity $[\eta]$ is actually independent of g over the g range from 1 to 10 sec^{-1} (plateau region).

If the viscosity of dilute polymer solutions is studied at rate gradients higher than 10–20 sec^{-1}, it is more convenient to use capillary viscometers whose operation is much simpler than that of rotary instruments. Their defect, however, is the nonlinear distribution of the liquid flow rate along the capillary radius. As a result, the rate gradient g varies from zero at the center of the capillary to a maximum value g_R at its wall.

For Newtonian fluids this variation is linear, and the rate gradient at a distance r from the capillary axis g_r is given by

$$g_r = (4Q/\pi R^4)\, r, \tag{3.53}$$

where R is the capillary radius and Q is the volume of the liquid flow per second. Q is dependent upon R, the capillary length L, the pressure difference p at its ends, and the viscosity of the liquid η according to Poiseuille's equation

$$Q = \pi p R^4/8\eta L. \tag{3.54}$$

Hence, for Newtonian fluids the viscosity η is determined according to Eq. (3.54) from the volume of the liquid flow per second Q and the parameters of the instrument $R, L,$ and p. The rate gradient may be characterized by one of its values averaged [according to Eq. (3.53)] either over the radius,

$$\overline{g_r} = 2Q/\pi R^3 = g_R/2, \tag{3.55}$$

or over the area S of the normal section of the capillary,

$$\overline{g_S} = 8Q/3\pi R^3 = 2g_R/3, \tag{3.56}$$

or over the volume of the liquid flow,

$$\overline{g_Q} = 32Q/15\pi R^3 = 8g_R/15, \tag{3.57}$$

where

$$g_R = 4Q/\pi R^3 \tag{3.58}$$

is the rate gradient in the wall layer of the capillary.

For both Newtonian and non-Newtonian fluids in any cylindrical layer of radius r the shear stress is given by

$$\Delta\tau_r = pr/2L \tag{3.59}$$

and, correspondingly, near the capillary walls it is given by

$$\Delta\tau_R = pR/2L. \tag{3.60}$$

However, since for non-Newtonian fluids the value of η in Eq. (2.14) is not constant, Eqs. (3.53)–(3.58) cannot be applied to them. For non-Newtonian fluids

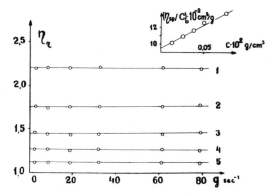

Fig. 3.27. Relative viscosity η_r versus rate gradient g for solutions of poly-(*para*-hydroxyphenylbenzoxazoleterephthalamide) in sulfuric acid. Numbers at the curves represent concentrations (c, g/100 cm³): 1) 0.085; 2) 0.052; 3) 0.038; 4) 0.025; 5) 0.0129. The dependence of $(\eta_{sp}/c)_{g \to 0}$ on c for the same sample is shown in the upper right-hand corner.

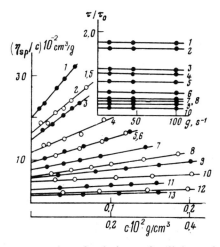

Fig. 3.28. Viscous properties of solutions of cellulose nitrate fractions at various shear rates g and concentrations c. Upper figure: relative viscosity $\eta_r = \tau/\tau_0$ versus g for solutions of high-molecular-weight fraction of cellulose nitrate ($M = 7.7 \cdot 10^5$) in cyclohexanone at solution concentrations, c, g/100 cm³: 1) 0.04; 2) 0.038; 3) 0.028; 4) 0.025; 5) 0.021; 6) 0.015; 7) 0.011; 8) 0.01; 9) 0.007; 10) 0.005. τ and τ_0 are the flow times of the solution and the solvent, respectively. Lower figure: dependence of $(\eta_{sp}/c)_{g \to 0}$ on c for solutions of a series of fractions of cellulose nitrate in ethyl acetate. The upper and lower concentration scales refer to fractions 1–10 and 11–13, respectively. The molecular weights of fractions $M_{sD} \cdot 10^{-5}$ are equal to: 1) 7.7; 2) 5.9; 3) 5.7; 4) 3.8; 5, 6) 3.04; 7) 2.45; 8) 1.94; 9) 1.80; 10) 1.60; 11) 1.00; 12) 0.70; 13) 0.50.

the value of η, determined from the experimental values of Q, p, R, and L according to Eq. (3.54), is only the "apparent" viscosity, and the value of g_R, determined according to Eq. (3.58), is only the "apparent" rate gradient near the capillary walls. The problem of the relationship between the true values of viscosity and rate gradient for non-Newtonian fluids has been solved by considering the conditions of their flow in the wall layer of the capillary [55]. It has been shown [55] that the true values of the rate gradient g_R* and the corresponding viscosity η_R* in the wall layer of a non-Newtonian fluid are determined by the expressions

$$g_R^* = g_R [(3/4) + (1/4)(d \ln Q/d \ln p)], \qquad (3.61)$$

$$(1/\eta_R^*) = (1/\eta_R)[(3/4) + (1/4)(d \ln Q/d \ln p)]. \qquad (3.62)$$

These expressions agree with the following general equation, analogous in form to Eq. (2.14):

$$g\eta = g^*\eta^* = \Delta\tau. \qquad (3.63)$$

Hence, to calculate the true values of g_R* and η_R* for a non-Newtonian liquid according to Eqs. (3.61) and (3.62) from experimental values of p and Q obtained in a capillary viscometer, the "apparent" values of g_R and η_R are calculated according to Eqs. (3.58) and (3.54), respectively, and $d \ln Q/d \ln p$ is found from the slope of the dependence of $\ln Q$ on $\ln p$.

Rate gradients g obtained in Ostwald capillary viscometers usually range from a few hundred to several thousand sec^{-1}. To extend this range toward lower g values, it is convenient to use a viscometer with a long spiral capillary schematically shown in Fig. 3.26. Several bulbs of different volumes V_1, V_2, ... at different heights h_1, h_2, ... make it possible to measure the flow rate at various pressures and, correspondingly, to determine viscosity at various values of g. Viscometric data obtained for an aromatic polyamide sample in sulfuric acid (Fig. 3.27) [56] and for fractions of cellulose nitrate in ethyl acetate and cyclohexanone (Fig. 3.28) [57] are taken as examples of the use of this type of viscometers. These data demonstrate that for these polymers, over the investigated range of rate gradients and concentrations, both η_r and $[\eta]$ are independent of shear rate (plateau region). Hence, the values of intrinsic viscosity obtained for these samples are virtually the values of $[\eta]_0 = \lim_{g \to 0}[\eta]$.

The possibility of reliable extrapolation of viscometric data to the conditions $g \to 0$ is of great importance when these data are used for the characterization of equilibrium hydrodynamic and conformational properties of rigid-chain polymers.

A decrease in $[\eta]$ with increasing g for cellulose nitrates has been revealed and studied quantitatively at much higher shear rates using capillary viscometers and manostats which ensure higher pressure drops at capillary ends [58, 59].

3.3. Interpretation of Shear Dependence of Intrinsic Viscosity

The theory of the shear dependence of intrinsic viscosity has been developed in the most complete form for solutions of rigid asymmetric particles [60–63]. This theory is based on Eq. (2.16) and the distribution function of orientations of asymmetric particles in a laminar flow according to Eqs. (2.21) and (2.22).

It has been shown that a decrease in $[\eta]$ with increasing shear rate is due to the orientation of the long axes of particles in the field direction and a subsequent increase in the degree of orientation with increasing g according to Eq. (2.22).

Moreover, as Kuhn has shown [62], at relatively low values of g and a low degree of orientation [determined by the parameter $\sigma = g/D_r$ in Eq. (2.22)] the value of $[\eta]$ is profoundly affected by friction losses caused by directed diffusive rotational motion of dissolved particles. Evaluation of these "diffusion" losses in combination with hydrodynamic losses for ellipsoids of revolution leads to the dependence of intrinsic viscosity $[\eta]$ on the orientation parameter σ at $\sigma \leq 2$ [63]

$$[\eta] = [\eta]_0 [1 - 0.026\sigma^2 (p - 1)/(p + 3)], \qquad (3.64)$$

where p is the degree of asymmetry of a spheroidal particle (see Chapter 2).

It follows from Eq. (3.64) that for a solution of rigid particles the gradient dependence of $[\eta]$ is possible only if the particles exhibit the asymmetry of shape ($p \neq 1$).

According to Eq. (3.64), the dependence of $[\eta]$ on σ at low σ has the shape of a parabola with slope equal to zero at $\sigma \to 0$. This conclusion corresponds to the fact that when the flow direction changes, the sign of the dependence of $[\eta]$ on g does not change. In actual practice this means that at relatively low g the experimental curve of the dependence of $[\eta]$ on g should have a plateau region. Similar results have been obtained using subsequent theories of the shear dependence of the viscosity of solutions of spheroidal particles [64, 65].

The asymmetry of shape is an indispensable condition for the dependence of $[\eta]$ on g not only for spheroidal particles, but also for a chain molecule if it is represented by a rigid (undeformable) spatial distribution of hydrodynamic resistance elements. This model has been used by Hearst and Tagami [66], who calculated the dependence of $[\eta]$ on g for a kinetically rigid wormlike necklace with cylindrical symmetrical bead distribution (see Chapter 2, Sections 4.3 and 4.4). For this model the "average" shape asymmetry decreases with increasing reduced chain length x from the value of $p = \infty$ (rodlike conformation) to $p = 1$ in the Gaussian range [Eq. (2.125)]. In accordance with this, at low x the authors of [66] have obtained the dependence of $[\eta]$ on g corresponding to the theory for elongated spheroids [63]. However, in the Gaussian range ($x \to \infty$) they could obtain the dependence of $[\eta]$ on g similar to Eq. (3.64) only after they had expanded the coordinates of the model along the axis of symmetry and thus "stretched" it up to the degree of asymmetry $p = (5/2)^{1/2}$.

The theoretical interpretation of the shear dependence of intrinsic viscosity for solutions of flexible-chain polymers is far from being so unequivocal.

Describing the hydrodynamic properties of a chain molecule by the elastic dumbbell model [67] without taking into account hydrodynamic interaction, or by the bead-and-spring subchain model [68] with different hydrodynamic interaction, Kuhn [67] and Zimm [68] have shown that for an ideally flexible Gaussian chain intrinsic viscosity should not depend on shear rate. This result has been interpreted qualitatively [63] as being due to the fact that a flexible Gaussian coil in flow is deformed and stretched. This should lead to an increase in viscosity, but at the same time the coil is oriented by the flow, and this decreases the viscosity of the solution. As a result of mutual compensation of these two effects, viscosity remains invariable.

However, experimental data show that for real chain molecules exhibiting relatively high equilibrium flexibility, such as the molecules of polyisoprene [69, 70], polyvinylacetate [71], polymethacrylates [72–74], polystyrene, and many others, intrinsic viscosity decreases with increasing shear rate. The higher the molecular weight, the more pronounced is this effect.

To explain this fact, Kuhn suggested that even if the equilibrium flexibility of the polymer chain is relatively high, it can exhibit a pronounced kinetic rigidity (see Chapter 1, Section 2) caused by hindrance of rotation about its valence bonds. To characterize kinetic rigidity quantitatively, Kuhn has introduced the concept of "internal viscosity" [67], the measure of resistance of the macromolecule to rapid changes in its conformation from external force effects. Internal viscosity is quantitatively determined by a force which counteracts coil deformation and is proportional to deformation rate. Hence, internal viscosity is manifested in a laminar flow in which the macromolecule is subjected to periodical forces of compression and stretching. Using the concept of inherent viscosity [77] in the framework of the elastic–viscous dumbbell model, Kuhn explained the dependence of $[\eta]$ on g for chain molecules. According to Kuhn, for the case of relatively high internal viscosity (as compared to that of the solvent used) the dependence of $[\eta]/[\eta]_0$ on g may be represented in the form [63]

$$[\eta]/[\eta]_0 = 1 - a_1 \beta_0^2 + a_2 \beta_0^4 - \dots, \qquad (3.65)$$

where a_1 and a_2 are the numerical coefficients ($a_1 = 1.86$) and the parameter β_0 is determined by the equation

$$\beta_0 = (M [\eta]_0 \eta_0 / RT) g. \qquad (3.66)$$

If Eqs. (2.27) or (2.96) are recalled, it can be clearly seen that the parameter β_0 differs from σ in Eq. (3.64) only in the numerical factor and, hence, the dependences of $[\eta]$ on g according to Eqs. (3.65) and (3.64) are similar.

Following Kuhn, Cerf has developed in detail the theory of internal viscosity based on the dynamic molecular Rouse–Zimm subchain model [68], supplemented by the evaluation of forces of internal friction [78, 79]. For very high internal viscosity, Cerf's theory yields a dependence of $[\eta]$ on β_0 differing from Eq. (3.65) only in numerical coefficients: $a_1 = 1.71$ in the absence of hydrodynamic interaction in the macromolecule, but is equal to 0.72 for strong hydrodynamic interaction.

For solutions of kinetically flexible chains exhibiting low internal viscosity, the value of $[\eta]/[\eta]_0$ in Cerf's theory is not only the function of the parameter β_0 but also (apart from hydrodynamic interaction) is dependent upon the ratio of internal viscosity of the molecule to the viscosity of the solvent η_0.

The problem of internal viscosity is important not only in the consideration of non-Newtonian flow, but also for understanding many problems of intramolecular dynamics and conformational changes in polymer chains. In particular, this refers to the problem of the significance of the mechanism of internal rotation related to the displacement of chain parts in the direction normal to its stretching: the mechanism in which internal viscosity (in contrast to the Kuhn–Cerf concept) is proportional to the viscosity of the solvent [84, 85]. Up to the present the concept of inherent viscosity has been the subject of animated discussion with different viewpoints coming into collision [80–91].

On the other hand, the phenomenon itself of non-Newtonian flow of dilute polymer solutions can be explained without using the concept of inherent viscosity if other concepts are applied.

Since instantaneous conformations of the Gaussian coil are not spherically symmetrical, it may be predicted that its hydrodynamic interactions in a chain molecule are anisotropic, attaining the maximum value in the normal direction [92–94]. As a result, the friction losses are at a minimum when the coil moves parallel to its length and at a maximum when it moves in the transverse direction. Therefore, the orientation and stretching of flexible random coil molecules in a laminar flow are accompanied by a decrease in $[\eta]$ with increasing g. At low g this dependence is quantitatively determined by an expression analogous to Eq. (3.65), in which the coefficient a_1 is of the order of magnitude of 0.01 [95].

The effect of hydrodynamic interaction changes in a molecular coil resulting from its inhomogeneous deformation in a laminar flow is probably more important for the gradient dependence of viscosity. The theory [95–97] shows that when the coil is stretched in a shear flow, the distances between the pairs of segments situated at the greatest distance from each other increase first. When the molecules rotate in a gradient field, these segments move in opposite directions, introducing "negative" contributions to hydrodynamic interaction, and thus increasing the total friction in the coil (see Chapter 2, Section 3.4.2). Hence, a decrease in these "negative" interactions with increasing distance between the interacting segments leads to a decrease in friction and intrinsic viscosity. At low g values, the dependence of $[\eta]$ on g due to this effect is quantitatively expressed by Eq. (3.65) where $a_1 = 0.15$.

If the effect of the thermodynamic strength of the solvent on the gradient dependence of viscosity is taken into account theoretically [98], it can be seen that the expansion of a flexible-chain polymer in a good solvent leads to lower values of $[\eta]/[\eta]_0$ at a given g value. However, the experimental data for polystyrene fractions obtained in solvents of different strengths can be plotted as a single curve of the dependence of $[\eta]/[\eta]_0$ on β_0 [76].

Some theories considering the deformation of a flexible polymer chain in a shear flow take into account a finite chain length. For this purpose Peterlin et al. [99–101, 97] have used the inverse Langevin function introduced by Kuhn [102] to characterize the limited deformability of the coil subjected to stretching force. In the works of Hearst et al. [103] limited chain deformation (or the constant value of its contour length) in a shear flow was ensured by the required functional dependence of the force constants of the chain on shear rate. This method has been used [103] by applying both the Rouse–Zimm bead-and-spring model [68] and the dynamic model for a polymer molecule "with local stiffness" [104, 105]. The evaluation of the limited deformability of a flexible chain in the theory of intrinsic viscosity showed that $[\eta]$ can decrease with increasing parameter β_0 (even if hydrodynamic interaction does not vary). The shorter the polymer chain (i.e., the smaller the number of segments it contains, $N = L/A$), the greater is this decrease. The curves of the dependence of $[\eta]/[\eta]_0$ on β_0 plotted at different values of N in these two theories [97, 103] virtually coincide not only qualitatively but also quantitatively.

Hence, it is possible to mention a number of effects which explain the existence of the shear dependence of intrinsic viscosity in solutions of flexible-chain polymers. However, at present there are no general concepts of the relative significance and probability of each of these effects and its application to a specific polymer. It should only be noted that all the foregoing theories are based on the concept of flexible-chain molecule deformation in flow as an indispensable condition for the dependence of $[\eta]$ on g. Hence, these theories and the effects discussed in them cannot be used for the interpretation of the shear dependence of viscosity in solutions of kinetically rigid chain molecules.

It might be expected that for real rigid-chain polymers the primary role in this phenomenon is played by the asphericity of molecular conformations and the related orientation of kinetically rigid chains in shear flow. In accordance with this, it is reasonable to discuss the data of the dependence of $[\eta]$ on g for these polymers from the standpoint of the Kuhn–Cerf theories [63, 78, 79] developed for molecules exhibiting high internal viscosity or with the aid of equations for rigid spheroidal particles essentially equivalent to these theories.

Taking into account the foregoing considerations, it is possible, for example, to evaluate the change in viscosity $1 - [\eta]/[\eta]_0$ which might be expected in a solution of a high-molecular-weight fraction of aromatic polyamide [56] (Fig. 3.27) at the maximum value of $g = 80$ sec^{-1}. The characteristic orientation angle for this polymer (see Chapter 6) was found [56] to be $[\chi/g] = 3 \cdot 10^{-4}$ sec, which corresponds to the rotational diffusion coefficient $D_r = (12[\chi/g])^{-1} = (1/36) \cdot 10^4$ sec^{-1} and (at $g = 80$ sec^{-1}), $\sigma = 0.3$. Hence, according to Eq. (3.64), if p is taken to be much greater than unity, one obtains $1 - [\eta]/[\eta]_0 \approx 3 \cdot 10^{-3}$ which is much less than the error in the experimental determination of this value. Similarly, for cellulose nitrate solutions (Fig. 3.28) for which $[\chi/g] = 5.3 \cdot 10^{-4}$ sec [57] according to Eq. (3.64) at $g = 100$ sec^{-1} one obtains $1 - [\eta]/[\eta]_0 = 1 \cdot 10^{-2}$, which is also within experimental error. Hence, the experimental results shown in Figs. 3.27 and 3.28 for two classes of typical rigid-chain polymers correspond to the theory of shear dependence of viscosity for aspherical kinetically rigid molecules.

4. HYDRODYNAMIC INVARIANT

4.1. Introduction

The reliable measurement of molecular weight of macromolecules is of major importance for the interpretation of data obtained by transport methods and for the use of these data for the characterization of their conformational properties. Frequently this problem may be solved by one of the "absolute" methods: light scattering or sedimentation–diffusion analysis of the polymer in solution. In the latter case we mean the combination of two hydrodynamic characteristics of a polymer molecule: the coefficient f of its translational friction (or the diffusion coefficient D) and its sedimentation rate in a centrifugal field (or the sedimentation constant s_0) according to Svedberg's equation (3.39).

However, as already indicated (Chapter 3, Section 2.5), it is not always possible to use the sedimentation method, and in these cases it is advisable to use a comparison of other hydrodynamic characteristics such as the friction coefficient f and intrinsic viscosity $[\eta]$ (here and below we mean by $[\eta]$ the value obtained under the conditions $g \to 0$).

Vallet, who probably was the first to discuss the problem of the relationship between f, $[\eta]$, and M [106, 107], measured D and $[\eta]$ for polystyrene fractions with different M values and showed the invariance of the parameter $(M[\eta])^{1/3}D\eta_0$ in a series of these fractions. Vallet explained this relationship by representing a polymer molecule as a compact sphere and applying to it Stokes' and Einstein's laws (see Chapter 2) for the description of diffusion and viscosity phenomena of polymer solutions.

Subsequently Mandelkern and Flory [108] confirmed the invariance of Vallet's parameter by representing a chain molecule by a hydrodynamically nondraining sphere with radius proportional to $\langle h^2 \rangle^{1/2}$. Proceeding from a comparison of Eqs. (2.84), (2.89), and (3.35), these authors [108] represented Vallet's invariant expression in the following form:

$$N_A \eta_0 (1 - \bar{v}\rho_0)^{-1} M^{-2/3} [\eta]^{1/3} s_0 = P^{-1} \Phi^{1/3}. \qquad (3.67)$$

Moreover, both P and Φ were considered by them to be universal constants for all polymers.

At the same time, Tsvetkov and Klenin [17], using the Debye–Bueche model of a partially drained sphere [109], have shown that the parameter

$$A_0 \equiv \eta_0 D \left(M\left[\eta\right]/100\right)^{1/3}/T = P^{-1}\Phi^{1/3}k/100^{1/3} \tag{3.68}$$

may be invariant over a wide molecular-weight range even when the molecular coil exhibits a marked draining effect, i.e., under the conditions when P and Φ do not remain constant separately if M is varied. These authors showed the invariance of A_0 by measuring D and $[\eta]$ for polystyrene fractions in dichloroethane over the molecular-weight range $(0.25–1.32)\cdot 10^6$ g/mole, and obtained the value of $A_0 = (3.42 \pm 0.06)\cdot 10^{-10}$ g cm² sec⁻² deg⁻¹ mole⁻¹ᐟ³

$A_0 = (3.42 \pm 0.06)\cdot 10^{-10}$ g cm² sec⁻² deg⁻¹ mole⁻¹ᐟ³ = $(3.42 \pm 0.06)\cdot 10^{-17}$ J deg⁻¹ mole⁻¹ᐟ³. This fact led to the conclusion that it is possible and desirable to determine the molecular weights of polymers according to viscometry and diffusion data of their solutions using Eq. (3.68) and the foregoing value of parameter A_0.

The problem of keeping the parameter A_0 constant when the molecular weight is varied is particularly important for rigid-chain polymers because their molecules are characterized by a considerable draining effect and in accordance with this both Φ and P vary with the chain length (see Chapter 2). This problem should first be considered from the standpoint of modern theories.

4.2. Theory

4.2.1. Dependence of A_0 on Chain Length. If a chain molecule is represented by a partially drained Gaussian necklace, then the parameter P, dependent upon hydrodynamic interaction by analogy with Eq. (2.84), is related to the friction coefficient f by the equation $P = f/\eta_0 n^{1/2} l_0$. A comparison of this equation with Eq. (2.80) gives

$$P = (6\pi^3)^{1/2} X/[1 + (8/3) X], \tag{3.69}$$

where the hydrodynamic interaction parameter X is determined from Eq. (2.81).

Expressing Φ according to Eqs. (2.90) and (2.91) and using Eqs. (3.69) and (3.68), one obtains

$$A_0 = (k/6\pi)(N_A/100)^{1/3} X^{-2/3}[1 + 8X/3](1 + X/C)^{-1/3}. \tag{3.70}$$

It follows from Eq. (3.70) that A_0 is a function of molecular weight since, according to Eq. (2.81), the parameter X is proportional to the square root of chain length. At high X, the limiting value of A is $A_\infty = \lim_{x \to 0} A_0 = (4k/9\pi)(N_A c/100)^{1/3}$. At $c = 3/2$ we have $A_\infty = 4\cdot 10^{-17}$ J deg⁻¹ mole⁻¹ᐟ³.

According to Eq. (3.70), at low X the parameter A_0 increases with decreasing X proportionally to $1/(X^{2/3})$. However, this increase becomes noticeable only at such small values of X where it is impossible to use Eq. (3.70) because its applicability is limited to the Gaussian range in which n and, correspondingly, X are relatively high.

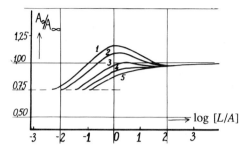

Fig. 3.29. Parameter A_0/A_∞ versus relative chain length L/A at various values of d/A according to the theory of a wormlike spherocylinder. Values of d/A: 1) 0.005; 2) 0.01; 3) 0.03; 4) 0.06; 5) 0.1.

TABLE 3.2. Relative Values of A_0/A_∞, the Hydrodynamic Parameter, at Various Values of Relative Length L/A and Diameter d/A of a Wormlike Chain

L/A	Values of $A_0/A_\infty = (P_\infty/P)(\Phi/\Phi_\infty)^{1/3}$ at various A/d values						
	2000	1000	200	100	100/3	100/6	10
10^6	0.998	0.999	0.999	0.999	0.999	0.999	1.000
10^5	0.995	0.996	0.996	0.997	0.997	0.998	0.999
10^4	0.997	0.998	0.990	0.990	0.991	0.994	0.996
10^3	0.976	0.978	0.978	0.978	0.978	0.983	0.989
500	0.977	0.977	0.975	0.974	0.973	0.978	0.985
200	0.989	0.985	0.976	0.972	0.967	0.971	0.978
100	1.008	1.000	0.982	0.975	0.965	0.965	0.971
70	1.024	1.013	0.989	0.979	0.965	0.962	0.967
50	1.043	1.029	0.997	0.985	0.967	0.960	0.962
20	1.115	1.090	1.033	1.011	0.977	0.956	0.948
10	1.186	1.151	1.070	1.037	0.987	0.953	0.934
7	1.225	1.184	1.090	1.051	0.991	0.951	0.926
5	1.261	1.215	1.108	1.052	0.992	0.948	0.917
2.278	1.339	1.282	1.144	1.084	0.990	0.934	0.897
2.0	1.362	1.300	1.154	1.092	0.995	0.938	0.899
1.0	1.382	1.312	1.149	1.079	0.973	0.910	0.868
0.6	1.380	1.306	1.133	1.060	0.950	0.886	0.844
0.4	1.362	1.286	1.109	1.035	0.923	0.861	0.821
0.2	1.310	1.231	1.051	0.977	0.870	0.813	0.778
0.1	1.242	1.162	0.983	0.912	0.816	0.768	0.758
0.05	1.167	1.088	0.914	0.850	0.768	—	—

Hence, in order to discuss the dependence of A_0 on chain length L over the entire possible range of L/A ratios it is necessary to use as molecular model a wormlike chain and characterize its hydrodynamic properties by a wormlike spherocylinder model.

According to Eq. (3.68) we have

$$A_0/A_\infty = (\Phi/\Phi_\infty)^{1/3}(P_\infty/P), \qquad (3.71)$$

where A_0, Φ, and P are functions of L/A and d/A, while A_∞, Φ_∞, and P_∞ are the values of A_0, Φ, and P corresponding to the conditions $L/A \to \infty$.

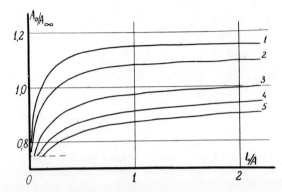

Fig. 3.30. Parameter A_0/A_∞ versus relative length L/A at various values of d/A. Numbers at the curves correspond to symbols in Fig. 3.29.

TABLE 3.3. Theoretical Values of A_∞ Calculated according to Eq. (3.68) at the Values of P_∞ Given and the Values of Φ_∞ Given in Table 2.1

$A_\infty \cdot 10^{10}$ g $\cdot$ cm^2 sec^{-2} deg$^{-1} \cdot$ mole$^{-1/3}$	P_∞	Reference in Chapter 2
4.1	5.11	[39]
4.0	5.11	[47]
3.83	5.11	[48]
3.82	5.11	[49]
3.75	5.11	[50]
3.79	5.11	[51]
3.51	5.11	[52]
3.13	5.99	[53]

P as a function of L/A and d/A for a wormlike spherocylinder is determined as follows: according to Eq. (2.84) we have

$$P = f/\eta_0 (LA)^{1/2} \equiv 3\pi (L/A)^{1/2} (3\pi\eta_0 L/f)^{-1}. \tag{3.72}$$

In Eq. (3.72) the term $3\pi\eta_0 L/f$ is a function of the values L/A and d/A determined using Eq. (2.130) at $L/A \leq 2.278$ and Eq. (2.132) at $L/A > 2.278$.

Similarly, the coefficient Φ as a function of L/A and d/A for a wormlike spherocylinder at $L/A > 2.278$ is determined by a combination of Eqs. (2.143) and (2.138). At $L/A \leq 2.278$, a combination of Eqs. (2.135) and (2.141) yields the expression for Φ:

$$\Phi = (2\pi N_A/45) (L/A)^{3/2} f (L/A)/F_\eta (p), \tag{3.73}$$

where $f(L/A)$ is defined by Eq. (2.136) and $F_\eta(p)$ is defined by Eq. (2.46).

The values of A_0/A_∞ calculated according to Eqs. (3.71)–(3.73) at various values of relative length L/A and relative diameter d/A of a wormlike chain are given in Table 3.2 and are shown in Figs. 3.29 and 3.30. At all d/A values the upper asymptotic limit predicted by the theory is equal to $A_\infty = 3.84 \cdot 10^{-17}$ J deg^{-1} mole$^{-1/3}$. At low L/A ratios the value of A_0/A_∞ also attains the limiting value $\lim_{L/A \to 0}(A_0/A_\infty) = 0.75$ obtained if in Eqs. (3.72), (2.130), and (3.73) the value of L/A is taken to be $L/A = d/A \to 0$. This corresponds to the value of $\lim_{L/A \to 0} A_0 = 2.88 \cdot 10^{-17}$ J deg^{-1} mole$^{-1/3}$.

The data presented in Table 3.2 and Figs. 3.29 and 3.30 show that, according to the theory of a wormlike spherocylinder, for all polymers in the range of $L/A > 20$ which exhibit the random coil conformation the parameter A_0 is invariant and close to the value of A_∞. For flexible-chain polymers ($d/A > 0.1$) at $L/A < 20$, when the chain length decreases, one should expect a monotonic decrease in A_0 to its lower limit (by 25%). For polymers exhibiting medium equilibrium chain rigidity ($0.1 > d/A > 0.03$) this decrease begins at lower values of L/A and is most pronounced at $L/A \leq 1$. Finally, for polymers with high chain rigidity the theory predicts a nonmonotonic dependence of A_0 on L/A in the range of short chains: at $L/A < 10$, when L decreases, the value of A_0 increases, attains a maximum at $L/A \approx 1$, and then decreases sharply up to the limiting value of $A_0/A_\infty = 0.75$. Hence, in the range of short chains ($L/A \leq 10$) where one might expect a change in A_0 with variation in L/A, according to the wormlike spherocylinder model a slightly higher value of A_0 corresponds to a more rigid polymer. This conclusion also agrees with the results obtained by Hearst et al. [105], who have calculated the values of A_0 for the range of values of $L/A \geq 10$.

4.2.2.   Numerical Value of A_∞.   It follows from Eq. (3.68) that the numerical value of A_∞ predicted by the theory depends on the theoretical values of P_∞ and Φ_∞. The value of $P_\infty = 3(6\pi^3)^{1/2}/8 = 5.11$ obtained first in the Kirkwood–Risemann theory (Chapter 2, Section 3.4.1) remained virtually invariable in all the subsequent theories apart from Zimm's computer simulation, in which P_∞ was found equal to 5.99. In contrast, the theoretical value of Φ_∞ varied greatly, as shown by the data presented in Table 2.1. In accordance with this, Table 3.3 gives the theoretical values of A_∞ calculated from the values of Φ_∞ listed in Table 2.1 and the generally accepted value of $P_\infty = 5.11$. There is only one exception: the last value in Table 3.3, Zimm's value $A_\infty = 3.13 \cdot 10^{-17}$ J deg^{-1} mole$^{-1/3}$, calculated from the values of Φ_∞ and P_∞ obtained by Zimm. The decrease in A_∞ in the series of values listed in Table 3.3 corresponds to a decrease in the theoretical values of Φ_∞ presented in Table 2.1 if they are considered in chronological order.

In principle, the excluded-volume effects can affect both the values of P_∞ and Φ_∞ and that of A_∞ because the expansion of the molecular coil in a good solvent is accompanied by a change in intramolecular hydrodynamic interaction. Although the results of theories developed for flexible-chain polymers are not always unequivocal, all of them lead to similar qualitative conclusions according to which both P_∞ and Φ_∞ decrease owing to the excluded-volume effect, and the change in Φ_∞ is more pronounced than that in P_∞ [110]. As a result, even if the combination of these values, $P_\infty^{-1}\Phi_\infty^{1/3}$, varies, these variations are so slight that even their sign cannot be reliably predicted by the present-day theories [111]. For rigid-chain polymers the effect of excluded volume upon A_∞ is particularly insignificant and can be neglected.

The values of P_∞, Φ_∞, and A_∞ discussed here refer to a polymer homogeneous with respect to molecular weight. It might be expected that the existence of molecular-weight polydispersity will lead to a decrease in Φ_∞ [112]. The parameter A_∞ also depends on polydispersity which can lead either to a decrease or an increase in A_∞, depending on how the experimental values of f, $[\eta]$, and M contained in Eq. (3.68) are averaged [113, 114]. However, the changes in A_∞ that might be expected for polymers with moderate polydispersity are not great [114], and in many cases may probably be neglected.

4.3. Experimental Data

4.3.1. Values of Φ and P. It follows from Eqs. (2.89) and (2.84) that, for separate experimental determinations of the coefficient Φ or P, two or three molecular characteristics should be measured. One of them should be the radius of gyration $\langle R^2 \rangle$ of the chain molecule, which in a θ-solvent for a polymer of relatively high molecular weight is $\langle R^2 \rangle = LA/6$. The measurement of $\langle R^2 \rangle$ carried out from the asymmetry of light scattering is the most difficult experimental procedure, always involving the possibility of great error and inaccuracy. At the same time, the light-scattering method allows the determination of the second value of molecular weight contained in Eq. (2.89). Finally, the third value that should be measured experimentally is either $[\eta]$ [if Φ is determined from Eq. (2.89)] or the sedimentation coefficient s_0 [if P is determined according to Eqs. (2.84) and (3.38)]. If P is determined using Eqs. (2.84) and (3.12), then apart from $\langle R^2 \rangle$ it is also necessary to measure the diffusion coefficient D, whereas the knowledge of M is not required.

Since the procedure of viscometric measurements is much simpler than the measurements of s or D, the majority of published experimental papers deal with the determination of Φ using Eq. (2.89). Moreover, in most cases the experiments have been carried out for polystyrene with a narrow molecular-weight distribution in a θ-solvent.

Some typical data obtained in these measurements are given in Table 3.4. The scattering of the data is very pronounced. However, they indicate that although the molecular weights of the samples are fairly high, the values of Φ are usually much lower than the theoretically predicted values of Φ_∞ (Table 2.1). The data presented in [118] do not indicate a unique dependence of Φ on the solvent strength.

The literature data on the experimental determination of P are very scarce. Some examples are listed in Table 3.5. Four out of five experimental values of P presented here exceed the value of 5.11 predicted by the theory, and only one of them is below the theoretical value. According to [118], the difference between the

TABLE 3.4. Some Experimental Values of the Coefficient Φ for Polystyrene in θ-Solvents (Cyclohexanone and Decalin) and in a Good Solvent (Tetrahydrobenzene)

$M_w \cdot 10^{-6}$ (g mole^{-1})	Solvent	$\Phi \cdot 10^{-23}$ (mole^{-1})	Reference
3.2	Cyclohexanone	1.85 †	[115]
4.0	»	1.76 †	[116]
0.62—1.6	»	2.6—2.9 †	[117]
0.62—4.4	Decalin	1.7—2.6 *	[117]
0.21—1.0	Cyclohexanone	1.8—2.4 †	[118]
0.34—2.9	»	2.4—2.6 †	[119]
27—44	»	1.5—2.0 †	[120]
1.2—4.6	Decalin	2.2—2.8 *	[121]
8.8—57	Cyclohexanone	2.2—2.4 †	[122, 123]
6.5	»	2.5±0.1 †	[124]
0.21—1.0	Tetrahydrobenzene	1.98—2.4 ‡	[118]

*In decalin, 20.4°C.
†In cyclohexanone, 35°C.
‡In tetrahydrobenzene, 25°C.

TABLE 3.5. Experimental Values of the Coefficient P for Some Polymer–Solvent Systems

Polymer	Solvent	P	References
Polystyrene	Methyl ethyl ketone–n-butanone, 25°C (θ-conditions)	4.7	[125]
Poly(methyl methacrylate)	Butyl chloride, 35°C (θ- conditions)	6.3	[126]
Polystyrene	Cyclohexane, 35°C (θ-conditions)	5.3	[118]
Polystyrene	Tetrahydrobenzene, 25°C (good solvent)	5.7	[118]
Poly-α-methylstyrene	Cyclohexane, 34.5°C (θ-conditions)	5.5	[127, 128]

values of P for polystyrene in a good solvent and a θ-solvent can be due to the same extent to the excluded-volume effect and to the scattering of data as a result of experimental error.

Hence, the problem of a reliable experimental determination of Φ and P is as yet unsolved and further detailed investigations are necessary.

If the experimental data listed in Tables 3.4 and 3.5 are compared to the theoretical data in Table 2.1, it can be seen that Zimm's method using computer simulation and avoiding the preaveraging of Oseen's tensor has led to a better agreement between theoretical and experimental values of Φ and P than the methods of other authors. However, if one takes into account the great scattering and insufficient reliability of the available experimental data and the difficulties involved in theoretical calculations, the acceptance of Zimm's data as the most reliable and preferable for use seems premature.

4.3.2. Values of the Parameter A_0. The differences in the experimental values of Φ and P obtained by different authors are not only due to the polydispersity of samples (by which they are often explained) but primarily to the errors in the measurement of molecular dimensions $\langle R^2 \rangle$ and $\langle h^2 \rangle$ from the asymmetry of light scattering in polymer solutions.

As already indicated, in order to calculate A_0 from experimental data it is necessary to know the values of M, $[\eta]$, and f, whereas the procedure of the determination of $\langle R^2 \rangle$ is excluded. Hence, in principle the precision and reliability of the experimental determination of A_0 greatly exceed those of separate determinations of Φ and P.

If M is determined according to Eq. (3.39) by a combination of diffusion and velocity sedimentation, then, replacing M contained in Eq. (3.68) by M_{sD} according to Eq. (3.39), one obtains

$$A_0 = \eta_0 \, (D/T)^{2/3} \, \{[\eta] \, s_0 R / 100 \, (1 - \bar{v}\rho)\}^{1/3}. \tag{3.74}$$

Hence, in this case the procedure of the determination of A_0 reduces to sedimentation–diffusion and viscometric measurements whose precision in the determination of D is of great importance since in Eq. (3.74) the exponent of D is twice as great as those at $[\eta]$ and s_0.

The main error in the sedimentation method is due to the measurement of the partial specific volume v determining the buoyancy factor $1 - v\rho$ in Eq. (3.74). A favorable point in using Eq. (3.74) is the fact that the buoyancy factor is contained in this equation with the exponent $1/3$, which decreases the error in the determination of v, whereas the diffusion coefficient contained in Eq. (3.74) with the exponent $2/3$ is measured fairly reliably (if the interferometric procedure is used). Hence, the sedimentation–diffusion method of the determination of A_0 should be considered to be the most reliable and, therefore, the preferred method. This method has been used in many experimental works in which the hydrodynamic parameter was determined.

A popular procedure known in the literature which gives experimental data used to calculate the parameter A_0 is a combination of the light-scattering method (for the determination of M_w) and velocity sedimentation (for the determination of f). Investigations in which light scattering and diffusion measurements are combined are infrequent. A combination of osmotic and diffusion measurements can be found in only a few papers in which low-molecular-weight samples (oligomers) have been investigated.

In a recent paper [111] the values of the parameter A_0 have been calculated using the published experimental data on the values of M, $[\eta]$, and f for a number of polymers. The values of A_0 for 200 polymer–solvent systems, including over 2000 fractions of various polymers, have been reported in this paper.

Some results of the analysis of these data [111] may be formulated as follows.

The most numerous data have been obtained for fractions of flexible-chain (mostly vinylic) polymers in "good" (not θ-) solvents. Comparison of the values of A_0 for 160 various polymer–solvent systems (including 750 fractions) does not indicate any systematic dependence of these values on the method used. No systematic changes in A_0 for separate fractions with variations in their molecular weight have been observed either.

Many investigations contain data obtained in θ-solvents (25 polymer–solvent systems with a total of 130 fractions). The mean values of the parameter A_0 for solutions of flexible-chain polymers in good and θ-solvents coincide and are equal to

$$A_0 = (3.2 \pm 0.2) \cdot 10^{-10} \ \text{g·cm}^2\text{·sec}^{-2}\text{·deg}^{-1}\text{·mole}^{-1/3}. \qquad (3.75)$$

These data show that the excluded-volume effects do not have any marked influence on the value of the parameter A_0.

Comparison of experimental values of A_0 obtained for narrow fractions and unfractionated samples suggests that for polymers exhibiting moderate polydispersity ($M_w/M_n < 2$) the dependence of A_0 on the width of molecular-weight distribution is slight.

The material considered in [111] also contains data obtained for polymers with moderate chain rigidity (ladder polysiloxanes, various cellulose derivatives, and other polysaccharides) and for polymers with very rigid chains (polyisocyanates, *para*-aromatic polyamides, etc.).

The whole complex of these experimental results suggests that for each polymer–solvent system over the entire range of molecular weights available in practice, the parameter A_0 determined by Eqs. (3.67), (3.68), and (3.74) is actually invariant. It is virtually independent of the chain length and the solvent strength and for moderately polydisperse samples it is also independent of their degree of polydispersity.

For all flexible-chain polymers and synthetic polymers with moderate chain rigidity the most reliable experimental value of A_0 is determined using Eq. (3.75). Accordingly, we have

$$\beta \equiv P^{-1}(\Phi/100)^{1/3} = 2 \ 3 \cdot 10^6. \qquad (3.76)$$

These experimental values are much lower than those predicted by modern analytical theories: $A_\infty = 3.8 \cdot 10^{-10} \ \text{g cm}^2 \ \text{sec}^2 \ \text{deg}^{-1} \text{mole}^{-1/3}$, $\beta = 2.71 \cdot 10^6$ (see Table 3.3).

For polymers with high chain rigidity, the mean experimental value of A_0 is

$$A_{0(\text{rigid})} = (3.8 \pm 0.4) \cdot 10^{-10} \text{ g·cm}^2\text{·sec}^{-2}\text{·deg}^{-1}\text{·mole}^{-1/3}, \qquad (3.77)$$

i.e., it coincides with the value of A_∞ predicted by analytical theories. However, this experimental value cannot be regarded as A_∞ because it has been obtained for short polymer chains in many cases at $L/A = 10$ or less. It is more logical to take as the experimental value of A_∞ that of A_0 determined from Eq. (3.75) since it is obtained at maximum chain lengths available experimentally. In accordance with this, the difference in the values of $A_{0,\text{rigid}}$ [according to Eq. (3.77)] and A_0 [according to Eq. (3.75)] should be regarded as the result of an increase in the A_0/A_∞ ratio for rigid-chain polymers ($d/A \ll 1$) at low L/A values according to the wormlike chain theory (Table 3.2 and Figs. 3.29 and 3.30).

Hence, when the foregoing results are used in practice to determine the molecular weights of polymers from the experimental values of $[\eta]$ and f and the known value of A_0 according to Eqs. (3.67) and (3.68), it is possible to recommend the value of A_0 from Eq. (3.75) for flexible-chain polymers and polymers with moderate chain rigidity and that from Eq. (3.77) for polymers with very rigid chains. According to the results of [111], the value $A_0 = (3.35 \pm 0.30) \cdot 10^{-10}$ g cm^2 sec^{-2} deg^{-1} mole$^{-1/3}$ should be used for cellulose derivatives and other polysaccharides, but it should be taken into account that this value is subject to great error.

As already indicated in Chapter 3, Sections 2.5 and 4.1, the method of using the invariant A_0 for the determination of molecular weights $M_{f\eta}$ is particularly important for polymers with high chain rigidity and low M values which, moreover, are soluble only in such specific solvents as concentrated sulfuric acid. A combination of all these properties virtually rules out the possibility of using the sedimentation method and makes the light-scattering method rather ineffective. Under these conditions the most efficient method is a combination of viscometry and diffusion measurements using Eq. (3.68) and the foregoing experimental values of A_0. This method is being widely used (see Chapter 4) and yields reasonable results.

The fact that the experimental value of A_∞ corresponds to the value $\beta = 2.3 \cdot 10^6$, rather than to that of $2.7 \cdot 10^6$ predicted by translational friction and viscosity theories, implies that the theoretical values $P_\infty = 5.11$ and $\Phi_\infty = 2.8 \cdot 10^{23}$ mole^{-1} are mutually *incompatible*. In other words, the interpretation of experimental data on diffusion (or sedimentation) and viscosity of solutions of the same polymer from the standpoint of a single theory of these processes (Chapter 2) cannot lead to coinciding values of chain rigidity (length of the Kuhn segment) determined from phenomena of translational friction and viscosity. The theories of transport processes in polymer solutions require further development. Its principal aims should be to eliminate the foregoing inherent contradiction and to ensure the theoretical value of A_∞ corresponding to experimental data. Zimm succeeded in this with the aid of computer simulation by decreasing the theoretical value of Φ_∞ and considerably increasing that of P_∞. Further investigations will show whether this is the only possible approach.

REFERENCES

1. T. Svedberg and K. O. Pedersen, The Ultracentrifuge, Oxford University Press, New York (1940).
2. H. K. Schachman, Ultracentrifugation in Biochemistry, Academic Press, New York (1959).
3. H. Fujita, Mathematical Theory of Sedimentation Analysis, Academic Press, New York (1962).
4. H. Fujita, Foundations of Ultracentrifugal Analysis, Wiley–Interscience, New York (1962).
5. J. W. Williams (ed.), Ultracentrifugal Analysis in Theory and Experiment, Academic Press, New York (1963).
6. H. G. Elias, Ultracentrifugen-Methoden, Beckman Instruments, Munchen (1961).
7. V. N. Tsvetkov, V. E. Eskin, and S. Ya. Frenkel, Structure of Macromolecules in Solutions, Nauka, Moscow (1964); National Lending Library for Science and Technology, Boston Spa, England (1971) [English translation: Butterworths, London (1970)].
8. C. H. Chervenka, A Manual of Methods for the Analytical Ultracentrifuge, Spinco Division of Beckman Instruments, Palo Alto, Calif. (1969).
9. E. Meyer (ed.), Polymer Molecular-Weight Determination Methods, Washington, D.C. (1973).
10. H. G. Elias, Macromolecules, Structure and Properties, Plenum, New York (1977).
11. P. P. Nefedov and P. N. Lavrenko, Transport Methods in Analytical Chemistry of Polymers [in Russian], Khimiya, Leningrad (1979).
12. A. Einstein, Annu. Rev. Phys., **17**, 549 (1905).
13. P. J. Flory, Principles of Polymer Chemistry, Cornell University Press, Ithaca (1953).
14. N. Gralen, Sedimentation and Diffusion Measurements on Cellulose, Almqvist and Wiksells, Uppsala (1944).
15. C. Beckmann and J. Rosenberg, Ann. N. Y. Acad. Sci., **46**, 209 (1945).
16. R. U. Stokes, Trans. Faraday Soc., **48**, 887 (1952).
17. V. N. Tsvetkov and S. I. Klenin, Dokl. Akad. Nauk SSSR, **88**, 49 (1953).
18. M. Daune, H. Benoit, and Ch. Sadron, J. Polym. Sci., **16**, 483 (1955).
19. M. Daune, L. Freund, and G. Scheibling, J. Chim. Phys., **57**, 924 (1957).
20. V. N. Tsvetkov, J. Eksp. Teor. Fiz., **21**, 701 (1951).
21. V. N. Tsvetkov and S. I. Klenin, Zh. Tekh. Fiz., **28**, 1019 (1958).
22. P. N. Lavrenko and O. V. Okatova, Vysokomol. Soedin., **A19**, 2640 (1977).
23. A. A. Lebedev, Tr. Gos. Opt. Inst. Leningr., **5**, 53 (1931).
24. V. N. Tsvetkov, Vysokomol. Soedin., **A9**, 1249 (1967).
25. V. N. Tsvetkov, V. S. Skazka, G. V. Tarasova, V. M. Yamshchikov, and S. Ya. Lyubina, Vysokomol. Soedin., **A10**, 74 (1968).
26. S. I. Klenin, V. N. Tsvetkov, and A. N. Cherkasov, Vysokomol. Soedin., **A9**, 1435 (1967).
27. A. N. Cherkasov, S. I. Klenin, and Yu. E. Eizner, Vysokomol. Soedin., **7**, 902 (1965).

28.   Yu. Eizner, A. N. Cherkasov, and S. I. Klenin, Vysokomol. Soedin.,
      A10, 1971 (1968).
29.   J. A. Fomin, Vysokomol. Soedin., **A15**, 1917 (1973).
30.   M. Wales and K. E. Van Holde, J. Polym. Sci., **14**, 81 (1954).
31.   M. Wales and S. J. Rehfeld, J. Polym. Sci., **62**, 179 (1962).
32.   I. Brandrup and E. H. Immergut (eds.), Polymer Handbook, Wiley–Inter-
      science, New York (1975).
33.   W. Archibald, J. Phys. Chem., **51**, 1024 (1947).
34.   V. N. Tsvetkov, Vysokomol. Soedin., **4**, 1575 (1962).
35.   V. N. Tsvetkov, V. S. Skazka, and N. A. Nikitin, Vysokomol. Soedin., **6**,
      69 (1964).
36.   V. N. Tsvetkov, V. S. Skazka, and P. N. Lavrenko, Vysokomol. Soedin.,
      A13, 2251 (1971).
37.   B. Nystrom, B. Porsch, and L. O. Sundelof, Eur. Polym. J., **13**, 683
      (1977).
38.   V. N. Tsvetkov, I. N. Shtennikova, M. G. Vitovskaya, E. I. Ryumtsev, T.
      V. Peker, Y. P. Getmantchuk, P. N. Lavrenko, and S. V. Bushin,
      Vysokomol. Soedin., **A16**, 566 (1974).
39.   P. N. Lavrenko, E. U. Urinov, and A. A. Gorbunov, Vysokomol.
      Soedin., **A18**, 859 (1976).
40.   M. G. Vitovskaya, P. N. Lavrenko, I. N. Shtennikova, A. A. Gorbunov,
      T. V. Peker, E. V. Korneeva, E. P. Astapenko, Y. P. Getmanchuk, and V.
      N. Tsvetkov, Vysokomol. Soedin., **A17**, 1917 (1976).
41.   N. V. Pogodina, P. N. Lavrenko, K. S. Pozhivilko, A. B. Melnikov, T.
      A. Kolobova, G. N. Martchenko, and V. N. Tsvetkov, Vysokomol.
      Soedin., **A24**, 331 (1982).
42.   S. V. Bushin, E. B. Lysenko, V. A. Cherkasov, K. P. Smirnov, S. A. Di-
      denko, G. N. Martchenko, and V. N. Tsvetkov, Vysokomol. Soedin.,
      A25, 1899 (1983).
43.   M. J. R. Cantov (ed.), Polymer Fractionation, Academic Press, New York
      (1967).
44.   B. L. Karger, L. R. Snyder, and C. Horwath, An Introduction to
      Separation Science, Wiley–Interscience, New York (1973).
45.   L. R. Snyder and J. J. Kirkland, Introduction to Modern Liquid
      Chromatography, Wiley, New York (1974).
46.   S. E. Bresler and S. Y. Frenkel, Zh. Tekh. Fiz., **23**, 1502 (1953).
47.   P. N. Lavrenko, A. A. Gorbunov, and E. U. Urinov, Vysokomol.
      Soedin., **A18**, 244 (1976).
48.   W. Kuhn and H. Kuhn, Helv. Chim. Acta, **28**(97), 1533 (1945); **29**, 72
      (1946).
49.   P. Eirich (ed.), Rheology, Academic Press, New York (1958).
50.   G. V. Vinogradov and A. J. Malkin, Rheology of Polymers [in Russian],
      Khimiya, Moscow (1977).
51.   E. Hatschek, The Viscosity of Liquids, G. Bell and Sons, Ltd., London
      (1928).
52.   E. Frey, D. Treves, and H. Eisenberg, J. Polym. Sci., **25**, 273 (1971).
53.   B. Zimm and D. Crothers, Proc. Natl. Acad. Sci. USA, **48**, 905 (1962).
54.   V. N. Tsvetkov and E. V. Korneeva, Vestn. Leningr. Gos. Univ., No. 22,
      Ser. Fiz. Khim., **4**(22), 75 (1965).
55.   R. Herzog and K. Weissenberg, Kolloid Z., **46**, 277 (1928).

56. V. N. Tsvetkov, N. V. Pogodina, and L. V. Starchenko, Eur. Polym. J., 19(9), 841 (1983).
57. N. V. Pogodina, K. S. Pozhivilko, A. B. Melnikov, S. A. Didenko, G. N. Martchenko, and V. N. Tsvetkov, Vysokomol. Soedin., A23, 2454 (1981).
58. S. Claesson and U. Lohmander, Makromol. Chem., 44–46, 461 (1961).
59. U. Lohmander and A. Svensson, Makromol. Chem., 65, 202 (1963).
60. G. Jeffery, Proc. R. Soc. London, A102, 161 (1922).
61. A. Peterlin, Z. Phys., 111, 232 (1938).
62. W. Kuhn and H. Kuhn, Helv. Chim. Acta, 28, 97 (1945).
63. W. Kuhn, H. Kuhn, and P. Buchner, Ergebn. Exakt. Naturwiss., 25, 1 (1951).
64. N. Saito, J. Phys. Soc. Jpn., 6, 297 (1951).
65. H. A. Scheraga, J. Chem. Phys., 23, 1526 (1955).
66. J. E. Hearst and Y. Tagami, J. Chem. Phys., 42, 4149 (1965).
67. W. Kuhn and H. Kuhn, Helv. Chim. Acta, 28, 1533 (1945).
68. B. Zimm, J. Chem. Phys., 24, 269 (1956).
69. M. Golub, J. Polym. Sci., 18(27), 156 (1955).
70. M. Golub, J. Phys. Chem., 60, 431 (1956); 61, 375 (1957).
71. S. Kapur and S. Gundiah, J. Polym. Sci., 26, 89 (1957); J. Colloid. Sci., 13, 170 (1958).
72. E. Wada, J. Polym. Sci., 14, 305 (1954).
73. O. V. Kallistov, Zh. Tekh. Fiz., 29, 70 (1959).
74. O. V. Kallistov and I. N. Shtennikova, Vysokomol. Soedin., 1, 842 (1959).
75. M. Copic, J. Chim. Phys., 53, 440 (1956); 54, 348 (1957).
76. T. Kotaka, H. Suzuki, and H. Inagaki, J. Chem. Phys., 45, 2770 (1966).
77. W. Kuhn and H. Kuhn, Helv. Chim. Acta, 29(1), 71; (3), 609; (4), 830 (1946).
78. R. Cerf, J. Phys. Radiat., 19, 122 (1958).
79. R. Cerf, Fortschr. Hochpolymer. Forsch., 1, 382 (1959).
80. R. Cerf, J. Chem. Phys., 66, 479 (1969).
81. H. C. Booij and P. H. Van Weichen, J. Chem. Phys., 52, 5056 (1970).
82. W. H. Stockmayer, Pure Appl. Chem., Suppl., Macromol. Chem., 8, 379 (1973).
83. E. R. Bazua and M. C. Williams, J. Chem. Phys., 59, 2858 (1973).
84. A. Peterlin, J. Polym. Sci., B10, 101 (1972).
85. A. Peterlin, J. Polym. Sci. Symp., 43, 187 (1973).
86. M. C. Williams and R. D. Zimmerman, Trans. Soc. Rheol., 17, 23 (1973).
87. R. Cerf, Adv. Chem. Phys., 33, 73 (1975).
88. W. H. Stockmayer, in: Molecular Fluids, R. Balian and G. Weill (eds.), Gordon and Breach (1976), p. 107.
89. R. Cerf, J. Phys., N4, 357 (1977).
90. De Gennes, J. Chem. Phys., 66, 5827 (1977).
91. Y. Y. Gotlieb and Y. Y. Svetlov, Vysokomol. Soedin., A22, 2442 (1980).
92. M. Copic, J. Chim. Phys., 53, 440 (1956).
93. Y. Ikeda, J. Phys. Soc. Jpn., 12, 378 (1957).
94. A. Peterlin and M. Copic, J. Appl. Phys., 27, 434 (1965).
95. A. Peterlin, J. Chem. Phys., 33, 1799 (1960).

96.   B. Zimm, Ann. N. Y. Acad. Sci., **89**(4), 670 (1961).
97.   A. Peterlin, Pure Appl. Chem., **12**, 563 (1966).
98.   M. Fixman, J. Chem. Phys., **45**, 793 (1966).
99.   A. Peterlin, Makromol. Chem., **44–46**, 338 (1961).
100. A. Peterlin, Polymer, **2**, 257 (1961).
101. C. Reinhold and A. Peterlin, J. Chem. Phys., **44**, 4333 (1966).
102. W. Kuhn and F. Grun, Kolloid Z., **101**, 248 (1942).
103. I. Noda and J. E. Hearst, J. Chem. Phys., **54**, 2342 (1971).
104. R. A. Harris and J. E. Hearst, J. Chem. Phys., **44**, 2595 (1966); **45**, 3106 (1966); **46**, 398 (1967).
105. J. E. Hearst, E. Beals, and R. A. Harris, J. Chem. Phys., **48**, 5371 (1968).
106. G. Vallet, J. Chem. Phys., **47**(7–8), 649 (1950).
107. G. Vallet, C. R. Acad. Sci., **230**(14), 1353 (1950).
108. L. Mandelkern and P. J. Flory, J. Chem. Phys., **20**, 212 (1952).
109. P. Debye and A. M. Bueche, J. Chem. Phys., **16**, 573 (1948).
110. H. Yamakawa, Modern Theory of Polymer Solutions, Harper and Row, New York (1971).
111. V. N. Tsvetkov, P. N. Lavrenko, and S. V. Bushin, Usp. Khim., **51**, 1699 (1982).
112. S. Newmann, W. R. Krigbaum, C. Laugier, and P. J. Flory, J. Polym. Sci., **14**, 451 (1954).
113. V. Petrus, Collect. Czech. Chem. Commun., **33**, 119 (1968).
114. V. P. Budtov, Vysokomol. Soedin., **A18**, 2606 (1976).
115. W. R. Krigbaum and D. K. Karpenter, J. Phys. Chem., **59**, 1166 (1955).
116. D. McIntyre, A. Wims, L. C. Williams, and L. Mandelkern, J. Phys. Chem., **66**, 1932 (1962).
117. G. C. Berry, J. Chem. Phys., **44**, 4550 (1966); **46**, 4887 (1967).
119. A. Yamamoto, M. Fujii, J. Tanaka, and H. Yamakawa, Polym. J., **2**, 799 (1971).
120. D. McIntyre, L. J. Fetters, and E. Slagowski, Science, **176**, 1041 (1972).
121. M. Fukuda, M. Fukutomi, Y. Kato, and H. Hashimoto, J. Polym. Sci., Polym. Phys. Ed., **12**, 871 (1974).
122. Y. Miyaki, Y. Einaga, and H. Fujita, Macromolecules, **11**, 1180 (1978).
123. Y. Einaga, Y. Miyaki, and H. Fujita, J. Polym. Sci., Polym. Phys. Ed., **17**, 2103 (1979).
124. Y. Miyaki, Y. Einaga, H. Fujita, and M. Fukuda, Macromolecules, **13**, 588 (1980).
125. J. Oth and V. Desreux, Bull. Soc. Chim. Belg., **66**, 303 (1957).
126. H. Lutje and G. Meyerhoff, Makromol. Chem., **68**, 180 (1963).
127. J. Noda, K. Mizutani, and T. Kato, Macromolecules, **10**, 618 (1977).

Chapter 4

HYDRODYNAMIC PROPERTIES.
EXPERIMENTAL DATA

1. Hydrodynamic Length and Diameter of a Polymer Chain

The length of the chain L of a polymer molecule is the main characteristic used in modern polymer theories. The information on this value is usually obtained from experimental values of the molecular weight M of the polymer investigated. However, the transition from the value of M to that of L requires the knowledge of an important structural characteristic of a chain molecule: the molecular weight per unit length $M_L = M/L = M_0/\lambda$, where M_0 is the molecular weight of the monomer unit and λ is the length of the monomer unit in the chain direction. The value of M_0 is determined by the chemical structure of the molecule and hence is usually known. The value of λ is not so unequivocally related to the chemical composition of the monomer unit and is determined mainly by its geometrical characteristics: the steric arrangement of atomic groups with respect to their nearest neighbors along the chain. Sometimes this information may be obtained from X-ray analysis data of polymers in bulk. However, these data are not always available and (which is more important) it is not always permissible to identify these data with the geometrical properties of the chain in solution, particularly in those cases when a specific secondary and ternary structure of the polymer molecule in solution can be formed.

Hence, the determination of the value of λ from data obtained directly from hydrodynamic investigations of a dilute polymer solution appears to be very important and desirable.

This determination can be carried out in practice only in those cases when we deal with polymers exhibiting relatively high rigidity and a molecular conformation close to rodlike in the range of the molecular weights actually used.

As has been shown in Chapter 2, for a homologous series of rodlike molecules of fairly large length ($L/d \gg 1$) Eqs. (2.54) and (2.55) should be valid.

147

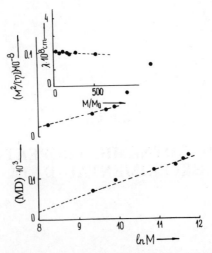

Fig. 4.1. Experimental dependences of MD and $M^2/[\eta]$ on $\ln M$ and that of λ_D [according to Eq. (4.1)] on $Z = (M/M_0)$ for poly(butyl isocyanate) fractions in tetrachloromethane [1, 2].

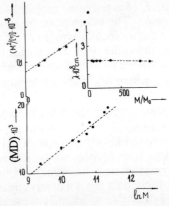

Fig. 4.2. The same dependences as in Fig. 4.1 for poly-(chlorohexyl isocyanate) fractions in tetrachloromethane [3].

According to these equations, the values of M/f and $M^2/[\eta]$ exhibit a linear dependence on the logarithm of chain length L (or molecular weight M). These dependences are shown in Fig. 2.4 for spheroid and spherocylinder models by asymptotic broken straight lines. According to Eqs. (2.54) and (2.55), the slopes of the straight lines representing the dependences of M/f and $M^2/[\eta]$ on $\ln M$ are proportional to M_L and M_L^3 respectively, and hence, in principle, may be used to determine M_L and λ from the experimental data on M and f or M and $[\eta]$. Moreover, in principle, it is also possible to determine the chain diameter d from the position of these straight lines in the chosen system of coordinates.

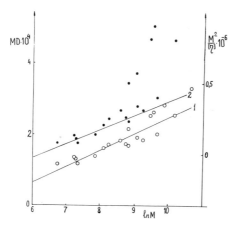

Fig. 4.3. Experimental dependences of MD (open circles) and $M^2/[\eta]$ (filled circles) on $\ln M$ for poly(para-hydroxyphenyl-benzoxazole terephthalamide) in sulfuric acid [2, 4].

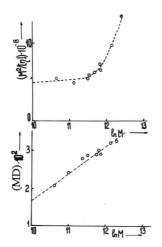

Fig. 4.4. Dependence of MD and $M^2/[\eta]$ on $\ln M$ for poly-(γ-benzyl-L-glutamate) fractions in dimethylformamide and di-chloroethane [5].

The actual possibilities of applying this method to various polymers are demonstrated with the aid of experimental data presented below.

In the following part of this book, the values of the molecular weight of the polymer M are (unless otherwise indicated) the values of M_{sD} obtained by sedimentation–diffusion analysis according to Svedberg's equation (3.39). All physical values are expressed in cgs units.

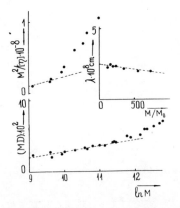

Fig. 4.5. Experimental dependences of MD and $M^2/[\eta]$ on $\ln M$ and that of λ_D [according to Eq. (4.1)] on M/M_0 for ladder poly(phenyl siloxane) fractions in benzene [6–9].

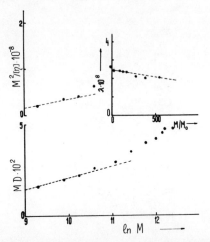

Fig. 4.6. Experimental dependences of MD and $M^2/[\eta]$ on $\ln M$ and that of λ_D [according to Eq. (4.1)] on M/M_0 for fractions of ladder poly(phenylalkyl siloxane) in butylacetate [10, 11].

The values of $MD = (M/f) \cdot kT$ and $M^2/[\eta]$ are plotted in Figs. 4.1 and 4.2 as functions of $\ln M$ for fractions of poly(butyl isocyanate) and poly(chlorohexyl isocyanate), respectively.

A linear portion can be seen in both the MD and the $M^2/[\eta]$ curves. However, in the latter case this portion is much shorter. The upward deviation of the experimental curves from the straight line at high values of $\ln M$ is a manifestation of chain flexibility (i.e., the deviation of its conformation from the rodlike conforma-

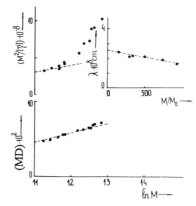

Fig. 4.7. Experimental dependences of MD and $M^2/[\eta]$ on $\ln M$ and that of λ_D [according to Eq. (4.1)] on M/M_0 for fractions of ladder poly(chlorophenyl siloxane)s [12, 13].

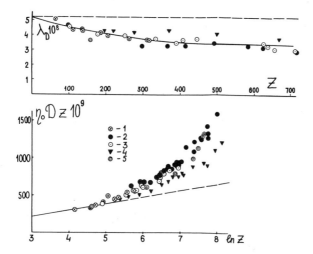

Fig. 4.8. Dependence of $\eta_0 DZ$ on $\ln Z$ (Z is the degree of polymerization) for fractions of a series of cellulose esters according to: 1) [14]; 2) [15]; 3) [16]; 4) [17]; 5) [18]. Dependence of λ_D [according to Eq. (4.1)] on Z for the same cellulose ethers and esters.

tion) with increasing length. Hence, the above differences in diffusion and viscosity curves confirm the fact that viscosity (and rotational friction) is much more sensitive to chain coiling (bending) than translational friction.

This relationship is also observed in experimental curves for a rigid-chain aromatic poly(*para*-hydroxyphenylbenzoxazole terephthalamide) shown in Fig. 4.3 and for helical molecules of poly-γ-benzyl-L-glutamate (Fig. 4.4).

The differences in experimental curves of translational friction and viscosity are still greater for less-rigid molecules of ladder polysiloxanes (Figs. 4.5–4.7) and cellulose derivatives (Figs. 4.8 and 4.9). For these polymers over the molecular

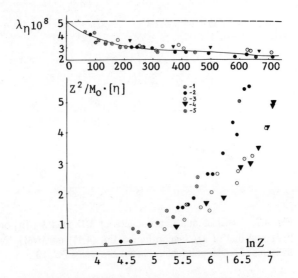

Fig. 4.9. Dependence of $Z^2/M_0[\eta]$ on $\ln Z$ and that of λ_η [according to Eq. (4.2)] on Z for the same cellulose ether and ester fractions as in Fig. 4.8.

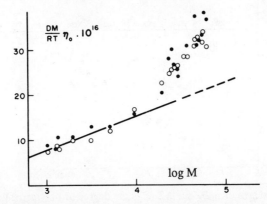

Fig. 4.10. Dependence of $DM\eta_0/RT$ on $\log M$ for fractions of aromatic poly(amide benzimidazole) in sulfuric acid (•) and dimethylacetamide (○).

weight range investigated a linear portion on the plots of $M^2/[\eta]$ or $Z^2/[\eta]$ (where Z is the degree of polymerization) is entirely absent and it is possible to determine λ from the slope of this portion only if the curves of the dependence of DM (or DZ) on $\ln M$ (or $\ln Z$) are used.

Consequently, the experimental data on translational friction (diffusion) of molecules should be used to obtain information about the structural characteristics of the polymer chain, the λ and d values, whereas viscosimetric data should be considered from the very beginning to be less suitable for this purpose.

TABLE 4.1. Length of a Monomer Unit λ in Chain Direction and Chain Diameter d according to Hydrodynamic Data in the Range of Rodlike Conformations

Polymer	$\lambda \cdot 10^8$, cm	$d \cdot 10^8$, cm
Poly(butyl isocyanate)	2 ± 0.1	10 ± 2
Poly(chlorohexyl isocyanate)	2 ± 0.1	12 ± 2
Poly(γ-benzyl-L-glutamate)	1.6 ± 0.2	25 ± 5
Ladder polysiloxanes		
Poly(phenyl siloxane)	2.5 ± 0.1	10 ± 2
Poly(chlorophenyl siloxane)	2.5 ± 0.1	12 ± 2
Poly(phenylisobutyl siloxane)	2.5 ± 0.1	14 ± 2
Poly(phenylisohexyl siloxane)	2.5 ± 0.1	14 ± 2
Poly(methylbutene siloxane)	2.5 ± 0.1	12 ± 2
Cellulose ethers and esters		
Cellulose nitrate, nitrogen 10.7%	5.1 ± 0.1	12 ± 2
Cellulose nitrate, nitrogen 12.1%	5.1 ± 0.1	12 ± 2
Cellulose nitrate, nitrogen 12.4%	5.1 ± 0.1	12 ± 2
Cellulose nitrate, nitrogen 13.4%	5.1 ± 0.1	12 ± 2
Cellulose carbanilate	5.1 ± 0.1	12 ± 2
Cellulose benzoate	5.1 ± 0.1	17 ± 4
Cellulose monophenylacetate	5.1 ± 0.1	17 ± 4
Cellulose diphenylphosphonocarbamate	5.1 ± 0.1	17 ± 4
Aromatic polyamides		
Poly(amide benzimidazole)	18 ± 1	10 ± 3
Poly(para-phenylbenzoxazole terephthalamide)	18 ± 2	8 ± 2

This statement may also be illustrated if Eqs. (2.54) and (2.55) are represented in the form of Eqs. (4.1) and (4.2), respectively:

$$\lambda_D = (1/3\pi\eta_0)\,(f/Z)\,[\ln(Z\lambda/d) + 0.38], \tag{4.1}$$

$$\lambda_\eta = (45M_0/2\pi N_A)^{1/3}\,([\eta]/Z^2)^{1/3}\,[\ln(Z\lambda/d) - 0.7], \tag{4.2}$$

where the numerical constants $\gamma_f = 0.38$ and $\gamma_\eta = -0.7$ are chosen in accordance with the spherocylinder molecular model.

If the value of λ/d in the logarithmic term in Eq. (4.1) is assumed to be equal to the value obtained from the plot of the dependence of M/f on $\ln M$ [according to Eq. (2.54)], as has been done in Figs. 4.1–4.8, then the value of λ_D on the left side of Eq. (4.1) may be calculated from the experimental value of f on Z for any value of Z according to Eq. (4.1). Under the same conditions λ_η may be calculated according to Eq. (4.2) for any Z value from the experimental values of $[\eta]$ and Z. The values of λ_D calculated in this manner are plotted in Figs. 4.1, 4.2, and 4.5–4.8 as a function of Z, and the dependence of λ_η on Z is shown in Fig. 4.9. These plots show that the curves $\lambda_D = \lambda(Z)$ may be extrapolated to the conditions $Z \to 0$. The intercepts of these curves with the ordinate correspond to the true value of λ.

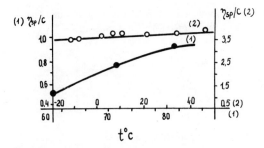

Fig. 4.11. Temperature dependence of intrinsic viscosity [η] of poly-(methacryloylphenyl) ester of cetyloxybenzoic acid in two solvents: 1) in dioxane (thermodynamically poor solvent); 2) in THF (good solvent).

TABLE 4.2. Temperature Coefficients of Intrinsic Viscosity $\delta = d(\ln[\eta])/dT$ for Some Polymers in Solutions

No.	Polymer	$M \cdot 10^{-5}$	Solvent	$[\eta] \cdot 10^{-2}$ at 293°K	Temp. range, K	$\delta \cdot 10^2$, K^{-1}
1	Poly(methacryloyl phenyl) ester of cetyloxybenzoic acid	18.5	Tetrahydro-furan	3.08	273—313	+0.2
			Dioxane	0.5 (at 333 K)	333—353	+2.9
2	Cellulose carbanilate	11.5	Dioxane	9.5	288—348	—1.1
3	Cellulose benzoate	1.9—3.9	Dioxane	2.8—4.5	300—330	—0.3
4	Cellulose monophenyl-acetate	2.3—37	Benzene	1.3—7.0	285—315	—0.4
5	Cyanoethylcellulose	1.38	Acetone	2.9	279—323	—0.6
6	Poly(butyl isocyanate)	2.5	Toluene	6.9	288—363	—1.0
7	Poly(amide benzimid-azole)	0.53	Sulfuric acid	7.2	283—333	—0.44
8	Poly(metaphenylene isophthalamide)	0.30	Sulfuric acid	0.85	273—353	—0.20
		1.70	Dimethyl-acetamide	2.84	283—339	—0.21
		3.05	Dimethyl-acetamide +3% LiCl	4.56	273—332	—0.16
		0.55	The same	1.5	273—332	—0.22
		0.37	The same	0.95	273—332	—0.21
		0.24	The same	0.65	273—332	—0.13
		0.155	The same	0.36	273—332	—0.19
9	Mesogenic aromatic polyester (PE–10)*	0.54	Dichloro-acetic acid	1.74	280—323	—0.24
10	Para–meta-aromatic polyether	0.15	Dichloro-acetic acid	1.43	288—329	—0.5

*Structure data given in Chapter 7, Section 10.2.2.

In contrast, the extrapolation of the curve $\lambda_\eta = \lambda(Z)$ in Fig. 4.9 to $Z \to 0$ is more or less arbitrary and cannot yield a reliable λ value.

It can be seen in Figs. 4.1–4.9 that the slope of the straight line representing MD vs. $\ln M$ at low M values cannot be determined here with high precision. In fact, the straight lines in these figures are plotted in accordance with the values of λ known from the chemical structure of the chain or X-ray diffraction data listed in Table 4.1. Thus, these values of λ correspond to cis-configurations of a polyiso-cyanate chain (see Fig. 7.48) and a ladder polysiloxane (Fig. 1.4), and the trans-conformation of a polyglucoside chain of cellulose ethers and esters (see Fig. 4.38). However, it may be assumed that the experimental points in Figs. 4.1–4.8 are in reasonable agreement with these data. The determination of the values of hydrodynamic chain diameter d from the λ values obtained and the position of straight lines in Figs. 4.1–4.8 is even less precise. These values are also given in Table 4.1.

Nevertheless, the difference in the order of magnitude between these values and those that might be expected on the basis of the chemical structure is not very great. The dependence of DM on $\log M$ for fractions of aromatic polyamido-benzimidazole shown in Fig. 4.10 [19] shows the same conclusion. The values of λ and d obtained for this polymer are also given in Table 4.1.

The agreement between the values of λ and d obtained from hydrodynamic data and molecular structural formulas justifies the description of hydrodynamic properties of a polymer chain based on a model of a wormlike cylinder with length L and the diameter d close to the contour length and the diameter of a real chain molecule.

For flexible-chain polymers ($A/d < 10$) the ranges of molecular weights in which the molecule exhibits the conformation of a weakly bending rod ($L/A < 1$) correspond to the values of $L/d < 10$. As can be seen from Fig. 2.4, for these val-ues of L/d the hydrodynamic characteristics f and $[\eta]$ are described by solid curves 1–4 and do not correspond to asymptotic straight lines 1'–4' or to Eqs. (2.54), (2.55), (4.1), or (4.2). Hence, the foregoing method of the determination of λ and d cannot be applied to flexible-chain polymers.

2. EFFECT OF LONG- AND SHORT-RANGE INTERACTION ON THE SIZE AND HYDRODYNAMIC PROPERTIES OF A POLYMER MOLECULE

2.1. Excluded-Volume and Draining Effects

As already mentioned, the upward deviation of experimental points in Figs. 4.1–4.9 is a manifestation of chain flexibility and hence, in principle, it can be used for the quantitative determination of flexibility by applying a theory chosen among those considered in Chapter 2. Usually this determination is based on the analysis of the dependence of hydrodynamic characteristics $[\eta]$, D, or s_0 on chain length L. However, a correct selection of the theory for analysis of experimental data is of great importance since the dependences of experimental values of $[\eta]$, D, or s_0 on L

exhibiting an external resemblance may have different physical natures and hence should be discussed in terms of different theories.

Thus, if for a homologous series of polymer fractions in a solvent the [η] value increases proportionally to M^a, where $a > 0.5$ and D decreases proportionally to M^{-b}, where $b > 0.5$, this dependence may be caused by either the excluded-volume effect (nonideality of the solvent) or the non-Gaussian behavior of polymer molecules due to their high rigidity and insufficient chain length (draining effect). In the former case the experimental data used for the determination of equilibrium chain rigidity should be interpreted using Eqs. (2.148) and (2.149) and sometimes Eq. (2.152). In the latter case, if excluded-volume effects are absent, the theories considered in Sections 3, 4, and 6 of Chapter 2 should be used.

In any case, the problem of selecting the theory for the interpretation of hydrodynamic behavior of polymer molecules in a dilute solution may be solved unequivocally if the investigations are carried out under θ-conditions (i.e., in the absence of volume effects) when the solvent and the experimental temperature are appropriately chosen. However, in practice this opportunity is seldom provided, particularly when polymers with high chain rigidity are studied, since their solubility is very limited.

As already shown (Section 1.5; Tables 1.1–1.3), even in good solvents the excluded-volume effects virtually do not affect molecular dimensions for polymers with relatively high equilibrium rigidity and, hence, their hydrodynamic properties correspond to those prevailing under θ-conditions. In practice it is possible to distinguish a typical rigid-chain polymer (with a high A value) from a flexible-chain polymer by comparing the intrinsic viscosities of two samples of equal molecular weights: for a rigid-chain polymer [η] is higher by one order of magnitude than for a flexible-chain polymer. Hence, high viscosity at low molecular weight (naturally, in the absence of specific effects, such as polyelectrolytic expansion) virtually implies that the excluded-volume effect may be neglected in the discussion of hydrodynamic properties of polymer molecules.

However, if polymers with moderate chain rigidity are studied, the intrinsic viscosities of solutions are not so high as in the preceding case, and the problem of the role played by long-range interactions (volume effects) in molecules in the formation of their conformations and in their hydrodynamic behavior requires additional consideration.

For this purpose it is useful to study the temperature dependence of intrinsic viscosity of a polymer solution because this dependence reflects either an increase or decrease in the size of the molecule with variation in solution temperature.

2.2. Temperature Dependence of Viscosity

It is known that, for polymers whose molecular dimensions are sensitive to long-range effects, intrinsic viscosity in poor solvents (under conditions similar to θ-conditions) increases drastically with increasing temperature T. With improving

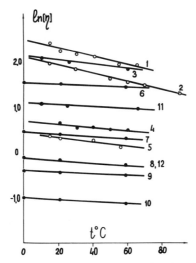

Fig. 4.12. Temperature dependence of intrinsic viscosity for solutions of some polymers. 1) Cellulose tricarbanilate, $M = 1.15 \cdot 10^6$, in dioxane; 2) poly(butyl isocyanate), $M = 2.5 \cdot 10^5$, in toluene; 3) poly(amidebenzimidazole), $M = 52 \cdot 10^3$, in 100% H_2SO_4; 4) mesogenic aromatic polyester (PE–10), $M = 54 \cdot 10^3$, in dichloroacetic acid; 5) para–meta-aromatic polyester, $M = 15 \cdot 10^3$, in dichloroacetic acid; 6–10) poly(m-phenylene isophthalamide) in DMAA + 3% LiCl at $M = 305 \cdot 10^3$ (6), $55 \cdot 10^3$ (7), $37 \cdot 10^3$ (8), $24 \cdot 10^3$ (9), and $15.3 \cdot 10^3$ (10); 11) poly(m-phenyl-isophthalamide), $M = 172 \cdot 10^3$, in DMAA; 12) poly(m-phenylene isophthalamide), $M = 30 \cdot 10^3$, in 96% H_2SO_4.

solvent strength the high positive temperature coefficient of viscosity $\delta = d \ln [\eta]/dT$ for these polymers decreases, and in thermodynamically good solvents becomes very low ($\approx 10^{-4}$). This behavior is typical for flexible-chain polymers.

These properties can be seen in Fig. 4.11, which shows the temperature dependence of intrinsic viscosity for solutions of poly(methacryloylphenyl) ester of cetyloxybenzoic acid in a poor (dioxane) and a good (tetrahydrofuran) solvent [20]. The corresponding values of the temperature coefficient of viscosity δ are given in Table 4.2. In both solvents the δ value is positive and in dioxane it is very high.

However, there are polymers for which the temperature dependence of intrinsic viscosity is inverse to that shown in Fig. 4.11; for these polymers [η] decreases with increasing temperature. A classical example of these polymers are cellulose derivatives for which the temperature coefficient of viscosity $\delta = d \ln [\eta]/dT$ is very high and negative [21]. Figure 4.12 shows, as an example, the temperature dependence of ln [η] for cellulose carbanilate solutions in dioxane [22]. The corresponding value of the temperature coefficient δ is given in Table 4.2, just as for some other cellulose esters and ethers [84, 85].

Similar thermoviscous properties have also been found for solutions of many other polymers: poly(alkyl isocyanate)s [23], various aromatic polyamides [24], and aromatic polyesters. The data obtained for some of these polymers are shown in Fig. 4.12 and listed in Table 4.2. For all these samples the coefficients δ are negative. The absolute value of δ is particularly high for poly(butyl isocyanate), a polymer with high equilibrium chain rigidity. For poly(*meta*-phenylene isophthalamide) [25], which was studied in greatest detail [25], the values of δ are negative, close to each other in various solvents, and virtually independent of the molecular weight of the fractions investigated.

Flory has considered the main features of the temperature dependence of viscosity for solutions of cellulose ethers and esters [21]. Evidently, similar considerations may be applied to other polymers with a negative coefficient δ. A qualitative discussion of this problem may be based on Eq. (2.141) if an additional factor, α_η^3, is introduced. This factor takes into account the change in the viscosity of solution due to the variation in molecular dimensions caused by long-range interaction (excluded-volume effect)

$$[\eta] = \Phi \frac{(LA)^{3/2}}{M} \alpha_\eta^3. \tag{4.3}$$

In this equation, as before, Φ is a function of chain parameters L, d, and A [Eq. (2.143)]. However, Φ should also depend on α_η, having a tendency to increase with decreasing α_η.

The size of a chain molecule (with given values of L, d, and M) in solution is determined by the rigidity parameter A (short-range interaction) and the expansion parameter α_η (due to excluded-volume effect–long-range interaction).

In all cases a temperature rise leads to a decrease in segment length A since the increase in the energy of thermal motion in the chain should decrease its skeletal rigidity maintained by short-range effects (rigidity of bond angles and chemical bonds, potentials limiting rotation about these bonds, local hydrogen bonds, etc.). According to Eq. (4.3), intrinsic viscosity $[\eta]$ should decrease, but its decrease is somewhat weakened by the simultaneous increase in Φ as a result of decreasing A according to Eq. (2.143). Hence, for polymers whose molecular dimensions in solution are determined only by the skeletal rigidity of the chain, one might expect negative temperature coefficients of viscosity δ with slightly lower absolute values than those of the parameter $A^{3/2}$.

The expansion coefficient α^3 contained in Eq. (4.3) is sensitive to temperature variations and increases with temperature for polymer solutions with a positive heat of dilution. This increase is very pronounced in thermodynamically pure solvents (near the θ–point) and less pronounced in good solvents. According to Eq. (4.3), these changes in α^3 should ensure a positive temperature dependence of viscosity which is somewhat weakened by a simultaneous decrease in Φ (due to a decrease in hydrodynamic interaction during chain uncoiling).

In principle, for a real polymer solution the temperature dependence of viscosity may be determined by a combination of both these mechanisms: the decrease in A and the increase in α with increasing temperature. However, taking into account the foregoing considerations, it may be assumed that a positive temperature coefficient of viscosity observed experimentally shows that the size of macromolecules is determined to a considerable extent by excluded-volume effects in the thermodynamically nonideal solution used.

In contrast, if the temperature coefficient of intrinsic viscosity of a polymer in solution can be reliably measured and is negative, then the major contribution to the formation of the conformational properties of its molecules is provided by equilibrium rigidity of the chain (short-range interaction), whereas the role played by volume effects (long-range interactions) is not important.

Hence, the value, and particularly the sign, of the coefficient δ may serve as a criterion for the correct choice of a theory adequately describing the hydrodynamic properties of the molecules of the polymer investigated.

For example, proceeding from this concept in the interpretation of experimental data for the polymers listed in Table 4.2, excluded-volume effects should be taken into account only for systems 1 and 2.

3. HYDRODYNAMIC CHARACTERISTICS OF COMB-SHAPED MOLECULES

The primary mechanism determining chain rigidity which is considered in the conformational statistics of flexible-chain polymers [36] is the interaction between side radicals of the polymer molecule.

Consequently, the effect of the size and structure of side groups on the conformation of polymer molecules and the rigidity of their main chain have been extensively studied experimentally [26].

Some data on this problem have been obtained in the investigation of hydrodynamic characteristics of molecules with a "comb-shaped" structure, i.e., polymer molecules with long (chain) side radicals. In particular, some polyvinyl esters have been used for this study: polyvinylcinnamate [27], polyvinyllaurate [28], and polycetylvinyl ester (polyvinylcetynate) [29].

3.1. Poly(alkyl acrylate)s and Poly(alkyl methacrylate)s

The most detailed investigations have been carried out for a poly(alkyl methacrylate) homologous series [30]

$$
\begin{array}{c}
CH_3 \\
| \\
-CH_2-C- \\
| \\
C=0 \\
| \\
O-C_nH_{2n+1}
\end{array}
$$

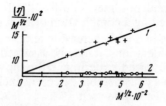

Fig. 4.13. Dependence of $[\eta]/M^{1/2}$ on $M^{1/2}$ for poly(vinyl cinnamate) fractions in bromoform (good solvent) (1) and methyl ethyl ketone (θ-solvent) (2).

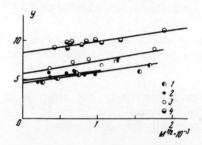

Fig. 4.14. Dependence of $y = (f/M^{1/2})(\eta_0 P_\infty)^{-1} \cdot M_L^{1/2} \cdot 10^4$ on $M^{1/2}$ for fractions of PA–1 (1) and PA–4 (2) in methyl ethyl ketone, and PA–8 (3) and PA–18 (4) in heptane [35].

TABLE 4.3. Hydrodynamic and Conformational Characteristics of Poly(n-alkyl acrylate)s with Side Chains of Various Length ($n = 1, 4, 8, 18$ is the number of carbon atoms in the alkyl group)*

Polymer	Solvent	$K_\eta \cdot 10^2$	$K_D \cdot 10^4$	$K_s \cdot 10^{15}$	a	b	$A_D \cdot 10^8$, cm	$A_\eta \cdot 10^8$, cm
PA-1	Methyl ethyl ketone	0.80	7.64	12.4	0.72	0.60	21.5	14.4
PA-4	Methyl ethyl ketone	0.89	4.67	5.0	0.73	0.55	25	22.5
PA-8	Heptane	1.02	5.2	6.6	0.67	0.56	34	27
PA-18	Heptane	1.02	5.2	6.6	0.67	0.56	72	46

*All data are given in cm, g, s.

up to poly(cetyl methacrylate) ($n = 16$) and in a poly(alkyl acrylate) series [31–35]

$$-CH_2-\underset{\underset{\underset{O-C_nH_{2n+1}}{|}}{\overset{\overset{C=O}{|}}{C}}}{\overset{\overset{H}{|}}{C}}-$$

up to poly(octadecyl acrylate) ($n = 18$).

The measurements of viscosity, diffusion, and sedimentation in solutions of these polymers have been carried out using a large number of fractions with a narrow molecular-weight distribution ($M_w/M_n < 1.2$).

The study of the temperature dependence of viscosity using various solvents has shown that the molecular dimensions of these polymers in solution are sensitive to the thermodynamic strength of the solvent, i.e., to excluded-volume effects. This can be seen in Fig. 4.13, which shows the dependence of $[\eta]/M^{1/2}$ on $M^{1/2}$ for poly(vinyl cinnamate) (PVC) fractions in bromoform (a good solvent) and methyl ethyl ketone (θ-solvent).

For all polymers of this type the molecular weight dependence of hydrodynamic characteristics $[\eta]$, D, and s_0 (sedimentation constant) may be expressed in the form of standard equations

$$[\eta] = K_\eta M^a, \quad D = K_D M^{-b}, \quad s_0 = K_s M^{1-b}, \tag{4.4}$$

where K_η, K_D, K_s, a, and b are constant over a wide molecular-weight range. The values of these constants for some poly(alkyl acrylate)s are given in Table 4.3. Values of exponents a and b higher than 0.5 reflect the influence of long-range interactions in molecules in thermodynamically good solvents.

Accordingly, the hydrodynamic properties of the comb-shaped macromolecules investigated are discussed in terms of theories accounting for excluded-volume effects, and the unperturbed size of a polymer chain is determined by extrapolation to the range of $M \to 0$ according to Eqs. (2.148) and (2.149).

The straight lines in Fig. 4.13 are an example of this extrapolation; they permit the determination of the length of the Kuhn segment A from viscometric data according to their intercept with the ordinate.

Another example is shown in Fig. 4.14 using data on the translational friction of molecules of some poly(alkyl acrylate)s and extrapolating these data to $M \to 0$ according to Eq. (2.148). Here the values of $y \equiv (f/M^{1/2})(\eta_0 P_\infty)^{-1}(M/L)^{1/2} \cdot 10^4$ cm$^{1/2}$ are shown as a function of $M^{1/2}$. In this figure the ordinate intercepts of the straight lines plotted according to Eq. (2.148) are equal to $A^{1/2} \cdot 10^4$, where A is the length of the Kuhn segment characterizing the rigidity of the main chain.

The lengths of the Kuhn segments obtained from translational friction (sedimentation and diffusion) A_D and viscometry A_η are given in Table 4.3 for some polyalkylacrylates. The values of both A_D and A_η increase with increasing alkyl side chain length.

The experimental data available on the dependence of the Kuhn segment length of the main chain on side chain length (number n of carbon atoms in the alkyl group) for polyacrylates and polymethacrylates are shown in Fig. 4.15. These data indicate that the rigidity of the main chain of comb-shaped molecules increases with increasing side chain length and at $n = 18$ exceeds that of typical flexible-chain polymer molecules by a factor of 2 or 3.

It is noteworthy that the values of A_D obtained from the data on translation friction exceed those of A_η determined from viscometric data. The reason for this difference is the fact that the values of A_D were calculated according to Eq. (2.148) using the value of $P_\infty = 5.11$, whereas the value of $\Phi_\infty = 2.8 \cdot 10^{23}$ mole^{-1} was used in the calculations of A_η according to Eq. (2.149). As shown in Section 4.3, these theoretical values of P_∞ and Φ_∞ are mutually incompatible and cannot lead to coinciding values of A_D and A_η.

The interaction between the side chains of a comb-shaped polymer is manifested in a decrease in the flexibility of its main chain with increasing side chain length. This interaction may be regarded as usual steric hindrances whose significance increases with the length of interacting side radicals. The interaction between side groups acquires a specific character in comb-shaped molecules whose side chains exhibit a structure characteristic of molecules capable of forming the liquid-crystalline phase [20, 38].

3.2. Molecules with Mesogenic Side Groups

Hydrodynamic characteristics in dilute solution have been investigated in greatest detail for fractions of polymers of some methacryloyl phenyl compounds (Table 4.4).

Figure 4.16 shows the molecular-weight dependence of intrinsic viscosity $[\eta]$ and diffusion coefficient D for solutions of fractions of poly(methacryloyl phenyl) ester of cetylhydroxybenzoic acid (No. 1 in Table 4.4) in tetrachloromethane (a θ-solvent). In accordance with θ-conditions, the experimental dependences of $1/D$ and $[\eta]$ on $M^{1/2}$ are represented by straight lines drawn from the origin, i.e., they correspond to Eqs. (4.4) with the value of exponents a and $b = 0.5$. This implies that over the entire molecular-weight range investigated up to $M = 8 \cdot 10^6$ the hydrodynamic properties of the molecules of this polymer correspond to those of nondrained Gaussian coils.

Hence, Eqs. (2.134) and (2.145) may be used in discussing these properties; the second term on the right side of these equations is equal to zero. Taking into account this condition, the equations for a nondraining Gaussian coil are obtained from Eqs. (2.134) and (2.145):

$$A_D = \frac{M_L}{(P_\infty \eta_0)^2}\left(\frac{kT}{DM^{1/2}}\right)^2, \quad A_\eta = \frac{M_L}{\Phi_\infty^{2/3}}\left(\frac{[\eta]}{M^{1/2}}\right)^{2/3}, \tag{4.5}$$

$$d_D = A_D/2.87, \quad d_\eta = A_\eta/2.13. \tag{4.6}$$

Applying Eqs. (4.5), it is possible to determine the length of the segment, A_D or A_η, from the slopes of straight lines 1 or 2 in Fig. 4.16, respectively. Equations (4.6) allow the determination of the chain diameter, d_D or d_η, from the values of A_D

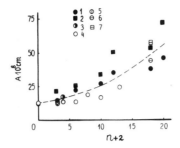

Fig. 4.15. Length of the Kuhn segment of the main chain of a comb-shaped molecule versus the number $n + 2$ of valence bonds in the side chain. Poly(alkyl acrylate)s (1–3) and poly-(alkyl methacrylate)s (4–7) according to viscometric data (1, 3–6) and translational friction (2, 7). Experimental points have been obtained: 1, 2) according to the data in {31, 34, 35]; 3, 5) according to the data in [36]; 4) according to the data in [37]; 6, 7) according to the data in [30].

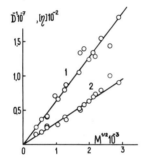

Fig. 4.16. Dependence of $1/D$ (1) and $[\eta]$ (2) on $M^{1/2}$ for fractions of poly(methacryl-oylphenyl) ester of cetyloxybenzoic acid in tetrachloromethane [39].

and A_η, respectively. The calculated values of A_D, A_η, d_D, and d_η are given in Table 4.4. The values of $P_\infty = 5.11$, $\Phi_\infty = 2.86 \cdot 10^{23}$ mole^{-1}, and $M_L = M_0/\lambda$, where $\lambda = 2.5 \cdot 10^{-8}$ cm, were used in calculations.

Similar investigations have been carried out [41] for fractions of poly-methacryloyloxybenzoic ester of p-(n-nonyloxy)phenol (No. 2 in Table 4.4) in tetrachloromethane which is also a θ-solvent for this polymer. Consequently, the dependences of D^{-2}, $s_0{}^2$, and $[\eta]^2$ on M for this polymer are represented by straight lines starting from the origin (Fig. 4.17), and molecular characteristics are calculated from experimental data using Eqs. (4.5) and (4.6).

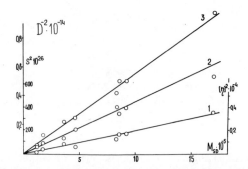

Fig. 4.17. Dependence of D^{-2} (1), s_0^2 (2), and $[\eta]^2$ (3) on molecular weight for fractions of poly(methacryloyloxybenzoate nonyloxyphenol) in tetrachloromethane [41].

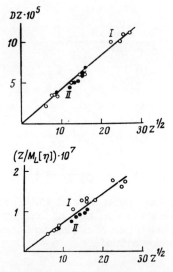

Fig. 4.18. Dependence of DZ (Z is the degree of polymerization) on $Z^{1/2}$ and that of $Z/M_L[\eta]$ on $Z^{1/2}$ for fractions of polyacryloylazoxybenzenes in tetrachloromethane [42]. I) Sample 3 in Table 4.4; II) sample 4 in Table 4.4.

Tetrachloroethane is a poor (but not a θ-) solvent for polyacryloylazoxybenzenes (Nos. 3 and 4 in Table 4.4) whose fractions have been studied in this solvent [42]. The measurements were carried out in a relatively low-molecular-weight range and as a result the excluded-volume effects were not very pronounced and the hydrodynamic behavior of molecules corresponded to the properties of nondraining Gaussian coils (Fig. 4.18). Therefore, the molecular characteristics were calculated using Eqs. (4.5) and (4.6).

TABLE 4.4. Hydrodynamic and Conformational Characteristics of Molecules of Polymethacryloyl $\left(CH_3-\overset{|}{\underset{CH_2}{C}}-\overset{|}{C}-\overset{\parallel}{\underset{O}{C}}-O-\bigcirc-R \right)$ (Nos. 1, 2, and 5) and Polyacryloyl $\left(H-\overset{|}{\underset{CH_2}{C}}-\overset{\parallel}{\underset{O}{C}}-O-\bigcirc-R \right)$ (Nos. 3 and 4) Derivatives with Mesogenic Side Groups*

No.	Radical R	M_0	Range of $M_{sD} \cdot 10^{-5}$	Solvent	$K_\eta \cdot 10^2$	$K_D \cdot 10^4$	$K_s \cdot 10^{15}$	a	b	$A_D \cdot 10^8$, cm	$A_\eta \cdot 10^8$, cm	$d_D \cdot 10^8$, cm	$d_\eta \cdot 10^8$, cm	References
1	$-O-\overset{O}{\overset{\parallel}{C}}-\bigcirc-O-C_{16}H_{33}$	522	1.4—84	Tetra-chloro-methane	3.15	1.58	—	0.5	0.5	55	49	20	23	[39, 40]
2	$-\overset{O}{\overset{\parallel}{C}}-O-\bigcirc-O-C_9H_{19}$	424	1.2—17	The same	4.46	1.33	2.1	0.5	0.5	63	49	22	23	[41]
3	$-N=N-\bigcirc-H$	267	0.1—2	Tetra-chloro-ethane	7.70	0.69	0.53	0.5	0.5	60	45	21	21	[42]
4	$-N=N-\bigcirc-C_7H_{15}$	324	0.3—0.8	The same	6.70	0.71	0.79	0.5	0.5	60	45	21	21	[42]
5	$-\overset{O}{\overset{\parallel}{C}}-O-C_{16}H_{33}$	430	0.2—5	Tetra-chloro-methane	0.33	0.33	8.1	0.8	0.6	72	54	25	25	[43]

*All data are given in cm, g. s.

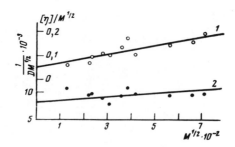

Fig. 4.19. Dependence of $[\eta]/M^{1/2}$ (1) and $1/DM^{1/2}$ on $M^{1/2}$ for fractions of the polycetyl ester of methacryloyloxybenzoic acid in dichloroethane (sample 5 in Table 4.4).

The fractions of the last sample (No. 5) listed in Table 4.4, the cetyl ester of methacryloyloxybenzoic acid, have been investigated [43] in tetrachloromethane, a good solvent for this polymer. Hence, in this case, the excluded-volume effect was quite distinctly displayed in the high values of exponents $a > 0.5$ and $b > 0.5$ in Eqs. (4.4). In accordance with this, the molecular parameters characterizing the unperturbed size of the molecules were determined by extrapolation of the dependences of $1/D\cdot M^{1/2}$ and $[\eta]/M^{1/2}$ on $M^{1/2}$ to the range of low M values according to Eqs. (2.148) and (2.149). The corresponding plot is shown in Fig. 4.19, and the values of A_D and A_η calculated according to Eqs. (4.5) are given in Table 4.4. This table also lists the values of chain diameter, d_D and d_η, calculated according to Eqs. (4.6).

For all the polymers listed in Table 4.4, the calculated values of A_D are higher than those of A_η. As already indicated, this fact is due to the numerical values of Φ_∞ and P_∞ used in this case. Both A_D and A_η exceed by a factor of 3–4 the corresponding values for typical flexible-chain polymers with no bulky side groups. This is an indication of the specific interaction between the mesogenic side groups of the molecule which increases the equilibrium rigidity of the main chain. The diameter of a wormlike chain describing the hydrodynamic properties of the molecule for all the polymers listed in Table 4.4 is close to $20\cdot10^{-8}$ cm. This value seems reasonable if the length of the side groups of investigated comb-shaped molecules is taken into account. It is noteworthy that for all the molecules listed in Table 4.4 the effect of hydrodynamic draining is virtually absent (under θ-conditions $a = b = 0.5$) in spite of a considerable equilibrium rigidity of the main chain ($A \approx 60\cdot10^{-8}$ cm). This is due to the fact that coil draining is determined by the A/d ratio rather than by segment length, and this ratio is relatively low for the comb-shaped molecules investigated.

Similar hydrodynamic properties have also been detected for the molecules of some graft copolymers.

3.3. Graft Copolymers

3.3.1. Chemical Structure of the Molecule.
The synthesis and study of the properties of graft copolymers is an important and extensively

investigated field of polymer science. This is largely due to the fact that the synthesis of graft copolymers provides wide possibilities for the modification of structure and physical properties of macromolecular materials. Moreover, it is also important to study the size, structure, and configuration properties of molecules of graft copolymers because these copolymers may serve as good models for branched-chain molecules. If the relative lengths of the main (L_A) and the grafted (L_B) chains and the degree of grafting is varied, it is possible to synthesize macromolecules with a given branch distribution and thus to vary their structure from "star-shaped" ($L_B > L_A$) to "comb-shaped" ($L_A \gg L_B$). The use of "living" [44] polystyrene chains initiated by organo-alkali catalysts for grafting onto polyalkyl-methacrylate chains was found to be particularly useful in this respect [45, 46].

The structural and conformational properties of the copolymer molecules obtained by this procedure have been investigated in solutions by hydrodynamic and optical methods which make it possible to prove the graft structure of these copolymers [47–51]. Thus, for the poly(methyl methacrylate)–polystyrene copolymer the structure of the molecule may be represented in the following form:

$$
\begin{array}{c}
\text{CH}_3 \ \text{H} \quad \text{CH}_3 \ \text{H} \quad \text{CH}_3 \qquad \text{CH}_3 \\
\mid \quad \mid \qquad \mid \quad \mid \qquad \mid \qquad\qquad \mid \\
\cdots\!-\!\text{C}\!-\!\text{C}\!-\!-\!\text{C}\!-\!\text{C}\!-\!-\!\text{C}\!-\!\cdots\!-\!-\!\text{C}\!-\!\cdots \\
\mid \quad \mid \qquad \mid \quad \mid \qquad \mid \qquad\qquad \mid \\
\text{O}\!=\!\text{C} \quad \text{H} \ \text{O}\!=\!\text{C} \quad \text{H} \quad \text{C}\!=\!\text{O} \qquad \text{C}\!=\!\text{O} \\
\mid \qquad\quad \mid \qquad\quad \mid \qquad\qquad\quad \mid \\
\text{O} \qquad \text{H}\!-\!\text{C}\!-\!\text{C}_6\text{H}_5 \ \text{O} \qquad \text{H}\!-\!\text{C}\!-\!\text{C}_6\text{H}_5 \\
\mid \qquad\quad \mid \qquad\qquad \mid \qquad\qquad\quad \mid \\
\text{CH}_3 \quad \text{H}\!-\!\text{C}\!-\!\text{H} \quad \text{CH}_3 \quad \text{H}\!-\!\text{C}\!-\!\text{H} \\
\qquad\qquad \mid \qquad\qquad\qquad\qquad \mid \\
\qquad\qquad \vdots \qquad\qquad\qquad\qquad \vdots
\end{array}
$$

The main characteristics of both initial copolymerized components are the length L_A (or molecular weight M_A) of the poly(methyl methacrylate) chain becoming the main chain of the graft copolymer and the length L_B (or molecular weight M_B) of each grafted polystyrene chain.

The graft copolymer is synthesized by grafting "living" polystyrene (PS) with a certain chain length L_B and a molecular-weight distribution as narrow as possible. To obtain a comb-shaped copolymer, the poly(methyl methacrylate) (PMMA) onto which polystyrene is grafted should be of relatively high molecular weight and (for subsequent fractionation) should have a broad molecular-weight distribution. The graft copolymer is fractionated by molecular weight, and each fraction is characterized by its molecular weight M (using standard methods) and the molar content of styrene in the copolymer $x_p = P_B/(P_B + P_A)$, where P_B is the degree of polymerization of the grafted polystyrene chain and P_A is that of the PMMA chain segment located between two neighboring graft points. The value of x_p characterizing the degree of grafting is determined from experimental data on the refractometry of solutions of PS, PMMA, and their copolymers. In this case, the use of a polarizing diffusiometer [52] is very effective (see Chapter 3, Figs. 3.14 and 3.15). Moreover, x_p may be determined by elemental analysis for the content of methoxy groups in the graft copolymer molecule. The molecular weight M_A of the main (PMMA) chain of each fraction of the graft copolymer is determined according to the molecular weight of the fraction from the relationship $M_A = M(1 - x_p)$. The length of the main chain L_A may be determined from the value of M_A. However, in this case it is necessary to take into account the correction for the molecular-weight decrease of those PMMA monomer units onto which the other monomer is grafted: as a result of grafting, this unit loses the methoxy group $-O-CH_3$.

The molecular characteristics of copolymer fractions determined from the data from hydrodynamic and optical investigations of their dilute solutions may differ greatly depending on the relative chain lengths of the components and the degree of grafting [47–51].

3.3.2. Hydrodynamic Properties and Chain Rigidity. Systematic quantitative data may be obtained from the investigations of a homologous series of fractions with a constant composition x_p and invariable length of grafted chains L_B (or M_B). If in this case, for all fractions $L_B \ll L_A$, this system may be regarded as a homologous series of "comb-shaped" molecules with similar structures but different molecular weights, varying as a result of changes in the length of the main chain. Although the structure of a "comb-shaped" graft copolymer is not as regular as that of the comb-shaped homopolymer [poly(cetyl methacrylate)], owing to random distribution of graft points along the main chain and a certain molecular-weight polydispersity of grafted chains, the hydrodynamic properties of this molecule may be described using a wormlike chain model [53].

Under actual conditions of the synthesis and fractionation of a graft copolymer, the fractions obtained may differ slightly in the degree of grafting and, hence, in the values of x_p. Therefore, for graft copolymers, it is desirable to use the equations of the hydrodynamics of long wormlike chains (2.134) and (2.145) in the following form:

$$\frac{\eta_0 LD}{kT} = \frac{1}{P_\infty} \left(\frac{L}{A}\right)^{1/2} + \frac{1}{3\pi} \left[\ln\left(\frac{A}{d}\right) - 1.0561\right], \qquad (4.7)$$

$$\frac{L}{M_L |\eta|} = \frac{1}{\Phi_\infty A} \left(\frac{L}{A}\right)^{1/2} + \frac{2.2}{\Phi_\infty A} \left[\ln\left(\frac{A}{d}\right) - 0.755\right], \qquad (4.8)$$

where L is the length of the backbone chain of the graft copolymer $L = L_A = M_A/M_L = M(1 - x_p)/M_L$, and M_L is the average molecular weight of the PMMA chain per unit chain length determined by taking into account the loss of some of the methoxy groups during grafting.

Figure 4.20 shows the experimental dependence of LD on $L^{1/2}$ for fractions of a graft copolymer obtained by grafting polystyrene chains with degree of polymerization $P_B = 16$ at average degree of grafting corresponding to the equation $P_B = 9P_A$. This means that a PS chain is grafted not less than one per each second PMMA monomer unit. This ratio of the length of the side chain to that of the main chain segment between the neighboring side chains approximately corresponds to the same ratio in a comb-shaped chain of poly(cetyl methacrylate) (PCMA) although the side chains are twice as long for a graft copolymer than for PCMA.

The data shown in Fig. 4.20 cover the range of molecular weights M_{sD} of the fractions investigated with degree of polymerization of the main chain from 850 to

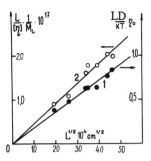

Fig. 4.20. Dependence of $\eta_0 LD/kT$ on $L^{1/2}$ for fractions of a PMMA–PS graft copolymer in butyl acetate (1) and dependence of L/M_L [η] on $L^{1/2}$ for the same fractions in bromoform (2) [54]. L is the length of the main chain. The M_L value averaged over fractions is $3\cdot10^{10}$ daltons cm^{-1}, $x_p \approx$ 0.9. Molecular weight of the chain of grafted polystyrene M_B is 1600 g/mole.

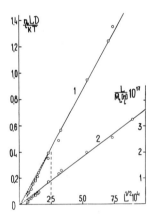

Fig. 4.21. Dependence of $\eta_0 LD/kT$ on $L^{1/2}$ for fractions of PMMA–PC graft copolymer in butyl acetate (1) and dependence of L/M_L [η] on $L^{1/2}$ for the same fractions in bromoform (2) [55]. L is the length of the main chain. The M_L value averaged over fractions is $3\cdot10^{10}$ daltons cm^{-1}, $x_p \approx$ 0.9, and $M_B = 2500$ g/mole.

140 PMMA monomer units. Hence, the lengths of the main chain of the molecule of even the lowest-molecular-weight fraction are higher by one order of magnitude than those of the side chains and hence all the molecules investigated may be considered comb-shaped.

The dependences of DL and $L/M_L[\eta]$ on $L^{1/2}$ shown in Fig. 4.20 are approximated by straight lines passing through the origin and therefore correspond to the

hydrodynamic properties of Gaussian chains exhibiting strong hydrodynamic interaction without a marked excluded-volume effect.

The slope of straight line 1 and Eq. (4.7) at $P_\infty = 5.11$ yield the length of the
Kuhn segment $A_D = 130 \cdot 10^{-8}$ cm. When Eq. (4.8) and the value of $\Phi_\infty = 2.86 \cdot 10^{23}$
mole^{-1} are used, the slope of straight line 2 yields $A_\eta = 80 \cdot 10^{-8}$ cm. From these
values of A_D and A_η it is possible to calculate, using Eq. (4.6), the hydrodynamic
diameter of this comb-shaped graft copolymer: $d_D = 45 \cdot 10^{-8}$ cm and $d_\eta = 40 \cdot 10^{-8}$
cm. Figure 4.21 illustrates, using the same system of coordinates as Fig. 4.20, the
results of sedimentation–diffusion and viscometric investigations of solutions of
fractions of another graft copolymer, PMMA–PS [55]. In this case, the degree of
polymerization of the grafted PS, P_B, is 25, and the average degree of grafting is
slightly lower than in the previous case, so that the P_B/P_A ratio = 9 remains the
same. The geometry of the "monomer unit" of this graft copolymer is similar to
that of the poly(cetyl methacrylate) unit, but on a scale three times greater. The
range of molecular weights M_{sD} of fractions shown in Fig. 4.21 is very wide and
corresponds to a degree of polymerization of the main chain from 3000 to 20
PMMA monomer units. The length of the main chain L_A of the molecules of low-
molecular-weight fractions of this series is comparable to that of the grafted side
chains L_B. For fractions of the lowest molecular weight the value of L_A is even less
than L_B. Hence, the molecules of these fractions are not comb-shaped but are more
like that of a "star" or a "hedgehog." If the condition $L_A/L_B \geq 10$ is assumed to be
the criterion for the comb-shaped structure of the molecule, then in the series
considered the fractions with $L_A^{1/2} = L^{1/2} \geq 25 \cdot 10^{-4}$ cm$^{1/2}$ satisfy this condition. The
lower limit of this range of L in Fig. 4.21 is shown by a broken straight line.

In the range of lengths L in which the molecules of a graft copolymer are
comb-shaped, the experimental points fit the straight lines passing through the
origin. This corresponds to the hydrodynamic properties of nondraining Gaussian
coils without excluded-volume effects. The slopes of these straight lines and Eqs.
(4.7), (4.8), and (4.6) yield the length of the Kuhn segment $A_D = 140 \cdot 10^{-8}$ cm,
$A_\eta = 100 \cdot 10^{-8}$ cm, and the chain diameter $d_D = 47 \cdot 10^{-8}$ cm, $d_\eta = 47 \cdot 10^{-8}$ cm.

Both the rigidity of the main chain and the hydrodynamic diameter of the investigated comb-shaped molecules of graft copolymers are much higher than those
of the comb-shaped poly(alkyl methacrylate)s considered above. This seems
natural if the longer length and much greater bulk of the interacting side chains of
polystyrene compared to the alkyl side chains of comb-shaped poly(alkyl methacrylate)s are taken into account.

The points in Fig. 4.21 representing the data for low-molecular-weight fractions markedly deviate from the straight line passing through the origin, and fall on
the curves with slope increasing with decreasing L.

This means that, in the low-molecular-weight range for the graft copolymers
considered here, the dependences of D, s_0, and $[\eta]$ on M are described by Eqs. (4.4)
in which the exponents a and b are *less than 0.5*. These dependences are not
characteristic of linear chain molecules and may be found only for nondraining
globular particles exhibiting a slight asymmetry of shape [Chapter 2, Eqs. (2.7) and

TABLE 4.5. Hydrodynamic and Conformational Characteristics of Molecules of Some Stepladder Polymers in Dilute Solutions*

Polymer	Radical R	Range of $M_{sD} \cdot 10^{-5}$	Solvent	$K_\eta \cdot 10^2$	$K_D \cdot 10^4$	$K_s \cdot 10^{15}$	a	b	$A_D \cdot 10^8$, cm	$A_\eta \cdot 10^8$, cm	$d_D \cdot 10^8$, cm	$d_\eta \cdot 10^8$, cm
Poly-N-substituted maleimides	Isobutyl [57] —C₄H₉	1.9—34	Butyl acetate	2.20	3.40	5.0	0.65	0.59	54	30	18	14
	Tolyl [59] ⟨○⟩—CH₃	0.4—5.6	Dimethyl-form-amide	0.63	2.88	3.65	0.7	0.58	40	30	14	14
	Dimethylphenyl [60] ⟨○⟩—CH₃ CH₃	0.4—2.0	The same	5.3	1.52	1.90	0.5	0.5	42	28	15	13
Polyacenaphthylene [61]	—	0.1—1	Benzene	6.5	1.8	2.3	0.5	0.5	38	25	13	12
		1—10	»	0.74	3.0	4.4	0.7	0.56	—	—	—	—

*All data are given in cm, g, s.

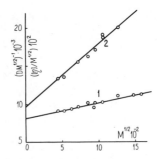

Fig. 4.22. Dependence of $1/DM^{1/2}$ (1) and $[\eta]/M^{1/2}$ on $M^{1/2}$ for poly(isobutyl maleimide) fractions in butyl acetate [57].

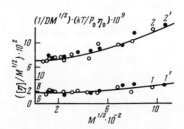

Fig. 4.23. Dependence of $1/DM^{1/2}$ (1) and $[\eta]/M^{1/2}$ (2) on $M^{1/2}$ for fractions of two samples of polyacenaphthylene in benzene. 1, 2) First sample; 1', 2') second sample [61].

(2.18)]. Naturally, this hydrodynamic behavior is revealed for the molecules of graft copolymers with the structure of a compact "star" or "hedgehog."

4. STEPLADDER POLYMERS

In principle, the interaction between neighboring side groups of the polymer chain may be intensified by the formation of true chemical (valence) bonds between these groups rather than by the increase in their length and bulk. In this case, the combination of this additional bond of two side groups and the valence bond of the main chain leads to the formation of a cyclic structure which virtually completely rules out the possibility of rotation about the "cyclized" valence bond of the main chain. If additional bonds are formed between all pairs of neighboring side groups, a "ladder" polymer appears, and if they are formed between a few pairs only, the polymer is called a "partial ladder" or a "stepladder" polymer.

Polymers with a regular "stepladder" structure may also be obtained without the formation of additional bonds by choosing a monomer exhibiting the appropriate structure. The following polymers may be used as examples: polyindene, polyimides of maleic and citraconic acids, polyacenaphthylene, etc. Five-membered rings rigidly bonded to the main chain rule out the possibility of rotation about each second bond of the main chain of these polymers (structural formulas in

Table 4.5). These polymers are fairly well soluble in common organic solvents and hence it is possible to investigate the hydrodynamic properties of their molecules in dilute solutions by sedimentation, diffusion, and viscometry using well-characterized narrow fractions.

The investigations of N-substituted poly(isobutyl maleinimide)s [56–58] over a wide molecular-weight range have revealed a marked influence of excluded-volume effects on the size of the molecules of these polymers in butyl acetate: the exponents a and b in Eqs. (4.4) are greater than 0.5 (Table 4.5). Hence, standard dependences have been plotted according to Eqs. (2.148) and (2.149) for determination of the unperturbed size of the molecules (Fig. 4.22).

In the study of aromatically substituted polymaleimides over a molecular-weight range of ten orders of magnitude, the influence of excluded-volume effects on the size of their molecules in solutions has also been observed. In accordance with this, for poly(tolyl maleimide) in dimethylformamide [59] the exponents a and b are greater than 0.5 (Table 4.5) and the determination of unperturbed molecular dimensions requires plots similar to those shown in Fig. 4.22.

For poly(dimethylphenyl maleimide), dimethylformamide is not a θ-solvent. However, this polymer has been investigated [60] over a narrow and relatively (compared to others) low-molecular-weight range in which excluded-volume effects have not been noticeable ($a = b = 0.5$) and, correspondingly, the dependences similar to those plotted in Fig. 4.22 have the shape of straight lines parallel to the abscissa.

Polyacenaphthylene has been obtained as a series of fractions covering a wide (a hundred orders of magnitude) molecular-weight range [61]. This allowed the observation of differences in the molecular-weight dependences of hydrodynamic characteristics of its benzene solutions in two ranges: a low- and a high-molecular-weight range. In the former we have the exponents $a = b = 0.5$, and in the latter both a and b are greater than 0.5 (Table 4.5). This dependence of the exponents a and b on M, in which a and b increase with increasing molecular weight, is rather unusual for linear chain molecules. It corresponds to the hydrodynamic properties of nondraining coils in a thermodynamically good solvent and may be detected only in the investigations carried out over a wide molecular-weight range, including low molecular weights. In this case, the long-range interaction in molecules at low M values virtually cannot be detected but becomes very noticeable with fractions of increasing molecular weight. This is illustrated in Fig. 4.23, which shows the dependences of $1/DM^{1/2}$ and $[\eta]/M^{1/2}$ on M for polyacenaphthylene fractions in benzene. These dependences are represented by curves whose slope decreases with decreasing molecular weight and at low M values becomes almost equal to zero.

If the equilibrium rigidity parameters A_D and A_η are to be calculated from the ordinate intercepts of the straight lines in Figs. 4.22 and 4.23 according to Eq. (4.5), the value of $M_L = M_0/\lambda$ should be known because the product $A \cdot \lambda$ is found directly from hydrodynamic data. For most carbon chain polyvinyl polymers discussed in previous sections, it is possible to take $\lambda = 2.5 \cdot 10^{-8}$ cm (simple trans-chain), whereas for the stepladder structures considered here this possibility is not evident. The elucidation of the problem of the values of λ for these polymers requires additional consideration of configurational properties of their molecules using optical anisotropy (Chapter 6, Section 2.3). For all the polymers listed in Table 4.5, in order to calculate the values of A_D and A_η it is assumed that $\lambda = 2.5 \cdot 10^{-8}$ cm, as in the case of a trans-chain. If the cis–trans-structure is assumed to be the most probable structure, the values of A increase by 20%.

As in all the other cases, in Table 4.5 the value of A_D is greater than that of A_η for the same sample in accordance with the accepted values of $\Phi_\infty = 2.86 \cdot 10^{23}$ and $P_\infty = 5.11$.

For all the polymers listed in Table 4.5 the parameter of equilibrium rigidity A is at least twice as high as that for typical flexible-chain polymers and does not vary greatly with the change in the structure of side groups. This illustrates the fact that for all these polymers high chain rigidity is due to the overall mechanism, i.e., to the stepladder structure for which rotation about one-half of all the bonds of the main chain is virtually completely hindered. The increase in equilibrium rigidity of the chain expected here may be qualitatively evaluated by applying, for example, Eq. (1.15) and interpreting the increasing rigidity in cyclization as the result of a reduction in the number of rotating bonds in the chain by a factor of two with the simultaneous increase in the length of each chain by the same factor. Moreover, the l/λ ratio in Eq. (1.15) remains invariable, whereas A increases by a factor of two. However, this evaluation cannot lay claim to quantitative reliability because complete cyclization of one-half of all rotating bonds should lead to more profound changes in the structure and equilibrium conformation of the chain and may greatly change the rotation hindrance about the noncyclized bonds.

The hydrodynamic diameters d_D and d_η of molecules given in Table 4.5 and calculated according to Eqs. (4.6) using the known values of A_D and A_η are relatively large. This was to be expected if we recall both the large bulk of side radicals and the rigidity of their attachment to the main chain of the stepladder molecule.

5. LADDER POLYSILOXANES

Attempts to obtain ladder or partial ladder polymers are usually based on the synthesis of aromatic–heterocyclic chains in which cyclic structures are separated by single links. These polymers may be based on benzimidazole [62], imide [63], imidazopyrrolone [64], benzobisoxazole [65], and other [66] heterocyclic structures. When the number of single links in the chain decreases, its partial ladder structure approaches the ladder structure and, if single links are absent, it becomes entirely a ladder structure as, for example, for the BBL polymer [67] or for polyquinolines.

Some literature data are available on the investigation of molecular properties of aromatic ladder polymers in dilute solutions [69, 70]. However, on a large scale these investigations are difficult and not always reliable because these polymers are poorly soluble.

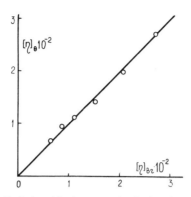

Fig. 4.24. Relationship between intrinsic viscosities $[\eta]_{Br}$ and $[\eta]_{\theta}$ of fractions of ladder polyphenylsiloxane (M_{sD} from $1.4 \cdot 10^5$ to $1.1 \cdot 10^6$) in bromoform (good solvent) and a benzene–butyl acetate (60:40) mixture (θ-solvent) [6, 73].

Well-soluble ladder polymers have been synthesized on the basis of double-strand siloxane chains with different side radical structures (polysilsesquioxanes) [71, 72]. This made it possible to carry out the investigations of hydrodynamic properties of these polymers in dilute solutions with the aid of standard transport methods (Chapter 3) by using a series of carefully prepared fractions over a wide molecular-weight range [73–80].

The structure of the monomer unit of ladder polysiloxanes may be represented by the following scheme:

$$
\begin{array}{c}
R_1 \\
| \\
-\mathrm{Si-O-} \\
| \\
\mathrm{O} \\
| \\
-\mathrm{Si-O-} \\
| \\
R_2
\end{array}
$$

where the radicals R_1 and R_2 may be either aromatic or aliphatic (Table 4.6).

The structure of the double-strand polyphenylsilsesquioxane chain is shown in Fig. 1.4. The proposed cis-syndiotactic structure [71] of this polymer is in good agreement with the length of the monomer unit $\lambda = 2.5 \cdot 10^{-8}$ cm determined from the translational friction of its molecules in solution (Figs. 4.5–4.7 and Table 4.1).

5.1. Molecular-Weight Dependence of Translational Friction and Viscosity

Numerous investigations of the dependence of D, s_0, and $[\eta]$ of solutions of ladder polysiloxanes on the molecular weight of the fractions have shown that over a limited molecular-weight range these dependences may be represented to a certain approximation by Eqs. (4.4) with the constant values of K_η, K_D, K_S, a, and b. In this case, the exponents a and b greatly exceed 0.5 and often attain the value of 1.0 (Table 4.6). These high values of a and b cannot be accounted for by excluded-volume effects and indicate the influence of the draining effect in the macromolecules.

The same conclusion may be drawn from a comparison of hydrodynamic characteristics of molecules in thermodynamically good and θ-solvents. This can be seen in Fig. 4.24, which shows the relationship between the intrinsic viscosities of a series of fractions of ladder polyphenylsiloxanes in a good (bromoform) and a θ-solvent [6, 73]. The values of $[\eta]_{Br}$ and $[\eta]_\theta$ virtually coincide for all fractions, which implies the absence of a marked effect of solvent thermodynamic strength and, correspondingly, the absence of effects of long-range interaction on the size of macromolecules in solution. Similar data have also been obtained for other ladder polysiloxanes [10].

The study of the dependence of $[\eta]$, D, and s_0 on M over a wide molecular-weight range shows that on a double logarithmic scale these dependences are actually represented by curves and hence cannot be expressed by Eqs. (4.4) with a and b values constant over the entire molecular-weight range investigated. This can be seen in Fig. 4.25, which shows, as an example, the corresponding dependences for fractions of ladder polychlorophenylsiloxane in benzene [12]. To a first approximation, however, these dependences may be described using the form of Eqs. (4.4) but taking different values of the coefficients K, a, and b in different ranges of M. Thus, the entire molecular-weight range shown in Fig. 4.25 is tentatively divided into two ranges in which coefficients K, a, and b have the values given in Table 4.6 (No. 6).

Similar results have also been obtained for other ladder polysiloxanes when their hydrodynamic properties have been studied over a relatively wide molecular-weight range. Some of these results (viscometric characteristics) are listed in Table 4.6 (Nos. 3, 10, and 11). For all ladder polysiloxanes the high exponents a and b in Eqs. (4.4) decrease with increasing molecular weight. This dependence is opposite (inverse) to that which might be expected with a marked influence of excluded-volume effect on the size of the molecules investigated (see, e.g., Fig. 4.23). In contrast, this dependence completely corresponds to an increase in hydrodynamic interaction (decrease in draining) in polymer chains with increasing molecular weight and the corresponding change in conformation from a weakly bending rod to a more compact Gaussian coil.

Hence, both the low sensitivity of hydrodynamic characteristics to the thermodynamic strength of the solvent and their molecular-weight dependence (high values of exponents a and b and their decrease with increasing M) observed in solutions of ladder polysiloxanes show convincingly that the hydrodynamic properties of their molecules are not determined by the excluded-volume effects but, rather, by skeletal rigidity and the corresponding hydrodynamic interaction in chains.

TABLE 4.6. Hydrodynamic and Conformational Characteristics of Molecules of Ladder Polysiloxanes in Dilute Solutions*

No.	Side group	Solvent	Range of $M \cdot 10^{-5}$	$K_\eta \cdot 10^4$	$K_D \cdot 10^4$	$K_s \cdot 10^{15}$	a	b	$A_D \cdot 10^8$, cm	$A_\eta \cdot 10^8$ cm	$d_D \cdot 10^8$, cm	$d_\eta \cdot 10^8$, cm	References
1	Phenyl I, $R_1 = R_2 = C_6H_5$	Benzene Bromoform θ-solvent	0.36—10 0.36—10 1.4—10	24 24 24	6.2 — —	9.2 — —	0.85 0.85 0.85	0.63 — —	200 — —	135 135 135	6—10 — —	— — —	[6, 73]
2	Phenyl II, $R_1 = R_2 = C_6H_5$	Benzene	0.4—4.2	1.3	6.1	7.2	1.10	0.62	140	—	6—10	—	[7, 75]
3	Phenyl III, $R_1 = R_2 = C_6H_5$	Benzene Bromoform	0.17—0.6 1.0—3.0	7.7 7.6	3.0 —	3.7 —	0.9 0.7	0.56 —	80 —	60 [74] —	— —	6 [74]	[7, 75]
4	Phenyl IV, $R_1 = R_2 = C_6H_5$	Benzene	0.10—4.7	16	9.3	12	0.9	0.66	300	320	7.7	—	[8]
5	Phenyl V, $R_1 = R_2 = C_6H_5$	Benzene	0.26—7.5	1.2	19.5	23	1.1	0.72	280	—	7	—	[9]
6	Chlorophenyl, $R_1 = R_2 = C_6H_4Cl$	Benzene	0.5—3.0 5—16	0.23 94	16 4.4	28 7.9	1.2 0.75	0.72 0.60	300 —	160 —	10 —	— —	[12]
7	Dichlorophenyl, $R_1 = R_2 = C_6H_3Cl_2$	Benzene	0.42—5.4	1.35	12.6	24.8	1.05	0.68	250	190	17	—	[13]
8	3-Methyl-1-butene I, $R_1 = R_2 =$ $=-CH=CH-CH(CH_3)CH_3$	Butylacetate Benzene	0.9—6.0 0.9—6.0	54 54	11 —	11 —	0.88 0.88	0.69 —	220 —	180 180	— —	— —	[10, 75]
9	3-Methyl-1-butene II, $R_1 = R_2 =$ $=-CH=CH-CH(CH_3)CH_3$	Butylacetate	0.005—7.4	—	—	—	—	—	240	170	9	—	[76]
10	Phenylisobutyl (1:1), $R_1 = C_6H_5$, $R_2 = -CH_2-CH(CH_3)CH_3$	Butylacetate Benzene [77]	0.14—1.6 1.6—23 0.3—3.2 3.2—23	12.6 7.0 11.35 397	3.31 0.525 — —	3.39 0.955 0.955 —	0.9 0.57 0.91 0.63	0.58 0.50 — —	96 — — —	84 — 100 [74] —	13 — — —	12 [74]	[11, 77]
11	Phenylisohexyl (1:1), $R_1 = C_6H_5$, $R_2 = -(CH_2)_3-CH(CH_3)CH_3$	Butylacetate Benzene	0.06—3.9	63.1	3.31	3.39	0.9	0.58	130	—	12	—	[11, 77]

*All data are given in cm, g, s.

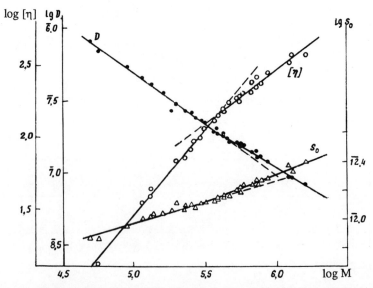

Fig. 4.25. Dependence of $\log[\eta]$, $\log D$, and $\log s_0$ on $\log M$ for ladder poly(chlorophenyl siloxane) fractions in benzene [12].

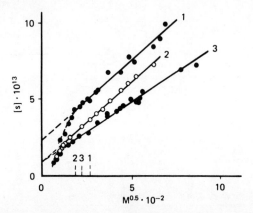

Fig. 4.26. Sedimentation coefficient $[s] = s_0$ versus $M^{0.5}$ for ladder polysiloxanes: 1) poly(phenyl siloxane) in benzene [8] (No. 4 in Table 4.6); 2) poly(phenyl isohexylsiloxane) in butyl acetate [11] (No. 10 in Table 4.6); 3) poly(3-methyl–1-butylsiloxane) in butyl acetate [10] (No. 8 in Table 4.6). Broken lines intersecting the abscissa give the chain length $L = 2.278$ Å for polymers 1, 2, and 3.

Hence, the theories of wormlike chains (wormlike spherocylinders) neglecting the excluded-volume effects will be used in the quantitative discussion of hydrodynamic properties of ladder polysiloxanes in relation to the conformational and structural characteristics of their molecules.

5.2. Parameters of Rigidity and Hydrodynamic Interaction

The theories relating the hydrodynamic properties of wormlike chains to their size and conformation are based on general equations (2.130), (2.132), (2.135), and (2.137). To solve specific problems these equations may be used in a particular form selected depending on the molecular-weight range and equilibrium rigidity of the polymer investigated.

5.2.1. High-Molecular-Weight Range.
If the hydrodynamic properties of molecules of the polymer investigated can be adequately expressed using a wormlike chain model, then at relatively high values of L (or M) these properties are described by Eqs. (2.134) for translational friction and by Eq. (2.145) for viscosity.

5.2.1.1. Translational friction.
According to Eq. (2.134) or (4.7) at $L/A \gg 1$ the dependence of $\eta_0 M/f$ on $M^{1/2}$ is represented by a straight line whose slope

$$\frac{\partial\,(\eta_0 M/f)}{\partial\,(M^{1/2})} = \frac{1}{P_\infty}\left(\frac{M_0}{A\lambda}\right)^{1/2} \tag{4.9}$$

allows the determination of A/M_L, whereas the intercept $(\eta_0 M/f)_0$ with the ordinate allows the determination of the chain diameter d from the equation

$$\ln(A/d) = 1.056 + (\eta_0 M/f)_0\, 3\pi\lambda/M_0. \tag{4.10}$$

Since $\eta_0 M/f = \eta_0 MD/kT = s_0\eta_0 N_A/(1 - \bar{v}\rho)$, analogous straight lines also determine the dependences of MD or s_0 on $M^{1/2}$. If the intercept is equal to zero, Eq. (4.10) becomes the first equation of Eqs. (4.6).

This relationship is valid for all ladder polysiloxanes investigated at medium and high M values. Figure 4.26 demonstrates this for some of them, showing the dependence of the sedimentation constant $s_0 \equiv [s]$ on $M^{1/2}$. Other examples may be found in the above-cited references to original papers.

It should be noted that the linear dependence of MD or s_0 on $M^{1/2}$ is retained for ladder polymers over the entire high-molecular-weight range experimentally available. This fact implies that even for fractions of the highest molecular weight the excluded-volume effects do not appreciably affect the translational friction of molecules of these polymers [compare with Eq. (2.152) in which ε should be taken equal to zero for agreement with experimental data].

In the low-molecular-weight range all curves in Fig. 4.26 exhibit a deviation from linearity, which corresponds to the theoretical predictions of Eqs. (2.130) and (2.131).

It should be noted that Eq. (2.134), predicting the linear dependence of M/f on $M^{1/2}$, is obtained from the general equation (2.132) when the terms d/A and A/L are neglected, i.e., under the condition $L/A \gg 1$. However, a more detailed analysis of Eq. (2.132) shows that Eq. (2.134) should be valid virtually over the entire range of values of $L/A \geq 2.278$. This theoretical prediction is also experimentally confirmed, in particular, by the data in Fig. 4.26, in which the deviation from the linear shape of the curves is observed only in the range $L/A < 2.2$. Hence, Eq. (2.134) [just as Eq. (2.111)] is more universal in the sense that it may be used over a wide molecular-weight range.

The lengths of the Kuhn segments A_D determined from the slopes of straight lines in Fig. 4.26 and from similar plots for other polymers obtained using Eq. (4.8) with $\lambda = 2.5 \cdot 10^{-8}$ cm are given in Table 4.6. The values of the hydrodynamic diameters d_D calculated from the ordinate intercepts and the values of A_D according to Eq. (4.9) are also listed here.

5.2.1.2. Viscometric data. According to Eqs. (2.137) and (2.137a), at $L/A > 2.278$ the dependence of $M/[\eta]$ on $M^{1/2}$ for a wormlike chain has the shape of the curves shown in Fig. 2.10, whose asymptotic limits at $L/A \gg 1$ are straight lines [74] determined by Eq. (2.145).

In those cases when ladder polysiloxane fractions are investigated in a high-molecular-weight range, the experimental data confirm the linear dependence of $M/[\eta]$ on $M^{1/2}$. This can be seen in Fig. 4.27, showing the dependence of $M/[\eta]$ on $M^{1/2}$ for fractions of ladder poly(phenyl siloxane) [74] and ladder poly(phenylbutyl) siloxane [74].

The slopes of the straight lines in Fig. 4.27, given by

$$\partial (M/[\eta])/\partial (M^{0.5}) = (M_L/A)^{3/2}/\Phi_\infty = (M_0/\lambda A)^{3/2}/\Phi_\infty, \qquad (4.11)$$

allow the determination of the Kuhn segment A_η. The values of A_η calculated at $\lambda = 2.5 \cdot 10^{-8}$ cm and $\Phi_\infty = 2.87 \cdot 10^{23}$ mole^{-1} for Nos. 3 and 10 are given in Table 4.6.

According to Eq. (2.145), the intercepts $(M/[\eta])_0$ of straight lines in Fig. 4.27 with the ordinate are related to the value of A/d by the equation

$$\ln (A/d) = 0.755 + (M/[\eta])_0 (\Phi_\infty/2.2) A/M_L^2. \qquad (4.12)$$

Equation (4.12) at $(M/[\eta])_0 = 0$ is equivalent to the second equation of Eqs. (4.6). Using Eq. (4.12) and the experimental values of $(M/[\eta])_0$ and A_η it is possible to calculate the chain diameter d_η. The values of d_η obtained for Nos. 3 and 10 are given in Table 4.6. They agree satisfactorily with those of d_D obtained from translational friction data.

As already mentioned (Chapter 2), the curves in Fig. 2.10 demonstrate the fact that the theoretical equation (2.145) is valid for the range of $L/A > 20$. This conclusion is confirmed by available experimental data, in particular those shown in Fig. 4.27. Hence, if a series of polymer fractions in a wide molecular-weight range is investigated, the plotting of the dependence of $M/[\eta]$ on $M^{1/2}$ is a convenient method of the determination of A and d for a polymer chain from the experimental viscometric data for high-molecular-weight polymer fractions (at $L/A > 20$).

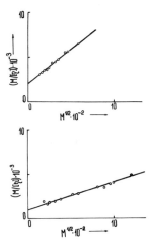

Fig. 4.27. Dependence of $M/[\eta]$ on $M^{1/2}$ for ladder poly(phenyl siloxane) in bromoform [74] (above) (No. 3 in Table 4.6) and ladder poly(phenyl isobutylsiloxane) in benzene [74] (below) (No. 10 in Table 4.6).

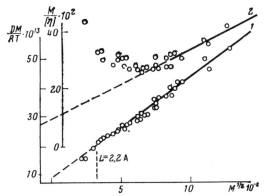

Fig. 4.28. Dependence of DM/RT (1) and $M/[\eta]$ (2) on $M^{1/2}$ for fractions of ladder poly(chlorophenyl siloxane) in benzene [12]. The point at which $L/A = 2.2$ is marked on the abscissa.

However, the theory predicts the deviation of the dependence $M/[\eta] = f(M)$ from the linear dependence with decreasing chain length at $L/A < 20$, as is shown by curves in Fig. 2.10. This theoretical prediction is confirmed by experimental data shown in Fig. 4.28. In this figure the dependences of MD and $M/[\eta]$ on $M^{1/2}$ are shown for fractions of the same polymer ladder polychlorophenylsiloxane [12]. The points corresponding to the experimental data on the diffusion, D, fit a straight line over the entire investigated range of M in which $L/A > 2.2$, whereas those representing viscometric data deviate from the straight line and form a curve concave upward, which is in qualitative agreement with the curves shown in Fig. 2.10.

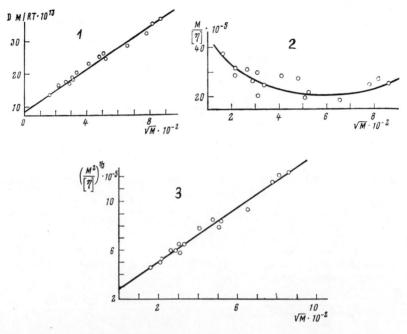

Fig. 4.29. Dependence of DM/RT (1), $M/[\eta]$ (2), and $(M^2/[\eta])^{1/3}$ (3) on $M^{1/2}$ for ladder poly(phenyl siloxane) fractions in benzene [9].

Hence, in order to use viscometric data in this range of molecular weights for the determination of the chain parameters A and d, it is necessary to compare experimental data with theoretical dependences (2.137)–(2.139) shown only in the functional form, and this requires the use of computers and complicates the experimenter's work.

In this situation it is desirable to find an analytical expression in the form of a simple linear relationship between viscometric data and molecular weight over the wider range of L/A. This expression has been obtained [9] using, first, a linear dependence of M/f on $M^{1/2}$ according to Eq. (2.134) and, second, the invariance of the hydrodynamic parameter A_0 [according to Eq. (3.68)]. Comparison of Eqs. (2.134) and (3.68) yields

$$(4.65A_0/k)(M^2/[\eta])^{1/3} = P_\infty^{-1}(M_L/A)^{1/2}M^{1/2} + (M_L/3\pi)[\ln(A/d) - 1.056]. \qquad (4.13)$$

Equation (4.13) predicts a linear dependence of $(M^2/[\eta])^{1/3}$ on $M^{1/2}$ over the entire molecular-weight range in which the dependence of M/f on $M^{1/2}$ is linear. The validity of this prediction can be seen in Fig. 4.29, which shows the dependences of MD, $M/[\eta]$, and $(M^2/[\eta])^{1/3}$ on $M^{1/2}$ for ladder poly(phenyl siloxane) fractions in benzene [9].

The length of segment A may be determined from the slope of the straight line plotted in Fig. 4.13, and the intercept $[4.65A_0/(M^2/[\eta])^{1/3}]_0$ of this line with the ordinate permits the calculation of the chain diameter d using an expression analogous to Eq. (4.10).   Naturally, the values of A and d obtained in this case will correspond to those of A_D and d_D since they are calculated using the experimentally justified value of A_0 and the value of P_∞ rather than that of Φ_∞. However, the use of Eq. (4.13) is important in those cases when only the data on viscometry and molecular weights determined by an independent method (such as light scattering) are available and the data on diffusion and sedimentation are absent.

5.2.2. Low-Molecular-Weight Range.

5.2.2.1. Translational friction. At low molecular weights the greater the length (or molecular weight) of a polymer molecule, the greater is the deviation of the chain conformation from a rodlike conformation. In accordance with this, the flexibility of the chain leads to the deviation of its hydrodynamic behavior from the properties determined by Eqs. (2.54) or (4.1). This deviation can be seen in Figs. 4.5–4.7 as a decrease in λ_D – the "apparent" length of the monomer unit with increasing degree of polymerization $Z = M/M_0$.

In the theory of wormlike chains at $L/A \leq 2.278$, flexibility is taken into account in Eq. (2.130) by terms depending on L/A. If we restrict ourselves to a linear approximation with respect to L/A and the condition $d \ll L$, then, according to Eq. (2.130), it is possible to write, instead of Eq. (4.1),

$$\lambda = (1/3\pi\eta_0)\,(f/Z)\,[\ln(Z\lambda/d) + 0.386 + 0.1667\,(L/A)]. \qquad (4.14)$$

Hence, according to Eqs. (4.1) and (4.14), the ratio of the length of the monomer unit $\lambda = 2.5 \cdot 10^{-8}$ cm of a ladder polymer to its "apparent" value λ_D (determined by the curves in Figs. 4.5-4.7 as a function of Z) is given by

$$\lambda/\lambda_D = 1 + 0.1667\,(Z\lambda/A)\,[\ln(Z\lambda/d) + 0.386]. \qquad (4.15)$$

Using the dependence of λ_D on $Z - M/M_0$ in Figs. 4.5–4.7 and Eq. (4.15), it is possible to determine the length of the Kuhn segment A.

A similar evaluation may be carried out if Eq. (2.130) for a thin chain ($d/L \ll 1$) is represented in the form

$$1/f - 1/f_{\text{rod}} \equiv D/kT - [\ln(L/d) + 0.386]/3\pi\eta_0 L$$
$$= (0.1667/3\pi\eta_0 A)\,[1 + 0.067\,(L/A) + 0.0067\,(L/A)^2 + \cdots], \qquad (4.16)$$

where f is the friction coefficient determined experimentally for the polymer molecule investigated, and f_{rod} is the friction coefficient of a thin rod of the same length L and diameter d. Equation (4.16) shows that at low L/A ratios the value of $1/A$, the inverse value of the Kuhn segment, represents directly the difference in the mobilities ($1/f$) and ($1/f_{\text{rod}}$) of a bending wormlike chain and a straight rod having the same dimensions. Hence, if f is determined experimentally at low L/A ratios and f_{rod} is calculated from the known values of L and d, it is possible to evaluate A according to Eq. (4.16).

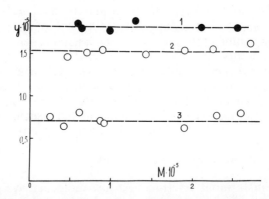

Fig. 4.30. Values of $y = 3\pi\eta_0[(1/f) - (1/f_{rod})]$ for molecules of ladder polysiloxanes of various molecular weights M obtained from the experimental data on D and M using Eq. (4.16). It was assumed in calculations that $\lambda = 2.5\cdot10^{-8}$ cm and $d = 10^{-7}$ cm. 1) Poly(phenylisobutyl siloxane) [11]; 2) poly(phenylisohexyl siloxane) [11]; 3) poly(phenyl siloxane) IV [8].

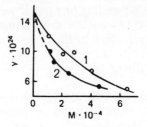

Fig. 4.31. Dependence of y [according to Eq. (4.17)] on M. Points correspond to the experimental values of y according to the values of M, $[\eta]$, M_0, λ, and d for fractions of ladder poly-(methylbutene siloxane) (1) and ladder poly(phenylisohexyl siloxane) (2). Curves represent the theoretical values of $\lambda^3 f(L/A)$, where $f(L/A)$ is calculated according to Eq. (2.136) at the values of $A = A_\eta$ listed in Table 4.7.

The points in Fig. 4.30 indicate the experimental values of $y \equiv 3\pi\eta_0[(1/f) - (1/f_{rod})]$ for molecules of some ladder polysiloxanes with various molecular weights M. According to Eq. (4.16), the extrapolation of the values of y to low M values gives the value of $0.1667/A$ and, correspondingly, the length of the Kuhn segment A. The values of A_D obtained by these methods according to Eqs. (4.15) and (4.16) are given in Table 4.7. They agree in order of magnitude with those given in Table 4.6, although the precision of the determination of A at low M values is lower than that at high M. This is mainly due to both the errors in the determination of molecular weights, s_0, and D for low-molecular-weight fractions, as well as the necessity for using the value of d generally known only approximately.

TABLE 4.7. Lengths of the Kuhn Segment for Molecules of Ladder Polysiloxanes Obtained from Translational Friction (A_D) and Viscometry (A_η) of Their Dilute Solutions at Low Molecular Weights

Polymer	$A_D \cdot 10^8$, cm	$A_\eta \cdot 10^8$, cm
Phenylsiloxane I	180 [6]	—
Phenylsiloxane II	120 [7]	—
Phenylsiloxane III	100 [7]	—
Phenylsiloxane IV	230 [8]	320 [8]
Phenylisobutylsiloxane	95 [11]	—
Phenylisohexylsiloxane	110 [11]	100 [80]
3-Methyl-1-butenesiloxane	180 [76]	220 [80]
Chlorophenylsiloxane	160 [12]	—
Dichlorophenylsiloxane	160 [13]	—

5.2.2.2. Viscometric data. In the range of $L/A \leq 2.278$, the relationship between the intrinsic viscosity $[\eta]$ and the molecular characteristics of a wormlike chain is described by Eq. (2.135). If the condition $d/L \ll 1$ is fulfilled over the entire molecular-weight range (for the ladder polymers discussed here it is fulfilled even for fractions of the lowest molecular weight), then the function $F_\eta(p)$ contained in Eq. (2.135) may be represented by its asymptotic limit according to Eq. (2.48). Under these conditions, Eq. (2.135) may be written in the following form:

$$y \equiv ([\eta]/M^2)\,(45M_0^3/2\pi N_A)\,[\ln\,(M\lambda/M_0 d) - 0.697] = \lambda^3 f\,(L/A), \qquad (4.17)$$

where $f(L/A)$ is determined by the series (2.136).

For a polymer for which the values of M_0, λ, and d are known, the left-hand side, y, of Eq. (4.17) is determined by the experimental values of $[\eta]$ and M. If y is plotted as a function of M, a curve which intersects the ordinate at a point $y_{M \to 0} = \lambda^3$ is obtained. The overall shape of this curve corresponds to the function $f(L/A)$. A comparison of the experimental curve of the dependence $y = y(M)$ and the theoretical curve $f(L/A)$ makes it possible to determine the length of segment A of the polymer investigated. Figure 4.31 shows the results of viscometric measurements of low-molecular-weight fractions of two ladder polysiloxanes [80], and the values of A_η are given in Table 4.7.

As in the methods using translational friction, in the viscometry of ladder polysiloxanes the precision of the determinations of the value of A_η in the low-molecular-weight range is lower than that in the moderate- and high-molecular-weight ranges.

As for other polymers, for ladder polysiloxanes, according to the data obtained at high M, A_η is generally less than A_D (Table 4.6). This is due to the values of P_∞ and Φ_∞ used here. In the low-molecular-weight range this relationship is not observed (Table 4.7) becuase in this range the values of A_D and A_η are determined without using the coefficients P_∞ and Φ_∞.

TABLE 4.8. Coefficients K_η, K_D, K_S, a, and b in Mark–Kuhn Equations (4.4) for Cellulose Esters and Ethers in Dilute Solutions (Temperature, 295°K)*

No.	Cellulose ethers and esters	Degree of substitution	Range of $M_{sD} \cdot 10^{-5}$ g/mole	Solvent	$K_\eta \cdot 10^4$	$K_D \cdot 10^4$	$K_S \cdot 10^{15}$	a	b	References
1	Butyrate , R: —CO—C$_3$H$_7$	2.9	0.63—17	Methyl ethyl ketone Tetrachloroethane Bromoform	1.2 1.9 1.8	4.8 — 1.6	6.4 — —	0.81 0.71 0.71	0.60 — 0.60	[86]
2	Benzoate , R: —CO—C$_6$H$_5$	2.2	0.51—6.6	Dioxane	300	1.3	1.75	0.75	0.60	[87]
3	Monophenyl acetate , R: —CO—CH$_2$—C$_6$H$_5$	2.2	0.70—7.0 7.0—50	Benzene »	50 900	4.6 4.6	4.8 4.8	0.85 0.60	0.58 0.58	[87]
4	Carbanilate [18] , R: —CO—NH—C$_6$H$_5$	2.3	0.20—9.2	Dioxane Ethyl acetate Benzophenone	10 7.8 3	— 7.8 —	— 12 —	1.0 1.0 1.0	— 0.64 —	[18]
5	Dimethylphosphone carbamate , R: $\overset{O}{\overset{\|}{\text{—CO—NH—P—(OCH}_3)_2}}$	2.0	2—24	0.2 M NaCl in water	98.2	5.5	5.02	0.8	0.65	[88]
6	Diphenylphosphone carbamate , R: $\overset{O}{\overset{\|}{\text{—CO—NH—P—(OC}_6\text{H}_5)_2}}$	2.2	0.30—16.7	Dioxane	14	3.0	3.0	0.9	0.63	[87]
7	Ethyl cellulose , R: —C$_2$H$_5$	2.27	0.05—0.60	Ethyl acetate Dioxane	100 100	4.8 —	4.8 —	0.92 0.92	0.62 —	[89]

#	Substance			Solvent						Ref.
8	Ethyl cellulose , R: —C₂H₅	2.75	0.11—0.80	Acetone Benzene	15.1 13.4	13.5 —	16.6 —	1.05 1.07	0.69 —	[90]
9	Cyanoethylcellulose , R: —C₂H₄—C≡N	2.6	0.26—3.3	Acetone Cyclohexanone	38 22	12 —	19 —	0.95 1.00	0.67 —	[85]
10	Cellulose nitrate , R: —N(=O)(=O)	2.7	0.4—3.0 $(M_{D\eta})$** 3.0—7.7 $(M_{D\eta})$**	Ethyl acetate Cyclohexanone Ethyl acetate Cyclohexanone	38 39.8 460 485	4.94 — 4.94 —	— — — —	0.99 0.98 0.79 0.77	0.63 — 0.63 —	[91]
11	Nitrate	2.4	0.25—2.8	Ethyl acetate	13.3	10.4	22.4	1.07	0.69	[16]
12	Nitrate	2.3	0.2—3.0 3.0—8.1 0.6—3.0 3.0—8.1	Ethyl acetate Ethyl acetate Cyclohexanone Cyclohexanone	98.5 — 28.8 1560	3.9 3.9 — —	6.8 6.8 — —	0.87 0.64 0.96 0.64	0.6 0.6 — —	[15]
13	Nitrate	2.0	0.05—0.50	Ethyl acetate	14.3	6.68	13.1	1.04	0.64	[92]
14	Nitrate	1.9	0.16—0.77	Methyl ethyl ketone Dioxane	120 140	0.43 —	9.6 —	0.86 0.86	0.60 —	[14]
15	Nitrate	1.14	0.67—5.7	Dimethylacetamide +6% LiCl	104	0.61	0.96	0.83	0.61	[16]
16	Cellulose, R: H	0	0.02—7.0	Cadoxene (cadmium–ethylene diamine complex)	450	0.72	1.8	0.74	0.58	[93]

* All data are given in cm, g, s.

** $M_{D\eta}$ are calculated from values of D and η using Eq. (3.68) with $A_0 = 3.5 \cdot 10^{-10}$ g·cm²·s⁻²·deg⁻¹·mole⁻¹/³.

5.3. Ladder Structure and Chain Rigidity

The hydrodynamic properties of dilute solutions and molecular characteristics of polysiloxanes described above show that the equilibrium rigidity of molecules of these polymers exceeds by one order of magnitude that of typical flexible-chain polymers. This fact is of great importance because it is the principal confirmation of the ladder structure of these polymers.

The main structural chain parameters $\lambda = 2.5 \cdot 10^{-8}$ cm and $d = (10 \pm 2) \cdot 10^{-8}$ cm, determined from the hydrodynamic characteristics of their molecules, correspond to the cis-syndiotactic chain structure shown in Fig. 1.4 and agree with the data of x-ray analysis of the polymer in bulk [81].

The various polymer samples listed in Tables 4.6 and 4.7 differ markedly in their equilibrium rigidity; the value of A is mainly within the range of (100–300)$\cdot 10^{-8}$ cm. These differences are not caused by the different structure of substituting side groups because they are observed to the same extent for poly-(phenyl siloxane)s I–V, which have side groups of the same structure.

Experimental data show that the values of A may differ depending on the method used for the chemical synthesis of the ladder structure. Thus, if in anionic polymerization a compound of the "tetrol" $(C_6H_5OHSiO)_4$ type is used as the initial unit [82], it is possible to obtain ladder poly(phenyl siloxane) IV with molecules whose equilibrium rigidity greatly exceeds [8] that of poly(phenyl siloxane)s I–III obtained by the anionic polymerization of products of the complete hydrolysis of phenyltrichlorosilane $(C_6H_5Si)_8O_{12}$ [71, 72]. However, even when the general schemes of chemical synthesis coincide, chain molecules may differ in their equilibrium rigidity as, for example, for poly(phenyl siloxane)s I–III.

These differences show that the factors relating to some defects in the ladder structure of these siloxanes play a certain role in the mechanism of flexibility of their double-strand chains. In fact, in polymerization some hydroxyl groups in a polycyclic compound may remain unreacted and, as a result, the ladder structure in these chain parts remains incomplete, and the polymer becomes a stepladder polymer.

The presence of some defects in the structure of the polymers investigated is also indicated by their A values which normally do not exceed $3 \cdot 10^{-6}$ cm. In fact, the flexibility mechanism of an ideal ladder chain with no defects should differ greatly from that of linear chain molecules arising from rotation about valence bonds without deformation of these bonds and bond angles. Since in an ideal ladder structure this rotation is impossible, the flexibility of a ladder molecule should be considered to be the result of microdeformations of this structure (i.e., bond angles and valence bonds) during thermal motion of the polymer chain. Taking into account the very high energetic barriers preventing this type of deformation, higher values of A should be expected for the molecules of ideal ladder polymers than are actually observed for the ladder polysiloxanes investigated.

However, although the defects of the ladder structure greatly determine the equilibrium rigidity of these polymers, it can be easily seen [8] that the relative number of defective units in their chains is small: it does not exceed 2 or 3% for the most flexible chains.

6. CELLULOSE DERIVATIVES

6.1. Molecular-Weight Dependence of $[\eta]$, D, and s_0

As already indicated (Chapter 1), cellulose derivatives have been known for a long time as polymers with high chain rigidity. In the hydrodynamic properties of their molecules in solutions this rigidity is shown mainly by the high intrinsic viscosity and relatively high a and b exponents in Eqs. (4.4).

The relevant data have been considered in the literature [83, 84]. They are supplemented by Table 4.8, which gives more up-to-date results. The table lists the hydrodynamic characteristics of some cellulose ethers and esters with different structures of the substituting groups R in the glucose ring and the degree of substitution γ.

All the data given in Table 4.8 have been obtained using fractions with a polydispersity parameter usually not exceeding the value of $M_w/M_n \leq 1.3$.

As for ladder polysiloxanes, for solutions of cellulose derivatives the dependences of $[\eta]$, D, and s_0 on M in the form of Eqs. (4.4) are fulfilled at constant values of coefficients and exponents only over a limited molecular-weight range. When the dependence of $\log[\eta]$ on $\log M$ is investigated over a wide range of M values, a deviation from linearity may be observed. This deviation corresponds to a decrease in the exponent a with increasing M. For cellulose nitrates this relationship was observed a long time ago [84]. In accordance with this, for some samples listed in Table 4.8 (Nos. 3, 10, and 12) the values of a are given for two ranges into which the entire range of molecular weights of the fractions investigated is tentatively divided.

The decrease in a with increasing M implies a decrease in draining of the molecule with increasing chain length. This relationship is not consistent with the hydrodynamic properties of a polymer molecule whose conformation is sensitive to excluded-volume effects. Hence, the hydrodynamic characteristics of cellulose derivatives in solutions and their molecular-weight dependence are in good agreement with the conformational properties of their molecules, discussed in Section 1.6.2.1 of this chapter.

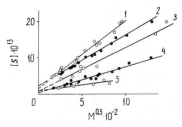

Fig. 4.32. Sedimentation coefficients $s_0 \equiv [s]$ versus $M^{1/2}$ for some cellulose esters. 1) Cellulose carbanilate in ethyl acetate [18]; 2) cellulose butyrate in methyl ethyl ketone [86]; 3) cellulose monophenylacetate in benzene [87]; 4) cellulose diphenylphosphonocarbamate in dioxane [87]; 5) cellulose benzoate in dioxane [87].

Hence, in the quantitative consideration of hydrodynamic properties of cellulose ethers and esters, just as for ladder polysiloxanes, it is quite reasonable to compare the experimental data and the result of theoretical calculations based on Eqs. (2.130), (2.132), (2.135), and (2.137).

6.2. Hydrodynamic Properties and Chain Rigidity

6.2.1. Translational Friction. As for ladder polymers, the experimental data on translational friction for cellulose derivatives are in good agreement with the linear dependence of M/f on $M^{1/2}$ predicted by Eqs. (2.134) and (4.7) over the entire range of $L/A \geq 2.3$. This can be seen in Fig. 4.32, which shows the dependence of the sedimentation coefficient $s_0 \equiv [s]$ on $M^{1/2}$ for solutions of fractions of some cellulose esters. Similar linear dependences have been obtained for all other samples listed in Table 4.8.

The lengths of segments A_D and chain diameters d_D for cellulose derivatives have been determined from the slopes of the straight lines of the dependence of M/f on $M^{1/2}$ and their ordinate intercepts using Eqs. (4.9) and (4.10) (at $\lambda = 5.15 \cdot 10^{-8}$ cm). These values are given in Table 4.9.

6.2.2. Viscometric Data. In those cases when the polymer fractions occupy a wide interval in the high-molecular-weight range, the dependence of $M/[\eta]$ on $M^{1/2}$ is linear in accordance with the theoretical equation (2.145). This is shown in Fig. 4.33, in which the corresponding dependence is presented for samples 1, 3, and 13 in Tables 4.8 and 4.9. The slopes of the straight lines in combination with Eq. (4.11) lead to the values of A_η, whereas the intercepts with the ordinate in combination with Eq. (4.12) lead to those of d_η (see Table 4.9).

If the molecular weights of the fractions are not sufficiently high to correspond to the linear portions of theoretical curves plotted in Fig. 2.10, then the experimental dependences of $M/[\eta]$ on $M^{1/2}$ for these fractions are represented by curves which cannot be used in practice for the determination of the rigidity param-

TABLE 4.9. Length of the Kuhn Segment A and Chain Diameter d of Cellulose Ether and Ester Molecules according to the Data of Translational Friction (A_D and d_D) and Viscometry (A_η and d_η) and Degrees of Hindrance to Intramolecular Rotation σ_D and σ_η Corresponding to the Values of A_D and A_η

No.	Polymer, degree of substitution	Solvent	$A_D \cdot 10^8$, cm	$d_D \cdot 10^8$, cm	$A_\eta \cdot 10^8$, cm	$d_\eta \cdot 10^8$, cm	σ_D	σ_η
1	Cellulose butyrate , $\gamma = 2.9$	Methyl ethyl ketone	260	10	200	7.0	4.8	4.2
2	Cellulose benzoate , $\gamma = 2.2$	Dioxane	240	20	180	40	4.6	4.0
3	Cellulose monophenyl acetate, $\gamma = 2.2$	Benzene	240	20	110	40	4.6	3.1
4	Cellulose carbanilate, $\gamma = 2.3$	Ethyl acetate	190	16	154	13	4.1	3.7
5	Cellulose dimethyl phosphone carbamate, $\gamma = 2.0$	0.2 M NaCl in water	180	10	165	9.0	4.0	3.8
6	Cellulose diphenyl phosphone carbamate, $\gamma = 2.2$	Dioxane	240	20	—	—	4.6	—
7	Ethyl cellulose, $\gamma = 2.27$	Ethyl acetate	180	12	175	12	4.0	3.9
8	Ethyl cellulose, $\gamma = 2.75$	Acetone	205	11	168	9.5	4.3	3.9
9	Cyanoethylcellulose, $\gamma = 2.6$	Acetone	330	11	250	8.3	5.4	4.7
10	Cellulose nitrate, $\gamma = 2.7$	Ethyl acetate	300	22	254	18.6	5.2	4.7
11	Cellulose nitrate, $\gamma = 2.4$	Ethyl acetate	260	12	236	11	4.8	4.5
12	Cellulose nitrate, $\gamma = 2.3$	Ethyl acetate	200	10	160	8	4.2	3.4
13	Cellulose nitrate, $\gamma = 2.0$	Ethyl acetate	130	12	102	9.5	3.4	3.0
14	Cellulose nitrate, $\gamma = 1.9$	Methyl ethyl ketone	150	14	113	10.5	3.6	3.2
15	Cellulose nitrate, $\gamma = 1.14$	Dimethylacetamide + 6% LiCl	150	12	106	8.5	3.6	3.0
16	Cellulose, $\gamma = 0$	Cadoxene	97	10	76	8.0	3.0	2.6

Note. Polymer samples in Table 4.9 are the same as in Table 4.8 and have identical numbers.

eters A_η and the diameters d. This can be seen in Figs. 4.34–4.36, which show the dependences of $M/[\eta]$ on $M^{1/2}$ for some cellulose ethers and esters.

Curves 1 and 2 in Fig. 4.37 represent the dependence of $M/[\eta]$ on $M^{1/2}$ for fractions of two cellulose nitrate samples differing in degree of substitution and, correspondingly, in chain rigidity. It is clearly seen that for a more flexible polymer (curve 2) some points fall on the linear portion of the curve, thus permitting a linear extrapolation to $M^{1/2} \to 0$. For a more rigid sample (curve 1) in the same M range (which corresponds to lower L/A ratios than those for sample 2) all the points fit a curve which has no linear portion, and this prevents carrying out linear extrapolation of data to the range $M^{1/2} \to 0$.

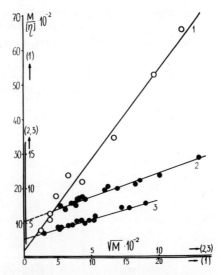

Fig. 4.33. Dependence of $M/[\eta]$ on $M^{1/2}$ for fractions of cellulose esters and ethers. 1) Cellulose monophenylacetate in benzene (No. 3 in Table 4.8); 2) cellulose butyrate in methyl ethyl ketone (No. 1 in Table 4.8); 3) cellulose nitrate in ethyl acetate (No. 13 in Table 4.8).

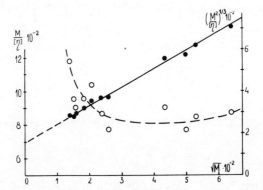

Fig. 4.34. Dependence of $M/[\eta]$ on $M^{1/2}$ (open circles) and that of $(M^2/[\eta])^{1/3}$ on $M^{1/2}$ (filled circles) for fractions of cellulose carbanilate in ethyl acetate (No. 4 in Table 4.8).

In these cases, in order to use experimental viscometric data for the determination of the parameters A_η and d_η it is possible to apply the plot of the dependence of $(M^2/[\eta])^{1/3}$ on $M^{1/2}$ which, according to Eq. (4.13), should be represented by a straight line over a wide molecular-weight range. These plots are shown in Figs. 4.34–4.37. They indicate that in all cases the experimental dependences of $(M^2/[\eta])^{1/3}$ on $M^{1/2}$ for cellulose ethers and esters are actually linear over the

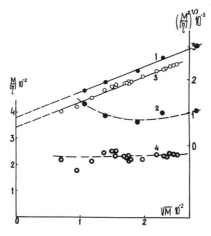

Fig. 4.35. Dependences of $M/[\eta]$ (curves 2 and 4) and $(M^2/[\eta])^{1/3}$ (curves 1 and 3) on $M^{1/2}$ for fractions of ethyl cellulose in acetone (curves 1 and 2, No. 8 in Table 4.8) and ethyl cellulose in ethyl acetate (curves 3 and 4, No. 7 in Table 4.8).

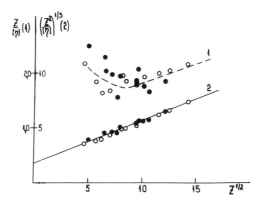

Fig. 4.36. Dependence of $Z/[\eta]$ on $Z^{1/2}$ (curve 1) and $(Z^2/[\eta])^{1/3}$ on $Z^{1/2}$ (curve 2) for fractions of cellulose nitrate in ethyl acetate (No. 13 in Table 4.8). Z is the degree of polymerization. Open and filled circles refer to samples differing in the degrees of substitution by 7%.

entire molecular-weight range investigated. The slopes of the straight lines plotted in these figures and their ordinate intercepts may be used to determine the values of A and d for the molecules being investigated according to Eq.(4.13). Evidently these values of A and d are those of A_D and d_D since the value of $P_\infty = 5.11$ and the parameter A_0 determined for the corresponding polymer–solvent system are used in Eq. (4.13).

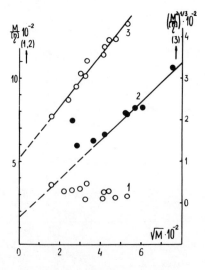

Fig. 4.37. Dependence of $M/[\eta]$ on $M^{1/2}$ (curves 1 and 2) and $(M^2/[\eta])^{1/3}$ on $M^{1/2}$ (curve 3) for fractions of cellulose nitrates. 1, 3) Sample with degree of substitution $\gamma = 2.4$ in ethyl acetate (No. 11 in Table 4.8); 2) sample with degree of substitution $\gamma = 1.14$ in dimethyl acetamide + 6% LiCl (No. 15 in Table 4.8).

On the other hand, if Eq. (3.68) is applied, Eq. (4.13) becomes

$$(M^2/[\eta])^{1/3} = \Phi_\infty^{-1/3}(M_L/A)^{1/2} M^{1/2} + (k/4.65A_0)(M_L/3\pi)[\ln(A/d) - 1.056]. \qquad (4.18)$$

In this form Eq. (4.13) may be used to interpret the experimental data concerning the dependence of $(M^2/[\eta])^{1/3}$ on $M^{1/2}$.

In this case, by using the slope of the straight line representing this dependence

$$\partial\{(M^2/[\eta])^{1/3}\}/\partial(M^{1/2}) = (M_0/\lambda A)^{1/2}/\Phi_\infty^{1/3}, \qquad (4.19)$$

and the theoretical value of $\Phi_\infty = 2.87 \cdot 10^{23}$ mole^{-1} it is possible to determine the parameter of equilibrium flexibility A_η.

The ordinate intercept $[(M^2/[\eta])^{1/3}]_0$ of the straight line, the value of A_η, and that of A_0 known for a given polymer–solvent system may be used to determine the chain diameter d_η from the equation

$$\ln(A/d) = 1.056 + (4.65A_0/k)(3\pi\lambda/M_0)[(M^2/[\eta])^{1/3}]_0. \qquad (4.20)$$

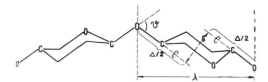

Fig. 4.38. Conformation of cellulose chain; $\lambda = 5.15 \cdot 10^{-8}$ cm is the projection of a monomer unit on the chain direction, $\Delta/2 = 2.65 \cdot 10^{-8}$ cm is the length of the effective unit about which rotation is possible, and $\delta = 1.42 \cdot 10^{-8}$ cm is the length of the effective unit about which rotation is impossible.

It can be seen that when the experimental data on M and $[\eta]$ are used, the length of the segment A_η determined according to Eq. (4.19) will differ from that of the segment A_D determined according to the slope of the straight line in Eq. (4.13). Their ratio is given by

$$A_\eta / A_D = (A_0 / A_\infty)^2, \qquad (4.21)$$

where $A_\infty = 3.8 \cdot 10^{-10}$ g cm^2 s^{-2} deg^{-1} mole^{-1} is the "theoretical" value of the hydrodynamic parameter (Table 3.3) and A_0 is its experimental value for a given polymer–solvent system.

The values of A_η and d_η obtained for cellulose ethers and esters using Eqs. (4.19) and (4.20) are given in Table 4.9.

6.3. Equilibrium Rigidity and Chain Structure

The data listed in Table 4.9 show that with very few exceptions the values of hydrodynamic diameter of chains fall in the range $(10-20) \cdot 10^{-8}$ cm. Taking into account the great error in the determinations of this value, it might be assumed that these values of d are in reasonable agreement with the geometric properties of cellulose ether and ester molecules.

Although the lengths of the Kuhn segments obtained by hydrodynamic methods and listed in Table 4.9 are different for different cellulose esters and ethers, they are close to each other in order of magnitude and higher by one order than those typical of flexible-chain polymers.

This result shows again that the high rigidity of molecules of various cellulose esters and ethers is their general property, determined mainly by the structure of the polyglucoside chain. There are no reasons for regarding this property as a specific feature of only some (nitrate and carbanilate) cellulose esters and ethers due to the peculiarities of their structure, as is sometimes done in the literature (see, e.g., [94]).

In those few cases when it is possible to compare the length of the Kuhn segment obtained by hydrodynamic methods with that determined from light scattering (see Chapter 1), they are found to be in satisfactory agreement. Moreover, it may be concluded that the values of A_D are slightly preferable to those of A_η and, hence, that of $P_\infty = 5.11$ is in better agreement with experimental data than $\Phi_\infty = 2.87 \cdot 10^{23}$ mole^{-1}.

Comparison of the A values for cellulose nitrates having different values of γ might suggest that the rigidity of nitroether molecules in ethyl acetate increases with increasing degree of substitution, as has already been reported [94–96]. However, these comparison should not be carried out using different solvents because it is well known [94] that the molecular sizes of the same cellulose esters and ethers may greatly differ in different solvents.

The chain flexibility of cellulose and its derivatives arises from a more or less hindered rotation of glucose rings about two bridge bonds, $O–C_1$ and $O–C_4$, between the neighboring rings.

The conformation of a part of the cellulose chain including the identity period is shown in Fig. 4.38 in its classical form. A real chain may be replaced by an equivalent chain with each unit consisting of two parallel bonds $\Delta/2$ about which rotation is possible, and one δ bond (normal to the former two), about which rotation is impossible. For this chain, if it is sufficiently long to be a Gaussian chain, and if rotation is completely free, the mean-square end-to-end distance $\langle h^2 \rangle_f$ is given by [98]

$$\langle h^2 \rangle_f = Z \left[\delta^2 + \Delta^2 (1 + \cos \vartheta)/(1 - \cos \vartheta) \right], \tag{4.22}$$

where $\vartheta = \pi - \theta$, θ is the bond angle at the bridge oxygen, and Z is the number of glucose rings in the chain (degree of polymerization).

The projection λ of the monomer unit on the chain direction (half identity period) is evidently given by

$$\lambda = \delta \sin (\vartheta/2) + \Delta \cos (\vartheta/2) \tag{4.23}$$

to give the length of the Kuhn segment for unhindered rotation

$$A_f = \langle h^2 \rangle_f / Z\lambda = [\delta^2 + \Delta^2 (1 + \cos \vartheta)/(1 - \cos \vartheta)]/[\delta \sin (\vartheta/2) + \Delta \cos (\vartheta/2)]. \tag{4.24}$$

Correspondingly, the number of monomer units in a segment for free rotation is given by

$$\begin{aligned} S_f = A_f/\lambda &= [\delta^2 + \Delta^2 (1 + \cos \vartheta)/(1 - \cos \vartheta)]/[\delta \sin (\vartheta/2) + \Delta \cos (\vartheta/2)]^2 \\ &= [(\delta/\Delta)^2 + (1 + \cos \vartheta)/(1 - \cos \vartheta)]/[\cos (\vartheta/2) + (\delta/\Delta) \sin (\vartheta/2)]^2. \end{aligned} \tag{4.25}$$

For a cellulose chain the identity period 2λ is $10.3 \cdot 10^{-8}$ cm and the distance between the neighboring oxygen bridging atoms O_1 and O_4 is $5.5 \cdot 10^{-8}$ cm [94]. These values correspond to $\Delta = 5.31 \cdot 10^{-8}$ cm, $\delta = 1.45 \cdot 10^{-8}$ cm, and $\vartheta = 71.5°$. Using these values, we obtain, according to Eqs. (4.24) and (4.25), the values of $s_f = 2.13$ and $A_f = 11 \cdot 10^{-8}$ cm for the chains of cellulose and cellulose ethers and esters.

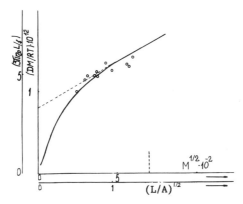

Fig. 4.39. Dependence of DM/RT on $M^{1/2}$ for fractions of poly(γ-benzyl-L-glutamate) in dimethylformamide [5]. The solid curve is the theoretical dependence plotted according to Eqs. (2.130) and (2.132) at $M_L = 109 \cdot 10^8$ daltons/cm, $M_0 = 218$, $\lambda = 2 \cdot 10^{-8}$ cm, $A = 2000 \cdot 10^{-8}$ cm, and $d = (15-20) \cdot 10^{-8}$ cm. The points represent experimental data. The point at which $L/A = 2.27$ is marked on the abscissa.

Comparison of this value of A_f with the experimental data given in Table 4.9 allows the determination of the degree of hindrance to rotation $\sigma = (A/A_f)^{1/2}$ in the chains of cellulose derivatives. The values of σ are listed in the two last columns of Table 4.9.

For cellulose derivatives the values of σ greatly exceed not only those characteristic of typical flexible-chain polymers [83] but also those σ for comb-shaped molecules (which follow from the data given in Tables 4.3 and 4.4). This fact shows that the interaction between the side groups of a polyglucoside chain hindering the rotation about valence bonds of the main chain does not reduce to common steric hindrance as in the case of most flexible-chain polymers. Evidently, a great role in this interaction is played by hydrogen bonds which cyclize the chain to a great extent and make its structure a more or less ladder-type structure.

This conclusion is supported by a drastic change in the size and hydrodynamic characteristics of molecules of cellulose derivatives in solution when one solvent is replaced by another, as well as by high negative temperature coefficients of molecular size and intrinsic viscosity (Table 4.2). Both a change in solvent strength and an increase in temperature lead to the breaking of intramolecular hydrogen bonds, thus decreasing the potentials hindering internal rotation in a polyglucoside chain.

7. POLYPEPTIDES

Synthetic polypeptides are a classical example of polymer molecules whose rigid regular conformation is retained by strong intramolecular hydrogen bonds

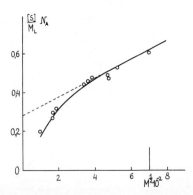

Fig. 4.40. Dependence of $[s]N_A/M_L \equiv L/f(1 - \bar{v}\rho)$ on $M^{1/2}$ for PBLG fractions in dimethylformamide [113]. The curve represents the theoretical dependence plotted according to Eqs. (2.130) and (2.132) at $M_0 = 218$, $\lambda = 1.5 \cdot 10^{-8}$ cm, $d = 30 \cdot 10^{-8}$ cm, and $A = 2000 \cdot 10^{-8}$ cm.

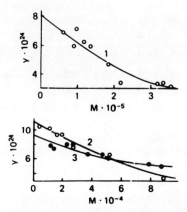

Fig. 4.41. Dependence of y [left side of Eq. (4.17)] on M. 1) PBLG fractions in dimethylformamide [80]. The values of λ and d used in plotting y are the same as those shown in Fig. 4.39. 2) Poly(butyl isocyanate) fractions in carbon tetrachloride [80]. The λ/d value is taken to be 0.1. 3) Poly(chlorohexyl isocyanate) fractions in carbon tetrachloride [80]. The λ/d value is taken to be 0.1.

formed between oxygen and hydrogen atoms of different amide groups –CONH– forming a segment of the polypeptide chain:

$$
\begin{array}{cc}
H & R \\
| & | \\
-N-C-C- \\
\parallel \ | \\
O \ \ H
\end{array}
$$

Hydrogen bonds lead to the formation of a molecule with a Pauling–Corie rigid helical structure [99] which is stable both in the polymer bulk and in some

TABLE 4.10. Length of the Kuhn Segment and Diameter d of the Chain of Some Polypeptides, Polyisocyanates, and DNA according to the Data of Translational Friction (A_D and d_D) and Viscometry (A_η and d_η)

Polymer	Monomer unit	Solvent	$A_D \cdot 10^8$, cm	$d_D \cdot 10^8$, cm	$A_\eta \cdot 10^8$, cm	$d_\eta \cdot 10^8$, cm	References
Poly(γ-benzyl-L-glutamate)	(structure shown)	Dimethylformamide	2000 2000	15 30	1640 —	20 —	[5, 80] [113]
Poly(butyl isocyanate)	(structure shown)	Tetrachloromethane	1000	9	900 800	— 14	[1, 80] [114]
Poly(chlorohexyl isocyanate)	(structure shown)	Tetrachloromethane	420	13	580	—	[3, 80]
Poly(tolyl isocyanate)	(structure shown)	Acetone	16	—	13	—	[115]
DNA	Mononucleotide	Water, 0.2 M NaCl	900	30	—	—	[117]

"helixogenic" solvents but is destroyed in strong solvents capable of breaking intramolecular hydrogen bonds.

Although many papers deal with the investigation of conformational characteristics of polypeptides in dilute solutions (see, e.g., [100, 112]), quantitative data on the equilibrium rigidity of their molecules are not numerous. The most complete information about the morphology of molecules has been obtained for poly(γ-benzyl-L-glutamate) (PBLG), a synthetic polypeptide with side radical R exhibiting the structure $(CH_2)_2$–$COCH_2$–C_6H_5.

Very high exponents a in the equations of viscosity and molecular weight obtained already in the pioneer papers of Doty et al. have suggested the "rodlike" shape of the molecules of synthetic polypeptides in helixogenic solvents [101, 102].

More recently, the study of dielectric]103, 104] and hydrodynamic [105] properties of dilute solutions of polypeptides as a function of their molecular weight has shown that the helical structures of their molecules are not absolutely rigid, and with increasing M the conformation of molecules deviates from the linear conformation.

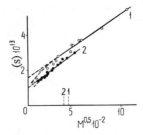

Fig. 4.42. Sedimentation coefficient [*s*] versus $M^{1/2}$ for poly(butyl isocyanate) [1] (1) and poly(chlorohexyl isocyanate) [3] (2) fractions in tetrachloromethane. Broken line intersecting the abscissa corresponds to $L/A = 2.27$.

However, in order to determine this deviation more or less reliably and thus to evaluate the equilibrium rigidity of the chain with the aid of hydrodynamic methods, it is necessary to use a series of polymer samples or fractions over a relatively wide molecular-weight range. This requirement is determined by the fact that the rigidity of the polypeptide chain in the helical conformation is high and the deviation of its shape and hydrodynamic properties from those of a straight rod is noticeable only at relatively high *M* values.

Figure 4.39 shows the dependence of *MD* on $M^{1/2}$ for fractions of poly(γ-benzyl-L-glutamate) (PBLG) in dimethylformamide (a helixogenic solvent) [5].

Although the points are greatly scattered (as a result of the difficulty of measuring low sedimentation coefficients s_0), they fall near the theoretical curve, and this makes it possible to evaluate the rigidity of the PBLG chain $A = 2000$ Å. This value is close to that obtained by diffusion in another paper [105], and is the highest of all the values known for various rigid-chain molecules.

Although the experimental points in Fig. 4.39 are located in the molecular-weight range up to $4 \cdot 10^5$, all of them correspond to values of *L/A* less than 2.27. However, even in this range of *L/A* the experimental dependence of *M/f* on $M^{1/2}$ may be partly approximated by a straight line, which indicates that Eq. (2.134) is "universal." This has already been mentioned (Chapter 4, Section 5.2.1.1).

Figure 4.40 shows similar data for PBLG obtained from sedimentation–diffusion measurements in a more recent work [113] covering a wider molecular-weight range (from $1 \cdot 10^4$ to $5 \cdot 10^5$). Comparison of experimental and theoretical data leads to the same value of the rigidity parameter *A* as in [5], but at somewhat different values of structural chain parameters λ and *d*. These differences may be caused not only by natural experimental errors but also by actually different conformational states (such as the degree of despiralization) of the molecules studied in these two papers.

The rigidity of the PBLG chain may also be evaluated from viscometric data using the plot of the function $y = \lambda^3 f(L/A)$ according to Eq. (4.17). This function plotted at the value of $\lambda/d = 0.1$ is shown in Fig. 4.41 (curve 1) and this plot leads to the segment length $A = 1640 \cdot 10^{-8}$ cm. This value agrees within experimental error with the value of A obtained from translational friction (Fig. 4.39).

It should be noted that the precision of the determination of equilibrium rigidity of PBLG molecules from their hydrodynamic characteristics (Table 4.10) is relatively low. This is due not only to specific experimental difficulties (such as a weak dependence of s_0 on M) but even to a greater extent to the necessity of using the theoretical dependences for a weakly bending rod determined by Eqs. (2.130) and (2.135) valid in the range of L/A less than 2.27. The important role played in these equations by the chain diameter, which may be evaluated only with great error, is an unfavorable factor.

8. POLYISOCYANATES

Polyisocyanates $\left(\begin{array}{cc} O & R \\ \parallel & \mid \\ -C-N- \end{array} \right)$ are polyamides whose chain consists entirely of amide groups (nylon 1) [106, 107]. Moreover, in these groups the hydrogen atom at the nitrogen atom is replaced by an aliphatic (alkyl isocyanate) or aromatic (phenyl or tolyl isocyanate) radical R. This point is of great importance because it rules out the possibility of the formation of intramolecular hydrogen bonds and rigid ordered secondary structures, in contrast to polypeptide molecules.

The second feature of polyisocyanate structure is also of great importance: the amide groups are not separated along the chain by methylene or other groups as is the case with common polyamides. As a result, great hindrance to rotation about the quasi-conjugated C–N bond of the amide group becomes characteristic of *all* the bonds of the main polyisocyanate chain, which should lead to high rigidity in their molecules. This has been confirmed even in the initial investigations of the properties of solutions of poly(alkyl isocyanate)s highly soluble in nonpolar solvents. Measurements have shown the high intrinsic viscosity of their solutions [106–109] and tremendously high dipole moments of their molecules [110]. This has been regarded as an indication of the rodlike shape of these molecules.

At the same time it has been found that the dipole moments of poly(alkyl isocyanate) molecules reflect monomer unit decrease with molecular weight [110, 111]. In early investigations this fact was interpreted by the hypothesis of random "head-to-head" and "head-to-tail" monomer attachments [110]. However, it was soon shown [111] that actually the change in both dielectric and hydrodynamic properties of poly(alkyl isocyanate) solutions as a function of molecular weight is an indication of their chain flexibility, and a quantitative description of these phenomena is possible using a wormlike chain model. This concept is generally adopted at present.

Some results of investigations of conformational properties of poly(alkyl isocyanate)s by the light-scattering method have been reported in Chapter 1, Section 6.2.2. Here we will consider their hydrodynamic properties.

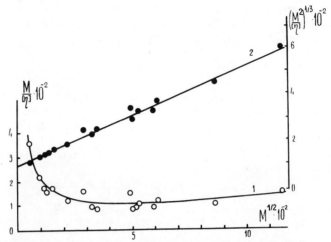

Fig. 4.43. Dependence of $M/[\eta]$ (1) and $(M^2/[\eta])^{1/3}$ on $M^{1/2}$ for poly(butyl isocyanate) fractions in tetrachloromethane according to the data in [1].

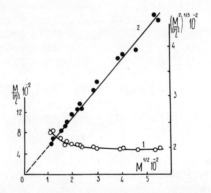

Fig. 4.44. Dependence of $M/[\eta]$ (1) and $(M^2/[\eta])^{1/3}$ on $M^{1/2}$ for poly(chlorohexyl isocyanate) fractions in tetrachloromethane according to the data in [3].

8.1. Translational Friction, Viscosity, and Chain Rigidity

8.1.1. Poly(alkyl isocyanate)s. In contrast to polypeptides, the conformational properties and chain rigidity of poly(alkyl isocyanate)s are not determined by intramolecular hydrogen bonds, but rather by the primary chemical structure: the rigidity of valence bonds and bond angles and conjugation, i.e., the resonance interaction in the chain. This fact ensures a greater stability in the conformational characteristics of poly(alkyl isocyanate)s in solution than polypeptides and provides more favorable conditions for their quantitative investigations.

Another important feature is the fact that the chemical synthesis of polyiso-cyanates makes it possible to obtain polymers of higher molecular weight than in the case of polypeptides. As a result, polymer molecules with a much greater reduced chain length L/A than that of polypeptides are available to the experimenter. This allows the investigation of hydrodynamic properties of polyisocyanates over a wide range of L/A ratios in which chain conformation varies from a slightly bending rod to a random coil.

Hence, in order to determine the rigidity parameter A and the chain diameter d according to the diffusion–sedimentation data, the canonical plots based on Eqs. (2.134) or (4.7) may be used. These plots are shown in Fig. 4.42 for poly(butyl isocyanate) (R: $-C_4H_9$) and poly(chlorohexyl isocyanate) (R: $-C_6H_{11}Cl$). It can be clearly seen that experiments provide a sufficient number of data (points) in the range of L/A greater than 2.27 for plotting a linear dependence of the sedimentation constant [s] on $M^{1/2}$. The values of A_D and d_D obtained from the slopes and intercepts of these straight lines (at $\lambda = 2\cdot10^{-8}$ cm) are given in Table 4.10.

However, according to viscometric data, the molecular-weight range at high M is insufficient for obtaining a linear dependence of $M/[\eta]$ on $M^{1/2}$ [according to Eqs. (2.145) and (4.8)] as is shown in Figs. 4.43 and 4.44. Hence, in discussing viscometric data in the range of L/A greater than 2.27, the plot of $(M^2/[\eta])^{1/3}$ vs. $M^{1/2}$ is used. According to Eq. (4.18), this plot gives a linear dependence, also shown in Figs. 4.43 and 4.44. The values of A_η and d_η calculated from the slopes and intercepts of these straight lines (at $\Phi_\infty = 2.87\cdot10^{23}$ mole^{-1}) are given in Table 4.10.

The rigidity of poly(alkyl isocyanate) chains may also be evaluated from vis-cometric data obtained in the range of L/A less than 2.27 by plotting the dependence of the value of y on M according to Eq. (4.17). This plot, shown in Fig. 4.41 for poly(butyl and chlorohexyl isocyanate) [80], leads to the Kuhn segment lengths listed in Table 4.10.

The data given in Table 4.10 show that, although the rigidity of poly(alkyl isocyanate) chains determined by the segment length A is less than that of helical polypeptides of the PBLG type, it exceeds that of all other known polymers whose flexibility is to some extent determined by rotation about the valence bonds of the chain. This result is not surprising if one takes into account that the flexibility mechanism of poly(alkyl isocyanate)s is of a specific nature and should be due considerably to the deformation of the planar trans- or cis-structure of the amide groups in the chain.

A marked difference in the rigidity of poly(butyl-) and poly(chlorohexyl iso-cyanate)s differing only in the structure of the side groups suggests that during polymerization a negligible amount of "defective" units may be formed having a polyacetal structure $\left(\begin{array}{c} -C-O- \\ \parallel \\ N-R \end{array} \right)$ and hence being an additional source of chain flexibility.

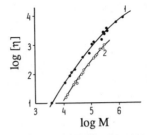

Fig. 4.45. Experimental dependence of $\log[\eta]$ on $\log M$ for poly(butyl isocyanate) (1) and poly(chlorohexyl isocyanate) fractions (2) in tetrachloromethane.

8.1.2. Aromatic Polyisocyanates. If an alkyl side radical R is replaced by an aromatic radical, the properties of polyisocyanate molecules undergo a drastic change. Poly(phenyl isocyanate) is insoluble not only in nonpolar but also in many polar solvents. Poly(tolyl isocyanate) (R: $-C_6H_4-CH_3$) was found to be soluble in many organic solvents, but in some nonpolar solvents (dioxane) a strong tendency of its molecules to form supermolecular structures and gels has been detected. Consequently, quantitative investigations of the molecular properties of these polymers have been carried out using polar solvents – bromoform [111] and acetone [115] – in which these phenomena were absent.

The investigation of hydrodynamic properties of a series of poly(tolyl isocyanate) fractions in acetone has shown that these properties differ markedly from those observed in solutions of poly(alkyl isocyanate)s.

For poly(tolyl isocyanate) in acetone the dependence of $[\eta]$, S, and s_0 on molecular weight may be described by the Mark–Kuhn equatins (4.4) containing coefficients constant over a wide molecular-weight range and equal to $K_\eta = 1.7 \cdot 10^{-3}$ g^{-1}, $K_D = 1.8 \cdot 10^{-3}$ cm^2 s^{-1}, $K_s = 2.2 \cdot 10^{-14}$ s, $a = 0.81$, and $b = 0.63$.

Both the values of the exponents a and b in poly(tolyl isocyanate) solutions, which are much lower than for poly(alkyl isocyanate) solutions, as well as the absolute values of $[\eta]$, which are lower by *one order of magnitude* (at equal M), show that poly(tolyl isocyanate) molecules in acetone behave as typical flexible-chain polymers. In this case, the deviation of the exponents a and b from 0.5 is not due to the draining of the molecules but, rather, to the excluded-volume effects which should be taken into account in the evaluation of chain rigidity according to hydrodynamic data.

In accordance with this, the plotting of the dependences of $1/DM^{1/2}$ and $[\eta]/M^{1/2}$ on $M^{1/2}$ according to Eqs. (2.148) and (2.149) allows for the determination of unperturbed dimensions and rigidity parameters A_D and $A\eta$ for poly(tolyl isocyanate) molecules. The values of A_D and $A\eta$ given in Table 4.10 are characteristic of typical flexible-chain polymers.

Hence, although the chemical structures of the chains of poly(alkyl and tolyl isocyanate)s are similar, their conformational characteristics and chain rigidity differ

TABLE 4.11. Values of Exponents in the Mark–Kuhn Equation Calculated [114] from the Theories of Intrinsic Viscosity of Wormlike Spherocylinders according to Eqs. (2.135), (2.137), and (4.26)

L/A	d/A						
	$5\cdot10^{-4}$	10^{-3}	$5\cdot10^{-3}$	10^{-2}	$3\cdot10^{-2}$	$6\cdot10^{-2}$	10^{-1}
10^6	0.508	0.507	0.505	0.504	0.503	0.502	0.502
10^5	0.523	0.521	0.516	0.514	0.511	0.507	0.505
10^4	0.566	0.560	0.546	0.541	0.532	0.507	0.515
10^3	0.656	0.645	0.618	0.605	0.583	0.564	0.546
10^2	0.780	0.770	0.736	0.716	0.678	0.656	0.630
70	0.802	0.792	0.759	0.738	0.697	0.677	0.651
50	0.825	0.815	0.803	0.761	0.718	0.699	0.674
20	0.911	0.901	0.869	0.847	0.799	0.774	0.754
10	1.017	1.006	0.972	0.951	0.901	0.856	0.830
7	1.088	1.076	1.039	1.019	0.969	0.908	0.871
5	1.160	1.147	1.108	1.088	1.037	0.962	0.908
2.278	1.292	1.280	1.235	1.202	1.131	1.075	0.938
2	1.407	1.390	1.334	1.295	1.202	1.111	1.021
1	1.527	1.507	1.432	1.379	1.248	1.119	0.990
0.6	1.618	1.594	1.500	1.432	1.263	1.096	0.932
0.4	1.676	1.647	1.534	1.452	1.244	1.040	0.856
0.2	1.723	1.684	1.527	1.410	1.116	0.865	0.715
0.1	1.720	1.667	1.446	1.279	0.901	0.713	
0.05	1.685	1.611	1.296	1.068	0.730		
0.03	1.639	1.544	1.140	0.888			
0.01	1.477	1.309	0.782				
0.005	1.311	1.083					
0.001	0.785						

Note. The maximum values of a attained in a homologous series of molecules (constant d/A) upon variation of their reduced length L/A are boxed in.

markedly. The reason for this difference may be the weakening of the resonance interaction between the nitrogen atoms and the carbonyl of the amide groups in poly(tolyl isocyanate) molecules due to the presence of aromatic rings in these polymers. This suggestion is confirmed by IR spectroscopy [116].

8.2. Applicability of the Mark–Kuhn Equations

As for all other rigid-chain polymers, for poly(alkyl isocyanate) solutions the dependence of intrinsic viscosity [η] and diffusion coefficient D on M may be only approximately expressed by Mark–Kuhn equations (4.4) with constant values of K_η, K_D, a, and b over a relatively narrow range of M.

In fact, for poly(alkyl isocyanate) solutions the changes in a or b in Eqs. (4.4) with M are even more pronounced than, for example, for ladder siloxanes and cellulose derivatives because their molecules are more rigid. These changes are easily observed by plotting the experimental dependence of log [η] on log M. This can be seen in Fig. 4.45, which shows this dependence for fractions of poly(butyl and chlorohexyl isocyanate) in tetrachloromethane. The decrease in the slope of the curves shown in this figure, and the corresponding decrease in the exponent a in Eq. (4.4) with increasing M, reflect the change in chain conformation from a bending rod to a random coil with increasing chain length L.

The type of the dependence of exponents a or b on the structural (L and d) and rigidity (A) characteristics of the chain may be predicted quantitatively if the chain is represented by a wormlike spherocylinder, and the theoretical equations describing the hydrodynamic properties of this model are used.

Thus, in the case of viscosity, the exponent a is found from the equation

$$a = (M/[\eta])\,(d\,[\eta]/dM),\qquad(4.26)$$

where $[\eta]$ as a function of M, L, d, and A is determined according to Eqs. (2.135) and (2.137) over the entire possible range of L/A ratios.

The exponents a calculated in this manner at various values of reduced diameter d/A and reduced length L/A of the wormlike chain are listed in Table 4.11 [114]. At high L/A ratios the exponent a is close to 0.5, which corresponds to the hydrodynamics of a nondraining Gaussian coil. In a homologous series of fractions of one polymer (constant d/A value) the exponent a increases with decreasing chain length L/A in accordance with a decrease in the degree of chain coiling and an increase in draining. The maximum value of a is attained in the range of L/A ratios in which the hydrodynamic properties of molecules of this series are closest to those of a thin rod. As the chain length decreases further, the exponent a also decreases, because, in this case, the d/L ratio increases and the hydrodynamic properties of the molecule become increasingly different from those of a thin rod. The lower the d/A ratio, i.e., the stiffer the molecules of the homologous series, the lower are the values of L/A at which the maximum value of a is attained and the higher is this maximum value. In Table 4.11 the maximum values of a for each homologous series are boxed in.

The data listed in Table 4.11 are illustrated in Fig. 4.46, in which the dependence of the exponent a on the reduced chain length L/A is shown for three homologous series of polymers differing in the degree of rigidity characterized by the values of d/A. It can be seen that for a wormlike spherocylinder the theoretical dependence of log $[\eta]$ on log M is represented by an S-shaped curve with inflection point corresponding to the maximum value of a (Fig. 4.47).

Figure 4.46 and Table 4.11 show that a maximum value of a equal to 1.7 and higher might be expected only for polymers whose rigidity is characterized by a parameter A/d less than 10^3. Moreover, for these polymers $a = 1.7$ is attained in the range of L/A less than 0.2. However, this very high value of A/d is not actually observed for any known polymers, even for those exhibiting the highest equilibrium rigidity of the chain. Thus, for poly(γ-benzyl-L-glutamate) in the helical conformation we have $A/d \approx 10^2$ and for native DNA $A/d \approx 30$ (Table 4.10). Hence, the statement that for real rodlike molecules the value a is 1.7 and higher, which is sometimes encountered, is not in agreement with modern theory.

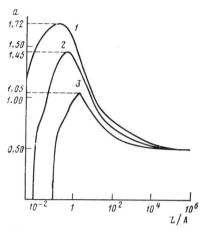

Fig. 4.46. Theoretical dependences of exponent a in the Mark–Kuhn equation on reduced length L/A of a wormlike spherocylinder at various values of reduced diameter d/A: 1) $5 \cdot 10^{-4}$; 2) $1 \cdot 10^{-2}$; 3) $1 \cdot 10^{-1}$.

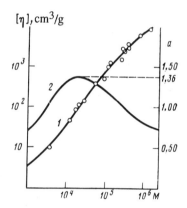

Fig. 4.47. Molecular-weight dependence of intrinsic viscosity $[\eta]$ (1) and exponent a (2) for poly(butyl isocyanate) fractions in tetrachloromethane. The points refer to experimental data [1]. The curves represent theoretical dependences in accordance with Eqs. (2.135) and (2.137) at $A = 800 \cdot 10^{-8}$ cm, $d = 14 \cdot 10^{-8}$ cm, and $M_L = 5.08 \cdot 10^9$ daltons/cm.

The theoretical dependence of $[\eta]$ on M is compared with the experimental data for poly(butyl isocyanate) (on a double logarithmic scale) in Fig. 4.47. The experimental points are in good agreement with the S-shaped theoretical curve plotted at $A = 800 \cdot 10^{-8}$ cm, $d = 14 \cdot 10^{-8}$ cm, and $\lambda = 2 \cdot 10^{-8}$ cm. In this case the dependence of the exponent a on M exhibits a maximum equal to 1.36.

For polymers with moderate rigidity, such as cellulose and its derivatives, for which the A/d value ranges from 15 to 30 and the L/A value varies from a few units to several tens, the theory predicts (according to Table 4.11) values of a close to unity, which agrees with the foregoing experimental data. This agreement may be regarded as another confirmation of the concept that for cellulose derivatives the high value of a is due to the effect of hydrodynamic interaction rather than to volume effects which are not of great importance for solutions of rigid-chain polymers, and hence are not taken into account in the theories considered here.

From the foregoing it may be concluded that, for rigid-chain polymers, the Mark–Kuhn equation with a constant value of a is a rough approximation valid only in a narrow range of M values. Moreover, the value of the exponent a is profoundly affected by the transverse dimension of the chain, and this fact should be taken into account in the analysis of experimental data. The Mark–Kuhn equation is a purely empirical dependence and cannot serve as a basis for far-reaching conclusions concerning the structure of the polymer investigated, although this is not always taken into account in its applications.

9. AROMATIC POLYAMIDES

Aromatic polyamides are of great practical importance in the preparation of strong thermally stable materials with a high modulus [38, 118, 119]. In contrast to polyisocyanates, in the molecules of aromatic polyamides the amide groups are separated along the chain by phenyl rings or other aromatic groups. As a result, single valence bonds appear about which rotation is possible. Hence, in principle it may be expected that the equilibrium flexibility of the chains of aromatic polyamides should exceed that of poly(alkyl isocyanate) chains. However, it may differ greatly for different representatives of this class of polymers, depending on the structure of the aromatic groups separating the amide groups and their type of insertion in the chain.

Although considerable progress has been attained in the synthesis of these polymers in recent years, the data on the hydrodynamic and conformational properties of their molecules in dilute solutions are very scarce. This is due, first, to the poor solubility of aromatic polyimides: most of them are soluble only in concentrated strong inorganic acids and only a few of these polymers may be dissolved in polar organic (often amide) solvents. In the latter few cases there is a favorable possibility of using transport methods to the full extent (including sedimentation) for the characterization of hydrodynamic and conformational properties of these polymers.

9.1. Poly(*meta*-phenylene isophthalamide) (PmPhIPhA)

9.1.1. Main Properties. PmPhIPhA is an aromatic polyamide with amide groups separated along the chain by phenyl rings in the meta position (Table 4.12). The monomer unit of PmPhIPhA has a relatively simple chemical structure. This facilitates the interpretation of experimental data on the size and rigidity of its molecules from the standpoint of conformational statistics of polymer chains.

PmPhIPhA readily dissolves (molecularly) not only in concentrated sulfuric acid but also in dimethylacetamide (DMA) (in particular, with the addition of LiCl)

TABLE 4.12. Coefficients K_η, K_D, K_S, a, and b in Mark–Kuhn Equations (4.4), Kuhn Segments A_D and A_η, and Chain Diameters d_D and d_η for Some Aromatic Polyamides and Polyesters according to the Data of Sedimentation, Diffusion, and Viscometry of Their Dilute Solutions*

Repeat unit of polymer	Range of $M \cdot 10^{-3}$	Solvent	$K_\eta \cdot 10^4$	$K_D \cdot 10^4$	$K_S \cdot 10^{15}$	a	b	$A_D \cdot 10^8$, cm	$d_D \cdot 10^8$, cm	$A_\eta \cdot 10^8$, cm	$d_\eta \cdot 10^8$, cm	References
PmPhIPhA	4—300	DMA + 3% LiCl	130	0.71	0.88	0.84	0.57	47	3.5	41	3.5	[122]
	20—250	96% H_2SO_4	113	—	—	0.84	—	—	—	45	—	[123]
PABI	1—10	DMA	50	2.8	—	1.11	0.70	—	—	—	—	[19]
		98% H_2SO_4	36	0.25	—	1.13	0.76	—	—	—	—	
	10—60	DMA	1400	1.6	2.7	0.77	0.64	260	5	130	5	
		98% H_2SO_4	1300	0.054	—	0.76	0.59	260	5	130	5	
PAH	3.5—20	DMS	10.7	1.87	—	1.25	0.714	—	—	—	—	[128—130]
	20—50	DMS	365	0.42	—	0.9	0.565	160	3.5	190	9	
PPBA	7—18	96% H_2SO_4	3.39	—	—	1.4	—	—	—	650	6	[136]
ACPE	2—25	TCE	353	0.828	—	0.83	0.61	160	13	160	13	[135]

$x_1 =$

$x_2 =$

*All data are given in cm, g, s.

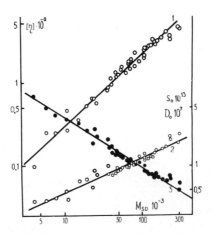

Fig. 4.48. Dependence of $[\eta]$ (1), s_0 (2), and D (3) on M_{sD} for PmPhIPhA fractions in dimethylacetamide + 3% LiCl [121, 122].

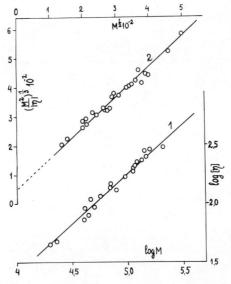

Fig. 4.49. Dependence of $\log[\eta]$ on $\log M$ (1) and $(M^2/[\eta])^{1/3}$ on $M^{1/2}$ for PmPhIPhA fractions in 96% sulfuric acid [123].

[120]. This facilitates its fractionation and makes it possible to study the hydrodynamic properties of its molecules with the aid of sedimentation–diffusion analysis and viscometry using relatively well-characterized fractions.

On the other hand, since PmPhIPhA can dissolve in strong organic acids without degradation, the hydrodynamic (and optical) properties of its molecules in these solvents may be compared to those of other polymers insoluble in organic

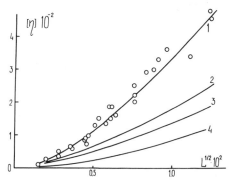

Fig. 4.50. Dependence of $[\eta]$ on $L^{1/2}$ for: 1) PmPhIPhA fractions in DMA + 3% LiCl [122] and for solutions of some flexible-chain polymers in good solvents; 2) polyethylene in trichlorobenzene ($a = 0.725$) [124]; 3) polystyrene in benzene ($a = 0.72$) [125]; 4) polymethylsiloxane in toluene ($a = 0.84$) [126].

solvents. This method of comparison is useful in the study of conformational characteristics of polymers soluble only in strong acids.

Owing to these features of PmPhIPhA, the hydrodynamic characteristics of its molecules have been studied in greater detail [120–123] than those of other aromatic polyamides.

The molecular-weight dependences of the hydrodynamic characteristics of PmPhIPhA in DMA and sulfuric acid are shown in Figs. 4.48 and 4.49. The dependences of $[\eta]$, D, and s_0 on M are represented on a logarithmic scale. This fact in itself indicates that PmPhIPhA chains are moderately rigid. The coefficients K_η, K_D, K_s, a, and b characterizing these dependences are given in Table 4.12.

9.1.2. Excluded-Volume and Draining Effects. The exponents a and b for PmPhIPhA solutions are not so high as to enable us to solve the problem of the relative contribution of draining and excluded-volume effects to the hydrodynamic behavior of molecules according to the difference between their values and 0.5 [as in the case of poly(alkyl isocyanate)s]. Therefore, an opinion exists in the literature that PmPhIPhA molecules in solutions are considered to be hydrodynamically nondraining coils. In this case the difference of the a and b exponents from 0.5 is ascribed to volume effects, and these effects are excluded [120] by using the extrapolation equations (2.148) and (2.149). However, a more careful consideration of the problem shows that this viewpoint is open to question.

Although the values of the exponent a and b for PmPhIPhA are close to those for flexible-chain polymers, the intrinsic viscosity of its solutions greatly exceeds that for typical flexible-chain polymers at equal chain length L. This can be seen in Fig. 4.50, which shows the dependence of $[\eta]$ on $L^{1/2}$ for PmPhIPhA in DMA (+ 3% of LiCl) and for some flexible-chain polymers in thermodynamically good solvents. The values of $[\eta]$ for these polymers have been calculated from the Mark–Kuhn equations reported in [124–126]. Higher values of $[\eta]$ (at equal L) for PmPhIPhA solutions than for flexible-chain polymers indicate that the conformation of their molecules is more expanded and the draining effects have a great influence on the hydrodynamic properties of these molecules. This is also confirmed by low values of the sedimentation coefficient s_0 at much higher M values [122].

The type of dependences of $[\eta]/M^{1/2}$ and $1/DM^{1/2}$ on $M^{1/2}$ [according to Eqs. (2.148) and (2.149)] for PmPhIPhA solutions lead to the same conclusion. These dependences (Fig. 4.51) differ greatly from those of polymers whose hydrodynamic characteristics are profoundly affected by excluded-volume effects as is shown, for example, in Fig. 4.23. The curves shown in Fig. 4.23 are concave upward in accordance with the fact that the significance of excluded volume effects leading to the "non-Gaussian" chain type increases with *increasing* molecular weight (chain length). In contrast, the curves in Fig. 4.51 are concave downward. This means that they illustrate the deviation from "Gaussian" properties of molecules which increases with their *decreasing* length. This behavior is observed only for those molecules whose hydrodynamic properties are determined by their skeleton rigidity and chain draining (increasing with decreasing M) rather than by volume effects. Accordingly, in principle the extrapolation of the curves plotted in Fig. 4.51 to the range of $M \to 0$ cannot lead to the "unperturbed dimensions" of the chain and cannot be used for the determination of its equilibrium rigidity.

Hence, the interpretation of experimental data on the hydrodynamic properties of PmPhIPhA molecules in the investigated M range should be carried out just as for other rigid-chain polymers using the equations of hydrodynamics of wormlike chains without taking into account excluded volume effects. The negative temperature dependence of $[\eta]$ for PmPhIPhA solutions (Table 4.2) may also serve as an indirect justification of this viewpoint.

9.1.3. Hydrodynamic Characteristics and Chain Rigidity.

Figure 4.52 (curve 1) shows the dependence of $\eta_0 MD/RT$ on $M^{1/2}$ for PmPhIPhA fractions in DMA (3% LiCl). Over the entire range of molecular weights up to $3 \cdot 10^5$ this dependence is represented by a canonical straight line in accordance with Eqs. (2.134) or (4.7), and shows no tendency to curving which might be expected if excluded-volume effects had a marked influence [Eq. (2.152)].

The slope of this straight line and its intercept with the ordinate have been used to calculate the length of the Kuhn segment A_D and the chain diameter d_D given in Table 4.12 according to Eqs. (4.9) and (4.10) at the values of $M_0/\lambda = M_L = 19.1 \cdot 10^8$ dalton cm^{-1} and $\eta_0 = 1.94 \cdot 10^{-2}$ poise.

Similar plots carried out according to Eq. (4.18) using viscometric data also lead to a linear dependence of $(M^2/[\eta])^{1/3}$ on $M^{1/2}$ and, according to Eqs. (4.19) and (4.20), make it possible to determine A_η and d_η [122], whose values are given in Table 4.12.

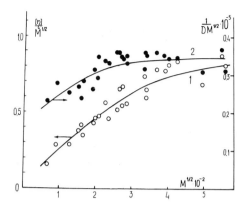

Fig. 4.51. Dependence of $[\eta]/M^{1/2}$ (1) and $1/DM^{1/2}$ (2) on $M^{1/2}$ for PmPhIPhA fractions in DMA (+3% LiCl) [122].

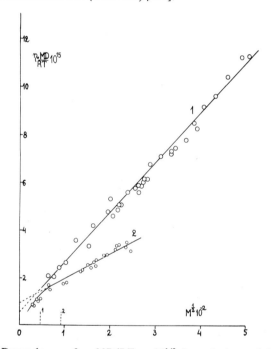

Fig. 4.52. Dependence of $\eta_0 MD/RT$ on $M^{1/2}$ for solutions of fractions of PmPhIPhA in DMA + 3% LiCl [122] (1) and PABI in DMA and in 98% sulfuric acid [19] (2). Broken lines on the abscissa show the points at which $L/A = 2.27$ for polymers 1 and 2, respectively.

The value of A_D for PmPhIPhA molecules in DMA exceeds that of A_η only slightly. If Eq. (4.21) is taken into account, this is explained by the fact that the experimental value of $A_0 = 3.55 \cdot 10^{-10}$ g cm^2 sec^{-2} deg^{-1} [122] for this polymer is close to its "theoretical" value $3.8 \cdot 10^{-10}$ (cgs).

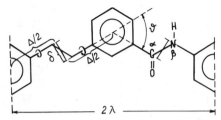

Fig. 4.53. Conformation of the extended chain of PmPhIPhA with the trans-structure of the amide group. λ is the projection of the monomer unit on the chain direction. Angles α and β at the carbon and nitrogen atoms of the amide group are assumed to be equal.

Although the length of the Kuhn segment for the PmPhIPhA molecules $A = (41\text{--}45)\cdot10^{-8}$ cm is relatively low, the A/d ratio, which is greater than 10 for this polymer, is considerably higher than for typical flexible-chain polymers and is close to that of such polymers as, for example, cellulose derivatives. This accounts for the fact that the draining effect typical of all rigid-chain polymers is distinctly indicated by the hydrodynamic behavior of the PmPhIPhA molecules.

It has been pointed out above that a comparative study of hydrodynamic characteristics of PmPhIPhA molecules in two solvents – DMA and sulfuric acid – is very useful. The data of viscometric measurements of solutions of PmPhIPhA in sulfuric acid (96%) are shown in Fig. 4.49 [123]. The polymer fractions used in these measurements have been previously studied in DMA and characterized according to their molecular weights M_{sD} by sedimentation and diffusion in this solvent. The values of the coefficients K_η and a obtained in sulfuric acid solutions according to curve 1 in Fig. 4.49 and listed in Table 4.12 are close to those of K_η and a obtained for PmPhIPhA solutions in DMA (3% LiCl). This shows that the PmPhIPhA molecular conformations in these two solvents are similar.

The plot of the dependence of $(M^2/[\eta])^{1/3}$ on $M^{1/2}$ for solutions in 96% sulfuric acid is straight line 2 (Fig. 4.49), whose slope determines, according to Eq. (4.19), the value of A_η given in Table 4.49. The intercept of curve 2 with the ordinate is not used for the determination of d_η according to Eq. (4.20) because the value of the parameter A_0 for the PmPhIPhA–sulfuric acid system is unknown. The values of A_η obtained for PmPhIPhA in DMA (3% LiCl) and sulfuric acid (96%) coincide within experimental error in accordance with similar molecular conformations of PmPhIPhA in these solvents.

It should be borne in mind that this coincidence may be absent if the concentration of sulfuric acid is changed because the size of aromatic polyamide molecules in sulfuric acid varies with the concentration of this acid and the higher the chain flexibility, the more pronounced is this effect [19].

9.1.4.   Chain Conformation and Rigidity. The experimentally determined length of the Kuhn segment A should be compared to the value of A which may be evaluated proceeding from the structure of the monomer unit of the PmPhIPhA chain. The part of this chain including the identity period (two "monomer" units) is shown in Fig. 4.53.

The amide groups in the chains of aromatic polyamides exhibit a planar trans-structure [127] excluding the possibility of rotation about the C–N bond. Hence, the main flexibility mechanism of the PmPhIPhA chain is the rotation about the meta-aromatic N–Ph and C–Ph bonds whose direction varies by the angle ϑ per distance of one monomer unit (length Δ) along the chain (length λ). Moreover, this moving along the chain is related to the displacement of the rotation axis in the direction normal to it by the distance δ. Hence each monomer unit of a real PmPhIPhA chain may be replaced by a combination of two parallel bonds $\Delta/2$ about which rotation takes place, and one bond δ normal to them about which no rotation occurs. In this case the directions of the $\Delta/2$ bonds in neighboring units differ by the angle ϑ. This "equivalent" chain is identical to an equivalent chain used for the characterization of the conformation of the cellulose chain (Fig. 4.38), and its statistical size and flexibility are determined by the same equations (4.22)–(4.25).

If the angles α and β at the carbon and nitrogen atoms of the amide group in the PmPhIPhA chain (Fig. 4.53) are assumed to be equal, then we have $\vartheta = 60°$, $\Delta = 6.45 \cdot 10^{-8}$ cm, and $\delta/\Delta = 0.2$. Hence, according to Eqs. (4.24) and (4.25) for the unhindered rotation in the chain we have $A_f = 20 \cdot 10^{-8}$ cm and $s_f = 3.3$. A comparison of this value with the experimental values of A_η and A_D given in Table 4.12 shows that the degree of hindrance to intramolecular rotation in the PmPhIPhA chain $\sigma = (A/A_f)^{1/2} \approx 1.5$. This value is not only many times lower than that of σ in the chains of cellulose derivatives (Table 4.9), but is also lower than those of σ characteristic of typical flexible-chain polymers [83]. Low hindrance to rotation in the chains of aromatic polyamides may be associated with their specific molecular structure. In these chains the neighboring rotating bonds are separated by aromatic rings and amide groups, which decrease steric hindrances and facilitate rotation in the chain.

Hence, the values of A_D and A_η obtained from experimental data using the wormlike chain theory, taking into account draining effects [Eqs. (4.7) and (4.18)], are in reasonable agreement with that of A_f evaluated from the data on the structure of the PmPhIPhA chain.

On the other hand, this agreement cannot be obtained if the experimental data on the hydrodynamic properties of PmPhIPhA are interpreted from the standpoint of the theory using the excluded volume effects and neglecting the draining effects. Thus, if the curves plotted in Fig. 4.51 are extrapolated to the range of $M \to 0$, it is possible to obtain $([\eta]/M^{1/2})_{M \to 0} = 0.05$ and $(DM^{1/2})_{M \to 0}^{-1} = 0.16 \cdot 10^5$. According to Eqs. (2.149) and (2.148), these values yield $A_\eta = 6 \cdot 10^{-8}$ cm and $A_D = 9 \cdot 10^{-8}$ cm. These values are lower by a factor of 2 or 3 than the value of $A_f = 20$ Å for these

216

CHAPTER 4

PmPhIPhA chains with free rotation. In other words, this result corresponds to the degree of hindrance $\sigma' \approx 0.5$, which has no physical sense.

9.2. Polyamidobenzimidazole [19]

An aromatic polyamidobenzimidazole (PABI) with the structure of the repeat unit given in Table 4.12 may serve as another example of an aromatic polyamide well soluble in sulfuric acid and some organic solvents (DMA). The application of a solvent with such low viscosity as DMA makes it possible to use sedimentation–diffusion analysis for the determination of molecular weights M_{sD} of PABI fractions in the range of M greater than $20 \cdot 10^3$. For fractions in the range of M less than $20 \cdot 10^3$, the sedimentation coefficients are too low for reliable measurements. Hence, for low-molecular-weight fractions, the values of $M_{D\eta}$ have been determined using Eq. (3.68) and the experimental values of D, $[\eta]$, and the parameter A_0 for PABI equal to $3.55 \cdot 10^{-10}$ g cm² sec⁻² deg⁻¹ mole⁻¹ᐟ³. The A_0 value was determined for high-molecular-weight fractions from the data on their sedimentation, diffusion, and viscometry.

The molecular-weight dependences of translational friction and viscosity for the PABI molecules in DMA and in 98% H_2SO_4 are characterized by the coefficients K_η, K_D, K_s, a, and b in Eqs. (4.4) listed in Table 4.12 [19]. Both the change in these coefficients with the variation in M and the high absolute values of viscosity for PABI solutions show unequivocally the high rigidity of its molecules and permit a reliable interpretation of experimental data from the standpoint of the theory of hydrodynamic properties of wormlike chains.

In accordance with this, Fig. 4.52 shows the dependence of $\eta_0 \, MD/RT$ on $M^{1/2}$ for PABI in DMA and in 98% H_2SO_4 (curve 2). The slope of the linear part of this curve is much shallower than that of straight line 1 for PmPhIPhA, which shows that the PmPhIPhA molecules are more rigid than those of PmPhIPhA. The values of A and d obtained from the slope and the intercept of the linear part of curve 2 using the values of $M_L = 20 \cdot 10^8$ dalton cm⁻¹ and Eqs. (4.7) are given in Table 4.12. However, the values of the parameter A for PABI in DMA and H_2SO_4 do not differ, and hence the molecular conformations of PABI in these two solvents are similar.

According to the data listed in Table 4.12, the equilibrium rigidity of the PABI chains is three or four times higher than that of the PmPhIPhA chains. This difference results from the difference in molecular structure of these two polymers. The

main source of the flexibility of the PABI chain is the benzimidazole ring ,

changing the direction of the rotation axis in the PABI chain by an angle ~30°, which is twice as small as the corresponding angle in the PmPhIPhA chain. Moreover, the distance between benzimidazole rings in the PABI chain is larger by a factor of three than that between meta-aromatic rings in the PmPhIPhA chain. If only these two structural features are taken into account, the number of monomer

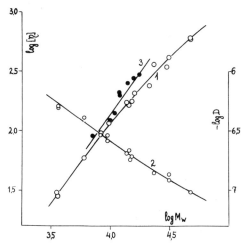

Fig. 4.54. Dependence of $\log [\eta]$ (1) and $\log D$ on $\log M_w$ (2) for PAH samples in dimethylsulfoxide [128–130], and $\log [\eta]$ on $\log M_w$ (3) for PBA samples in 96% H_2SO_4.

units S_f in the Kuhn segment and the segment length A_f with free rotation in the PABI chain may be calculated using Eqs. (4.22)–(4.25), in which it is assumed that $\vartheta = 30°$; $\vartheta/\Delta = 0.133$, and $\lambda = 17.8 \cdot 10^{-8}$ cm. Under these conditions, the values of $S_f = 14$ and $A_f = 250 \cdot 10^{-8}$ cm are obtained. If the experimental value of A_D in Table 4.12 is taken into account, this value leads to the diffusion parameter $\sigma \sim 1$, which appears to be too low. We will also consider this problem when dealing with optical properties of these polymers (Chapter 6, Section 2.8.2).

9.3. Polyamidehydrazide

Although polyamidehydrazide (PAH) (see Table 4.12) is soluble in sulfuric acid, it is rapidly degraded, and hence quantitative investigations of the hydrodynamic properties of this polymer are impossible if sulfuric acid is used as solvent. However, it readily dissolves in some polar organic solvents (DMAA and dimethylsulfoxide), in which the hydrodynamic characteristics of this polymer may be investigated by sedimentation and diffusion analysis and by viscometry [128, 129].

The molecular weights M_w of some PAH samples have been measured by light scattering of their solutions in dimethylsulfoxide (DMS) [130], and these values of M_w are used for the interpretation of hydrodynamic properties of these samples in the same solvent.

Figure 4.54 shows on a logarithmic scale the dependence of intrinsic viscosity and the diffusion coefficient of PAH on molecular weight M_w. The dependences of $\log [\eta]$ and $\log D$ on $\log M_w$ are represented by curves which may be approximated by the Mark–Kuhn linear dependences for narrow M_w ranges. These ranges, and the corresponding values of the coefficients K_η, K_D, a, and b, are

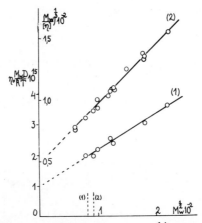

Fig. 4.55. Dependences of $\eta_0 M_w D/RT$ on $M_w^{1/2}$ (1) and $(M_w^2/[\eta])^{1/3}$ on $M^{1/2}$ (2) for PAH samples in dimethylsulfoxide [128–130]. Abscissa shows points where $L/A = 2.2$.

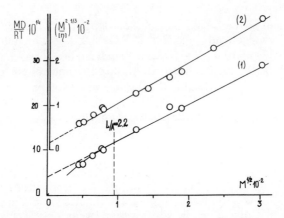

Fig. 4.56. Dependences of MD/RT (1) and $(M^2/[\eta])^{1/3}$ (2) on $M^{1/2}$ for fractions of an aromatic copolyester (ACPE) in TCE.

given in Table 4.12. A considerable curvature of the plots in Fig. 4.54 and high values for the exponent a indicate that PAH chains are fairly rigid.

This fact is confirmed by the character of the dependence of $M_w D$ on $M_w^{1/2}$ and that of $(M_w^2/[\eta])^{1/3}$ on $M^{1/2}$ shown in Fig. 4.55 for PAH solutions in DMS. The values of A and d given in Table 4.12 have been determined from the slopes and intercepts of these straight lines using Eqs. (4.7) and (4.18), and the values of $M_L = 19 \cdot 10^8$ dalton cm^{-1} and $\eta_0 = 2 \cdot 10^{-2}$ poise (viscosity of DMS at 20°C). The differences in these values of A_η and A_D reflect the errors due to experimental procedures. These values are close to that of A obtained by light scattering [130].

Hence, according to the results of hydrodynamic investigations, the rigidities of PAH and PABI molecules are similar. In the PAH chain all phenyl rings are in the para position and, hence, in contrast to the PABI molecules there are no units regularly changing rotation direction axes along the chain. However, a specific source of flexibility in the PAH chains may be hydrazide bonds increasing the probability for an amide group to adopt the cis-conformation [128].

9.4. Poly(*para*-benzamide)

Poly(*para*-benzamide) (PBA) is a "classical" aromatic polyamide, and the unique properties of these polymers – the ability of forming the mesophase in concentrated solutions, and hence the possibility of using them for the preparation of ultrahigh modulus fibers – have been found first for this polymer.

However, quantitative information on hydrodynamic characteristics and the rigidity of PBA molecules is more scarce and less reliable than for other aromatic polyamides considered above.

The first reason for this is the fact that molecular solutions of PBA can be obtained only in strong inorganic acids. Although it can also dissolve in dimethylacetamide (for technical purposes), molecular association is observed in these solutions, and hence they are unsuitable for the study of hydrodynamic and conformational properties of the dissolved polymer [140].

The molecular weights, M_w, of a number of PBA samples have been determined by light scattering of their solutions in 96% sulfuric acid [136]; intrinsic viscosities have also been measured for these solutions. The dependence of $[\eta]$ on M_w plotted on a logarithmic scale is shown in Fig. 4.54 (curve 3). Over the relatively narrow M_w range available this dependence is approximated by the Mark–Kuhn straight line with parameters K_η and a given in Table 4.12. The interpretation of this dependence according to Eq. (2.135) leads to the values of the length of the Kuhn segment A_η and the chain diameter d_η also listed in Table 4.12. Although for the above reasons the precision of this value of A_η is not high, it is possible to conclude that the PBA molecules are characterized by high equilibrium rigidity exceeding that of other aromatic polyamides listed in Table 4.12. These results have been confirmed by French researchers [137].

10. AROMATIC POLYESTERS

The chain structure of aromatic polyesters is similar to that of aromatic polyamides: in the simplest the ester groups alternate with phenyl rings: $-OCOC_6H_4-$. Taking into account a greater probability of the planar trans-conformation of the ester group [36] and the analogy with aromatic polyamides [127], the equilibrium chain rigidity of para-aromatic polyesters may be expected to be relatively high [131, 132]. This is indirectly confirmed by the tendency to form the liquid-crystal state of substances whose molecules contain combinations of the ester groups with para-aromatic rings [133]. However, quantitative data on the rigidity of molecules of a typical para-aromatic polyester ($-OCOC_6H_4-$) have not been available before because this polymer is soluble only in strong inorganic acids (H_2SO_4) and is immediately degraded in these acids.

In order to obtain soluble polyesters, copolymer compositions with the chain structure $-X-OCO-C_6H_4OCO-$ have been synthesized [134]. In this case the fragment X exhibits the structure X_1 or X_2 listed in Table 4.12. The main sources of chain flexibility are evidently fragments X_2 containing groups $-C(CH_3)_2-$, which ensure a change in the direction of rotational axes by the tetrahedral angle. This aromatic copolyester (ACPE) containing in the chain fragments X_1 and X_2 in relative amounts m:n = 7:3, respectively, readily dissolves in tetrachloroethane (TCE). Hence, it has been possible to fractionate this copolymer and to study the hydrodynamic characteristics of its fractions in TCE by sedimentation, diffusion, and viscometry [135].

In the molecular-weight range investigated the dependences of $[\eta]$ and D on M for ACPE in TCE may be represented by the Mark–Kuhn equations with the coefficients K_η, K_D, a, and b listed in Table 4.12. The exponents a and b differ greatly from 0.5, and taking into account the low molecular weights of the fractions investigated this indicates that the ACPE molecules exhibit considerable draining and allows the interpretation of their hydrodynamic properties using Eqs. (4.7) and (4.18).

In accordance with this, Fig. 4.56 shows the dependences of MD and $(M^2/[\eta])^{1/3}$ on $M^{1/2}$ for ACPE fractions in TCE. The values of A and d calculated from the slopes and intercepts of the curves plotted in Fig. 4.56 using Eqs. (4.7) and (4.18) and the values of $M_L = 26 \cdot 10^8$ dalton cm^{-1} are given in Table 4.12. This value of A is close to those for aromatic polyamides listed in Table 4.12. However, it should be expected that the equilibrium rigidity of an aromatic polyester containing no flexible X_2 fragments should be much higher. The data concerning this problem may be obtained by additional investigations of these polymers using flow birefringence.

11. AROMATIC POLYMERS:
VISCOMETRIC AND DIFFUSION DATA

11.1. Molecular-Weight Determination

As already mentioned, many aromatic polymers which are of major practical importance are soluble *only* in strong inorganic acids, mainly in concentrated H_2SO_4.

The possibilities of experimental investigations of hydrodynamic properties of these polymers in dilute solutions are much less favorable than in the foregoing cases because the application of sedimentation analysis to these polymers is virtually ruled out (see Chapter 3, Section 2.5).

In the investigation of these polymers the experimenter is forced to restrict himself to viscometric and diffusion measurements. Although these measurements provide quantitative information on the rotational and translational friction of molecules, the determination of the molecular weight of the polymer presents a dif-

TABLE 4.13. Molecular Weights $M_{D\eta}$, Coefficients K_η, K_D, K_s, a, and b in Mark–Kuhn Equations, Kuhn Segments A_D, and Chain Diameters d_D for Some Aromatic Polymers according to Diffusion and Viscometric Data for Their Dilute Solutions in 96% Sulfuric Acid *

No.	Repeat unit of polymer	Range of $M_{D\eta} \cdot 10^{-3}$	$K_\eta \cdot 10^4$	$K_D \cdot 10^4$	a	b	$M_L \cdot 10^{-8}$, dalton cm^{-1}	$A_D \cdot 10^8$, cm	$d_D \cdot 10^8$, cm	References
1	Poly(naphthoylene benzimidazole) (PNBI)	2.4–800	500	0.032	0.5	0.5	25	39	13.5	[139]
2	Poly(meta-hydroxyphenylbenzoxazole terephthalamide) (PmHPhBOTPhA)	1–42	12	0.032	0.76	0.60	20	96	16	[4]
3	Poly(para-phenylene oxadiazole) (PPhOD)	1–5	12.7	17.7	1.23	0.70	20	—	—	[138]
		5–50	411	10.3	0.825	0.63	—	100	6.5	
4	Poly(para-hydroxyphenylbenzoxazole terephthalamide) (PpHPhBOTPhA)	1–5	21	0.245	1.25	0.77	20	—	—	[4]
		5–42	60	0.132	1.13	0.70	—	320	13	
5	Para-phenylene terephthalamide–benzamide copolymer (PTBC)	2–10	14.1	0.195	1.35	0.75	20	320	16	
		10–40	316	0.195	1.0	0.75	—	—	—	

*All data are given in cm, g, s.

Note: In the calculations of $M_{D\eta}$ the value of the hydrodynamic parameter is assumed to be $A_0 = 3.6 \cdot 10^{-10}$ erg deg^{-1} mole$^{-1/3}$.

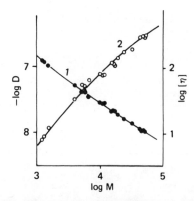

Fig. 4.57. Dependence of the diffusion coefficient D (1) and intrinsic viscosity $[\eta]$ (2) on molecular weight $M_{D\eta}$ for PPhOD samples in 96% H_2SO_4 [138].

ficult problem because, in this case, the traditional measurements of M_{sD} are impossible.

Under these conditions, the possibility of using the hydrodynamic invariant A_0 (the theory and experimental values of which are presented in Chapter 3, Section 4) becomes very important.

In principle, the molecular weight of the polymer may be calculated [see Eq. (3.68)] from the experimental values of D, $[\eta]$, and A_0 according to the equation

$$M_{D\eta} = A_0^3 \left(\frac{T}{\eta_0} \right)^3 \frac{100}{|\eta| D^3}.$$
(4.27)

It has been shown in Chapter 3 for flexible-chain polymers over a wide molecular-weight range of practical importance (i.e., in the range of $L/A > 20$) that the value of A_0 is really invariant. Hence, Eq. (4.27) with the recommended experimental value of A_0 may doubtless be used for the calculation of a reliable value of $M_{D\eta}$.

For polymers with high chain rigidity the problem of the determination of $M_{D\eta}$ is not so simple if it is borne in mind that, according to theoretical predictions (Figs. 3.29 and 3.30 and Table 3.2), for each polymer in the range of $L/A < 20$ (i.e., in the range of M usual for rigid-chain polymers) the value of A_0 varies with L (i.e., with M) and the character of these changes is different for polymers with different chain rigidities (with different d/A ratios). Hence, for a homologous series of polymer fractions or samples in the range of $L/A < 20$, in principle the calculation of $M_{D\eta}$ according to Eq. (4.27) at the assumed constant value of A_0 may lead to an incorrect dependence of the molecular characteristics D and $[\eta]$ on molecular weight.

Fortunately, in many cases of practical importance the situation is not so un-favorable. The most favorable situation (according to theoretical predictions) is ob-served for a wide class of polymers with moderate rigidity for which $d/A \approx 0.03$. As can be seen in Table 3.2 and Figs. 3.29 and 3.30, in this case the value of A_0 remains constant over the entire range of L/A ratios from 1 to ∞ to within 2–3%.

Although for more rigid polymers ($d/A \approx 0.01$) in the range of $0.5 < L/A < 5$ of practical importance for these polymers, the theoretical value of A_0 exceeds that of A_∞ by 6–8%, but varies with L by not more than 2–3%. Since these changes are within the precision of experimental measurements, they may be neglected.

The results of the calculation of the absolute value of $M_{D\eta}$ from experimental values of $[\eta]$ and D according to Eq. (4.27) are markedly dependent on the value of A_0 used for this purpose, because this value is contained in Eq. (4.27) with the ex-ponent 3. For a newly investigated rigid-chain polymer (soluble only in H_2SO_4) this value is usually unknown. In these cases it is reasonable to assume a value of A_0 equal to that of another polymer with a similar structure determined from visco-metric and diffusion–sedimentation measurements using Eq. (3.74). If the experi-mental values of A_0 for poly(*meta*-phenylene isophthalamide) and poly(amide ben-zimidazole) considered above are taken into account, it is reasonable to assume for many aromatic polymers the value of $A_0 = (3.5–3.6) \cdot 10^{-10}$ g cm^2 s^{-2} deg^{-1} mole$^{-1/3}$.

In particular, this interpretation of experimental data has been used in the in-vestigations of diffusion and viscosity of some samples of poly(*para*-phenylene oxadiazole) (PPhOD, Table 4.13) in 96% sulfuric acid and has made it possible to characterize these polymers by both their molecular weights and the equilibrium rigidities of their molecules [138].

Figure 4.57 shows on a logarithmic scale the dependences of $[\eta]$ and D on M for PPhOD, where the values of $M \equiv M_{D\eta}$ are calculated according to Eq. (3.74) from experimental values of D and $[\eta]$ at $A_0 = 3.6 \cdot 10^{-10}$ g cm^2 s^{-2} deg^{-1} mole$^{-1/3}$ and at $T = 299$ K. The points show the experimental values of $[\eta]$ and D and the curves represent the theoretical dependences calculated from general equations for worm-like chains, Eqs. (2.130)–(2.132) and (2.135)–(2.140), at the values of chain pa-rameters M_L, d, and A given in Table 4.13. Satisfactory agreement between ex-perimental points and theoretical curves at reasonable values of M_L, d, and A shows that it is possible to determine the molecular weight $M_{D\eta}$ of a rigid-chain polymer from the measured values of D and $[\eta]$.

Molecular weights $M_{D\eta}$ have also been determined for some PBA samples in 96% sulfuric acid [140]. The values obtained agree with those of M_w measured for the same polymers by light scattering [136].

The curves shown in Fig. 4.57 may be approximated by two linear Mark–Kuhn dependences for two molecular-weight ranges. The corresponding coeffi-cients K_η and a for PPhOD and some other aromatic polymers are given in Table 4.13.

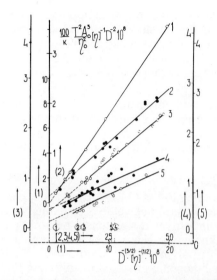

Fig. 4.58. Dependence of $100 \cdot T^2 A_0{}^3/(k[\eta]\eta_0{}^2 D^2)$ on $1/(D^{3/2}[\eta]^{1/2})$ for solutions of aromatic polymers in 96% sulfuric acid: 1) poly(naphthoylene benzimidazole); 2) poly(*meta*-hydroxyphenylbenzoxazole terephthalamide); 3) poly(*para*-phenylene oxadiazole); 4) poly(*para*-hydroxyphenylbenzoxazole terephthalamide); 5) copolymer of *para*-phenyleneterephthalamide and *para*-benzamide.

11.2. Determination of the Equilibrium Rigidity of the Chain

If the molecular weights $M_{D\eta}$ of a homologous series of samples (or fractions) are known, usual plots based on Eqs. (4.7) or (4.18) may be applied to the determination of the chain parameters A and d of polymers if their molecular weights are fairly high.

However, if a researcher is interested mainly in the evaluation of chain rigidity of this homologous series of polymers, it is possible to avoid the calculation of $M_{D\eta}$ excluding M from Eqs. (4.7) or (4.18) by substitution of Eq. (4.27) into these equations. Then we will obtain Eqs. (4.28) and (4.29) instead of Eqs. (4.7) and (4.18), respectively:

$$\frac{100}{k}\frac{T^2 A_0^3}{\eta_0^2}\left(\frac{1}{|\eta| D^2}\right) = \frac{10}{P_\infty}\left(\frac{M_L}{A}\right)^{1/2}\left(\frac{A_0 T}{\eta_0}\right)^{3/2}\left(\frac{1}{|\eta| D^3}\right)^{1/2} + \frac{M_L}{3\pi}\left(\ln\frac{A}{d} - 1.056\right), \quad (4.28)$$

$$\frac{100}{k}\frac{T^2 A_0^3}{\eta_0^2}\left(\frac{1}{|\eta| D^2}\right) = \frac{10 A_0}{k}\left(\frac{100}{\Phi_\infty}\right)^{1/3}\left(\frac{M_L}{A}\right)^{1/2}\left(\frac{A_0 T}{\eta_0}\right)^{3/2}\left(\frac{1}{|\eta| D^3}\right)^{1/2}$$
$$+ \frac{M_L}{3\pi}\left(\ln\frac{A}{d} - 1.056\right). \quad (4.29)$$

Plotting the left side of these equations as a function of $([\eta]D^3)^{-1/2}$ at a constant chosen value of A_0 we should obtain a straight line whose slope and intercept make it possible to determine A and the diameter D, respectively. These plots (Fig. 4.58) are based on the experimental values of D and $[\eta]$ for homologous series of some aromatic polymers dissolved in 96% sulfuric acid. For all these polymers the linear portions of the curves are sufficiently large to enable us to use Eqs. (4.28) and (4.29).

Naturally, Eqs. (4.28) and (4.29) are equivalent if Eq. (3.69) is taken into account, which relates the values of A_0, P_∞, and Φ_∞ to each other.

However, the theoretical value of P_∞ should be used in the experimental interpretation of the straight line plotted according to Eq. (4.28), whereas if Eq. (4.29) is applied, the theoretical value of Φ_∞ should be used. Consequently, Eq. (4.28) will yield the values of A_D and d_D, whereas Eq. (4.29) will give those of $A\eta$ and d_η. According to Eq. (4.21), the former and latter values should differ.

Table 4.13 lists the values of A_D and d_D obtained from the straight lines plotted in Fig. 4.58 using Eq. (4.28). The chain rigidities of the polymers listed in Table 4.13 differ greatly.

PNBI (sample 1) is found to be the most flexible, although its monomer unit is the most bulky and exhibits the quasi-ladder structure. The high flexibility of this polymer is due both to the presence of an oxygen bond between neighboring monomer units and, probably, to incomplete imidization during its synthesis. As a result, the ladder structure of its unit remains incomplete [139]. Hence, the hydrodynamic behavior of this polymer corresponds to the properties of a nondraining Gaussian coil over the very wide molecular-weight range investigated. Similar properties have also been found for a polymer containing no oxygen atom in the bond between neighboring monomer units [139, 141].

The molecules of samples 4 and 5 in Table 4.13 exhibit the highest rigidity. This seems quite natural for sample 5 because it is a completely para-aromatic polyamide containing no phenyl rings in the meta-position (in contrast to PmPhIPhA, Table 4.12). The rigidity of the PpHPhBOTPhA chain (Table 4.13, No. 4) greatly exceeds that of the PABI chain (Table 4.12), although the structures of these two polymers are similar. This difference is probably due to the presence of an OH substituent in the phenyl ring of the PpHPhBOTPhA monomer unit, which may lead to increasing rotation hindrance of the neighboring benzoxazole heteroring in the chain and thus favor an increase in its rigidity.

Samples 2 and 3 in Table 4.13 exhibit a much lower rigidity than samples 4 and 5. This demonstrates the increase in chain flexibility when phenyl rings in the meta-position (No. 2) or other rings changing the direction of the rotation axis of the chain (No. 3) are inserted in it.

Figure 4.58 clearly shows that the precision of the chain rigidity determination according to hydrodynamic properties is profoundly affected by the location and width of the molecular-weight range used in the experiment. The wider this range and the more it extends to that of $L/A > 2.8$, the more reliable is the value of A obtained, and the greater is the precision of its determination. In this sense the most favorable experimental conditions exist for the study of sample 1, and are satisfactory for samples 2 and 3, and more unfavorable for samples 4 and 5.

However, in all cases it should be borne in mind that these experimental values of A are tentative to a certain extent because the values of A_η and A_D are known to be different, and at present the choice between them should be considered to be arbitrary.

As to the hydrodynamic diameters d of polymer chains, it is possible to evaluate their order of magnitude only, as can be seen from the data presented in Table 4.13 and all the foregoing data.

REFERENCES

1. V. N. Tsvetkov, I. N. Shtennikova, M. G. Vitovskaya, E. I. Rjumtsev, T. V. Pecker, Y. P. Getmanchuk, P. N. Lavrenko, and S. V. Bushin, Vysokomol. Soedin., A16, 566 (1974).
2. V. N. Tsvetkov, Vysokomol. Soedin., A25, 1571 (1983).
3. M. G. Vitovskaya, P. N. Lavrenko, I. N. Shtennikova, A. A. Gorbunov, T. V. Pecker, E. V. Korneeva, E. P. Astapenko, Y. P. Getmanchuk, and V. N. Tsvetkov, Vysokomol. Soedin., A17, 1917 (1975).
4. V. B. Novakovski, P. N. Lavrenko, and V. N. Tsvetkov, Eur. Polym. J., 19, 831 (1983).
5. V. N. Tsvetkov, Y. V. Mitin, I. N. Shtennikova, V. R. Glushenkova, G. V. Tarasova, V. S. Skazka, and N. A. Nikitin, Vysokomol. Soedin., 7, 1098 (1965).
6. V. N. Tsvetkov, K. A. Andrianov, I. N. Shtennikova, G. I. Okhrimenko, L. N. Andreeva, G. A. Fomin, and V. I. Pakhomov, Vysokomol. Soedin., A10, 547 (1968).
7. V. N. Tsvetkov, K. A. Andrianov, G. I. Okhrimenko, I. N. Shtennikova, G. A. Fomin, M. G. Vitovskaya, V. I. Pakhomov, A. A. Yarosh, and D. N. Andreev, Vysokomol. Soedin., A12, 1892 (1970).
8. K. A. Andrianov, S. V. Bushin, M. G. Vitovskaya, V. N. Yemelyanov, P. N. Lavrenko, N. N. Makarova, A. M. Musafarov, V. Y. Nikolaev, G. F. Kolbina, I. N. Shtennikova, and V. N. Tsvetkov, Vysokomol. Soedin., A19, 469 (1977).
9. S. V. Bushin, V. N. Tsvetkov, Y. B. Lysenko, and V. N. Yemelyanov, Vysokomol. Soedin., A23, 2494 (1981).
10. V. N. Tsvetkov, K. A. Andrianov, M. G. Vitovskaya, N. N. Makarova, E. N. Zakharova, S. V. Bushin, and P. N. Lavrenko, Vysokomol. Soedin., A14, 369 (1972).
11. M. G. Vitovskaya, E. P. Astapenko, S. V. Bushin, V. S. Skaska, V. M. Yamschikov, N. N. Makatova, K. A. Andrianov, and V. N. Tsvetkov, Vysokomol. Soedin., A15, 2549 (1973).
12. M. G. Vitovskaya, P. N. Lavrenko, A. A. Gorbunov, S. V. Bushin, N. N. Makarova, K. A. Andrianov, and V. N. Tsvetkov, Vysokomol. Soedin., B17, 593 (1975); A24, 2101 (1982).
13. K. A. Andrianov, M. G. Vitovskaya, S. V. Bushin, V. N. Yemelyanov, A. M. Musafarov, D. Y. Zvankin, and V. N. Tsvetkov, Vysokomol. Soedin., A22, 1277 (1978).

14. N. V. Pogodina, K. S. Pojivilko, N. P. Evlampieva, A. B. Melnikov, S. V. Bushin, S. A. Didenko, G. N. Marchenko, and V. N. Tsvetkov, Vysokomol. Soedin., **A23**, 1252 (1981).
15. N. V. Pogodina, P. N. Lavrenko, K. S. Pojivilko, A. B. Melnikov, T. A. Kolobova, G. N. Marchenko, and V. N. Tsvetkov, Vysokomol. Soedin., **A24**, 332 (1982).
16. S. V. Bushin, E. B. Lisenko, V. A. Cherkasov, K. P. Smirnov, S. A. Didenko, G. N. Marchenko, and V. N. Tsvetkov, Vysokomol. Soedin., **A25**, 1899 (1983).
17. G. M. Pavlov, A. F. Koslov, G. N. Marchenko, and V. N. Tsvetkov, Vysokomol. Soedin., **B24**, 234 (1982).
18. L. N. Andreeva, P. N. Lavrenko, E. U. Urinov, L. I. Kutsenko, and V. N. Tsvetkov, Vysokomol. Soedin., **B17**, 326 (1975).
19. M. G. Vitovskaya, P. N. Lavrenko, O. V. Okatova, E. P. Astapenko, V. B. Novakovsky, S. V. Bushin, and V. N. Tsvetkov, Eur. Polym. J., **18**, 583 (1982).
20. V. N. Tsvetkov, I. N. Shtennikova, E. I. Rjumtsev, G. F. Kolbina, I. I. Konstantinov, Y. B. Amerik, and B. A. Krentsel, Vysokomol. Soedin., **A11**, 2528 (1969).
21. P. J. Flory, O. K. Spurr, and D. K. Carpenter, J. Polym. Sci., **27**, 231 (1958).
22. P. N. Lavrenko, E. U. Urinov, L. N. Andreeva, K. J. Linov, H. Dautzenberg, and B. Philipp, Vysokomol. Soedin., **A18**, 2579 (1976).
23. A. I. Grigoryev, V. N. Zgonnik, O. Z. Korotkina, and V. E. Eskin, Vysokomol. Soedin., **B17**, 884 (1975).
24. M. G. Vitovskaya, P. N. Lavrenko, O. V. Okatova, E. P. Astapenko, V. B. Novakovsky, S. V. Bushin, and V. N. Tsvetkov, Eur. Polym. J., **18**, 583 (1982).
25. V. N. Tsvetkov, P. N. Lavrenko, G. M. Pavlov, S. V. Bushin, E. P. Astapenko, A. A. Boikov, N. A. Shildyaeva, S. A. Didenko, and B. F. Malichenko, Vysokomol. Soedin., **A24**, 2343 (1982).
26. V. N. Tsvetkov, V. E. Eskin, and S. Y. Frenkel, Structure of Macromolecules in Solution, Butterworths, London (1970).
27. V. N. Tsvetkov, E. N. Sakharova, G. A. Fomin, and P. N. Lavrenko, Vysokomol. Soedin., **A14**, 1956 (1972).
28. V. N. Tsvetkov, E. V. Korneeva, P. N. Lavrenko, D. Hardi, and K. Nitrai, Vysokomol. Soedin., **B13**, 426 (1971).
29. E. V. Korneeva, V. N. Tsvetkov, and P. N. Lavrenko, Vysokomol. Soedin., **A12**, 1369 (1970).
30. V. N. Tsvetkov, D. Hardi, I. N. Shtennikova, E. V. Korneeva, G. F. Pirogova, and K. Nitrai, Vysokomol. Soedin., **A11**, 349 (1969).
31. V. N. Tsvetkov, L. N. Andreeva, E. V. Korneeva, P. N. Lavrenko, N. A. Platé, V. P. Shibaev, and B. S. Petrukhin, Vysokomol. Soedin., **A13**, 2226 (1971).
32. V. N. Tsvetkov, E. V. Korneeva, I. N. Shtennikova, P. N. Lavrenko, G. F. Kolbina, D. Hardi, and K. Nitrai, Vysokomol. Soedin., **A14**, 427 (1972).
33. V. N. Tsvetkov, L. N. Andreeva, E. V. Korneeva, and P. N. Lavrenko, Dokl. Akad. Nauk SSSR, **205**, 895 (1972).

34. V. N. Tsvetkov, L. N. Andreeva, E. V. Korneeva, P. N. Lavrenko, N. A. Platé, V. P. Shibaev, and B. S. Petrukhin, Vysokomol. Soedin., **A14**, 1737 (1972).
35. L. N. Andreeva, A. A. Gorbunov, S. A. Didenko, E. V. Korneeva, P. N. Lavrenko, N. A. Platé, and V. P. Shibaev, Vysokomol. Soedin., **B15**, 209 (1973).
36. P. J. Flory, Statistical Mechanics of Chain Molecules, Wiley–Interscience, New York (1969).
37. S. Chinai and R. Juzzi, J. Polym. Sci., **41**, 475 (1959).
38. A. Blumstein (ed.), Liquid-Crystalline Order in Polymers, Academic Press, New York (1978).
39. V. N. Tsvetkov, I. N. Shtennikova, E. I. Rjumtsev, E. V. Korneeva, and J. F. Kolbina, Vysokomol. Soedin., **A15**, 2158 (1973).
40. V. N. Tsvetkov, E. I. Riumtsev, I. N. Shtennikova, E. V. Korneeva, B. A. Krentsel, and Yu. B. Amerik, Eur. Polym. J., **9**, 481 (1973).
41. S. V. Bushin, E. V. Korneeva, I. I. Konstantinov, Yu. B. Amerik, S. A. Didenko, I. N. Shtennikova, and V. N. Tsvetkov, Vysokomol. Soedin., **A24**, 1469 (1982).
42. G. M. Pavlov, E. V. Korneeva, T. A. Garmonova, D. Hardy, and K. Nitrai, Vysokomol. Soedin., **A20**, 1634 (1978).
43. V. N. Tsvetkov, I. N. Shtennikova, E. I. Riumtsev, N. V. Pogodina, G. F. Kolbina, E. V. Korneeva, P. N. Lavrenko, O. V. Okatova, Yu. B. Amerik, and A. A. Baturin, Vysokomol. Soedin., **A18**, 2016 (1976).
44. M. Szwarc, M. Levy, and R. Milkowitch, J. Am. Chem. Soc., **78**, 2656 (1956).
45. S. P. Mitsengendler, G. A. Andreeva, K. I. Sokolova, and A. A. Korotkov, Vysokomol. Soedin., **4**, 1366 (1962).
46. P. Rempp, V. I. Volkov, J. Parrod, and Ch. Sadron, Vysokomol. Soedin., **2**, 1521 (1960); Bull. Soc. Chim. Fr., **5**, 919 (1960).
47. V. N. Tsvetkov, S. Y. Magarik, S. I. Klenin, and V. E. Eskin, Vysokomol. Soedin., **5**, 3 (1963).
48. V. N. Tsvetkov, S. I. Klenin, and S. Y. Magarik, Vysokomol. Soedin., **6**, 400 (1964).
49. I. A. Baranovskaya, S. I. Klenin, S. Y. Magarik, V. N. Tsvetkov, and V. E. Eskin, Vysokomol. Soedin., **7**, 878, 884 (1965).
50. T. Kadirov, A. N. Cherkasov, I. A. Baranovskaya, V. E. Eskin, S. I. Klenin, S. Y. Magarik, and V. N. Tsvetkov, Vysokomol. Soedin., **A9**, 2094 (1967).
51. V. N. Tsvetkov, G. A. Andreeva, I. A. Baranovskaya, S. I. Klenin, S. Y. Magarik, and V. E. Eskin, J. Polym. Sci., **C16**, 239 (1967).
52. G. A. Fomin, Vysokomol. Soedin., **A15**, 1917 (1973).
53. V. N. Tsvetkov, S. Y. Magarik, T. Kadirov, and G. A. Andreeva, Vysokomol. Soedin., **A10**, 943 (1968).
54. V. N. Tsvetkov, L. N. Andreeva, S. Y. Magarik, P. N. Lavrenko, and G. A. Fomin, Vysokomol. Soedin., **A13**, 2011 (1971).
55. G. M. Pavlov, S. Y. Magarik, G. A. Fomin, V. M. Yamschikov, G. A. Andreeva, V. S. Skaska, I. G. Kirillova, and V. N. Tsvetkov, Vysokomol. Soedin., **A15**, 1696 (1973).
56. M. G. Vitkovskaya, V. N. Tsvetkov, L. I. Godunova, and T. V. Sheremeteva, Vysokomol. Soedin., **A9**, 1682 (1967).

57. V. N. Tsvetkov, G. A. Fomin, P. N. Lavrenko, I. N. Shtennikova, T. V. Sheremeteva, and L. I. Godunova, Vysokomol. Soedin., **A10**, 903 (1968).
58. T. V. Sheremeteva, G. N. Larina, V. N. Tsvetkov, and I. N. Shtennikova, J. Polym. Sci., **C22**, 185 (1968).
59. V. N. Tsvetkov, N. N. Kupriyanova, G. V. Tarasova, P. N. Lavrenko, and I. I. Migunova, Vysokomol. Soedin., **A12**, 1974 (1970).
60. V. N. Tsvetkov, G. V. Tarasova, E. L. Vinogradov, N. N. Kupriyanova, V. M. Yamschikov, V. S. Skaska, V. S. Ivanov, V. K. Smirnova, and I. I. Migunova, Vysokomol. Soedin., **A13**, 620 (1971).
61. V. N. Tsvetkov, M. G. Vitovskaya, P. N. Lavrenko, E. N. Zakharova, I. F. Gavrilenko, and N. N. Stefanovskaya, Vysokomol. Soedin., **A13**, 2532 (1971).
62. H. Vogel and C. S. Marvel, J. Polym. Sci., **50**, 511 (1961).
63. C. E. Sroog, A. L. Endrey, S. V. Abramo, C. E. Berr, W. M. Edwards, and K. L. Olivier, J. Polym. Sci., **A3**, 1373 (1965).
64. V. L. Bell and G. F. Pezdirtz, J. Polym. Sci., **B3**, 977 (1965).
65. J. F. Wolfe and F. E. Arnold, Macromolecules, **14**, 909 (1981).
66. J. F. Wolfe, B. H. Loo, and F. E. Arnold, Macromolecules, **14**, 915 (1981).
67. A. J. Sicree, F. E. Arnold, and R. L. Van Dausen, J. Polym. Sci., Chem. Ed., **12**, 265 (1974).
68. P. D. Sybert, W. H. Beever, and J. K. Stille, Macromolecules, **14**, 493 (1981).
69. G. C. Berry, Discuss. Faraday Soc., **49**, 121 (1970).
70. G. C. Berry, J. Polym. Sci., Polym. Sympos., **N65**, 143 (1978).
71. J. F. Brown, J. Polym. Sci., **CN1**, 83 (1963).
72. K. An Andrianov, Vysokomol. Soedin., **7**, 1477 (1965), **A11**, 1362 (1969); **A13**, 253 (1971).
73. V. N. Tsvetkov, K. A. Andrianov, E. L. Vinogradov, I. N. Shtennikova, S. E. Yakushkina, and V. I. Pakhomov, J. Polym. Sci., **C23**, 385 (1968).
74. V. N. Tsvetkov and A. V. Lezov, Vysokomol. Soedin., **A26**, 494 (1984).
75. V. N. Tsvetkov, K. A. Andrianov, G. I. Okhrimenko, and M. G. Vitovskaya, Eur. Polym. J., **7**, 1215 (1971).
76. V. N. Tsvetkov, K. A. Andrianov, M. G. Vitovskaya, N. N. Makarova, S. V. Bushin, E. N. Zakharova, A. A. Gorbunov, and P. N. Lavrenko, Vysokomol. Soedin., **A15**, 872 (1973).
77. V. N. Tsvetkov, K. A. Andrianov, E. N. Zakharova, T. A. Rotinyan, and N. N. Makarova, Vysokomol. Soedin., **A16**, 1792 (1974).
78. V. N. Tsvetkov, Makromol. Chem., **160**, 1 (1972).
79. V. N. Tsvetkov, K. A. Andrianov, N. N. Makarova, M. G. Vitovskaya, E. I. Rjumtsev, and I. N. Shtennikova, Eur. Polym. J., **9**, 27 (1973).
80. M. G. Vitovskaya and V. N. Tsvetkov, Eur. Polym. J., **12**, 251 (1976).
81. K. A. Andrianov, G. L. Slonimsky, Y. V. Genin, V. I. Gerasimov, N. Y. Levin, N. N. Makarova, and D. Y. Tsvankin, Dokl. Akad. Nauk SSSR, **187**, 1285 (1969).
82. K. A. Andrianov, S. A. Pavlova, I. I. Tverdohlebova, V. N. Yemelyanov, T. A. Larina, and A. Yu. Rabkina, Vysokomol. Soedin., **A14**, 2246 (1972).

83. J. Brandrup and E. H. Immergut (eds.), Polymer Handbook, Wiley, New York (1975).
84. M. Marx-Figini and G. V. Schulz, Makromol. Chem., **54**, 102 (1962).
85. V. N. Tsvetkov, P. N. Lavrenko, L. N. Andreeva, A. I. Mashoshin, O. V. Okatova, O. I. Mikrjukova, and L. I. Kutzenko, Eur. Polym. J., **20**, 823 (1984).
86. S. Ya. Lyubina, S. I. Klenin, I. A. Strelina, A. V. Troitskaya, V. I. Kurlyankina, and V. N. Tsvetkov, Vysokomol. Soedin., **A15**, 691 (1973).
87. E. V. Korneeva, P. N. Lavrenko, E. Urinov, A. K. Khripunov, L. I. Kutsenko, and V. N. Tsvetkov, Vysokomol. Soedin., **A21**, 1547 (1979).
88. E. N. Zakharova, L. I. Kutsenko, V. N. Tsvetkov, V. S. Skazka, G. V. Tarasova, and V. M. Yamschikov, Vestn. Leningr. Gos. Univ., **N16**, 55 (1970).
89. E. I. Rjumtsev, F. M. Aliev, M. G. Vitovskaya, E. U. Urinov, and V. N. Tsvetkov, Vysokomol. Soedin., **A17**, 2676 (1975).
90. G. Meyerhoff and N. Sutterlin, Makromol. Chem., **87**, 258 (1965).
91. N. V. Pogodina, K. S. Pozhivilko, A. B. Melnikov, S. A. Didenko, G. N. Martchenko, and V. N. Tsvetkov, Vysokomol. Soedin., **23**, 2454 (1981).
92. N. V. Pogodina, A. B. Melnikov, O. I. Mikrjukova, S. A. Didenko, and G. N. Marchenko, Vysokomol. Soedin., **A26**, 2515 (1984).
93. S. Ya. Lyubina, S. I. Klenin, I. A. Strelina, A. V. Troitskaya, A. K. Khripunov, and E. U. Urinov, Vysokomol. Soedin., **A19**, 244 (1977).
94. M. Bikales and L. Segal (eds.), Cellulose and Cellulose Derivatives. Part 4, Wiley–Interscience, New York (1971).
95. G. V. Schulz and E. Penzel, Makromol. Chem., **112**, 260 (1968).
96. E. Penzel and G. V. Schulz, Makromol. Chem., **113**, 64 (1968).
97. K. H. Mayer and L. Misch, Helv. Chim. Acta, **20**, 232 (1937).
98. H. Benoit, J. Polym. Sci., **3**, 376 (1948).
99. L. Pauling, R. V. Corey, and H. R. Branson, Proc. Natl. Acad. Sci. USA, **37**, 205, 235, 241 (1951).
100. A. Teramoto and H. Fujita, Adv. Polym. Sci., **18**, 65 (1975).
101. P. Doty, A. Holtzer, J. Brandbury, and E. Blout, J. Am. Chem. Soc., **76**, 4493 (1954); **78**, 947 (1956).
102. J. Lang, J. Am. Chem. Soc., **80**, 1783, 5139 (1958).
103. V. Luzzati, M. Cesari, G. Spach, F. Masson, and J. Vincent, J. Mol. Biol., **3**, 566 (1961).
104. J. Marchal and E. Marchal, Vysokomol. Soedin., **3**, 561 (1964).
105. J. Spach, L. Freund, M. Daune, and H. Benoit, J. Mol. Biol., **7**, 468 (1963).
106. V. E. Shashoua, J. Am. Chem. Soc., **81**, 3156 (1959).
107. V. E. Shashoua, W. Sweeny, and R. F. Tietz, J. Am. Chem. Soc., **82**, 886 (1960).
108. W. Burchard, Makromol. Chem., **67**, 182 (1963).
109. N. S. Schneider, S. Furusaki, and R. W. Lenz, J. Polym. Sci., **A3**, 933 (1965).
110. H. Yu, A. J. Bur, and L. J. Fetters, J. Chem. Phys., **44**, 2568 (1966).
111. V. N. Tsvetkov, I. N. Shtennikova, E. I. Rjumtsev, L. N. Andreeva, Yu. P. Getmanchuk, Yu. L. Spirin, and R. I. Dryagileva, Vysokomol. Soedin., **A10**, 2132 (1968).

112. S. Itou, N. Nishioka, T. Norisuye, and A. Teramoto, Macromolecules, **14**, 904 (1981).
113. P. N. Lavrenko , Dissertation, Institute of Macromolecular Compounds, Academy of Sciences of the USSR, Leningrad (1987).
114. I. P. Kolomietz and V. N. Tsvetkov, Vysokomol. Soedin., **B25**, 813 (1983).
115. M. G. Vitovskaya, I. N. Shtennikova, E. P. Astapenko, and T. V. Pecker, Vysokomol. Soedin., **A17**, 1161 (1975).
116. B. Z. Volchek and V. N. Nikitin, Dokl. Akad. Nauk SSSR, **205**, 622 (1972).
117. H. B. Gray, V. A. Bloomfield, and J. E. Hearst, J. Chem. Phys., **16**, 1493 (1967).
118. L. B. Sokolov, Thermostable Aromatic Polyamides, Khimiya, Moscow (1975).
119. P. W. Morgan, J. Polym. Sci., C, Polymer Sympos., No. 65, 1 (1978).
120. I. K. Nekrasov, Vysokomol. Soedin., **A13**, 1707 (1971).
121. M. G. Vitovskaya, E. P. Astapenko, V. I. Nikolaev, S. A. Didenko, and V. N. Tsvetkov, Vysokomol. Soedin., **A18**, 691 (1976).
122. V. N. Tsvetkov, P. N. Lavrenko, G. M. Pavlov, S. V. Bushin, E. P. Astapenko, A. A. Boikov, N. A. Shildaeva, S. A. Didenko, and B. F. Malichenko, Vysokomol. Soedin., **A24**, 2343 (1982).
123. N. A. Mikhailova, "Flow birefringence in solutions of rigid-chain aromatic polymers," Thesis, Leningrad University (1982).
124. H. L. Wagner and C. Λ. Hoeve, J. Polym. Sci., **A2**, 1189 (1973).
125. G. Meyerhoff, Z. Phys. Chem., **4**, 336 (1955).
126. H. Takimoto, G. Forbes, and R. Laudenschlanger, J. Appl. Polym. Sci., **5**, 153 (1961).
127. V. N. Tsvetkov, Eur. Polym. J., **12**, 867 (1976); Macromolecules, **11**, 306 (1978).
128. V. N. Tsvetkov and N. A. Mikhailova, Vysokomol. Soedin., **A20**, 191 (1978).
129. M. G. Vitovskaya, P. N. Lavrenko, E. P. Astapenko, O. V. Okatova, and V. N. Tsvetkov, Vysokomol. Soedin., **A20**, 320 (1978).
130. V. N. Tsvetkov and S. O. Tsepelevich, Eur. Polym. J., **19**, 267 (1983).
131. F. Laupretre, L. Monnerie, and B. Fayolle, J. Polym. Sci., Polym. Phys. Ed., **18**, 2243 (1980).
132. B. Erman, P. J. Flory, and J. P. Hummel, Macromolecules, **13**, 484 (1980).
133. D. Demus, H. Demus, and H. Zaschke, "Flussige Kristalle in Tabellen," Verlag Grundstoff Industrie, Leipzig (1974).
134. Z. Edlinski and D. Sek, Eur. Polym. J., **7**, 827 (1971).
135. V. N. Tsvetkov, L. N. Andreeva, S. V. Bushin, A. I. Mashoshin, V. A. Edlinksi, and D. Sek, Eur. Polym. J., **20**, 373 (1984).
136. M. G. Vitovskaya, S. V. Bushin, V. D. Kalmikova, A. V. Volokhina, G. I. Kudrjavtzev, and V. N. Tsvetkov, Vysokomol. Soedin., **B18**, 588 (1976).
137. M. Arpin and C. Strazielle, Polymer, **18**, 591 (1977); Makromol. Chem., **179**, 1261 (1978).
138. V. N. Tsvetkov, V. B. Novakovski, N. A. Mikhailova, A. V. Volokhina, and A. B. Raskina, Vysokomol. Soedin., **A22**, 133 (1980).

139. V. N. Tsvetkov, V. B. Novakovski, V. V. Korshak, A. L. Rusanov, A. M. Berlin, and S. H. Fidler, Vysokomol. Soedin., **A27**, 86 (1985).
140. M. G. Vitovskaya, P. N. Lavrenko, and O. V. Okatova, Vysokomol. Soedin., **A19**, 1966 (1977).
141. G. C. Berry, Discuss. Faraday Soc., **N49**, 121 (1970).
142. J. R. Schaefgen, V. S. Foldi, F. M. Logullo, V. H. Good, L. W. Gulrich, and F. L. Killian, Polymer Prepr., **17**, No. 1, 69 (1976).

Chapter 5

FLOW BIREFRINGENCE.
THEORY

Flow birefringence is known to be a widely used method for the study of optical, conformational, and hydrodynamic properties of molecules of flexible-chain polymers in solutions. Experimental and theoretical data available on this problem are very extensive and have been discussed in many publications and reviews [1–12]. They make it possible to understand this phenomenon not only on a qualitative but also quantitative level.

The situation in the field of rigid-chain polymers is much less favorable. Experimental investigations are relatively few. This is probably due to both the above-mentioned experimental difficulties and the deficiency of polymer samples suitable for investigation, because the synthesis of rigid-chain molecules with a well-known chemical structure has been widely developed only in the last decade.

The number of theoretical papers on the dynamo-optical properties of rigid-chain molecules is even more limited, and the published papers deal essentially with the explanation of certain individual phenomena and do not allow a description of the whole complex of properties and experimental data from the standpoint of a single theory.

However, the analysis of experimental data available on flow birefringence and the use of some theoretical (sometimes semiempirical) concepts make it possible to draw important conclusions concerning optical, conformational, and structural characteristics of rigid-chain molecules [13]. Hence, we have sufficient reason for considering the method of flow birefringence as an effective and useful means of investigating rigid-chain polymers, in particular in those cases when other known methods prove to be ineffective.

As we have seen, according to their hydrodynamic properties in solutions, molecules of rigid-chain polymers occupy an intermediate position between abso-

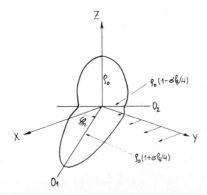

Fig. 5.1. Distribution of orientation $\rho(\varphi, \vartheta)$ in the flow plane XY and in the plane ZO_2; O_1 and O_2 are the two main optical directions in the flow plane. O_1 is the optical axis of an anisotropic solution, the direction of preferred orientation.

lutely rigid rodlike particles and typical flexible-chain polymers, exhibiting certain features of resemblance to both of them. A similar situation may be expected in the consideration of dynamo-optical properties of these molecules. Hence, it is desirable, first, to characterize briefly the main features of the phenomena of flow birefringence (FB) in solutions of both rigid nondeformable particles and flexible-chain molecules.

1. RIGID PARTICLES

The phenomenon of flow birefringence (FB) in a solution containing rigid particles asymmetrical in shape arises from their orientation in a gradient field and is similar in physical nature to phenomena determining the shear dependence of intrinsic viscosity in this solution. The FB theory has been developed in the most complete quantitative form for particles represented by ellipsoids of revolution (spheroids) with a uniaxial symmetry of optical properties and with coinciding axes of geometrical and optical ellipsoids [14].

1.1. Distribution Function for Orientations

The theory is based on Jeffery's equations (2.19a) and (2.19b) describing the rotational motion of spheroids in a laminar flow with a rate gradient g [Eq. (2.13)] in the coordinate system shown in Fig. 2.1. The "kinematic" orientation appearing here is characterized (with the evaluation of the opposing effect of Brownian rotational motion of particles) by the distribution function $\rho(\vartheta, \varphi)$ of spatial orientations of axes of spheroids. The function $\rho(\vartheta, \varphi)$ is determined by diffusion equation (2.21), whose solution under steady-state conditions ($g = $ const) in a weak flow has the form of Eq. (2.22). At $b_0 = 1$, Eq. (2.22) coincides with the function previously obtained by Boeder [15] for rodlike particles.

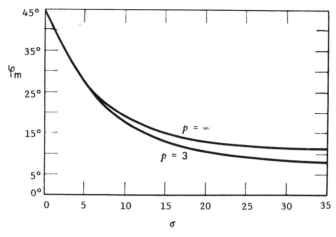

Fig. 5.2. Dependence of extinction angle φ_m on $\sigma = g/D_r$ for extended spheroidal particles at degrees of asymmetry $p = 3$ and $p = \infty$.

It follows from Eq. (2.22) that, in the direction normal to the flow plane ($\vartheta = 0$), the distribution function $\rho(\vartheta, \varphi)$ does not change in the shear field but remains equal to ρ_0, its value in the absence of the field. The flow plane XY is the plane of symmetry of the function $\rho(\vartheta, \varphi)$, since the replacement of ϑ by $\pi - \vartheta$ does not change Eq. (2.22).

The distribution function exhibits the greatest asymmetry in the XY flow plane (Fig. 5.1, $\vartheta = \pi/2$), where at low $\sigma \equiv gW/kT$ this function is represented by a figure resembling an ellipse (although not identical with it) with the semiaxes $a_1 = \rho_0(1 + \sigma b_0/4)$ and $a_2 = \rho_0(1 - \sigma b_0/4)$. Under these conditions ($\sigma \ll 1$) the distribution function $\rho(\varphi)$ may be represented by the equation

$$\rho(\varphi) = a_1 \cos^2 \varphi_1 + a_2 \sin^2 \varphi_1, \tag{5.1}$$

where $\varphi_1 = \varphi - \pi/4$ is the angle taken from the direction of preferred orientation O_1.

1.2. Extinction Angle

If the particles (molecules) in solution are optically anisotropic, their orientation in flow leads to the appearance of macroscopic optical anisotropy of the solution. This anisotropy corresponds to that of an optically biaxial crystal since the function $\rho(\varphi, \vartheta)$ in space (Fig. 5.1) is described by the surface of a body exhibiting triaxial symmetry. However, if the optical properties of a solution in a light beam parallel to the Z axis are studied, the picture observed corresponds to the distribution function in the flow plane and hence is characterized by two mutually perpendicular main directions O_1 and O_2. The direction O_1 is generally called the optical axis of solution, and the angle φ_m formed by the optical axis and the flow direction is termed the orientation or extinction angle.

When the term quadratic with respect to σ is used in Eq. (2.22), apart from the linear term, it can be easily seen that the orientation angle φ_m decreases with increasing shear rate g. Hence, if in a weak flow the direction of the preferred orientation coincides with that of the extension in a shear field, then the axis O_1 approaches the flow direction X with increasing σ. In a very strong flow, or at a very weak thermal motion (large particles), i.e., under the conditions $\delta \to \infty$, the hydrodynamic forces bring about a preferred orientation of particles in the flow direction and $\varphi_m \to 0$.

According to Eq. (2.22), at low σ the dependence of φ_m on σ may be approximated [15] by the equation

$$\tan 2\varphi_m = 6/\sigma = 6D_r/g = 6kT/gW \tag{5.2}$$

or by a more precise expression [14]:

$$\varphi_m = \frac{\pi}{4} - \frac{\sigma}{12}\left[1 - \frac{\sigma^2}{108}\left(1 + \frac{24}{35}b_0^2\right) + \cdots\right]. \tag{5.3}$$

It follows from Eq. (5.3) that with increasing σ the extinction angle becomes not only a function of σ but also of b_0.

It is impossible to use the function for the calculation of φ_m at high σ values in the form of Eq. (2.22); this function should be represented by expansion in a series of exponents of b_0[16]. The values of φ_m tabulated as the function of σ and the degrees of asymmetry of a spheroid p over the entire range of values $0 < \sigma < \infty$ and $1 < p < \infty$ have been obtained using this series with computer calculations.

Figure 5.2 shows the dependences of φ_m on σ for spheroidal particles with the ratio of axes $p = 3$ and $p = \infty$, according to tabulated data [17]. The asymptotic limit of both curves is the same, $\varphi_m \to 0$ at $\sigma \to \infty$. However, the higher p is, the slower the approach to this limit. It should be noted that at low σ the shape of curves $\varphi_m = \varphi_m(\sigma)$ is virtually independent of the degree of asymmetry of particles. Thus, at the values of $\sigma \leq 5$ or $45° - \varphi_m < 20°$ the curves corresponding to $p = 3$ and $p = \infty$ coincide to within 2%. This means that in this range of σ and φ_m values the extinction angle φ_m is related by a unique dependence to the shear rate g and the coefficient of rotational friction W of the particles investigated.

It is also very important that, according to Eq. (5.3) and Fig. 5.2, at low σ the dependence of φ_m on σ is virtually *linear*. Thus, at the value of $\sigma \approx 2$ and, correspondingly, at $45° - \varphi_m \approx 10°$ the deviation from linearity of the function $\varphi_m(\sigma)$ does not exceed 4%, which corresponds to the difference in the measured angle $\Delta\varphi_m \approx 0.4°$. At $\sigma \approx 1$ and, correspondingly, $45° - \varphi_m \approx 5°$, the deviation of $\varphi_m(\sigma)$ from linearity does not exceed 1% and, correspondingly, $\Delta\varphi_m \approx 0.05°$, which is much lower than the error in the experimental determination of angles φ_m. This means that when the experimental values of φ_m are in the range of $45° - \varphi_m < 10°$, it

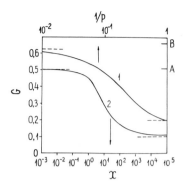

Fig. 5.3. Coefficient G as a function of molecular shape: 1) as a function of shape asymmetry $1/p$ of extended spheroidal particles according to Eq. (5.7); 2) as a function of $x = 2L/A$ for kinetically flexible chains with local rigidity according to the theory advanced in [89]. A and B are the values of G for kinetically rigid Gaussian coils at strong and weak hydrodynamic interactions, respectively [89].

is possible to reliably determine the initial slope of the dependence of φ_m on g and to calculate the intrinsic value of the orientation angle

$$[\chi/g] \equiv \lim_{\substack{g \to 0, \\ c \to 0}} (\chi/g) \equiv \lim_{\substack{g \to 0, \\ c \to 0}} [(\pi/4 - \varphi_m)/g]. \qquad (5.4)$$

This value is of major importance because, regardless of the optical properties of particles, according to Eqs. (5.2) and (5.3), it is uniquely related to the diffusion D_r and friction W coefficients for the rotation of the spheroid about the short axis

$$[\chi/g] = 1/12 D_r = W/12kT. \qquad (5.5)$$

Comparison of Eqs. (5.5) and (2.27) gives Eq. (5.6) relating the intrinsic orientation to the molecular weight M of a spheroidal particle and the intrinsic viscosity $[\eta]$ of solution

$$[\chi/g] = GM [\eta] \eta_0/RT, \qquad (5.6)$$

where the coefficient G is a function of p and is related to the function $F(p)$ in Eq. (2.27) by the equation

$$G = 1/12F (p) = f_0 (p)/12\nu (p), \qquad (5.7)$$

where the function $f_0(p)$ is determined by Eq. (2.11) and the function $\nu(p)$ is determined by Eqs. (2.24) and (2.25).

The dependence of G on $1/p$ according to Eq. (5.7) is shown in Fig. 5.3 (curve 1). When the asymmetry of the spheroid changes, G varies from the value of 0.625 for a rod ($p = \infty$) to 0.2 for a sphere ($p = 1$). At $p = 10$, $F(p) = 1/6$ and, accordingly, $G = 1/2$.

1.3. Value of Birefringence

When the light beams pass through the solution in the direction of the Z axis (Fig. 5.1), the value of the observed birefringence Δn is determined by the difference in refractive indices, n_1 and n_2, of two beams with electric vectors parallel to the two main directions, O_1 and O_2, respectively. The value of Δn depends on both the type of distribution function $\rho(\varphi, \vartheta)$ and the optical properties of dissolved particles or molecules.

In the theory considered here, the optical properties of a particle are characterized by two main polarizability coefficients, γ_1 and γ_2, i.e., they exhibit an axial symmetry, and the optical axis (γ_1) coincides with the axis of a spheroid describing the hydrodynamic properties of a particle (molecule).

If in a system of O_1O_2Z coordinates (Fig. 5.1) the axis of the molecule (γ_1) forms a polar angle ϑ with the Z axis, and the azimuth φ with the O_1 axis (in the O_1O_2 plane), then the difference between the polarizabilities $(\gamma_1 - \gamma_2)_0$ introduced by this molecule on the axes O_1 and O_2 is given by

$$(\gamma_1 - \gamma_2)_0 + (\gamma_1 - \gamma_2) \sin^2 \vartheta \, \cos 2\varphi. \tag{5.8}$$

Hence, the average (over all orientations) difference in the polarizabilities of the molecule in the two main directions, O_1 and O_2, of an optically anisotropic solution is given by

$$\overline{(\gamma_1 - \gamma_2)}_0 = \frac{1}{4\pi} \int_\Omega (\gamma_1 - \gamma_2) \, \rho(\varphi, \vartheta) \, d\Omega, \tag{5.9}$$

where $\Omega = \sin \vartheta_0 / \vartheta d\varphi$ is an element of the solid angle and $\rho(\varphi, \vartheta)$ is the distribution function of molecule orientations, which is expressed in a system of O_1O_2Z coordinates.

The function $\rho(\varphi, \vartheta)$ in Eq. (2.22) is transformed into the O_1O_2Z system if the XYZ system of coordinates is rotated by the angle φ_m about the Z axis (Fig. 5.1).

For an optically anisotropic molecular system in which the refractive index of the substance is related to the polarizability of molecules by the Lorentz–Lorenz equation, the difference between the two main refractive indices, $n_1 - n_2$, is given by

$$n_1 - n_2 \equiv \Delta n = \frac{2\pi N_A}{9} \frac{(n^2 + 2)^2}{n} \frac{c}{M} \overline{(\gamma_1 - \gamma_2)}_0, \tag{5.10}$$

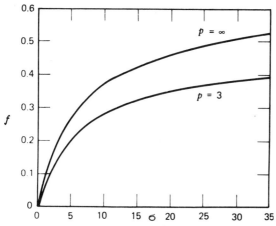

Fig. 5.4. Dependence of orientation factor $f(\sigma, b_0)$ on the parameter $\sigma = g/D_r$ for extended spheroidal particles at degrees of asymmetry $p = 3$ and $p = \infty$ [17].

where M is the molecular weight of dissolved molecules, c is the solution concentration in g/cm³, and n is the average refractive index of the solution.

The system of equations (5.8)–(5.10) leads to the expression for the value of FB caused by the orientation of dissolved molecules

$$\frac{\Delta N}{c} = \frac{2\pi N_A}{9} \frac{(n^2 + 2)^2}{n} \frac{\gamma_1 - \gamma_2}{M} f(\sigma, b_0), \tag{5.11}$$

where

$$f(\sigma, b_0) = (1/4\pi) \int_0^{2\pi} \int_0^{\pi} \rho(\varphi, \vartheta) \sin^3 \vartheta \cos 2\varphi \, d\varphi \, d\vartheta. \tag{5.12}$$

The function $f(\sigma, b_0)$ may be called the orientation factor because it characterizes the degree of orientation of axes of particles (molecules) in a laminar flow. The values of the orientation factor (σ, b_0) have been calculated and tabulated [17] over the entire range of σ values from 0 to ∞ using the distribution function $\rho(\varphi, \vartheta)$ represented in the form of a series of exponents of b_0 [16]. Figure 5.4 shows the dependences of (σ, b_0) on σ at two values of shape asymmetry of a spheroid p.

These dependences have the shape of curves whose initial slopes and asymptotic limits depend on the degree of asymmetry p of the particle increasing with it. At $p \to \infty$ (thin rod), the limiting value of (σ, b_0) is equal to 1.

At low σ the orientation factor may be represented [16] by the equation

$$f(\sigma, b_0) = \frac{\sigma b_0}{15} \left[1 - \frac{\sigma^2}{72} \left(1 + \frac{6b_0^2}{35} \right) + \cdots \right]. \tag{5.13}$$

It follows from this equation that in this range the function (σ, b_0) and, hence, FB are proportional to σ (or to g) and the factor of shape asymmetry b_0.

Equations (5.11) and (5.13) give, for the intrinsic value of FB, the equation

$$\lim_{\substack{g \to 0, \\ c \to 0}} \left(\frac{\Delta n}{g c \eta_0} \right) \equiv [n] = \frac{2\pi N_A^-}{135kT} \frac{(n^2 + 2)^2}{n} (\gamma_1 - \gamma_2) \frac{b_0 W}{M \eta_0} . \tag{5.14}$$

Using Eq. (2.27), it is possible to represent Eq. (5.14) in the following form:

$$\frac{[n]}{[\eta]} = \frac{4\pi}{45kT} \frac{(n^2 + 2)^2}{n} (\gamma_1 - \gamma_2) \frac{b_0}{6F(p)} . \tag{5.15}$$

Equation (5.15) shows that the ratio of intrinsic values of FB to the viscosity of solution may characterize the optical anisotropy $\gamma_1 - \gamma_2$ of dissolved molecules.

Equations (5.11), (5.13), and (5.14) contain two main factors determining the FB value: the optical factor $\gamma_1 - \gamma_2$ and the mechanical (hydrodynamic) factor (σ, b_0) which may be represented in various forms but is always proportional to the parameter of shape asymmetry b_0. This fact implies that the indispensable condition of FB in a solution of rigid particles (molecules) is not only their optical anisotropy, but also the asphericity of their shape where b_0 differs from zero.

1.4.   Low-Molecular-Weight Liquids

Although in Eqs. (5.11)–(5.15) the hydrodynamic properties of particles are described by a model of a macroscopic spheroid, these equations are intended for real molecules because the optical factor in these equations contains the internal field factor according to Lorentz–Lorenz.

In particular, Eq. (5.14), corresponding to the conditions $\sigma \ll 1$, may be used for the discussion of FB in low-molecular-weight liquids for which, according to the theory, Δn increases proportionally to g and the extinction angle φ_m is equal to $\pi/4$ at all g values accessible to experiments. In this case, if the theory is applied to a pure liquid rather than to a solution, the concentration c and the value of η_0 in Eq. (5.14) are replaced by ρ, the density of the substance, and η, its viscosity, respectively.

A more general model than that used in Eqs. (5.11)–(5.15) is a triaxial ellipsoid with optical axes γ_1, γ_2, and γ_3 and the corresponding geometrical axes a_1, a_2, and a_3. For this model the following expression may be used instead of Eq. (5.14):

$$\lim_{\substack{g \to 0, \\ c \to 0}} (\Delta n / g \eta_0 c) \equiv [n]$$

$$= \frac{\pi N_A}{135kT} \frac{(n^2 + 2)^2}{\eta_0 n M} [(\gamma_1 - \gamma_2) b_{12} W_3 + (\gamma_1 - \gamma_3) b_{13} W_2 + (\gamma_2 - \gamma_3) b_{23} W_1], \tag{5.16}$$

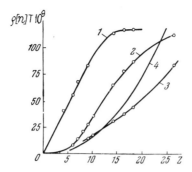

Fig. 5.5. Dependence of $[n]T\rho$ on the number Z of carbon atoms in the molecule for: 1) fatty acids; 2) normal alcohols; 3) normal alcohol solutions in cyclohexanol; 4) theoretical curve plotted according to Eq. (5.20) (on an arbitrary scale along the ordinate). ρ is the density of the liquid: for a solution $\rho = c$, i.e., it is the concentration in g/cm^3 [19–21].

where W_1, W_2, and W_3 are the friction coefficients of a triaxial ellipsoid rotating about axes 1, 2, and 3, respectively [24]. The coefficients of shape asymmetry of the ellipsoid b_{ik} are determined by the equation

$$b_{ik} = \frac{(a_i/a_k)^2 - 1}{(a_i/a_k)^2 + 1} = \frac{p_{ik}^2 - 1}{p_{ik}^2 + 1}. \tag{5.17}$$

If $\gamma_2 = \gamma_3$ and $a_2 = a_3$, the ellipsoid becomes a spheroid and Eq. (5.16) becomes Eq. (5.14), which is generally used for the interpretation of experimental data on FB.

Equation (5.16) is similar in form to Eq. (5.18) obtained for FB of low-molecular-weight liquids in the Raman–Krishnan theory [25].

$$\frac{\Delta n}{g\eta\rho} \equiv [n]$$

$$= \frac{2\pi}{45kT} \frac{(n^2 + 2)^2}{n\rho} \frac{(a_1 - a_2)(\gamma_1 - \gamma_2) + (a_1 - a_3)(\gamma_1 - \gamma_3) + (a_2 - a_3)(\gamma_2 - \gamma_3)}{a_1 + a_2 + a_3}. \tag{5.18}$$

In this theory, as in the foregoing theories, the FB of a pure liquid is considered to be the result of the orientation of molecules asymmetric in shape in the flow shear field. However, in contrast to the foregoing theories, in the Raman–Krishnan theory molecular orientation is caused by dilatation stresses directed at the angle $\pi/4$ to the flow direction rather than by the nonuniform rotation of the molecule in flow [Eqs. (2.19a) and (2.19b)]. Hence, in the theory [25] the orientation of molecules is *static*, and at all shear stresses the axis of this orientation forms an angle $\pi/4$ with the flow direction. Evidently, this theory cannot be applied to solutions of high-molecular-weight substances in which the extinction angle is a function of shear stress. However, for pure low-molecular-weight liquids this theory leads to the same qualitative results as those based on the concept of kinematic orientation of molecules.

In order to compare both theories under consideration, it is convenient to represent Eq. (5.18) in the form applicable to a molecular model with axial symmetry of optical and geometrical properties ($\gamma_2 = \gamma_3$ and $a_2 = a_3$). Then Eq. (5.18) becomes

$$[n]_\rho \equiv \frac{\Delta n}{g\eta} = \frac{4\pi}{45kT} \frac{(n^2 + 2)^2}{n} (\gamma_1 - \gamma_2) \frac{p - 1}{p + 2}, \qquad (5.19)$$

where, as previously, $p = a_1/a_2$. If Eqs. (2.10) and (2.20) are applied, Eq. (5.14) for a pure liquid may be represented in the form

$$[n]_\rho \equiv \frac{\Delta n}{g\eta} = \frac{4\pi}{45kT} \frac{(n^2 + 2)^2}{n} (\gamma_1 - \gamma_2) \frac{p^2 - 1}{p^2 + 1} \frac{f_0(p)}{6}, \qquad (5.20)$$

where $f_0(p)$ is the function determined from Eq. (2.11).

Hence, Eqs. (5.19) and (5.20) differ only in the type of dependence of the hydrodynamic factor on the shape asymmetry of the molecule.

It follows from Eqs. (5.19) and (5.20) that the sign of FB in a liquid is determined by that of the product of optical ($\gamma_1 - \gamma_2$) and hydrodynamic factors.

The available experimental data [18–22] show that FB is positive for all low-molecular-weight liquids. This fact implies that for the molecules of these liquids the signs of $\gamma_1 - \gamma_2$ and b_0 coincide. In other words, in low-molecular-weight substances the direction of the largest geometrical length of the molecule coincides with that of its greatest optical polarizability. This fact is one of the most important experimental confirmations of the general principles of the theory of induction interaction relating the anisotropy of polarizability of molecules to their shape [23].

However, it cannot be expected that the optical anisotropy $\gamma_1 - \gamma_2$ and geometrical characteristics b_0 and W of molecules of low-molecular-weight liquids obtained using Eqs. (5.19) and (5.20) according to FB data should be in quantitative agreement with the corresponding values determined from the optical properties of molecules in the gaseous phase and from their size, determined by x-ray analysis, in the solid phase. The application of the laws of hydrodynamics of macroscopic bodies to small molecules, as well as the reduction of the internal field in a liquid to the Lorentz–Lorenz isotropic field, is a rough approximation and hence may lead only to qualitatively reasonable results.

A comparison of experimental data with Eqs. (5.19) or (5.20) in the investigation of a homologous series of molecules whose shape and anisotropy change systematically (and in the known simple manner) in this series is more justifiable. These series are observed, for example, for normal hydrocarbons and their derivatives (alcohols and aliphatic acids). In this case, although the absolute value of the hydrodynamic factor in Eqs. (5.19) and (5.20) applied to low-molecular-weight liquids remains uncertain, its relative change with increasing chain length may be evaluated with greater reliability using the hydrodynamics of macroscopic particles.

Figure 5.5 shows the experimental values of $[n]T\rho$ as a function of z, the number of carbon atoms in the molecule for aliphatic acids, normal alcohols, and solutions of the latter in cyclohexanone [19–21]. Curve 4 in this figure represents (on an arbitrary scale along the ordinate) the theoretical dependence according to Eq. (5.20).

The absolute magnitudes of the experimental values of $[n]$ are much lower than those predicted by both Eq. (5.19) and, in particular, Eq. (5.20), which does not seem surprising in light of the foregoing considerations.

In the type of dependence of $[n]$ on chain length at low Z, experimental curve 1 for aliphatic acids is in better agreement with theoretical equation (5.19) than with Eq. (5.20). In contrast, the shape of experimental curves 2 and 3 for alcohols and their solutions is in better agreement with Eq. (5.20) than with Eq. (5.19). However, the general type of dependence of $[n]$ on Z shows, at least qualitatively, that the hydrodynamic and optical properties of low-molecular-weight normal hydrocarbons in solution and in bulk correspond to those of rigid rodlike particles.

In the investigated range of Z values the curve for acids lies above that for alcohols, which lies above that for their solutions. This may probably be ascribed to the effect of intermolecular interaction and molecular association increasing the effective value of the hydrodynamic parameter in Eqs. (5.19) and (5.20).

As the chain becomes longer, the experimental curves begin to show an inflection point and a tendency to saturation. This may be explained by a deviation in the shape of the hydrocarbon molecule from the planar trans-chain as a result of its natural flexibility.

Similar data have been obtained for n-alkanes [22].

1.5. Optical Anisotropy of Macroscopic Particles

For solutions containing particles whose size is much larger than that of solvent molecules, the optical properties are similar to those of a system of colloidal particles surrounded by a continuous medium of the solvent with the refractive index n_s. In this case, the optical field acting on the particle does not differ from the average macroscopic field in the medium, and the main refractive indices, n_x and n_y, of the anisotropic medium are not related to the corresponding average components, $\overline{\gamma_x}$ and $\overline{\gamma_y}$, of the polarizability of particles by the Lorentz–Lorenz equation but, rather, by the equations

$$n_x^2 - 1 = 4\pi\,(N_A c/M)\,\overline{\gamma}_x, \quad n_y^2 - 1 = 4\pi\,(N_A c/M)\,\overline{\gamma}_y. \tag{5.21}$$

Hence, if it is assumed that $n_x + n_y \approx 2n$, the following equation is obtained instead of Eq. (5.11):

$$\Delta n/c = (2\pi N_A/n)\,[(\gamma_1 - \gamma_2)/M]\,f\,(\sigma,\ b_0), \tag{5.22}$$

where $f(\sigma, b_0)$ does not differ from the corresponding function in Eq. (5.11).

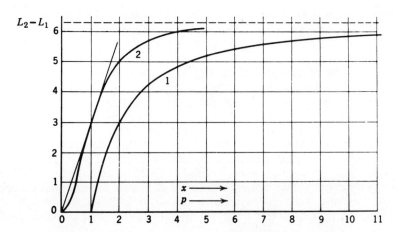

Fig. 5.6. Dependence of the optical factor of form anisotropy $L_2 - L_1 = 4\pi e$: 1) on the axial ratio p of the spheroid; 2) on the ratio $x = h/\langle h^2 \rangle^{1/2}$ for a Gaussian chain.

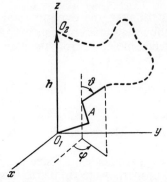

Fig. 5.7. Orientation of a segment of a freely jointed chain in a system of XYZ coordinates in which the Z axis coincides with the vector $O_1O_2 = h$ joining the chain ends. ϑ is the angle formed by the segment axis and the vector h: A and φ are the length and the azimuth of the segment, respectively.

    In the general case the difference between the main polarizabilities, $\gamma_1 - \gamma_2$, of a macroscopic (colloidal) particle may be the sum of two terms: the intrinsic anisotropy of the particle $(\gamma_1 - \gamma_2)_i$ and the anisotropy of its shape $(\gamma_1 - \gamma_2)_f$. Intrinsic anisotropy is due to the ordered structure of the substance of the molecule (the existence of orientational order of valence bonds and atomic groups composing the particle, the crystal structure of this substance, etc.). When the size of the particle is relatively large, a certain refractive index may be ascribed to the substance of the particle. For an anisotropic particle the value of this index is different in different directions (n_1, n_2, and n_3).

The shape anisotropy is characteristic of particles whose geometrical shape exhibits no spherical symmetry with the refractive index differing from that of the solvent, n_s. In particular, as Maxwell has shown [26], a particle having the shape of a spheroid with coinciding geometrical and dielectric axes, located in a medium with the refractive index n_s, and situated in a uniform electric field, is polarized uniformly but anisotropically.

Hence, for an extended spheroid ($p > 1$) the total difference between the two main polarizabilities of the particle, $\gamma_1 - \gamma_2$, is given by [14]

$$\gamma_1 - \gamma_2 = (g_1 - g_2)v = (\gamma_1 - \gamma_2)_i + (\gamma_1 - \gamma_2)_f$$

$$= \frac{(3n_s^2)^2}{4\pi} \frac{(n_1^2 - n_2^2) + e(n_1^2 - n_s^2)(n_2^2 - n_s^2)/n_s^2}{[(n_1^2 + 2n_s^2) - 2e(n_1^2 - n_s^2)][(n_2^2 + 2n_s^2) + e(n_2^2 - n_s^2)]} v. \tag{5.23}$$

where v is the volume of the spheroid, g_1 and g_2 are the main specific polarizabilities of the particle, and e is the factor of shape asymmetry of the spheroid varying from 0 to 1/2 when the degree of its asymmetry p changes from 1 to ∞. The dependence of e on p is expressed by

$$e = \left[2p^2 + 4 - \frac{3p}{\sqrt{p^2-1}} \ln \frac{p + \sqrt{p^2-1}}{p - \sqrt{p^2-1}}\right] \bigg/ 4(p^2 - 1). \tag{5.24}$$

Curve 1 in Fig. 5.6 represents the dependence of the parameter $L_2 - L_1 = 4\pi e$ on p according to Eq. (5.24). Curve 1 shows a rapid increase in the asymmetry factor e with increasing shape asymmetry p and with subsequent approach of the curve $e(p)$ to its asymptotic limit $e = 1/2$.

According to Eq. (5.23), intrinsic anisotropy disappears at $n_1 = n_2$. The shape anisotropy disappears at $e = 0$, i.e., for a spherical particle, and also for a particle of any shape if the refractive index of the solvent n_s is equal to n_1 or n_2. At all other values of n_s the anisotropy of shape is positive and, according to Eq. (5.23), the dependence of $(\gamma_1 - \gamma_2)_f$ on n_s is described by a curve similar to a parabola with a minimum at $n_s^2 \approx (n_1^2 + 2n_2^2)/3$. The value of FB determined by Eq. (5.22) exhibits a similar dependence on the optical properties of the system.

2. FLEXIBLE CHAIN MOLECULES

It has already been pointed out in Chapter 1 that a chain of freely jointed statistical segments – a Gaussian chain – is often used as a model for describing the properties of chain molecules. In fact, this model is in good agreement with the molecular conformation of real flexible-chain polymers in θ-solvents.

In the theoretical consideration of optical properties of molecules and FB in flexible-chain polymer solutions, the freely jointed Gaussian chain also serves as the main model.

2.1. Optical Anisotropy of a Chain Molecule

2.1.1. Intrinsic Anisotropy. The theory of anisotropy of chain molecules has been developed in a classical paper by Kuhn and Grün [27], who used for this purpose a freely jointed chain model.

It is assumed in the theory that each statistical segment (of length A) is characterized by two main optical polarizabilities α_1 and α_2, and the axis of optical polarizability (α_1) of the segment coincides with its geometrical axis A.

A system of N anisotropic segments comprising the freely jointed chain produces the anisotropy of the whole molecule. However, in contrast to the mean-square anisotropy δ^2 [Eq. (1.50)], depending on the angles between the axes of various segments, the anisotropy of the molecule manifested in FB is determined by the type of segment orientation with respect to the direction fixed in the molecule, and assumed to be its axis. In the Kuhn theory the vector $\mathbf{h}$ joining the chain ends is assumed to be this axis. Hence, to calculate the difference in the two main polarizabilities, $\gamma_1 - \gamma_2$, of a chain molecule in the direction of the vector $\mathbf{h}$ and in that normal to this vector, it is necessary to consider the segment distribution on the angles ϑ of their orientation with respect to h (Fig. 5.7).

Considering segment distribution on azimuths φ (Fig. 5.7) to be random, and using the general methods of statistical mechanics, Kuhn and Grün have shown that for a long chain ($N \gg 1$) the most probable distribution on the angles ϑ is determined by the function

$$\rho(\vartheta)\,d\vartheta = Ce^{q\cos\vartheta}\sin\vartheta d\vartheta. \tag{5.25}$$

Here the parameter q is determined by the equation

$$\overline{\cos\vartheta} = h/L = \mathscr{L}(q), \tag{5.26}$$

where L is the contour length of the chain and $\mathscr{L}(q)$ is the Langevin function, given by

$$\mathscr{L}(q) = \coth q - 1/q, \tag{5.27}$$

and $\overline{\cos\vartheta}$ is the cosine of the angle formed by a segment with the axis h averaged over all chain segments and all chain configurations corresponding to a given value of h.

Using function (5.25), it is easy to calculate the mean square $\overline{\cos^2\vartheta}$ given by

$$\overline{\cos^2\vartheta} = 1 - 2(h/L)/\mathscr{L}^*(h/L), \tag{5.28}$$

where $\mathscr{L}^*(h/L)$ is the "inverse" Langevin function determined by the series

$$\mathscr{L}^*(h/L) = 3h/L + (9/5)(h/L)^3 + (297/175)(h/L)^5 + (1539/875)(h/L)^7 + \ldots \tag{5.29}$$

Hence, both $\overline{\cos\vartheta}$ and $\overline{\cos^2\vartheta}$ in a freely jointed chain are unique functions of the h/L ratio. For a chain of a given length L the change in the end-to-end distance h is accompanied by a change in the degree of orientation of segments. At $h \to 0$ (ends coincide), according to Eqs. (5.29) and (5.28), $\overline{\cos^2\vartheta}$ is 1/3 and segment orientations are random. At $h \to L$ (completely extended chain), we have $\mathscr{L}^*(h/L) \to \infty$ and, correspondingly, $\overline{\cos^2\vartheta} \to 1$, i.e., all segments are oriented parallel to h.

The preferred orientation of an assembly of N segments in the direction h induces the anisotropy of the molecule "as a whole," and the difference in its main polarizabilities γ_1 (in the direction h) and γ_2 (normal to h) is evidently related to the anisotropy of the segment, $\alpha_1 - \alpha_2$, by the equation

$$\gamma_1 - \gamma_2 = N (\alpha_1 - \alpha_2)(\overline{3\cos^2\vartheta} - 1)/2 \tag{5.30}$$

or, applying Eq. (5.28),

$$\gamma_1 - \gamma_2 = N (\alpha_1 - \alpha_2)[1 - 3 (h/L)/\mathscr{L}^* (h/L)]. \tag{5.31}$$

It can easily be shown [28] that the expression

$$\mathscr{L}^* (h/L) = 3 (h/L) [1 - (2/5) (h/L)^2]/[1 - (h/L)^2] \tag{5.32}$$

is a good approximation for $\mathscr{L}^*(h/L)$.

Using Eq. (5.32) instead of Eq. (5.31), we obtain

$$\begin{aligned}\gamma_1 - \gamma_2 &= (3/5) N (\alpha_1 - \alpha_2) (h/L)^2/[1 - (2/5) (h/L)^2] \\ &= (3/5)(\alpha_1 - \alpha_2) (h^2/AL)/[1 - (2/5) (h/L)^2]. \end{aligned} \tag{5.33}$$

Equation (5.33), just as Eq. (5.31), is obeyed at all h values $0 \le h \le L$, i.e., it may be applied to a chain molecule at any degree of coiling h/L and any conformation varying from a completely extended conformation ($h = L$) to that with coinciding chain ends ($h = 0$).

According to Eq. (5.33) at $h = L$ we have

$$\gamma_1 - \gamma_2 = N (\alpha_1 - \alpha_2), \tag{5.34}$$

as is to be expected for a rodlike molecule.

In the range of coiled conformations in which $h/L \ll 1$, it follows from Eq. (5.33) that

$$\gamma_1 - \gamma_2 = (3/5) (\alpha_1 - \alpha_2) (h^2/AL) = \vartheta_i h^2/AL = \vartheta_i \frac{h^2}{\langle h^2 \rangle}. \tag{5.35}$$

For a Gaussian chain in the equilibrium state we have $h^2 = \langle h^2 \rangle = AL$ and, according to Eq. (5.35),

$$\langle \gamma_1 - \gamma_2 \rangle = (3/5)(\alpha_1 - \alpha_2) = \theta_i.$$ (5.36)

Hence, the optical anisotropy of the Gaussian chain averaged over all its conformations is 3/5 of that of the corresponding Kuhn segment.

When the Gaussian chain undergoes extension deformation, according to Eq. (5.35) its anisotropy increases proportionally to h^2, the square end-to-end distance.

2.1.2. Solvent Effect. Under real conditions, a polymer molecule in solution is surrounded by the solvent medium which affects the optical properties of the polymer chain.

If the refractive index of the solvent n_s differs from that of the dissolved polymer n_k, then additional anisotropy of the molecule appears as a result of mutual optical interaction of single chain parts.

In this case, the interaction between chain elements relatively distant along the chain (long-range optical interaction) and that between neighboring elements (short-range optical interaction) should be distinguished.

2.1.2.1. Anisotropy of macroform. Instantaneous coil conformations are aspherical [29–31]. As a result of nonspherical mass distribution in a coiled Gaussian chain, optical long-range interaction in a chain molecule leads to the anisotropy of the polarizing field inside the molecular coil and to the additional difference in the polarizabilities of the molecule $(\gamma_1 - \gamma_2)_f$, the anisotropy of macroform.

The simplest procedure for the evaluation of the anisotropy of macroform is the approximation of a molecular coil by a spheroid with an average optical density n_ω determined by the equation [32, 33]

$$n_\omega^2 = n_s^2 + (n_k^2 - n_s^2)/\omega.$$ (5.37)

where the value of $\omega = v\rho/m$ (m is the mass of the molecule and ρ is the density of dry polymer) shows the factor by which the volume v occupied by a molecule in solution is greater than that, m/ρ, in a dry polymer.

Under Eqs. (5.23) and (5.37) we obtain the equation for the anisotropy of macroform of a chain molecule in solution [34, 35]:

$$(\gamma_1 - \gamma_2)_f = \left(\frac{n_k^2 - n_s^2}{4\pi n_s \rho N_A}\right)^2 \frac{M^2}{v} 4\pi e = \theta_f 4\pi e,$$ (5.38)

which is always positive.

According to Kuhn [30, 31], the average shape asymmetry of a nondeformed molecular coil is determined by its average maximum $\langle H_1 \rangle$ and minimum $\langle H_2 \rangle$ dimensions (Fig. 2.6) related to the mean square end-to-end distance $\langle h^2 \rangle^{1/2}$ by the equations

$$\langle H_1 \rangle = 1.4 \langle h^2 \rangle^{1/2}, \quad \langle H_2 \rangle = 0.7 \langle h^2 \rangle^{1/2}. \tag{5.39}$$

Accordingly, the volume of the Gaussian coil is given by

$$v = \pi \langle H_1 \rangle \langle H_2 \rangle^2 / 6 = 0.36 \langle h^2 \rangle^{3/2} = 0.36 \, (LA)^{3/2}. \tag{5.40}$$

Hence, the axial ratio p of the coil may be expressed [36] as a function of the parameter $x \equiv h/\langle h^2 \rangle^{1/2} = h/(AL)^{1/2}$:

$$p = [1 + (3/2) \, x^2]^{3/4}. \tag{5.41}$$

Comparison of Eqs. (5.24) and (5.41) allows the calculation of the parameter $4\pi e$ at various x values. The dependence of $4\pi e$ on x obtained in this manner is shown by curve 2 in Fig. 5.6. This curve shows that in the range $0.5 \leq x \leq 1.5$ (i.e., at a slight deformation of molecules), we have $4\pi e \approx 3x$ and hence we obtain instead of Eq. (5.38) the following equation:

$$(\gamma_1 - \gamma_2)_{f, \, x < 3/2} \simeq 3\theta_f h / \langle h^2 \rangle^{1/2}. \tag{5.42}$$

The expressions for $(\gamma_1 - \gamma_2)_f$ analogous to Eq. (5.42) have also been obtained in [37–39], in which the optical interaction in a Gaussian coil has been expressed in terms of the theory of induced dipole interaction [23, 40].

At $x > 1.5$, Eq. (5.42) is inapplicable and curve 2 in Fig. 5.6 should be used.

2.1.2.2. Anisotropy of microform. Neighboring elements (monomer units) of the chain are arranged in a certain linear order with respect to each other. Hence, their optical interaction cannot be spherically symmetrical. This axially symmetrical short-range interaction in the chain leads to the appearance of local field anisotropy, just as the asphericity of form of the entire molecule leads to the appearance of average field anisotropy. The local field anisotropy depends on the microstructure of the chain and increases with its equilibrium rigidity (segment size and shape asymmetry). As a result of this effect, additional anisotropy of the segment polarizability and, hence that of the whole macromolecule, appears (microform anisotropy); it is also positive in sign.

The difference between the two main polarizabilities of the macromolecule (in the direction of the vector h and normal to it) corresponding to the microform effect is given [41, 42] by

$$(\gamma_1 - \gamma_2)_{fs} = \frac{3}{5} \left(\frac{n_k^2 - n_s^2}{4\pi n_s} \right)^2 \frac{M_0 S}{\rho N_A} 4\pi e_s \frac{h^2}{LA} = \theta_{fs} \frac{h^2}{LA}, \tag{5.43}$$

where M_0 is the molecular weight of the monomer unit, S is the number of monomer units in a Kuhn segment, and $4\pi e_s$ is the function determined by Eq. (5.24) and the curve in Fig. 5.6 where the axial ratio p characterizes the asymmetry of segment form.

The total difference in polarizabilities of a chain molecule is equal to the sum of three effects:

$$\gamma_1 - \gamma_2 = (\theta_i + \theta_{fs})(h^2/AL) + \theta_f/4\pi e, \qquad (5.44)$$

where θ_i, θ_f, and θ_{fs} are determined by Eqs. (5.36), (5.38), and (5.43), respectively.

Comparison of Eqs. (5.38), (5.40), and (5.43) readily shows that for a given macromolecule (with given L, M, and n_k) in a certain solvent (with given n_s) the relative part played by the macroform (θ_f) and microform (θ_{fs}) effect depends on the equilibrium rigidity parameter of the chain A. The value of θ_f is inversely proportional to v or $A^{3/2}$, whereas θ_{fs} is directly proportional to S or A. This means that the higher the equilibrium rigidity of the chain, the greater is θ_{fs} and the lower is θ_f in the total anisotropy of the molecule. Hence, the experimentally determined value of θ_{gs}/θ_f may serve as a measure of equilibrium rigidity of a chain molecule [43].

2.1.2.3. Anisotropic solvents. Experiments show that various solvents may affect the value of intrinsic anisotropy θ_i of the polymer chain, and this effect is not directly related to the difference in the values of n_k and n_s, i.e., this is not the macro- or microform effect. Thus, in solvents with strongly and anisotropically polarized molecules, the positive intrinsic anisotropy of the molecules of some polymers is higher than in solvents with optically isotropic molecules [9, 43–45]. To account for this fact, some authors have suggested that the anisotropic molecules of the solvent exhibit a preferred orientation of the axes of their greatest polarizability in the direction of dissolved polymer chains [46–48]. However, no proofs of the existence of this orientation obtained by a more direct method are available at present. But the dependence of intrinsic anisotropy of a polymer molecule on that of solvent molecules implies that matching solvents ($n_k = n_s$) with optically isotropic molecules should be used for the determination of θ_i.

2.2. Hydrodynamic Properties

Various molecular models used in discussing the hydrodynamic behavior of polymer molecules in the phenomena of translational and rotational friction and viscometry have been considered in Chapter 2.

In principle the same models may be used in FB theories as has been done in Section 1 of this chapter in discussing FB in solutions of rigid spheroidal particles.

However, if flexible-chain molecules are considered, it should be taken into account that both the intramolecular thermal motion of chain elements and chain deformation in the flow shear field may play an important part in the FB phenomenon in the solutions of these molecules. Hence, in contrast to Eq. (2.21), diffusion equations describing the behavior of a flexible-chain molecule in FB should contain terms in which the flexibility and intramolecular (micro-Brownian) motion of the chains are taken into account.

To describe the FB of flexible-chain polymers, the following molecular models, in which these factors are taken into account, have been used: a more primitive elastic dumbbell model [41, 49, 50] and a more realistic elastic necklace model [51].

2.2.1.   Elastic Dumbbell.   In the theory advanced by Kuhn, the hydrodynamic properties of a chain molecule are represented by a model for an elastic (deformable) dumbbell with length equal to h, the end-to-end distance. Hydrodynamic forces in flow are applied to each of the two end points (beads) of the dumbbell (to chain ends) and are proportional to chain length L, the viscosity of the solvent, and its rate with respect to the bead. The proportionality of these forces to L actually means that the theory does not take into account the hydrodynamic interaction between chain elements, and the true spatial arrangement of these elements is replaced by two points, the chain ends, which is doubtless a rough approximation. However, the adoption of this rough model leads to results which are in qualitative agreement with the data obtained using more rigorous theories, and the relative simplicity of the mathematical treatment makes it possible to obtain a clear physical picture of the FB phenomenon.

The chain ends are kept at a distance h apart by the entropy elastic force F acting on these ends and is, according to Kuhn, determined by the equation

$$F = -\frac{kT}{A}\,\mathcal{L}^*\,(h/L). \qquad (5.45)$$

At $h/L \ll 1$, according to Eq. (5.32), Eq. (5.44) becomes

$$F = -\frac{3kT}{AL}\,h = -\frac{3kT}{\langle h^2 \rangle}\,h. \qquad (5.45a)$$

A combination of the entropy force F and the diffusion motion of chain ends in a stationary solution leads to the equilibrium distribution of the end-to-end distance h determined by Eq. (1.2).

In a laminar flow, the distribution type changes as a result of nonuniform rotation of the dumbbells [see Eqs. (2.19)] and the rotational diffusion flow of molecules induced by this rotation. Moreover, an elastic dumbbell rotating in a laminar flow of a viscous solvent undergoes periodical tensions and compressions which are maximum in the directions to the flow $+\pi/4$ and $-\pi/4$, respectively. This leads to a flow of chain ends in the radial direction (parallel to h). Under these conditions the state of a flexible-chain molecule is determined by both the value of h (varying under the influence of viscous flow forces and diffusion of chain ends) and the orientation (φ, ϑ) of the vector $\mathbf{h}$ in space. Correspondingly, in contrast to Eq. (2.22), the chain-end distribution in the coordinate system of Fig. 2.1 is a function $\rho(h, \varphi, \vartheta)$ whose projection on the flow plane $(\vartheta = \pi/2)$ is represented by the equation [41, 49]

$$\rho\,(h,\ \varphi) = \frac{3}{2\pi\,\langle h^2 \rangle}\,\frac{1}{(1 + \beta_0^2)^{1/2}}\,\exp\left\{-\frac{3}{2}\,\frac{h^2}{\langle h^2 \rangle}\left[1 - \beta_0\,\frac{\sin 2\varphi + \beta_0 \cos 2\varphi}{1 + \beta_0^2}\right]\right\}, \qquad (5.46)$$

where β_0 is the hydrodynamic parameter determined from Eq. (3.66).

In a stationary solution ($\beta_0 = 0$), function (5.46) is equivalent to distribution (1.2) converted to two-dimensional space.

The function $\rho(\varphi)$ of the orientational distribution of molecular axes h in the flow plane (regardless of the value of h) is obtained by integration of Eq. (5.46) between $0 < h < \infty$:

$$\rho\,(\varphi) = \frac{1}{2\pi}\,\frac{1}{(1+\beta_0^2)^{1/2}}\,\frac{1}{1-\beta_0\,\dfrac{\sin 2\varphi + \beta_0\cos 2\varphi}{1+\beta_0^2}}\cdot \tag{5.47}$$

The preferred orientation angle φ_m is found from the condition of the maximum $\rho(\varphi)$ [i.e., the minimum in the denominator in Eq. (5.47)], which gives

$$\tan 2\varphi_m = 1/\beta_0. \tag{5.48}$$

Equation (5.48) coincides with Eq. (5.2) for rigid rodlike particles or spheroids with the degree of symmetry $p = 10$ for which we have, in Eq. (2.27), $F(p) = 1/6$.

2.2.2. Elastic Necklace.

2.2.2.1. Molecular model. The elastic necklace model, or the "bead-and-spring model" [52], is based on the Gaussian necklace shown in Fig. 2.7, and is used in Chapter 2 to describe the transport properties of polymer chains.

However, as already indicated, in order to describe these properties under steady-state conditions it is possible to use absolutely rigid models frozen in the most probable (or "average") conformations, whereas in the FB phenomenon it is necessary to take into account the kinetic chain flexibility and the thermal motion of chain parts (beads).

Hence, in the elastic necklace model, the freely jointed bonds between neighboring beads (Fig. 2.7) are not absolutely rigid. In the steady state their lengths l, according to the law of Gaussian chains (1.2), are distributed about average values $\langle l^2 \rangle^{1/2}$, which is the result of the action of diffusion forces and entropy elasticity forces F. The entropy forces F are determined by Eq. (5.45), in which h and $\langle h^2 \rangle$ are replaced by l and $\langle l^2 \rangle$, respectively.

When a flexible molecule is deformed in flow, the distance l between beads may vary, whereas $\langle l^2 \rangle$ is constant. Hence, the value of $3kT/\langle l^2 \rangle$ is equivalent to the force constant of an elastic spring representing the entropy restoring force F. Thus, each chain element between neighboring beads is a Gaussian "subchain" for which the hydrodynamic properties in flow are analogous to those of the Kuhn elastic dumbbell. However, since each chain contains $N + 1$ beads, its friction characteristics are calculated taking into account hydrodynamic interactions between beads according to Kirkwood's method (see Chapter 2).

In accordance with the foregoing considerations, when the model of an elastic necklace in a shear flow is used, the problem of the distribution function should be solved for $N + 1$ beads composing each chain rather than for chain ends (as in the case of an elastic dumbbell). This leads to a differential equation with $3(N + 1)$ independent variables. Using the Zimm method of normal coordinates [23], this equation is separated into $N + 1$ three-dimensional equations. The solution of this system yields the distribution function ρ in the form of the product $\rho = \Pi\rho_i$, $i = 1, 2, 3, \ldots, N$, where ρ_i is the distribution function characteristic of the ith motion mode of the polymer chain. Particular functions ρ_i resemble the distribution function for the elastic dumbbell model [50].

Hence, simultaneous motion of all chain segments (subchains) is described as the sum of a series of constituents (mobility modes). Each constituent represents the motion of only one chain part, and this motion ranges from the relative motion of chain ends ($i = 1$) to those of neighboring segments ($i = N$). A certain relaxation time τ_i corresponds to each mode and the totality of times τ_i forms the spectrum of relaxation times of intramolecular motions of the entire chain.

The spectrum of relaxation time of a polymer molecule may be studied by the method of oscillatory FB, in which a sinusoidal shear field is used [53–56], and the interpretation of experimental data is based on the elastic necklace theory [52, 54]. ·

Here the results of the theory are reported for the case of a steady flow (field frequency is zero).

2.2.2.2. Intrinsic viscosity and extinction angle. The application of the distribution function $\rho = \Pi\rho_i$ makes it possible to calculate the main hydrodynamic characteristics of a solution of molecules represented by a Gaussian elastic necklace: relaxation times τ_i, intrinsic viscosity $[\eta]$, and the angle of preferred orientation (extinction angle)

$$\tau_i = \frac{\langle l^2 \rangle \zeta}{6kT\lambda_i} = M\,[\eta]\,\eta_0 \bigg/ \left(RT\lambda_i \sum_{i=1}^{N} \frac{1}{\lambda_i} \right), \tag{5.49}$$

$$[\eta] = \frac{N_A \langle l^2 \rangle \zeta}{6M\eta_0} \sum_{i=1}^{N} \frac{1}{\lambda_i}, \tag{5.50}$$

$$\tan 2\varphi_m = \frac{\sum\limits_{i=1}^{N} \tau_i}{g \sum\limits_{i=1}^{N} \tau_i^2}, \tag{5.51}$$

where γ_i are the numerical coefficients (eigenvalues) characterizing the contribution of various motion modes to the phenomenon under investigation.

Final results that may be obtained from Eqs. (5.49)–(5.51) depend on the hydrodynamic interaction in the molecule determined by the parameter X in Eq. (2.81). For the model of a freely jointed elastic necklace discussed here, the effective bond length l_0 contained in Eq. (2.81) is $l_0 = \langle l^2 \rangle^{1/2} = A$, where A is the segment length and, correspondingly, the number of beads $n = N$.

In the absence of hydrodynamic interaction [51] ($x \to 0$, a draining coil), the coefficients γ_i are equal to $\gamma_i = \pi^2 i^2/N^2$ and, correspondingly, $\sum_{i=1}^{N} (1/\lambda_i) = N^2/6$ and $\sum_{i=1}^{N} (1/\lambda_i^2) = N^4/90$.

The substitution of these equations into Eqs. (5.49)–(5.51) gives

$$\tau_i = \langle l^2 \rangle \zeta N^2/(6kT\pi^2 i^2) = 6M\,[\eta]\,\eta_0/(\pi^2 i^2 RT), \tag{5.52}$$

$$[\eta] = N_A \zeta \langle l^2 \rangle N^2/(36\eta_0 M), \tag{5.53}$$

$$\tan 2\varphi_m = 1/C\beta_0. \tag{5.54}$$

The coefficient $1/c$ in Eq. (5.54) is 2.5 and the parameter β_0 has the same value as in Eqs. (5.48) and (3.66).

Equation (5.53) coincides with Debye's equation (2.87) for the intrinsic viscosity of a rigid Gaussian necklace without hydrodynamic interaction.

Equation (5.54) differs from Kuhn's equation (5.48) in the presence of a numerical coefficient C differing from unity. Although both equations describe the hydrodynamics of a freely draining chain molecule (without hydrodynamic interaction), Eq. (5.54) is based on a multisegmental model characterized by an array of relaxation times τ_i, whereas Eq. (5.48) corresponds to the hydrodynamics of a single-segment elastic dumbbell with one relaxation time τ. If a single relaxation time is assumed for the elastic necklace model and, correspondingly, in Eqs. (5.50) and (5.51) λ_i and τ_i are replaced by λ and τ, Eq. (5.50) becomes

$$\tau = M\,[\eta]\,\eta_0/RT, \tag{5.50a}$$

and Eq. (5.54) becomes Eq. (5.48).

Moreover, if Eqs. (5.50a) and (2.27) [with $F(p) = 1/6$] are compared, it can easily be seen that $\tau = 1/6D_r$, where D_r is the rotational diffusion coefficient for a rigid dumbbell (rod) rotating about the transverse axis. According to Eq. (5.52), relaxation time τ is greater by a factor of $\pi^2/6$ than relaxation time τ_1 for the first mode of the multisegmental model.

If the hydrodynamic interaction in the molecule is high ($X \gg 1$) [52], the coefficients λ_i in Eqs. (5.49)–(5.51) are functions of the parameter X in Eq. (2.81):

$$\lambda_i = \left(4X/\sqrt{2}\,N^2\right)\lambda_i' = 2\zeta\left(\sqrt{3\pi^3}\,\eta_0\langle l \rangle^{1/2}N^{3/2}\right)^{-1}\lambda_i', \tag{5.55}$$

where the coefficients λ_i' calculated by Zimm [57] are as follows: $\lambda_1' = 4.04$, $\lambda_2' = 12.79$, $\lambda_3' = 24.2$, and $\lambda_4' = 37.9$, and for high i values they are determined from the equation $\lambda_i' = (\pi^2 i^{3/2}/2)(1 - 1/2\pi i)$. The substitution of λ_i from Eq. (5.55) into Eqs. (5.49) and (5.50) gives

$$\tau_i = \frac{M\,[\eta]\,\eta_0}{RT}\left(\lambda_i' \sum_{i=1}^{N} \frac{1}{\lambda_i'}\right)^{-1}, \tag{5.56}$$

$$[\eta] = \frac{N_A\sqrt{3\pi^3}}{12}\frac{(N\langle l^2\rangle)^{3/2}}{M}\sum_{i=1}^{N}\frac{1}{\lambda_i'}. \tag{5.57}$$

Numerical summation yields [52] the values of the sums

$$\sum_{i=1}^{\infty}(1/\lambda_i') = 0.586, \quad \sum_{i=1}^{\infty}(1/\lambda_i'^2) = 0.0703. \tag{5.56a}$$

The substitution of the value of the first sum into Eq.(5.57) leads to Flory's equation (2.89) for a nondraining Gaussian coil with the coefficient Φ_∞ given by

$$\Phi_\infty = \frac{N_A\sqrt{3\pi^3}}{12}\cdot 0.586 = 2.84\cdot 10^{23}. \tag{5.58}$$

The substitution of the value of the first sum into Eq. (5.56) and the use of both sums in Eq. (5.51) for obtaining the extinction angle yields Eq. (5.54) with the value of the coefficient

$$1/C = (0.586)^2/0.0703 = 4.88.$$

Hence, if a dynamic model for a chain molecule, taking into account the kinetic flexibility and intramolecular motion of the Gaussian chain, is used in the theory of intrinsic viscosity, this theory leads to the same results as those based on the consideration of kinetically rigid chains "frozen" in conformations of the Gaussian coil. In this case, the differences in the numerical values of the coefficient Φ_∞ in Zimm's and Kirkwood's theories (Chapter 2) arise from the different calculations of hydrodynamic interaction in molecules rather than from the difference in their dynamic models. This fact justifies the neglect of intramolecular mobility of chain molecules assumed in the theories of their steady-state transport properties considered in Chapter 2.

Comparison of the values of coefficients $1/\lambda_1'$, $1/\lambda_2'$, ... contained in Eq. (5.57) with their sum $\sum_{i=1}^{\infty}(1/\lambda_i') = 0.586$ readily shows that the contribution of one first mode to intrinsic viscosity is about half the total effect, and the first five modes provide over 80% of the total losses for friction. This means that the losses for friction in a steady flow of a solution of flexible-chain molecules are mainly due to large-scale chain motion, whereas their small-scale deformations do not play an important part.

The evaluation of intramolecular motion and kinetic flexibility (deformability) of the Gaussian chain leads to much more important (cardinal) changes in the orientation angle of FB than in intrinsic viscosity. Thus, the replacement of a single-segment draining elastic dumbbell with one relaxation time by a multisegmental draining elastic necklace with an array of relaxation times τ_i increases the coefficient

$1/C$ in Eq. (5.54) from 1 to 5/2, and if the latter model takes into account strong hydrodynamic interaction, this coefficient increases to 4.88.

If chain draining is moderate, the values of the coefficient $1/C$ are intermediate between these two limits [58]. Moreover, deviation from Gaussian properties, as well as the coil expansion due to the excluded volume effects, decrease C because this expansion increases molecular draining [58]. Figure 5.8 shows the dependence of $1/C$ on the parameter of hydrodynamic interaction X. Different curves correspond to different values of the excluded-volume parameter ε determined by Eq. (1.36).

The coefficient C in Eq. (5.54) can be experimentally determined by plotting the dependence of $\tan 2\varphi_m$ on $1/\beta_0$. If a linear experimental dependence is obtained, it will mean that C is constant over the entire range of β_0 values.

However, it may be expected that some additional causes might lead to a change in the coefficient C with increasing parameter β_0. They are the inhomogeneous coil deformation in flow [59] and the finite value of the length of a soft chain molecule, which may be taken into account using the function $\mathscr{L}^*(h/L)$ in Eq. (5.44) [60, 61]. The theories of these effects are complex and contain several hypothetical suggestions on the possible deformation mechanism of chain molecules in the shear flow field, and at present cannot be checked by reliable experimental data. The few experimental data available show a decrease in the coefficient C with increasing β_0 [10]. However, it is difficult to give a unique quantitative interpretation of this effect, particularly because β_0 is a function of both the molecular dimensions (M and [η]) and the rate gradient g. Hence, it is more reliable to discuss the experimental data on extinction angles obtained under the conditions of limitingly low shear rates when the angles φ_m and $\chi = \pi/4 - \varphi_m$ may be considered to be linear functions of g. Under these conditions, the initial slope of the dependence $\chi = \chi(g)$ gives the characteristic orientation angle $[\chi/g]$ determined according to Eq. (5.4), and the coefficient C in Eq. (5.54) is equal to double the coefficient G in Eq. (5.6).

Hence, according to Eq. (5.48) for an elastic dumbbell, the coefficient G is equal to 0.5, and according to Eq. (5.49) for an elastic Gaussian necklace the value of G varies from 0.2 (draining coil) to 0.103 (nondraining coil).

For all the molecular models considered here (rigid spheroid, flexible dumbbell and flexible necklace), the extinction angle represents the rate at which FB relaxes when the shear field is no longer applied. However, for a monodisperse system of molecules homogeneous in their optical and hydrodynamic characteristics the extinction angle is completely determined by hydrodynamic properties of the system (dimensions, shape, and rigidity of particles) and is independent of the optical properties of the molecules being investigated.

This fact is of great practical importance because it allows the FB method to be adopted for the independent study of both the hydrodynamic properties of dissolved macromolecules according to measured FB orientations, and their optical characteristics according to the measurements of the value and sign of FB.

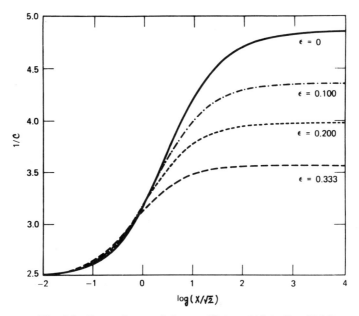

Fig. 5.8. Dependence of the coefficient $1/C$ in Eq. (5.54) on the parameter of hydrodynamic interaction $X/\sqrt{2} = \zeta N^{1/2}$. $(12\pi^3)^{-1/2} \cdot \eta_0 \langle l^2 \rangle^{1/2}$. Different curves correspond to different values of the excluded-volume parameter ε [58].

2.3. Value and Sign of FB

FB is a result of the orientation and deformation of optically anisotropic flexible-chain molecules in a shear field. The value and sign of this effect are strongly dependent on the optical properties of the solvent used.

2.3.1. FB of Solution in a Matching Solvent. In a matching solvent $(n_k = n_s)$ the form effect is absent and the anisotropy of the chain is its intrinsic anisotropy, defined by Eq. (5.35).

When the flexible dumbbell model is used, Eq. (5.35) is applied to the entire chain. For the elastic necklace model, Eq. (5.35) is applied to a single subchain and, accordingly, h and $\langle h^2 \rangle$ are replaced by l and $\langle l^2 \rangle$, respectively.

The excess FB of solution, Δn, over that of a pure solvent is calculated using (5.35) and the distribution functions of segments (subchains) $\rho = \Pi \rho_i$ (for the elastic necklace model) or the distribution function of chain ends [Eq. (5.46)] (for the elastic dumbbell model).

2.3.1.1. Intrinsic value of FB. In the range of weak flows ($g \to 0$), the theory predicts a linear dependence of Δn on g and if the elastic necklace model is used, this leads to the equation for the intrinsic value of FB [52]:

$$[n] \equiv \lim_{\substack{g \to 0, \\ c \to 0}} \left(\frac{\Delta n}{g c \eta_0} \right) = \frac{4\pi}{45kT} \frac{(n^2 + 2)^2}{n} (\alpha_1 - \alpha_2) \frac{N_A \langle l^2 \rangle \zeta}{6M\eta_0} \sum_{i=1}^{N} \frac{1}{\lambda_i}. \tag{5.59}$$

Comparison of Eqs. (5.59) and (5.50) yields an important relationship:

$$[n]/[\eta] = (4\pi/45kTn)(n^2 + 2)^2 (\alpha_1 - \alpha_2), \tag{5.60}$$

Equation (5.60) shows that the ratio of the intrinsic values of FB and viscosity in a matching solvent is a direct measure of the anisotropy of the segment, and the sign of $[n]$ coincides with that of $\alpha_1 - \alpha_2$.

Equation (5.60) has also been obtained on the basis of the elastic dumbbell model [41, 42]. It coincides with Eq. (5.15) for a rigid rodlike molecule whose anisotropy, $\gamma_1 - \gamma_2$, is equal to that of the segment, $\alpha_1 - \alpha_2$, of the Gaussian chain.

It is noteworthy that although, according to Eqs. (5.59) and (5.50), for a Gaussian chain both $[n]$ and $[\eta]$ depend on hydrodynamic interaction in the molecule, the $[n]/[\eta]$ ratio is independent of the degree of coil draining.

If it is assumed that solution concentration equally affects the hydrodynamic factor $N_A \langle l^2 \rangle \zeta \sum_{i=1}^{N} (1/\lambda_i)/6M\eta_0$ in Eqs. (5.50) and (5.59), then with the variation in concentration c the ratio

$$\left(\frac{\Delta n}{g c \eta_0} \right)_{g \to 0} \bigg/ \left(\frac{\eta - \eta_0}{\eta_0 c} \right)_{g \to 0} = \left[\frac{\Delta n}{g (\eta - \eta_0)} \right]_{g \to 0} \equiv \frac{\Delta n}{\Delta \tau} \tag{5.61}$$

should remain constant and equal to $[n]/[\eta]$.

Here, $\Delta n/\Delta \tau$, the ratio of the excess FB value (over that for the solvent) to the excess shearing stress $\Delta \tau = g(\eta - \eta_0)$ in a flowing solution, is the shear optical coefficient equal to one-half of the stress optical coefficient [64, 65]. The value of $\Delta n/\Delta \tau = [n]/[\eta]$ in a weak flow in a matching solvent actually remains constant over a wide range of variations in polymer concentration, as confirmed by numerous experimental data [9].

Moreover, the shear optical coefficient remains virtually invariable when the thermodynamic strength of the solvent varies and excluded-volume effects increase [66]. This fact implies that the flexible polymer coil under the influence of excluded-volume effects has no marked influence on the intrinsic segmental anisotropy of the chain, $\alpha_1 - \alpha_2$.

It follows from Eqs. (5.60) and (5.62) that the $\Delta n/\Delta \tau$ ratio is independent of the molecular weight of the polymer. This theoretical prediction is confirmed by numerous experimental data on FB in flexible-chain polymer solutions [9].

The measurement of the shear optical coefficient $\Delta n/\Delta\tau$ for a polymer in a matching solvent is the principal method for the experimental determination of intrinsic optical anisotropy of the chain. The anisotropy of the segment, $\alpha_1 - \alpha_2$, obtained by this method, is directly related to that of a monomer unit of the chain Δa by the equation

$$\alpha_1 - \alpha_2 = S\Delta a, \tag{5.62}$$

where S is the number of monomer units in the Kuhn segment.

The value and sign of the anisotropy of the monomer unit are sensitive indicators of its structure and may be very useful in the analysis of the molecular structure of the polymer [9, 13].

2.3.1.2.   Dependence of FB on shear rate.   With increasing shear rate g the deformation of the polymer chain develops and is manifested in a change in the end-to-end distance of the chain h and in those of subchains or segments l. Under these conditions, the theories based on the elastic necklace model and the elastic dumbbell model lead to a similar dependence

$$(\Delta n/g\eta_0 c)_{c \to 0} = [n](1 + \cot^2 2\varphi_m)^{1/2}, \tag{5.63}$$

where $[n]$ is determined according to Eq. (5.60) and $\cot 2\varphi_m$ is determined according to Eq. (5.54). At relatively high g values, Eq. (5.63) predicts a nonlinear dependence of Δn on shear rate. For the dumbbell model the nonlinearity is more pronounced than for the bead model, and for the latter it becomes weaker with increasing hydrodynamic interaction in accordance with different values of the coefficient C in Eq. (5.54). This theoretical result is qualitatively confirmed by experimental data for solutions of flexible-chain polymers of high molecular weight in matching solvents: Δn increases faster than proportionally to g [9, 67]. This is illustrated in Fig. 6.5, in which curve 1 represents the dependence of Δn on shear stress $\Delta\tau$ for polyisobutylene solutions in benzene (matching solvent).

It should be noted that the type of the dependence of Δn on g predicted by the theory [Eq. (5.63)] and observed experimentally in flexible-chain polymer solutions (upward curvature) differs greatly from the corresponding dependence [Eq. (5.11)] (Fig. 5.4) for rigid asymmetric molecules for which the curves $\Delta n = f(g)$ exhibit the downward curvature. These differences indicate that molecular deformation plays an important part in the FB phenomenon in flexible-chain polymer solutions at high values of g.

It must be borne in mind that the relationships predicted by Eq. (5.63) may be valid at such values of g (or, to be precise, of the parameter β_0) at which the end-to-end distance of the deformed chain is still much shorter than its outline length L ($h \ll L$). When the chain extension in flow is high, the theories taking into account the finite length of the entire chain [60] and of its segment [61] predict that the curves of the dependence $\Delta n = f(\beta_0)$ tend to the limit (saturation effect). However, the

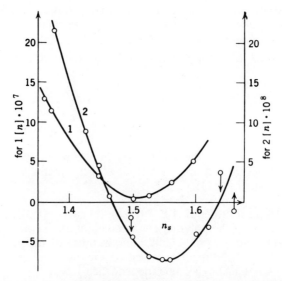

Fig. 5.9. Dependence of $[n]$ on n_s. Curve 1: for PMMA ($M = 4.2 \cdot 10^6$, $n_k = 1.50$) [68]. Curve 2: for PPTBPhMA ($M = 0.57 \cdot 10^6$, $n_k = 1.55$) [69].

shear stresses required for this effect are very high and are not attained in practice in the experimental investigations discussed in this book.

2.3.2. Influence of Form Effect on FB. In solutions in which the refractive index of the solvent n_s differs from that of the solute n_k, the relationships of FB are greatly complicated by the influence of the form effect. The observed FB is a sum of three effects caused by the intrinsic anisotropy of molecules, the anisotropy of their macroform, and that of their microform.

2.3.2.1. Intrinsic values of FB. At low shear rates ($g \to 0$) a chain molecule is virtually nondeformed by flow and the end-to-end distance h is equal to $\langle h^2 \rangle^{1/2}$, just as the end-to-end distance l for subchains in the necklace model is equal to $\langle l^2 \rangle^{1/2}$. Under these conditions, the factor dependent upon h is absent in Eqs. (5.35), (5.42), and (5.43), and Δn increases proportionally to shear stress, just as in the absence of the form effect. In this case, the intrinsic value of FB, $[n]$, or the shear optical coefficient, $[n]/[\eta]$, obtained by extrapolation to $g \to 0$ and $c \to 0$ may be represented as the sums of three terms

$$\frac{[n]}{|\eta|} = \frac{2\pi}{27kT} \frac{(n^2+2)^2}{n} \left[\theta_i + \theta_{fs} + \left(\frac{3}{2}\right)^2 \theta_f \right] = [n]_i/[\eta] + [n]_{fs}/[\eta] + [n]_f/[\eta]. \quad (5.64)$$

where the values of θ_i, θ_{fs}, and θ_f determined from Eqs. (5.36), (5.43), and (5.38), respectively, characterize the contributions to FB provided by the intrinsic anisotropy of molecules and the anisotropy of macro- and microform.

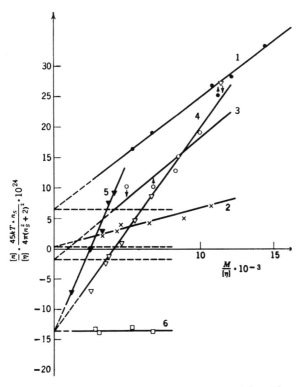

Fig. 5.10. Dependence of $([n]/[\eta])[45kTn/4\pi(n^2 + 2)^2]$ on $M/[\eta]$. 1) Polyisobutylene in hexane [70]; 2) poly(dimethylsiloxane) in toluene [71]; 3) poly(butyl methacrylate) in ethyl acetate [72]; 4) polystyrene in dioxane, $n_k - n_s = 0.18$ [73]; 5) polystyrene in methyl ethyl ketone, $n_k - n_s = 0.22$ [73]; 6) polystyrene in bromoform, $n_k - n_s \approx 0$ [73].

The $[n]_i/[\eta]$ ratio is defined by Eq. (5.60), where $\alpha_1 - \alpha_2$ is the intrinsic anisotropy of the segment.

According to Eqs. (5.43) and (5.64), we have

$$\frac{[n]_{fs}}{[\eta]} = \frac{(n^2 + 2)^2 \, (n_k^2 - n_s^2)^2}{45 RT \rho n_s^3} \, M_0 S e_s. \tag{5.65}$$

According to Eqs. (5.38), (5.40), and (2.89), the value of $[n]_f/[\eta]$ contained in Eq. (5.64) may be represented in the following form:

$$\frac{[n]_f}{[\eta]} = \frac{(n^2 + 2)^2 \, (n_k^2 - n_s^2)^2 \, 0.058\Phi}{\pi \rho^2 n^3 N_A RT} \, \frac{M}{[\eta]}, \tag{5.66}$$

where Φ is the Flory coefficient.

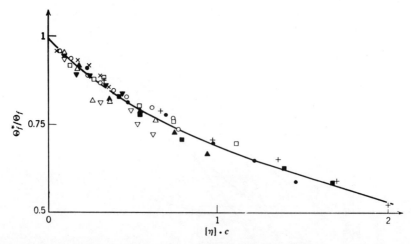

Fig. 5.11. Dependence of relative form anisotropy $\theta_f^*/\theta_f = [\Delta n/g(\eta - \eta_0)]/([n]_f/[\eta])$ on $[\eta]\cdot c$ for polyisobutylene fractions in hexane and poly(carboethoxyphenyl methacrylamide) in ethyl acetate [74].

Equations (5.64)–(5.66) show that the dependence of the shear optical coefficient on the refractive index of the solvent, n_s, has a parabolic shape. This dependence has actually been experimentally observed for all flexible-chain polymers investigated. Figure 5.9 shows as an example the dependence of the intrinsic value of FB, $[n]$, on n_s for poly(methyl methacrylate) (PMMA) and poly[(*para*-tertiary)-butylphenyl methacrylate] (PPTBPhMA) in various solvents. According to theoretical predictions, the experimental parabolic curves exhibit a minimum in a solvent for which $n_s = n_k$. Under these conditions, the value of $[n]$ corresponds to that of $[n]_i$, which is positive for PMMA and negative for PPTBPhMA. For flexible-chain polymers the dependence of $[n]$ on n_s is virtually determined by the term $[n]_f$ in Eq. (5.64) because, for these polymers, we have $[n]_{fs} \ll [n]_f$. Hence, a comparison of experimental curves in Fig. 5.9 with theoretical dependences enables us to evaluate the parameter p_0 of the shape asymmetry of a molecular coil in the equilibrium state. The resulting values are in reasonable agreement [9] with that of $p_0 \cong 2$ assumed in the theory [Eq. (5.41)].

Another characteristic indication of the form effect is the dependence of the shear optical coefficient on the molecular weight of the polymer. The values of θ_i and θ_{fs} contained in Eq. (5.64) represent the segmental anisotropy of the chain, and hence are independent of chain length and, correspondingly, of M. In contrast, the value of θ_f is determined by the optical properties of the coil as a whole and, according to Eqs. (5.38) and (5.40), for a Gaussian chain this value is proportional to $M^{1/2}$. In accordance with this, the $[n]_i/[\eta]$ and $[n]_{fs}/[\eta]$ ratios in Eqs. (5.60) and (5.65) are independent of M, whereas $[n]_f/[\eta]$ in Eq. (5.66) is proportional to $M/[\eta]$.

These relationships are confirmed by experimental data shown in Fig. 5.10. This figure represents the dependence of $[n]/[\eta]$ on $M/[\eta]$ for a series of polymer fractions in a certain solvent. In accordance with Eq. (5.64), the experimental points fit a straight line whose slope [according to Eq. (5.66)] characterizes the anisotropy of macroform, and the intercept with the ordinate gives the total value of $([n]_i + [n]_{fs})/[n]$.

The broken line passing through the intersection point parallel to the abscissa represents the dependence of $[n]/[\eta]$ on $M/[\eta]$ for the same polymer in the absence of the form effect. For polystyrene fractions the data obtained in three solvents are given. One of them is a matching solvent. Two inclined straight lines and the corresponding horizontal line intersect the ordinate at virtually the same point. This fact signifies that the form effect observed in flexible-chain polymer solutions may be practically considered as the macroform effect $[n]_f$, and the much smaller microform effect, $[n]_{fs}$, may be neglected. When Eqs. (5.38)–(5.42) are applied, the slope of straight lines in Fig. 5.10 allows the evaluation of $p_0 \approx 2$, and when Eq. (5.66) is applied, this slope leads to the value of the Flory coefficient [9] close to those obtained by other methods.

2.3.2.2. Dependence of FB on concentration and excluded-volume effect. As shown above, in matching solvents the shear optical coefficient $\Delta n /\Delta\tau$ determined from Eq. (5.61) is independent of solution concentration. In nonmatching solvents in which $n_k \neq n_s$, an increase in solution concentration decreases the difference between the average refractive index of the coil n_ω [Eq. (5.37)] and that of the medium. This should lead to a decrease in θ_f [Eq. (5.33)] and, correspondingly, to that in $(\Delta n /\Delta\tau)_f$ with increasing solution concentration. In very concentrated solutions with the chain entanglement and coil interpenetration, a complete disappearance of the macroform effect should be expected with increasing c.

The decrease in $(\Delta n /\Delta\tau)_f$ with increasing solution concentration is a general effect for all polymer–solvent systems and is distinctly revealed experimentally [9]. For polymers with negative intrinsic anisotropy, when concentration increases, this effect may be accompanied by a reversal in the sign of the observed FB from positive to negative [73].

The value of $[\eta]\cdot c$ characterizing the volume fraction in solution occupied by coils may serve as a parameter determining the concentration dependence θ_f for various polymer–solvent systems. Figure 5.11 shows the concentration dependence of relative anisotropy $\theta_f^*/\theta_f \equiv (\Delta n /\Delta\tau)_f/ [n]/[\eta])_f$ on $[\eta]\cdot c$ for solutions of fractions of two polymers. The points corresponding to fractions of two different samples with molecular weights ranging from $0.2\cdot10^6$ to $10\cdot10^6$ are concentrated close to one curve. This figure illustrates the universal importance of the parameter $[\eta]\cdot c$ in the concentration dependence of the form effect of flexible chains in solution.

The anisotropy of microform [Eq. (5.43)] also depends on the difference between the refractive indices of the polymers and the medium and, in principle, should also decrease with increasing solution concentration. This effect has actually been observed experimentally and investigated in detail [75]. For the quantitative investigation of both the anisotropy of the microform itself and its concentration dependence, in the case of a flexible-chain polymer, solutions of high concentration (gels or swollen polymers) should be used for which the macroform effect virtually no longer exists [9, 75].

The influence of excluded-volume effects on the anisotropy of macroform is also very specific.

Both the intrinsic anisotropy θ_i and the anisotropy of microform θ_{fs} are determined by the optical properties of a chain segment, and hence do not vary upon coil expansion under the influence of excluded-volume effects. Consequently, the shear optical coefficients, $[n]_i/[\eta]$ and $[n]_{fs}/[\eta]$, are independent of these effects.

In contrast, according to Eq. (5.38), an increase in coil volume V when the coil expands in a thermodynamically good solvent should lead to a decrease in θ_f. Since in this case the parameter of the shape asymmetry $4\pi e$ does not change [76], it should be expected, according to Eq. (5.66), that the shear optical coefficient $[n]_f/[\eta]$ decreases when the molecule expands as a result of volume effects. This is confirmed by experimental data [9, 77] showing that the excluded-volume effects greatly increasing the coil size of flexible-chain polymers do not lead to a marked change in the asymmetry of coil shape.

2.3.2.3. Dependence of FB on shear rate. According to Eq. (5.63), the coil deformation appearing with increasing shear rate g in a matching solvent leads to a nonlinear dependence of Δn on the parameter β_0. Moreover, according to Eq. (5.35), during deformation the intrinsic anisotropy of the chain (or of its segment in the elastic necklace model) increases proportionally to the *square* of the end-to-end distance, h^2 (l^2 in the elastic necklace model), or proportionally to the parameter $x^2 = h^2/\langle h^2 \rangle$ in Eq. (5.41). According to Eq. (5.43), in a solution in which the form effect exists ($n_k \neq n_s$) during chain deformation the anisotropy of microforms, $(\gamma_1 - \gamma_2)_{fs}$, also increases proportionally to x^2. Hence, the dependence of Δn_{fs} on the shear rate (or on the parameter β_0) does not differ from that of Δn_i on β_0 and is described by the same Eq. (5.63).

The relationships for the macroform effect are more complex.

According to Eq. (5.38), when the coil is deformed (extended) in the shear field, the anisotropy of macroform $(\gamma_1 - \gamma_2)_f$ changes proportionally to the parameter of shape asymmetry e. At relatively slight deformations, this leads to the dependence [Eq. (5.42)] according to which $(\gamma_1 - \gamma_2)_f$ increases proportionally to $x = h/\langle h^2 \rangle^{1/2}$, i.e., increases more slowly than $(\gamma_1 - \gamma_2)_i$ or $(\gamma_1 - \gamma_2)_{fs}$. Consequently, the dependence of Δn_f on the shear rate during chain deformation should be less sharp than for Δn_i or Δn_{fs}. This dependence may be found using Eqs. (5.24) and (5.41) and, correspondingly, curve 2 in Fig. 5.6 for the determination of $(\gamma_1 - \gamma_2)_f$ {Eq. (5.38) as a function of the parameter β_0. In the framework of the flexible dumbbell hydrodynamic model, the dependence of Δn_f on β_0 has been obtained

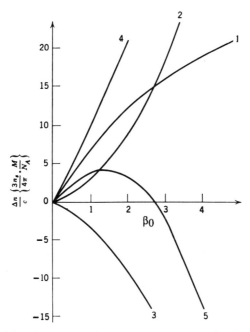

Fig. 5.12. FB versus shear rate parameter β_0 for flexible-chain molecules according to Eq. (5.67) at $C = 1$ (elastic dumbbell): 1) anisotropy of macroform: $[(n^2 + 2)/3]^2\theta_f = 3$; 2) intrinsic anisotropy and anisotropy of microform: $[(n^2 + 2)/3]^2(\theta_i + \theta_{fs}) = +2$; 3) $[(n^2 + 2)/3]^2(\theta_i + \theta_{fs}) = -2$; 4) total effect of 1 and 2; 5) total effect of 1 and 3.

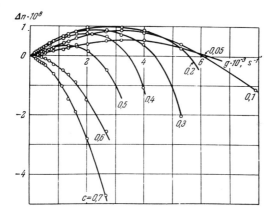

Fig. 5.13. Flow birefringences Δn versus rate gradient g for polystyrene solutions in dioxane [73]. Solution concentrations (in g/100 cm^3) are shown by numbers on the curves.

using distribution functions (5.46) and (5.48) [73]. Taking into account all three FB components, we obtain [42, 79]

$$\lim_{c \to 0} \left(\frac{\Delta n}{c} \right) \frac{3 n_s}{4 \pi} \left(\frac{3}{n_s^2 + 2} \right)^2 \frac{M}{N_A} = (\theta_i + \theta_{fs}) \, \beta \, \sqrt{1 + C^2 \beta_0^2} + \left(\frac{3}{2} \right)^2 \theta_f \Psi \, (\beta_0). \qquad (5.67)$$

where C has the same meaning as in Eq. (5.54). The function $\Psi(\beta_0)$ describes the dependence of the anisotropy of the chain macroform $(\gamma_1 - \gamma_2)_f$ on the parameter β_0. The values of this function are determined by the following series: $\beta_0 = 0$, $\Psi = \beta_0$; $\beta_0 = 1$, $\Psi = 1$; $\beta_0 = 3$, $\Psi = 2.4$; $b_0 = 4$, $\Psi = 2.85$; $\beta_0 \to \infty$, $\Psi = (4/3)\pi$. Hence, at values of $\beta_0 \leq 1$, a part of FB corresponding to the macroform effect increases proportionally to the shear rate, similarly to the parts Δn_i and Δn_{fs}. However, in contrast to these parts, as β_0 increases further, the curve of the dependence of Δn on β_0 tends to the limit in accordance with the dependence of the parameter of shape asymmetry e on p or x (Fig. 5.6).

The general shape of the dependence of Δn on β_0 described by Eq. (5.67) may be very different, depending on the value of θ_f and the value and sign of the sum $\theta_i + \theta_{fs}$. Thus, in some cases, for polymers exhibiting negative intrinsic anisotropy with increasing rate gradient, the FB sign may change from the positive to the negative. Figure 5.12 shows as an example the dependences of Δn on β_0 for the cases when $\theta_i + \theta_{fs}$ and θ_f coincide (curve 4) or are opposite (curve 5) in sign.

These theoretical dependences are actually in agreement with experimental results obtained for some polymers with negative intrinsic anisotropy (Figs. 5.13 and 5.14).

Figure 5.13 shows the dependence of Δn on g for polystyrene solutions ($M = 3.3 \cdot 10^6$) in dioxane. Different curves correspond to different concentrations, shown by numbers on the curves. The initial course of the curves ($g \to 0$) characterizes the value and sign of the total shear optical coefficient $(\Delta n / \Delta \tau)_{g \to 0}$, which is positive at low concentrations and negative at high concentrations, in accordance with the concentration dependence of the macroform effect (see Fig. 5.11).

For more dilute solutions for which the initial FB is positive, with increasing g the curve $\Delta n = f(g)$ attains a maximum and passes to the range of negative Δn values, in qualitative agreement with the theoretical curves plotted in Fig. 5.12.

The lower the solution concentration, i.e., the greater the relative contribution of the macroform effect to the observed birefringence, the higher is the shear stress at which the sign of FB is reversed.

For a quantitative comparison of the theoretical and experimental data, the latter should be extrapolated to $c \to 0$ and expressed as a function of the parameter β_0. Figure 5.14 shows these data for fractions of poly[(*para*-tertiary)butylphenyl methacrylate] in tetrachloromethane.

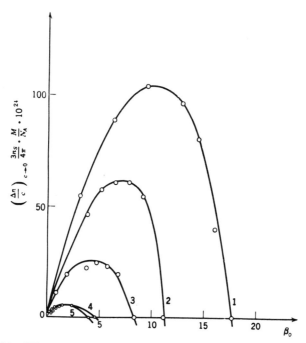

Fig. 5.14. FB versus the parameter $\beta_0 = M[\eta]\eta_{0g}/RT$ for fractions of poly(para-tert-butylphenyl methacrylate) in tetrachloromethane [80, 81]. 1) $M = 13.2\cdot10^6$; 2) $M = 7.4\cdot10^6$; 3) $M = 6.0\cdot10^6$; 4) $M = 2.2\cdot10^6$; 5) $M = 1.67\cdot10^6$.

The general shape of the theoretical curves shown in Fig. 5.12 is in agreement with the experimental data presented in Fig. 5.14. In particular, according to the theoretical requirements [Eq. (5.67)], for experimental curves it is observed that the higher the molecular weight M of the fraction, the higher are the values of the parameter β_0 at which the FB sign is reversed. However, quantitative coincidence between the theoretical and experimental curves is observed only in the range of low β_0 values in which their initial slopes are the same. Subsequent increase in the curvature of the experimental curves and their intercepts with the abscissa are located at higher β_0 values than for the theoretical curves. This discrepancy between the theory and experimental data is much less pronounced when the hydrodynamic properties of polymer molecules are described by the elastic necklace (rather than the elastic dumbbell) model and when the value of C corresponding to this model is used in Eq. (5.67). Moreover, the quantitative difference between experimental data and theoretical predictions may be due to the effect of kinetic rigidity (internal viscosity) of molecules which is not taken into account in the theory of ideally flexible chains upon which Eq. (5.67) is based.

2.3.2.4. Extinction angle. It has been mentioned above that the extinction angle in the FB phenomenon is the angle of the preferential orientation of molecules in flow, determined by their hydrodynamic characteristics and independent of their optical properties.

However, this rule is not always obeyed when the molecular system is inhomogeneous (polydisperse) with respect to the optical and hydrodynamic properties of molecules. This situation may occur, in particular, in a solution of flexible-chain molecules whose optical anisotropy is related to the macroform effect.

In all cases the extinction angle φ_m may be found from the condition that the tensor of optical polarizability of the medium calculated using a system of XYZ coordinates (Fig. 5.1 or Fig. 2.1) reduces to a diagonal form by the transformation of the coordinates to the O_1O_2Z system, i.e., by the rotation of the XYZ system about the Z axis by an angle φ_m (Fig. 5.1). In this case we have [2, 14]

$$\tan 2\varphi_m = 2n_{XY}^2/(n_{XX}^2 - n_{YY}^2), \qquad (5.68)$$

where n_{ij}^2 are the components of the tensor of refractive indices of the medium in the system of XYZ coordinates and characterize the optical polarizability of the medium along the i axis caused by the light wave field along the j axis.

The values of n_{ij}^2 contained in Eq. (5.68) may be calculated if the distribution function $\rho(\varphi, \vartheta)$ of molecular orientations in the XYZ system and the value of anisotropy of the molecule, $\gamma_1 - \gamma_2$, are known. Elementary calculations give

$$\tan 2\varphi_m = \frac{\int_{\Omega} (\gamma_1 - \gamma_2)\, \rho\, (\varphi,\, \vartheta)\, \sin^2\vartheta\, \sin 2\varphi\, d\Omega}{\int_{\Omega} (\gamma_1 - \gamma_2)\, \rho\, (\varphi,\, \vartheta)\, \sin^2\vartheta\, \cos 2\varphi\, d\Omega}\,, \qquad (5.69)$$

where the element of the solid angle $d\Omega = \sin\vartheta\, d\varphi\, d\vartheta$.

If the system is optically monodisperse, then the difference in the main polarizabilities $\gamma_1 - \gamma_2$ is identical for all molecules and is excluded from Eq. (5.69). Hence, the extinction angle depends only on the hydrodynamic characteristics of the molecule [contained in $\rho(\varphi, \vartheta)$].

For a system polydisperse with respect to optical and hydrodynamic properties of molecules, $\gamma_1 - \gamma_2$ is no longer a constant value; it is related to the hydrodynamic characteristics of molecules and cannot be excluded from Eq. (5.69). Hence, in this case, φ_m is a function not only of hydrodynamic but also of the optical properties of molecules.

If a polydisperse system is represented as a series of fractions, each of which leads to flow birefringence Δn_j with extinction angle φ_j, then the value of Δn and the extinction angle φ_m for the entire system are defined by the equations [2, 82]

$$\Delta n^2 = \left(\sum_j \Delta n_j \sin 2\varphi_j\right)^2 + \left(\sum_j \Delta n_j \cos 2\varphi_j\right)^2, \qquad (5.70)$$

$$\tan 2\varphi_m = \left(\sum_j \Delta n_j \sin 2\varphi_j\right) \Big/ \left(\sum_j \Delta n_j \cos 2\varphi_j\right) \tag{5.71}$$

A solution of flexible-chain molecules is always mechanically and optically polydisperse even if it exhibits no polymolecularity (an ideally narrow fraction). This is a polydispersity of shape because the conformational distribution exists in a molecular system (this distribution may be characterized, for example, by the value h). In a solution at rest this distribution is determined by Eq. (1.2) and in flow it is defined by function (5.46) for the dumbbell model or by the function $\rho = \Pi\rho_i$ for the necklace model.

The polydispersity with respect to conformations (h values) is accompanied by that related to the anisotropy of molecules because different values of $\gamma_1 - \gamma_2$ correspond to different values of h. According to Eq. (5.35), in the absence of the form effect the *sign* of anisotropy is *identical* for all molecules being determined by that of θ_i. When the form effect exists for a polymer with negative intrinsic anisotropy θ_i, the values of $\gamma_1 - \gamma_2$ different *in sign* may correspond to molecules in different conformations. This polydispersity of the sign of anisotropy should be particularly pronounced in the range of those values of $\theta_i + \theta_{fs}$, θ_f, $h^2/\langle h^2\rangle$, and e at which the first negative term in the right side of Eq. (5.44) is compensated by the second positive anisotropy, whereas less-extended molecules show positive anisotropy. According to Eq. (5.67), this occurs under the conditions when the measured values of Δn reverses its sign (Figs. 5.12–5.14) with increasing β_0.

The problem of the dependence of the extinction angle on β_0 may be discussed qualitatively if the true conformational distribution is replaced by a combination of two components providing contributions to FB, Δn_i, and Δn_f,

$$\Delta n_i/\Delta n_f = (\theta_i + \theta_{fs})\beta\sqrt{1+\beta_0^2}\Big/\left(\frac{3}{2}\right)^0 \theta_f\Psi(\beta_0) = x \tag{5.72}$$

with the corresponding extinction angles

$$\tan\varphi_{mi} = A_i/\beta_0, \quad \tan 2\varphi_{mf} = A_f/\beta_0, \tag{5.73}$$

where A_i and A_f are constants differing in value. Then, according to Eq. (5.71), the extinction angle φ_m for the total FB is determined from the condition

$$\tan 2\varphi_m = \frac{\Delta n_i \sin 2\varphi_{mi} + \Delta n_f \sin 2\varphi_{mf}}{\Delta n_i \cos 2\varphi_{mi} + \Delta n_f \cos 2\varphi_{mf}}. \tag{5.74}$$

If the value of the difference $\varphi_{mf} - \varphi_{mi} \equiv \delta$ is not high, then Eq. (5.74) becomes

$$\tan 2\varphi_m = \tan 2\varphi_0 + \frac{1-x}{1+x}\tan 2\delta, \tag{5.75}$$

where $\varphi_0 \equiv (\varphi_{mf} + \varphi_{mi})/2$ is the extinction angle which a system would have in the absence of polydispersity of the anisotropy of the molecules, and which characterizes their hydrodynamic properties.

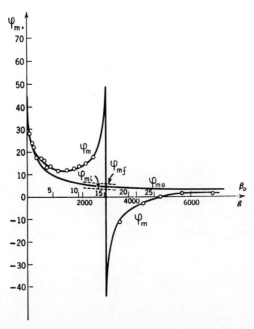

Fig. 5.15. Extinction angle versus shear rate g (or β_0) for the solution of a polymer with negative intrinsic anisotropy θ_i and the macroform effect θ_f [9]. The curves represent theoretical values according to Eqs. (5.72), (5.73), and (5.75) at the values of the parameters $(\theta_i + \theta_{fs})/\theta_f = -0.4$, $A_i = 0.5$, and $A_f = 0.6$. The points represent experimental data for poly(para-tert-butylphenyl methacrylate) in tetrachloromethane [81].

Consequently, for polymer solutions of the type considered here, the value of the extinction angle φ_m characterizes not only hydrodynamic but also, to a considerable extent, optical properties of the system. Moreover, a slight hydrodynamic polydispersity (low δ value) leads to very great changes in the shape of the experimental dependence $\varphi_m(\beta_0)$.

The anomalous dependence of the magnitude and sign of the extinction angle of FB on the rate gradient observed experimentally (Figs. 5.13-5.15) is the most direct and conclusive proof of the fact that at high shear rate FB in flexible-chain polymer solutions is largely due to the deformation of a soft molecular coil in the flow shear field.

The manifestation of the macroform effect in FB in the range of limiting weak shearing stresses ($\Delta\tau \to 0$) implies that instantaneous conformations of molecules of flexible-chain polymers under equilibrium conditions are aspherical and the dynamo-optical properties of their solutions are strongly dependent upon this asphericity.

Figure 5.15 shows a theoretical curve $\varphi_m(\beta_0)$ plotted according to Eq. (5.75) at the values of the optical and hydrodynamic parameters of molecules (dumbbell model) indicated in the caption. At $x = -1$, i.e., in the range of β_0 in which flow birefringence reverses its sign, the curve $\varphi_m(\beta_0)$ exhibits a break and sign reversal showing a sharply anomalous shape in this range.

The points in Fig. 5.15 representing the experimental data for a sample of poly(para-tertiary-butylphenyl methacrylate) in tetrachloromethane [81] are in good agreement with the theoretical curve.

2.4. Effect of Internal Viscosity on FB

In the FB theories considered above, based on the elastic dumbbell or the elastic necklace model, a chain molecule is assumed to be ideally flexible according to both equilibrium properties (Gaussian chain) and kinetic characteristics.

This assumption implies that when a chain is deformed by the action of friction forces in flow, the force opposing this deformation is the elastic entropy force increasing proportionally to deformation and determined by Eqs. (5.45) or (5.45'). According to Eq. (5.49), the deformation rate characterized by relaxation times τ_i is limited by the friction coefficient of the segment ζ depending on the solvent viscosity η_0.

However, as indicated above (Chapter 1, Section 2; Chapter 3, Section 3.3), the deformation of any real chain molecule is related to the transition over potential barriers oppositing the rotation of chain units and thus limiting the deformation *rate*.

This fact was first indicated by Kuhn [49], and it induced him to introduce the concept of internal viscosity in order to characterize the kinetic rigidity of the chain. According to Kuhn, if the end-to-end distance h (dumbbell model) changes under the influence of external forces at a rate dh/dt, then the oppositing internal viscosity force appears; this force is applied to chain ends and is proportional to the deformation rate:

$$\mathbf{F} = -B\,(d\mathbf{h}/dt). \qquad (5.76)$$

The coefficient B is the measure of internal viscosity (kinetic rigidity) of the chain. According to Kuhn, B is inversely proportional to the molecular weight (length) of a chain molecule since a change in conformation due to internal rotation may occur with equal probability at any chain element.

The concept of internal viscosity made it possible to explain the dependence of intrinsic viscosity of polymers on the shear rate (Chapter 3, Section 3.3), and was used by Kuhn in the theory of FB of polymer solutions.

Moreover, in the framework of the dumbbell model for a chain molecule it was shown that the relationship between the intrinsic values of FB and the solution viscosity determined according to Eq. (5.60) remains valid regardless of the internal viscosity of the molecule. In contrast, the internal viscosity of the chain leads to a considerable increase in intrinsic orientation $[\chi/g]$ [Eq. (5.4)]. Thus, according to Eq. (5.48), for a dumbbell model in the absence of internal viscosity (kinetically flexible chain), the coefficient G contained in Eq. (5.6) is equal to 1/2, whereas for an elastic dumbbell with limiting high internal viscosity this coefficient is three times as high: $G = 3/2$.

Subsequently, the concept of internal viscosity has been developed by Cerf, who applied it both to the problem of the shear dependence of intrinsic viscosity (Chapter 3, Section 3.3) and to the FB theory.

The original model for a chain molecule in Cerf's theory in its initial variation has been a viscoelastic sphere [83], subsequently replaced with the Rouse–Zimm elastic necklace model supplemented by taking into account the internal viscosity of the chain [7, 84]. In application to the multisegmental elastic necklace model the determination of the internal viscosity force is modified because the motion of the molecule does not reduce to the motion of both chain ends (as in the case of the elastic dumbbell) but rather is described by an array of N different eigenmodes. The deformation rate $U_{\text{def}, i}$ of the molecule caused by the motion of a chain element is assumed to be the rate of motion of this element in a coordinate system rotating at an angular rate ω together with the molecule. Hence, $U_{\text{def}, i}$ is equal to the difference between the total rate dx_i/dt of the change in the coordinate x_i of the element and the rate of this element $\omega \cdot x_i$ due to the rotation of the molecule as a whole (without deformation)

$$U_{\text{def}, i} = \partial \mathbf{x}_i / \partial t - \mathbf{\omega} \cdot \mathbf{x}_i. \tag{5.77}$$

Chain deformation related to the ith eigenmode of its motion induces the internal viscosity force F_i proportional to the deformation rate $U_{\text{def}, i}$. According to Cerf, similarly to Eq. (5.76) we have

$$F_i = -(i\varphi/N) U_{\text{def}, i}, \tag{5.78}$$

where, as previously, N is the number of beads in the necklace (and, correspondingly, the total number of eigenmodes) and φ is a constant value independent of molecular weight (and, consequently, of N) which characterizes the internal viscosity (kinetic rigidity) of the chain.

The eigenvalue

$$\varphi_i = i\varphi N \tag{5.79}$$

characterizes the resistance to molecular deformation corresponding to the ith eigenmode. The first eigenvalue, $\varphi_1 = \varphi/N$, corresponds to the change in the end-to-end distance for the entire molecule and characterizes the resistance to a rapid change in h. Similarly to the parameter B in Eq. (5.76), the value of φ_1 decreases

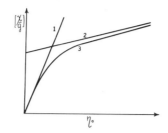

Fig. 5.16. Intrinsic orientation $[\chi/g]$ versus solvent viscosity: 1) high internal viscosity ($\varphi \gg \zeta$), orientational FB; 2) low internal viscosity ($\varphi \ll \zeta$), deformational FB; 3) dependence which, in accordance with Cerf, may be expected for the real chain molecule.

with increasing molecular weight and disappears for an infinitely long chain. The highest eigenvalue $\varphi_N = \varphi$ corresponding to the deformation of neighboring chain segments is independent of molecular weight. The higher the eigenmode index i, the shorter the chain segment N/i along which the conformational change occurs in the process of deformation, and hence the higher the resistance to a change in shape. Hence, the internal viscosity coefficient φ characterizes the resistance of the subchain to a rapid change in the shape of the molecule.

For a chain molecule in a viscous solvent the rate of conformational changes also depends on the friction coefficient of the segment ζ and, correspondingly, on the solvent viscosity η_0. Hence, φ/ζ is a characteristic parameter of the molecule. This ratio is equal to zero for an ideally flexible chain and increases with molecular stiffness.

The concept of internal viscosity has been somewhat modified in more recent papers [85–88] but these modifications have not essentially involved Cerf's results for the intrinsic values of FB and extinction angle.

The substitution of the internal viscosity forces F_i into the diffusion equation for the elastic necklace complicates this equation, and the final solution may be obtained only for intrinsic values of FB and extinction angles, and only in the limiting cases of low and high internal viscosity.

Similarly to the elastic dumbbell model, the intrinsic values of viscosity, $[\eta]$ and FB, $[n]$, and their ratio $[n]/[\eta]$, for the elastic necklace model are independent of the internal viscosity of the molecule. Hence, for an entirely rigid Gaussian coil, as for an entirely soft coil, the shear optical coefficient is independent of molecular weight and is determined by Eq. (5.60).

In contrast, in the elastic necklace theory the behavior of the intrinsic orientation $[\chi/g]$ is very strongly dependent on the internal viscosity of the molecule.

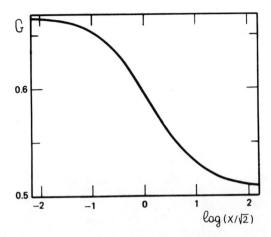

Fig. 5.17. Coefficient G according to Eq. (5.6) versus the parameter of hydrodynamic interaction $X/\sqrt{2} = \zeta N^{1/2}/\sqrt{12\pi^3}$. $\eta_0 \langle l^2 \rangle^{1/2}$ for a kinetically rigid Gaussian coil [89].

In this theory the condition of low internal viscosity is equivalent to that of high solvent viscosity η_0 (or low φ/ζ value). Under these conditions the intrinsic orientation for a Gaussian coil is the sum of two terms [7, 87]:

$$[\chi/g] = GM [\eta] \eta_0/RT + \alpha\varphi \langle h^2 \rangle/kT, \qquad (5.80)$$

where, as usual, $\langle h^2 \rangle$ is the mean-square end-to-end distance of the nondeformed Gaussian coil (molecule). The coefficients G and α are dependent on the hydrodynamic interaction in the chain.

The coefficient G is equivalent to its value for an elastic necklace without internal viscosity, i.e., $G = C/2$, where C is the coefficient in Eq. (5.54). In accordance with this, at weak hydrodynamic interaction G is equal to 0.2 and at strong hydrodynamic interaction G is equal to 0.103. The dependence of G on the parameter of hydrodynamic interaction X may be represented by the curve in Fig. 5.8.

The coefficient α in Eq. (5.80) has the value of 0.0062 at weak hydrodynamic interactions and 0.0045 at strong hydrodynamic interactions.

Equation (5.80) describing the properties of a kinetically flexible Gaussian coil weakly influenced by internal viscosity contains two terms, the first of which depends on the solvent viscosity η_0, and the second of which depends on the internal viscosity parameter φ and is independent of η_0.

Hence, the dependence of $[\chi/g]$ on η_0 according to Eq. (5.80) is described by a straight line with slope determined by a constant factor $GM[\eta]/RT$ in the first term, and the intercept with the ordinate determined by the second term (Fig. 5.16, straight line 2).

According to the theory of an elastic necklace with low internal viscosity, in a laminar flow a coil-shaped molecule rotates as a whole at a virtually constant angular rate $\omega = g/2$.

Under these conditions the observed FB is due to intramolecular motion and the deformation of chain segments rather than to the orientation of the chain as a whole, and hence may be regarded as deformational FB.

A Gaussian elastic necklace with high internal viscosity (a kinetically rigid chain) is not deformed in a laminar flow, but rather rotates at an angular rate which is dependent on the angular position of the molecule. Hence, FB in a solution of these molecules is a result of their kinematic orientation, just as for aspherical rigid particles (spheroids). In this case, according to Cerf's theory, the intrinsic orientation, $[\chi/g]$, is proportional to the parameter $M[\eta]\eta_0$ as for rigid ellipsoids {Eq. (5.6)]. Moreover, the coefficient G is dependent on hydrodynamic interaction in a kinetically rigid coil [89] and for the two limiting cases its values are $G = 0.51$ for a nondraining coil and $G = 0.67$ for a draining coil. The dependence of this coefficient G on the parameter of hydrodynamic interaction $X/\sqrt{2}$ is shown in Fig. 5.17.

Hence, at a constant $M[\eta]$ value the dependence of $[\chi/g]$ on η_0 for a kinetically rigid coil is described by straight line 1 (Fig. 5.16) passing through the origin and having a slope greater by a factor of three to six than straight line 2.

According to Cerf's theory, in principle, for a rigid chain molecule characterized by a finite value of the internal viscosity parameter φ, at high η_0 values (where $f/\varphi \gg 1$) FB is of a deformational nature and is determined by straight line 1 in Fig. 5.16. In contrast, in low-viscosity solvents (in which $f/\varphi \ll 1$) FB should be of an orientational nature and the dependence of $[\chi/g]$ on η_0 should be determined by line 1. Hence, the total dependence of $[\chi/g]$ on η_0 should be represented by curve 3 in Fig. 5.16. Consequently, according to Cerf, it is possible to detect the existence of kinetic rigidity for a flexible-chain molecule and to determine quantitatively the internal viscosity parameter φ by plotting the experimental dependence of $[\chi/g]$ on the solvent viscosity η_0. If the intercept of the tangent to the curve $[\chi/g] = f(\eta_0)$ with the ordinate is finite (as for curve 3 in Fig. 5.16), then the length of this intercept characterizes the kinetic rigidity of the chain and determines the value of φ. On the other hand, if over the entire range of solvent viscosities η_0 being investigated the experimental dependence of $[\chi/g]$ on η_0 is a straight line passing through the origin, this dependence may be characteristic of both a kinetically flexible chain [in Eq. (5.80) $\alpha = 0$] and a kinetically very rigid chain [Eq. (5.6)]. In this case, the choice between these two structures can be based on the slope of the straight line describing the dependence of $[\xi/g]$ on $M[\eta]\eta_0$, i.e., on the value of the coefficient G in Eqs. (5.6) and (5.80) which should be several times greater for a rigid than for a flexible chain.

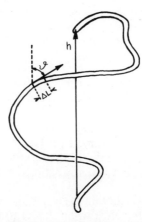

Fig. 5.18. Orientation of an element ΔL of a wormlike chain with respect to the vector **h**.

In a number of cases, the experimental investigations carried out many years ago in solutions of flexible-chain polymers [polystyrene, poly(methyl methacrylate)] revealed the dependences of intrinsic orientation $[\chi/g]$ on the viscosity of the solvent which corresponded to the theoretical curve shown in Fig. 5.16. However, the differences in the values of the internal viscosity parameter reported in different papers were too great to be able to impart to these values any quantitative importance. In some papers [92] the investigated FB was a result of the macroform effect which in itself involves difficulties in the interpretation of experimental data. Of considerable importance in all these measurements are errors due to difficulties in extrapolating experimental data on extinction angles to low shear rates and zero concentrations. These difficulties are particularly marked in the investigation of flexible-chain molecules of high molecular weight. Considerable errors may also be introduced by the polydispersity of the samples.

Hence, the possibility of carrying out experimental determinations of the internal viscosity parameter φ for kinetically flexible molecules from the intercepts of the curves shown in Fig. 5.16 with the ordinate has not yet been completely elucidated. Evidently, the sensitivity of this method is also decreased by the fact that, according to modern concepts [86, 88], the internal viscosity may depend on the viscosity of the solvent η_0, which levels off the differences in the behavior of the first and the second terms in Eq. (5.80).

This observation refers to the evaluation of the possibilities of quantitative experimental determinations of the internal viscosity of polymer chains and by no means concerns the problem of the fundamental relevance [95] of the concept of internal viscosity itself.

It is beyond doubt that relaxation times characteristic of intramolecular motions in a chain molecule should depend on the hindrance to internal rotations and should increase with this hindrance. In the description of steady-state friction characteristics of polymer molecules, intramolecular motions are not important, and the use of internal viscosity cannot give new results. However, in physical phenomena related to intramolecular chain motions (which include FB), the part played by the kinetic rigidity of the molecule is very important, and the results of the theory of internal viscosity are of great significance.

Unfortunately, at present the theory of internal viscosity has been developed only for two cases, the limiting low and limiting high internal viscosity, and it is impossible to use this theory for the characterization of dynamo-optical properties of polymer molecules in the range of intermediate values of their kinetic flexibility.

3. RIGID-CHAIN MOLECULES

The content of the preceding chapters has shown that the conformational and hydrodynamic (transport) properties of chain molecules with high equilibrium rigidity may be adequately described on the basis of the wormlike chain model.

Hence, it also seems desirable to use this model in the consideration of dynamo-optical properties of rigid-chain polymers.

As in the case of flexible-chain polymers, one of the main values determining the dynamo-optical properties of a solution containing rigid-chain polymers is their optical anisotropy. However, as we have seen, the optical anisotropy of a flexible-chain molecule is calculated in terms of the freely jointed chain model, whereas for a rigid-chain polymer this model should be replaced by a wormlike chain.

3.1. Optical Anisotropy of the Wormlike Chain

As indicated in Chapter 1, the anisotropy of the optical polarizability of the chain may be characterized by its mean-square anisotropy, δ^2 [Eq. (1.50)]. This value, determined by the squares of cosines of the angles between the chain elements (segments), may be calculated and expressed by Eq. (1.52) which is, for example, successfully used for the characterization of anisotropic light scattering by a polymer solution (see Chapter 1).

However, the value of δ^2 cannot be used to characterize FB because this value is a measure of the correlation between mutual orientations of chain segments, and is not related to the character of segment orientation with respect to a direction fixed in the molecule or in the shear field. In particular, the value of δ^2 cannot indicate the *sign* of anisotropy of a chain molecule, which is one of its main characteristics.

Hence, in the study of FB, as in the case of a Gaussian chain, in the molecule represented by a wormlike chain a fixed direction should be selected with respect to which the orientation of segments or chain elements is considered.

3.1.1. Anisotropy in a System of Coordinates Related to Chain Ends. The line h joining the ends of a wormlike chain (Fig. 5.18) may be taken as

the molecular axis, and the chain anisotropy may be characterized by the difference $\gamma_1 - \gamma_2$ between its main polarizabilities γ_1 and γ_2 in the direction h and that normal to h, respectively. Then the contribution $\Delta\gamma$ to the chain anisotropy, $\gamma_1 - \gamma_2$, of a chain element ΔL (characterized by the axial symmetry of its optical properties) is evidently given by

$$\Delta\gamma = \beta\Delta L (3\cos^2\vartheta - 1)/2, \tag{5.81}$$

where β is the anisotropy of unit length of the wormlike chain and ϑ is the angle formed by the element ΔL and the h axis.

In accordance with this, the total difference between two principal polarizabilities of the wormlike chain is given by

$$\gamma_1 - \gamma_2 = \beta L (3\,\overline{\cos^2\vartheta} - 1)/2, \tag{5.82}$$

where L is the chain length and $\cos^2\vartheta$ is the value averaged over all chain elements.

Equation (5.82) is quite analogous to Eq. (5.30) because the value $N(\alpha_1 - \alpha_2) = \beta L$ is the difference between two principal polarizabilities of a completely extended chain.

In the statistics of the freely jointed Kuhn chains the most probable orientational distribution of segments in the chain is determined by Eq. (5.25). Accordingly, Eq. (5.28) is obtained for $\cos^2\vartheta$ and the value of $\gamma_1 - \gamma_2$ is given by Eq. (5.33), which may be represented in the following form:

$$\gamma_1 - \gamma_2 = (3/5)\beta L (h/L)^2/[1 - (2/5)(h/L)^2]. \tag{5.83}$$

In contrast to the value $\cos^2\vartheta_{ik}$ contained in Eq. (1.50), a rigorous calculation of $\cos^2\vartheta$ in Eq. (5.82) for a wormlike chain is difficult and an approximate solution of this problem should be used. In this connection it should be borne in mind that Eq. (5.83), obtained for a freely jointed chain, is applicable to chain molecules exhibiting any conformations ranging from a straight rod ($h = L$) to a Gaussian coil ($h \ll L$). In this case the mechanism ensuring the chain end-to-end distance h (and consequently the h/L ratio) has not been taken into account in the derivation of Eqs. (5.31) and (5.83). Hence, it may be reasonably assumed [28] that Eq. (5.83) can be applied not only to flexible-chain molecules but also to wormlike chains for which a small degree of coiling (at low L/A values) is determined by the skeletal rigidity limiting the thermal mobility of chain elements along the entire length of the chain rather than by the external stretching force.

On this assumption, the anisotropy of a wormlike chain with contour length L and persistent length a, averaged over all conformations, may be determined using Eq. (5.83), if in this equation the value of h^2 is assumed to be averaged over all possible conformations of the chain in the equilibrium state, i.e.,

$$\langle \gamma_1 - \gamma_2 \rangle = (3/5)\, \beta L (\langle h^2 \rangle / L^2) / [1 - (2/5) \langle h^2 \rangle / L^2]. \tag{5.84}$$

According to Eq. (1.24), for a wormlike chain we have

$$\langle h^2 \rangle / L^2 = (2/x)[1 - (1 - e^{-x})/x], \tag{5.85}$$

where $x = 2L/a$.

The substitution of Eq. (5.85) into Eq. (5.84) transforms Eq. (5.84) into the form

$$\langle \gamma_1 - \gamma_2 \rangle = (3/5)\, \beta A f(x) \equiv (6/5)\, \beta L f(x)/x = \beta L [1 - (5/9)\, x \\ + (85/324)\, x^2 - (170/1458)\, x^3 + \cdots], \tag{5.86}$$

where

$$f(x) = [1 - (1 - e^{-x})/x]/\{1 - (4/5x)[1 - (1 - e^{-x})/x\}. \tag{5.87}$$

It follows from Eqs. (5.85)-(5.87) that the optical anisotropy of a wormlike chain in a coordinate system of the vector $\mathbf{h}$ at low x varies proportionally to chain length (as for a rodlike molecule), attaining in the Gaussian range ($x \to \infty$) the asymptotic limit $(3/5)\beta A$. This limit corresponds to the anisotropy of the freely jointed Gaussian chain according to Kuhn [27].

The dependence of $\langle \gamma_1 - \gamma_2 \rangle / \beta A$ on x plotted according to Eq. (5.86) is shown in Fig. 5.19 by curve 1.

The anisotropy of the wormlike chain in a coordinate system of the vector $\mathbf{h}$ has also been calculated in [96] with the application of other simplifying assumptions in the determination of $\cos^2 \vartheta$. The resulting dependence of $\langle \gamma_1 - \gamma_2 \rangle$ on x is shown by curve 2 in Fig. 5.19. In the range of $x \leq 3$, curve 2 is close to curve 1. However, as x increases further, curve 2 acquires a nonmonotonic shape; it attains a maximum and decreases to the limiting value of $0.5\beta A$, which is 5/6 of the limit for a Gaussian chain. This unrealistic shape of the curve appears to indicate the inadequacy of the approximations assumed in the calculation of $\overline{\cos^2 \vartheta}$.

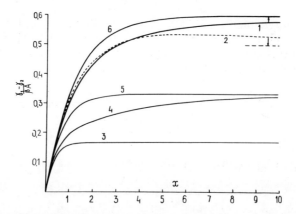

Fig. 5.19. Relative anisotropy of a wormlike chain $\langle\gamma_1 - \gamma_2\rangle/\beta L$ versus relative chain length $x = 2L/A$. In a system of coordinates of the vector **h**, 1) according to Eq. (5.86); 2) according to [96]; 3) in a system of coordinates of the first chain element according to Eq. (5.94); 4) in a system of coordinates of a chain element located at a distance of $0.1L$ from the chain extremity according to Eq. (5.95); 5) in a system of coordinates of a middle chain element according to Eq. (5.96); 6) according to Eq. (5.97).

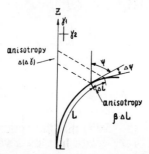

Fig. 5.20. Optical anisotropy of a wormlike chain in the system of coordinates of its first element.

3.1.2. Anisotropy in a System of Coordinates Related to a Chain Element. An alternative choice for the molecular system of coordinates used to calculate the chain anisotropy is the superposition of the origin on one of the chain elements [97, 98]. One axis of the system is directed along the axis of this element (along the tangent to the wormlike chain) and is assumed to be the axis of cylindrical symmetry in the spatial distribution of all other chain elements (the Z axis in Fig. 2.8).

The case when the Z axis coincides with the first (and extreme) chain element is shown in Fig. 5.20.

A chain region of length L consists of n elements, each of length ΔL, so that $L = n\Delta L$. Each element is characterized by axial symmetry of optical properties and exhibits the anisotropy $\beta \Delta L$, where β is the anisotropy per unit length of the chain.

The average anisotropy introduced by one element j in the direction (axis) of the neighboring element is evidently given by

$$\Delta \gamma_{j,\,j\pm 1} = \beta \Delta L K, \quad K = (3 \langle \cos^2 \Delta \psi \rangle - 1)/2, \tag{5.88}$$

where $\langle \cos^2 \Delta \psi \rangle$ is the square of cosines of the angle $\Delta \psi$ between the neighboring elements, averaged over all chain conformations.

In accordance with Eq. (5.88), the nth chain element of length ΔL removed along the chain contour from a chain beginning with the value L provides a contribution $\Delta(\Delta \gamma)$ to the anisotropy of the chain $\Delta \gamma \equiv \gamma_1 - \gamma_2$ in the coordinate system of its first element

$$\Delta(\Delta \gamma) = \beta \Delta L K^{(n-1)} = \beta \Delta L K^{\left(\frac{L}{\Delta L}-1\right)} = \beta \Delta L e^{\ln K \left(\frac{L}{\Delta L}-1\right)}. \tag{5.89}$$

At $\Delta L \to 0$, $\Delta \psi \to 0$, and $K \to 1$; Eq. (5.89) yields the differential form of the dependence of chain anisotropy $\Delta \gamma$ (in the coordinate system of the first element) on its length L:

$$d(\Delta \gamma)/dL = \beta \exp \left[\left(\frac{\ln K}{\Delta L}\right) L \right]$$

and, correspondingly, the integral form:

$$\Delta \gamma = \beta (\Delta L/\ln K) [e^{(\ln K/\Delta L)\, L} - 1]. \tag{5.90}$$

At a large length a wormlike chain becomes a Gaussian chain and, according to Eq. (5.90), its anisotropy is given by

$$\lim_{L \to \infty} \Delta \gamma = \Delta \gamma_\infty = -\beta (\Delta L/\ln K) \tag{5.91}$$

and hence,

$$\Delta \gamma = \Delta \gamma_\infty (1 - e^{-\beta L/\Delta \gamma_\infty}). \tag{5.92}$$

On the other hand, the curvature of the chain may be expressed by its persistent length a according to Eq. (1.20). A comparison of Eqs. (5.91) and (1.20) yields

$$\Delta \gamma_\infty = a \beta \ln \langle \cos \Delta \psi \rangle / \ln K,$$

and taking into account that $\lim_{\Delta \psi \to 0} [\ln \langle \cos \Delta \psi / \ln K] = 1/3$, we obtain

$$\Delta\gamma_\infty = (1/3)\,a\beta = (1/6)\,A\beta = (1/6)\,(\alpha_1 - \alpha_2), \tag{5.93}$$

where $\alpha_1 - \alpha_2$ is the anisotropy of the Kuhn segment.

A combination of Eqs. (5.92) and (5.93) gives the dependence on the chain length L of the difference in the polarizabilities γ_1 and γ_2 of a wormlike chain in the directions parallel and perpendicular to the Z axis in Fig. 5.20, respectively,

$$\langle\gamma_1 - \gamma_2\rangle = (1/6)\,\beta A\,(1 - e^{-3x}) = \beta L\,[1 - (3/2)\,x + (3/2)\,x^2 - (9/8)x^3 + \ldots], \tag{5.94}$$

where $x = 2L/A$.

Curve 3 in Fig. 5.19 describes the dependence of $\langle\gamma_1 - \gamma_2\rangle/A\beta$ on $x = 2L/A$ according to Eq. (5.94). The general shape of curve 3 corresponds to that of curve 1, but the limiting value of the anisotropy of the Gaussian chain in the coordinate system of its first element is only 5/18 of that in the coordinate system of the vector **h**.

This method can be readily used to show that the anisotropy of a wormlike chain (of length L), calculated in a system of coordinates of its element located at a distance L_1 from one of the chain ends, is given by

$$\langle\gamma_1 - \gamma_2\rangle = (1/6)\,A\beta\,[2 - e^{-3\,(L_1/L)\,x} - e^{-3\,[1-(L_1/L)\,x]}]. \tag{5.95}$$

The shape of the dependence of $\gamma_1 - \gamma_2$ on x can undergo a great change upon variation in the position of the selected element in the chain. All the curves $(\gamma_1 - \gamma_2)/\beta A = f(x)$ obtained by this method exhibit coinciding initial slopes of 0.5 and asymptotic limits twice larger than the limit of the curve described by Eq. (5.94) and obtained in the system of coordinates of the first chain element ($L_1 = 0$ or $L_1 = L$). However, the steepness of the curves determined from Eq. (5.95) may differ greatly. The steepness is maximum at $L_1 = L/2$, i.e., when the molecular axis coincides with the middle chain element. This is illustrated by curves 4 and 5 in Fig. 5.19, calculated according to Eq. (5.95), at $L_1 = 0.1L$ and $L_1 = 0.5L$, respectively.

According to Eq. (5.95), the anisotropy of the wormlike chain in a system of coordinates of the middle element ($L_1 = 0.5L$) is given by

$$\langle\gamma_1 - \gamma_2\rangle = \tfrac{1}{3}\beta A\left(1 - e^{-\frac{3}{2}x}\right) = \beta L\left(1 - \tfrac{3}{4}x + \tfrac{3}{4}x^2 - \ldots\right). \tag{5.96}$$

A comparison of curves 3, 4, and 5 with curve 1 in Fig. 5.19 shows that in the calculation of the anisotropy of the wormlike chain the application of different coordinate systems does not lead to a change in the general type of dependence of $\langle\gamma_1 - \gamma_2\rangle$ on x but the maximum value of the anisotropy of the Gaussian chain is obtained in the coordinate system of the vector **h**.

Taking this into account, it might be assumed that the general equation (5.92) also characterizes approximately the dependence of $\Delta\gamma$ on x in the coordinates of the vector $\mathbf{h}$ if in Eq. (5.92) the anisotropy of the Gaussian chain $\Delta\gamma_\infty$ is taken to be equal to its value in the system of coordinates of the vector $\mathbf{h}$. Accordingly, assuming $\Delta\gamma_\infty = (3/5)\beta A$ we obtain, from Eq. (5.92),

$$\langle\gamma_1 - \gamma_2\rangle = (3/5)\beta A\,(1 - e^{-5x/6}) = \beta L\left[1 - \frac{5}{12}x + \frac{25}{216}x^2 - \frac{125}{5184}x^3 + \cdots\right]. \quad (5.97)$$

According to Eq. (5.97), the dependence of $\langle\gamma_1 - \gamma_2\rangle/\beta A$ on x is described by curve 6 in Fig. 5.19. The difference between curves 6 and 1 is slight, and hence both of them may be used to express the dependence of the anisotropy of a wormlike chain on its length in the h system of coordinates.

The fact that the difference between the optical polarizabilities of the Gaussian chain in the coordinates of the vector $\mathbf{h}$ is maximum as compared to its values in other systems of molecular coordinates implies that the direction h is in fact the main direction in a chain molecule and may be definitely regarded as the axis of symmetry of its optical properties. Hence, the calculation of $\gamma_1 - \gamma_2$ in the coordinates of the vector $\mathbf{h}$ gives the most reliable characteristic of the optical anisotropy of the chain molecule.

3.1.3. Optical Form Effect.
The difference between the polarizabilities $\langle\gamma_1 - \gamma_2\rangle$ of a wormlike chain is dependent not only on the values of L and A determining its conformation, but also directly upon the parameter β determined by the optical properties of structural elements of the chain molecule. The value of β is related to the difference between the main polarizabilities $\Delta a \equiv a_\| - a_\perp$ of the monomer unit in the directions parallel $(a_\|)$ and perpendicular $(a_\perp)$ to the chain length: $\beta = \Delta a/\lambda$, where λ is the length of the monomer unit in the chain direction. The anisotropy of the monomer unit of a real chain molecule located in the solvent medium contains not only the intrinsic anisotropy Δa_i but also the microform anisotropy which may be expressed by the parameter θ_{fs} contained in Eq. (5.43).

For rigid-chain polymers in which the segment length A is much larger than the chain diameter d, the form coefficient in Eq. (5.43), e_s, is equal to 1/2, and using this equation it is possible to obtain the expression for the parameter β [9]:

$$\beta = (\Delta a_i + \Delta a_f)/\lambda = [\Delta a_i + (dn/dc)^2\,M_0/(2\pi N_A \bar{v})]/\lambda, \quad (5.98)$$

where dn/dc is the refractive index increment in the polymer–solvent system and $\bar{v}$ is the partial specific volume of the polymer in solution.

As for any chain molecule, for a wormlike chain the macroform effect is defined by Eq. (5.38). However, as already mentioned, for rigid-chain polymers for which the volume V of the coil is much larger than for flexible-chain molecules, this effect is low, and in most cases may be neglected.

3.2. Form Asymmetry of the Molecular Chain

As repeatedly emphasized in Section 1 of this chapter, the form asphericity of rigid particles or molecules is an indispensable condition for the appearance of FB in their solutions because only if this condition is obeyed can the kinematic orientation of particles in shear flow be possible. Therefore, all the expressions for the value of birefringence of rigid particles [e.g., Eqs. (5.11), (5.14), and (5.15)] contain in the hydrodynamic multiplier the factor of shape asymmetry b_0.

In the theories of FB of solutions of flexible-chain molecules their form asymmetry is discussed only in connection with the optical macroform effect, and in this case the factor of the form asymmetry e appears in the *optical* multiplier {e.g., in Eq. (5.38)]. Hydrodynamic properties of flexible chains in FB phenomena are described by either the elastic dumbbell model or the elastic necklace model. In both cases the kinetic unit undergoing the action of shear stress in flow is the dumbbell (the entire molecule or the subchain) whose hydrodynamic behavior is equivalent to that of a thin rod. Hence, in the hydrodynamic multiplier in equations describing FB of a solution of flexible chains it is automatically assumed that b_0 is equal to unity.

For a rigid-chain polymer described in the FB phenomenon by the wormlike chain model, the role played by the shape asymmetry of the molecule in its hydrodynamic behavior evidently depends on the kinetic flexibility of the chain. The higher the kinetic rigidity of the molecule, the closer are its hydrodynamic properties to those of an aspherical rigid particle (spheroid) and the greater is the importance of evaluating the parameter of shape asymmetry b_0 in the description of its hydrodynamic behavior in the FB phenomenon. For a wormlike chain the shape asymmetry evidently should depend on the chain length L because, with increasing L, the chain conformation undergoes a change from the straight rod to a Gaussian coil. In the rodlike conformation the asphericity is determined simply by the relation $L/d = p$, whereas the characteristic of the form asymmetry of the Gaussian coil is not so unequivocal.

3.2.1. Asphericity of the Gaussian Coil.

As already mentioned, the optical macroform effect long known [67] and considered in previous sections may serve as direct experimental evidence of the shape asymmetry of molecules of real flexible-chain polymers in solutions.

The asphericity of instantaneous conformations of chain molecules has also been shown in the experiments carried out with tagged polymers [99].

Various methods of theoretical characterization of the shape asymmetry of a chain molecule are possible.

Thus, to describe the instantaneous shape of a coil adopting a random conformation, Kuhn [30, 31] has proposed to measure the distance H_1 between the two most distant chain points and the distance H_2 between two such points in the direction normal to H_1 (Fig. 2.6). The averaging of the values of H_1 and H_2 over a large number of conformations gives the average values $\langle H_1 \rangle$ and $\langle H_2 \rangle$ related to

Eqs. (5.39), and leads to the average shape asymmetry of the nondeformed Gaussian coil $p = 2$.

In this method the coil shape is determined in the molecular system of coordinates which is selected for each individual conformation, and varies on passing to another conformation; it is not related to any fixed chain units.

This approach has been developed in modern theoretical papers by Stockmayer et al. [100-102] and others [103–110]. In these papers, computer simulation was used, and the shape of a flexible-chain molecule was characterized from the components of the radii of gyration R_1, R_2, and R_3 in the three main directions in the coil (fixed for each conformation). The calculations carried out for a linear chain, a macroring, and a starlike chain showed that the average instantaneous coil shape may be approximated by a triaxial ellipsoid with a considerable degree of asphericity. In particular, for a linear chain in the Gaussian coil conformation the ratio of squares of axes is $\langle R_2^3 \rangle : \langle R_2^2 \rangle : \langle R_1^2 \rangle = 11.7 : 2.7 : 1$. Hence, the degree of asymmetry of the coil p is

$$p = [2 \langle R_3^2 \rangle / (\langle R_2^2 \rangle + \langle R_1^2 \rangle)]^{1/2} \approx 2.5.$$

With a decreasing number of units in the chain the shape asymmetry increases. Thus, according to data reported in [106], where the Monte Carlo method was used to generate the configurations of the polymethylene chain, when the number of units decreased from 1000 to 20, the value of p increased from 2.6 to 3.3.

This relationship reflects the deviation of the conformation of a chain molecule from a Gaussian coil and its approach to a more extended shape with decreasing chain length.

Another possible method for the characterization of the shape asymmetry of a chain molecule is based on the choice of the main direction in the molecule related to the position of certain fixed chain elements. Thus, the Z axis in Fig. 2.3 tangent to the wormlike chain in its middle, i.e., related to the direction and position of the middle chain element, may be chosen as this direction. As we have seen, in this system of coordinates the shape asymmetry p is determined by Eq. (2.125), based on Hearst's calculation [111]. The dependence of p on the relative length $x = 2L/A$ of a wormlike chain is represented according to Eq. (2.125) by curve 1 in Fig. 5.21. In the range of $x \to 0$, we have $p \to \infty$, which adequately reflects the properties of the wormlike chain. However, as the chain length increases, the shape asymmetry drastically decreases, and in the range of the Gaussian coil ($x \to \infty$) we have $p = 1$, which corresponds to a spherically symmetrical spatial distribution of chain elements.

Similar results have been obtained in [106, 112], in which the conformations of flexible chains have been simulated by the Monte Carlo method. Furthermore, the shape asymmetry was characterized by three main components of the tensor of gyration of the chain. However, in contrast to Stockmayer's method [100], these components were determined and averaged in a system of coordinates related to the two neighboring bonds in the chain. Calculations have shown that in this definition

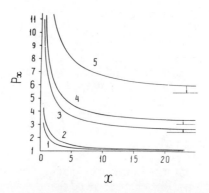

x

Fig. 5.21. Shape asymmetry p of a wormlike chain versus $x = 2L/A$. 1 – Curve plotted according to Eq. (2.125), 2–5 – curves plotted according to Eq. (5.101): 2) $p_\infty = 1$; 3) $p_\infty = 2.5$; 4) $p_\infty = 3.1$; 5) $p_\infty = 5.5$.

of form asymmetry the coil of any flexible-chain polymer of relatively high molecular weight is spherically symmetric.

Hence, the characterization of conformations of a flexible-chain molecule in a system of coordinates determined by the orientation of two selected chain units located close to each other (or by the orientation of an element of a wormlike chain) cannot give an adequate idea of the shape asymmetry of the coil.

However, it has long been known that the asymmetry of the Gaussian coil may be characterized in a system of coordinates related to two chain elements if these elements are located at chain ends. As Kuhn has shown [29], to evaluate the degree of asymmetry of the Gaussian coils, the end-to-end vector $\mathbf{h}$ may be chosen as the main direction, and the length h may be compared to transverse coil dimensions. In this case, transverse dimensions are determined by distances h_1 and h_2 from vector $\mathbf{h}$ relative to the chain points situated at the greatest distance from h in two mutually perpendicular directions normal to h. Calculations have shown that the ratio of squares of coil dimensions averaged over all conformations $\langle h^2 \rangle : \langle h_1^2 \rangle : \langle h_2^2 \rangle$ is $1 : (1/6) : (1/24)$. Hence, according to Kuhn, the direction of the maximum average dimensions of the Gaussian coil is the direction of the end-to-end vector $\mathbf{h}$, and the square of the degree of coil asymmetry is given by

$$p^2 = 2 \langle h^2 \rangle / (\langle h_1^2 \rangle + \langle h_2^2 \rangle) = 48/5 \text{ and } p = 3.1.$$

This result shows again that the ends of a chain molecule are "specific" chain points, and the end-to-end vector $\mathbf{h}$ may be regarded as the main direction in the molecule in characterizing its shape asymmetry.

It should be noted that the values of p predicted by the above-mentioned theories agree in order of magnitude with those obtained for flexible-chain molecules from the macroform effect in the FB phenomenon [9].

3.2.2. Shape Asymmetry of the Wormlike Chain. In connection with the results of the previous section, the shape asymmetry of a wormlike chain in a system of coordinates of the vector **h** should be evaluated, particularly because its optical anisotropy was determined in the same system.

For this purpose, a wormlike coil is represented by an extended body of revolution [97, 113, 115] for which the central radius of gyration $\langle R^2 \rangle$ is related to the longitudinal H and transverse Q dimensions by the general equation

$$\langle R^2 \rangle = \gamma H^2 + \delta Q^2, \tag{5.99}$$

where γ and δ are constant coefficients depending on the selected model. Thus, for a spheroid, $\gamma = 1/20$ and $\delta = 1/10$, whereas for a compact cylinder $\gamma = 1/12$ and $\delta = 1/8$. The shape asymmetry of the molecular model p is equal to H/Q.

Equation (5.99) is supplemented by the following conditions: 1) at all the values of $x = 2L/A$, relationships (1.24) and (1.25) characteristic of a wormlike chain are obeyed between $\langle h^2 \rangle$, $\langle R^2 \rangle$, and x; 2) at all values of x the major axis of the coil H is equal to $\alpha \langle h^2 \rangle^{1/2}$, where α is the constant coefficient; 3) at $x \to 0$ the shape asymmetry of a wormlike chain $p_x \to \infty$ (if the chain diameter d is relatively small) and, moreover, according to Eqs. (1.24) and (1.25), we have $\langle R^2 \rangle / \langle h^2 \rangle \to 1/12$; 4) at $x \to \infty$, we have $p_x \to p_\infty$, and according to Eq. (1.26), $\langle R^2 \rangle / \langle h^2 \rangle = 1/6$. If these conditions and Eq. (5.99) are employed, the following equation is obtained:

$$1/p_x^2 = (12^2/p_\infty^2)(\langle R^2 \rangle / \langle h^2 \rangle - 1/12). \tag{5.100}$$

If Eqs. (1.24) and (1.25) are taken into account, Eq. (5.100) is equivalent to the equation

$$\frac{1}{p_x^2} = \frac{1}{p_\infty^2} \left[\frac{2 - 6/x}{1 - (1 - e^{-x})/x} + \frac{12}{x^2} - 1 \right]. \tag{5.101}$$

Equation (5.101) relates the shape asymmetry of a wormlike chain p_x to the relative chain length x and the shape asymmetry p_∞ of the chain in the Gaussian coil conformation.

Curves 2–5 on Fig. 5.21 show the dependence of p_x on x according to Eq. (5.101) at various hypothetical values of p_∞.

Curve 2, corresponding to $p_\infty = 1$, is similar to curve 1, although it is obtained in different systems of coordinates: curve 1 is obtained in the central system (Fig. 2.8), and curve 2 in that related to the vector **h**. In both cases at $x \to 0$ we have $p \to \infty$, which corresponds to the conformation of a straight infinitely thin rod. With increasing chain length the shape asymmetry decreases sharply and, even for chains containing 3–4 Kuhn segments, virtually attains the asymptotic limit corresponding to a spherically symmetric segment distribution.

Curves 3, 4, and 5 in Fig. 5.21 also show a drastic decrease in the shape asymmetry of the molecule with increasing length, although in these cases the asymptotic limit is the Gaussian coil having an asymmetric shape.

Equation (5.101) satisfies the condition that, at $x \to 0$, $p_x \to \infty$, as for an infinitely thin rod. For a real molecule it is necessary to take into account the finite value of its diameter d [97, 114] because it is evident that the shape asymmetry cannot exceed the L/d value. At low x, when the chain conformation is close to a straight rod, $p \to L/d$.

In this range of x values with increasing L the degree of asymmetry does not decrease as in the plots of Fig. 5.21, but rather increases. However, it increases more slowly than L because, in this case, the degree of chain coiling $L/\langle h^2 \rangle^{1/2}$ increases and, correspondingly, the effective diameter of the hydrodynamically equivalent particle also slightly increases (as compared to d). If the increase in the effective diameter with increasing chain length is neglected at low x values, then the maximum possible degree of asymmetry p in this range of x for the wormlike chain may be assumed to be

$$p = \langle h^2 \rangle^{1/2}/d = (A/d)[(x - 1 + e^{-x})/2]^{1/2}.$$

This actually means that the shape symmetry parameter b_0 contained in the FB equations is expressed by the equation

$$b_0 = \frac{p_x^2 - 1}{p_x^2 + 1} \frac{p^2 - 1}{p^2 + 1} = \frac{p_x^2 - 1}{p_x^2 + 1} \frac{x - 1 + e^{-x} - 2(d/A)^2}{x - 1 + e^{-x} + 2(d/A)^2}, \tag{5.102}$$

where p_x is determined from Eq. (5.101).

With increasing chain length (i.e., x) the second multiplier in Eq. (5.102) approaches unity and the dependence of b_0 on x at high x is determined only from the first multiplier. In contrast, at low x the first multiplier is close to unity and the dependence of b_0 on x is determined from the second multiplier.

3.3. Intrinsic Value of Birefringence

3.3.1. Molecular Model.
The conformational and optical properties of a chain molecule considered above may be uniquely described using the wormlike chain model, while its dynamic behavior in a shear flow, and the corresponding FB, are determined to a great extent by an additional factor, the kinetic rigidity of the chain. According to the theory of internal viscosity [Eq. (5.79)], the kinetic rigidity of the chain determining its large-scale motion decreases with increasing chain length. Hence, in principle it might be expected that the change in the conformation of a wormlike chain from a straight rod to a Gaussian coil with increasing chain length should be accompanied by a decrease in its kinetic rigidity.

However, as we have seen, the results of the theory of internal viscosity refer only to the limiting cases of a kinetically flexible and a kinetically rigid Gaussian coil and only allow the determination of intrinsic values of orientation and flow birefringence of coiled molecules (weak flow range). Hence, we restrict ourselves to the consideration of FB of kinetically rigid wormlike chains for which over the entire range of investigated values of $x = 2L/A$ the kinetic rigidity is sufficiently high to ensure the orientational character of FB. In other words, it is assumed that the conformation of the chain remains unchanged during the time interval which is large compared with that required for the establishment of steady orientational distribution of molecules in shear flow.

The conformational and optical properties of the wormlike model are characterized by an axially symmetric distribution of chain elements in which the geometrical and optical axes of the molecule coincide (vector $\mathbf{h}$). The difference between the main polarizabilities $\langle \gamma_1 - \gamma_2 \rangle$ is determined by Eqs. (5.86) and (5.87) or (5.97), and the shape asymmetry is defined by Eq. (5.102).

The shape asymmetry of a kinetically rigid chain is the cause of its nonuniform rotation (with an orientation-dependent angular velocity) in flow, which leads to the kinematic orientation of axes of these anisotropic molecules. In the weak-flow range the distribution function $\rho(\varphi, \vartheta)$ coincides with that for spheroids and is defined by Eq. (2.22).

Accordingly, intrinsic birefringence for wormlike chains is expressed by an equation similar to Eq. (5.14):

$$[n] = \frac{2\pi N_A}{135 kT \eta_0 M} \frac{(n^2 + 2)^2}{n} \langle b_0 W (\gamma_1 - \gamma_2) \rangle, \qquad (5.103)$$

where all designations coincide with those used in Eq. (5.14). These equations differ in the presence of angle brackets $\langle \; \rangle$ in Eq. (5.103), which imply averaging over all chain conformations.

3.3.2. Conformational Polydispersity. The necessity of averaging shown in Eq. (5.103) follows from the fact that in a system of chain molecules conformational distribution (polydispersity) always occurs. The higher the kinetic rigidity of the chain and, hence, the greater the time of existence of any individual conformation, the more important is the evaluation of this distribution. The conformations of various kinetically rigid molecules in solution are virtually "frozen"; at any moment they differ for different molecules and each of them is characterized by certain values of $\gamma_1 - \gamma_2$, W, and b_0.

The values of $\langle \gamma_1 - \gamma_2 \rangle$ and $\langle b_0 \rangle$ averaged over all conformations for wormlike chains are determined by Eqs. (5.86), (5.97), and (5.101). The value of $\langle W \rangle$ can also be expressed (see below) by experimentally determined values. However, the transition from the average product $\langle b_0 W (\gamma_1 - \gamma_2) \rangle$ contained in Eq. (5.103) to the product of average values $\langle b_0 \rangle \langle W \rangle \langle \gamma_1 - \gamma_2 \rangle$ can be made only approximately on the basis of some qualitative considerations.

Thus, it is possible to make a simplifying assumption that all the conformations of molecules are highly aspherical, i.e., $p \gg 1$; then, correspondingly, $\langle b_0 \rangle = 1$ and the product $\langle W (\gamma_1 - \gamma_2) \rangle$ in Eq. (5.103) is averaged.

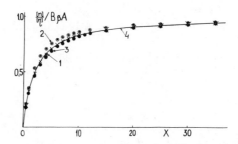

Fig. 5.22. Relative value of the shear optical coefficient $([n]/[\eta])/B\beta A$ versus relative chain length $x = 2L/A$ for a wormlike chain: 1) according to Eqs. (5.106), (1.24), (1.28), and (5.86); 2) according to Eqs. (5.106), (1.24), (1.28), and (5.97); 3) according to Eq. (5.110); 4) according to Eq. (5.111).

In the Gaussian range, when the chain conformation undergoes a change (i.e., h undergoes a change), according to Eq. (2.95), the value of W changes proportionally with h^2. Similarly, according to Eq. (5.35), for a Gaussian chain $\gamma_1 - \gamma_2$ is proportional to h^2. Hence, for a Gaussian coil the following equation is obeyed:

$$\langle W (\gamma_1 - \gamma_2) \rangle / (\langle W \rangle \langle \gamma_1 - \gamma_2 \rangle) = \langle h^4 \rangle / \langle h^2 \rangle^2. \tag{5.104}$$

It follows from Eq. (5.104) that for a Gaussian chain [see Eqs. (1.4) and (1.5)] we have

$$\langle W (\gamma_1 - \gamma_2) \rangle / \langle W \rangle \langle \gamma_1 - \gamma_2 \rangle = 5/3.$$

For a wormlike chain in the rodlike conformation ($x \to 0$), Eq. (5.104) is also obeyed because under these conditions conformational polydispersity is absent and $\langle h^4 \rangle / \langle h^2 \rangle^2 = 1$. Consequently, for a wormlike chain, Eq. (5.104) is obeyed for the limiting possible values of $x = 0$ and $x \to \infty$. It is natural to assume that Eq. (5.104) is also valid for all the intermediate values of x for which $\langle h^4 \rangle / \langle h^2 \rangle^2$ is defined by Eqs. (1.28) and (1.24).

If Eq. (5.104) is applied and $\langle b_0 \rangle$ is assumed to be unity, Eq. (5.103) becomes

$$[n] = \frac{2\pi N_A}{135\eta_0 kT} \frac{(n^2 + 2)^2}{n} \frac{\langle W \rangle}{M} \langle \gamma_1 - \gamma_2 \rangle \frac{\langle h^4 \rangle}{\langle h^2 \rangle^2}. \tag{5.105}$$

The rotational friction coefficient $\langle W \rangle$ is related to the intrinsic viscosity $[\eta]$ by a general relationship (2.27) valid for all molecular models at the corresponding value of coefficient $F(p)$ (see Chapter 2). At $\langle b_0 \rangle = 1$, $p = 10$, and $F(p) = 1/6$, if $\langle W \rangle$ is replaced according to Eq. (2.27), Eq. (5.105) becomes

$$[n]/[\eta] = B \langle \gamma_1 - \gamma_2 \rangle \langle h^4 \rangle / \langle h^2 \rangle^2, \tag{5.106}$$

where

$$B = (4\pi/45kT)(n^2 + 2)^2/n. \qquad (5.107)$$

According to Eq. (5.106), the limiting values of $[n]/[\eta]$ for wormlike chains are: a) at $x \to 0$, $([n]/[\eta])_0 = B\langle\gamma_1 - \gamma_2\rangle_{x\to 0}$ as for a thin rod, and 2) at $x \to 0$, $([n]/[\eta])_\infty = B(\alpha_1 - \alpha_2) = B\beta A$ as for a kinetically flexible Gaussian coil [Eq. (5.60)].

The dependence of the relative value of $([n]/[\eta])/([n]/[\eta])_\infty$ on x according to Eq. (5.106) is shown in Fig. 5.22 by points 1 and 2. In plotting this figure, the dependence of $\langle h^4\rangle/\langle h^2\rangle^2$ on x from Eqs. (1.28) and (1.24) is used and points 1 and 2 differ in the dependences of $\langle\gamma_1 - \gamma_2\rangle$ on x: points 1 and 2 correspond to Eqs. (5.86) and (5.97), respectively.

The similarity of curves 1 and 2 shows that Eqs. (5.86) and (5.97) are virtually identical.

Hence, the shear optical coefficient $[n]/[\eta]$ for a kinetically rigid wormlike chain is dependent on the chain length increasing at small x and attaining the limiting value in the Gaussian range where it is no longer dependent on chain length. According to Eq. (5.106), this change is entirely determined by the change in the optical anisotropy of the chain $\langle\gamma_1 - \gamma_2\rangle$ in accordance with curves 1 and 6 in Fig. 5.19. In this case the fact that the change in the shape of wormlike chains with increasing x should also affect the value of $[n]$ and $[n]/[\eta]$ in Eq. (5.106) is not taken into account because in Eq. (5.105) it was initially assumed that $b_0 = 1$.

If no *a priori* assumption is made that in Eq. (5.103) the shape coefficient $b_0 = 1$, it is necessary to consider the change occurring in its average value $\langle b_0\rangle$ upon variation of molecular conformation from rodlike to a Gaussian coil, taking into account that, in this case, the shape asymmetry of the wormlike chain p_x varies according to Eq. (5.101).

Moreover, the coefficient $F(p)$ in Eq. (2.27) also varies with the length of the wormlike chain and may be represented as a function of x in the following form [113, 116]:

$$F_x = 1/6 + 0.133/p_x. \qquad (5.108)$$

Further, it should be borne in mind that the deviation of instantaneous conformations of a rigid chain from the "average" conformation must lead to the deviation from average values not only for W and $\gamma_1 - \gamma_2$, which was taken into account in Eqs. (5.105) and (5.106), but also for the shape coefficient b_0, which was not taken into account in these equations. If it is assumed that when the conformations undergo a random change, the coefficient b_0 changes proportionally to h^2 as do the values of W and $\gamma_1 - \gamma_2$, then the product $b_0 W(\gamma_1 - \gamma_2)$ changes proportionally with h^6. Hence, the average value of this product in Eq. (5.103) may be represented as

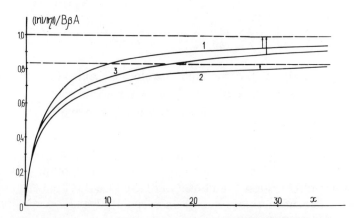

Fig. 5.23. Relative value of the shear optical coefficient $([n]/[\eta])/B\beta A$ versus relative chain length $x = 2L/A$ for a wormlike chain according to various theories: 1) according to Eq. (5.111); 2) according to the theory presented in [118, 119]; 3) according to the theory presented in [120].

$$\langle b_0 W (\gamma_1 - \gamma_2)\rangle = \langle b_0\rangle \langle W\rangle \langle \gamma_1 - \gamma_2\rangle \langle h^6\rangle/\langle h^2\rangle^3. \qquad (5.109)$$

Applying Eq. (5.109), replacing $\langle W\rangle/M$ in Eq. (5.103) according to (2.27), and expressing $\langle \gamma_1 - \gamma_2\rangle$ according to Eq. (5.97), we obtain, instead of Eq. (5.106), the expression

$$[n]/[\eta] = B\beta A (3/5) (\langle h^6\rangle/\langle h^2\rangle^3) \langle b_0\rangle (1 - e^{-5x/6}) (F_0/F_x), \qquad (5.110)$$

where $F_0 = 1/6$, F_x is expressed as a function of p_x according to Eq. (5.108), $\langle b_0\rangle = (p_x^2 - 1)/(p_x^2 + 1)$, and p_x is determined from Eq. (5.101). The $\langle h^6\rangle/\langle h^2\rangle^3$ ratio for a wormlike chain is defined by Eqs. (1.29) and (1.24).

The dependence of $([n]/[\eta])/B\beta A$ on x according to Eq. (5.110) at $p_\infty = 2$ (which corresponds to Kuhn's data [30, 31]) is shown by curve (points) 3 in Fig. 5.22. According to this curve, in the range of limiting small x (rodlike conformation) we have $([n]/[\eta])_{x\to 0} = B\beta L$ and in the Gaussian range the limiting value is $([n]/[\eta])_\infty = B\beta A$, which corresponds both to the initial slope and to the limit of curves 1 and 2 (for which it is assumed that $\langle b_0\rangle = 1$). In all the other ranges of x values the difference between curve 3 and curves 1 and 2 is slight, and when compared to experimental data, this difference is virtually within experimental error.

Hence, if conformational polydispersity is taken into account, the dynamo-optical properties of an assembly of rigid coiled molecules (for which $\langle p\rangle = 2$) approach those of molecules with a high shape asymmetry ($\langle p\rangle \to \infty$). The reason for this resemblance is the fact that in FB observed in a polydisperse system the main contribution is provided by the molecules adopting the most extended instantaneous conformations, since these conformations correspond to the highest values of W and $\gamma_1 - \gamma_2$, as well as to the highest values of b_0. Consequently, when theoretical curves are compared to experimental data, both Eq. (5.106) and Eq. (5.110) may be used with equal validity.

Equations (5.106) and (5.110) may be adequately approximated by the simple relationship (5.111) interpolating the dynamo-optical properties of kinetically rigid wormlike chains over the entire range of changes in their conformations from a straight thin ($d \ll L$) rod to a Gaussian coil

$$[n]/[\eta] = B\beta Ax/(x + 2). \tag{5.111}$$

The dependence corresponding to Eq. (5.111) is shown in Fig. 5.22 by curve 4.

The similarity between curve 4 and curves 13 in Fig. 5.22 ensures the validity of Eq. (5.111), and its simplicity is useful in the interpretation of experimental data on FB in solutions of kinetically rigid chain molecules.

The complete coincidence of Eq. (5.111) in the Gaussian range ($x \to \infty$) with Eq. (5.60) corresponds to the conclusions of the internal viscosity theory, that the shear optical coefficients for a kinetically flexible and a kinetically rigid Gaussian coil do not differ and are defined by the same equation (5.60).

The theory of the shear optical coefficient $[n]/[\eta]$ dependence on chain length L for molecules represented by a wormlike chain has been proposed also in [118, 119]. The calculations have been carried out without considering the dynamics of intramolecular motion, as for a kinetically rigid chain. However, the problem of the dependence of the shape asymmetry of the molecule on its length and the effect of this dependence on FB has not been considered; hence the multiplier b_0 is not introduced in the equations of the theory [118]. The resulting dependence of $([n]/[\eta])/B\beta A$ on x is represented by curve 2 in Fig. 5.23. The initial slope and the region of curve 2 in the range of low x coincide with curve 1 [determined from Eq. (5.111)] but the asymptotic limit of curve 2 is 5/6 that of curve 1. Accordingly, the shear optical coefficient of a kinetically rigid Gaussian coil is smaller in the same proportion as that for a kinetically flexible Gaussian chain [Eq. (5.60)]. This difference appears to be related to the fact that in the theory [118] the anisotropy of the Gaussian chain is assumed to be equal to $\beta A/2$ (curve 2 in Fig. 5.19 [96]) rather than to the value $(3/5)\beta A$ used in plotting curve 1 and corresponding to Eq. (5.36).

To express the dependence of $[n]/[\eta]$ on the molecular weight of the polymer M, Eq. (5.111) may be represented in the following form:

$$[n]/[\eta] = B\beta AM/(M + M_s), \tag{5.112}$$

where $M_s = M_0 S = AM_L$.

Here M_0, M_L, and M_s are the molecular weights of the monomer unit, the unit length of the chain, and the Kuhn segment, respectively, and S is the number of monomer units in the Kuhn segment.

According to Eq. (5.112), the dependence of $[n]/[\eta]$ on M is represented by a curve similar to curve 1 in Fig. 5.22. Its limiting value is $([n]/[\eta])_\infty = B\beta A$ and the initial slope is $[\partial([n]/[\eta])/\partial M]_{M\to 0} = B\beta/M_L$. Correspondingly, the ratio of the limit to the initial slope $AM_L = M_s$ is equal to the molecular weight of the Kuhn segment.

3.4. Intrinsic Orientation of FB

For kinetically rigid chain molecules the value and orientation of flow birefringence are determined by the character of the rotation of the molecule as a whole. Hence, for these molecules, as for any other rigid particles, intrinsic orientation $[\chi/g]$ is uniquely related to the rotational diffusion coefficient D_r and rotational friction coefficient W by Eq. (5.5).

Hence, Eq. (5.6) may also be valid for rigid-chain molecules, but the value of the coefficient G in this equation depends on the model used.

Thus, as we have seen, for a kinetically rigid chain in the Gaussian coil conformation the values of G are found to be 0.67 and 0.51 at weak and strong hydrodynamic interactions, respectively (Fig. 5.17) [89]. These values, denoted by symbols A and B in Fig. 5.3, are three to five times higher than the corresponding theoretical values of G for kinetically flexible Gaussian coils (Section 2.4). The reason for these very high values of G (and hence those of $[\chi/g]$ at equal $[\eta]$ and M values) for kinetically rigid coils as compared to kinetically flexible coils can be explained by the effect of conformational polydispersity. In fact, rigid chains exhibiting extended conformations are oriented at lower g values than rigid coiled chains, which should lead to an increase in the *initial* slope of the dependence of χ on g in a conformationally polydisperse assembly of rigid molecules.

On the other hand, if kinetically rigid chains are of the rodlike shape ($x \to 0$ for the wormlike model), then the assembly exhibits no conformational polydispersity and $G = 0.5$–0.63 for both a straight necklace and an extended spheroid (curve 1 in Fig. 5.3), which is close to the values of G for a rigid coil ($x \to \infty$ for the wormlike model), denoted by symbols A and B in Fig. 5.3.

Hence, with increasing length of kinetically rigid chains and the corresponding change in their conformation from a rodlike conformation to a Gaussian coil (change in x from 0 to ∞ in the wormlike model) the theoretical value of G is virtually invariable, remaining close to 0.5–0.6.

This conclusion can naturally be valid only if the kinetic rigidity (internal viscosity) of the chain does not vary markedly over the investigated range of changes in x. With increasing length of both flexible and rigid chains their hydrodynamic interaction (parameter X) increases, and this in itself should lead to a decrease in the coefficient G (Figs. 5.8 and 5.17). However, for a kinetically rigid wormlike chain (in which the average conformation also varies from a rod to a coil) this decrease in G is compensated by an increase in G due to increasing conformational polydispersity.

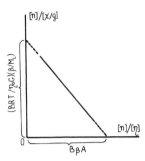

Fig. 5.24. Dependence of $[n]/[\chi/g]$ on $[n]/[\eta]$ for a kinetically rigid wormlike chain according to Eq. (5.114).

The measurement of the value of intrinsic orientation $[\chi/g]$ for kinetically rigid chains is the most direct experimental method used for the determination of the coefficient of rotational friction W (or diffusion D_r) according to Eq. (5.5).

Moreover, it is possible to evaluate the molecular weight of the polymer from experimental values of $[\eta]$ and $[\chi/g]$ using Eq. (5.6) if the value of the coefficient G for polymers of the class investigated is known. This method is of great practical importance, in particular in those cases when it is difficult to use other known methods (for example, for polymers soluble only in strong acids).

This method may also be used [117] in the study of the dependence of the shear optical coefficient $[n]/[\eta]$ on the chain length if the measurements of molecular weight M are replaced by those of intrinsic orientation $[\chi/g]$. For this purpose it is sufficient to write Eq. (5.6) in the form

$$[n]/[\chi/g] = (RT/GM\eta_0)([n]/[\eta]) \qquad (5.113)$$

and to exclude M from the system of equations (5.112) and (5.113). Then we obtain

$$[n]/[\chi/g] = [(RT)/(\eta_0 G)][(B\beta/M_L) - (1/AM_L)([n]/[\eta])]. \qquad (5.114)$$

If the coefficient G remains invariable in a homologous series of rigid-chain molecules over the molecular-weight range investigated, then Eq. (5.114) predicts a linear dependence of $[n]/[\chi/g]$ on $[n]/[\eta]$. A straight line corresponding to the function $[n]/[\chi/g] = f([n]/[\eta])$ forms the abscissa intercept $B\beta A$ and the ordinate intercept $(BRT/\eta_0 G)(\beta/M_L)$ (Fig. 5.24). Using experimental data on $[n]$, $[\eta]$, and $[\chi/g]$ it is possible to determine from the intercepts the values of β and A (if M_L is known) or the anisotropy of the monomer unit Δa and the number of monomer units S in a Kuhn segment.

3.5. Kinetically Flexible Wormlike Chains

The FB theory in solutions of chain molecules whose conformational properties correspond to those of a wormlike chain has been developed by Hearst et al. [120].

The theory is based on the model of a molecule whose dynamic properties in the isolated state are represented by the dynamics of an elastic string with the bending force constant α [121, 122]. A comparison of the bending energy with the curving of the string caused by its thermal motion shows that the string acquires the conformation of a wormlike chain determined by Eq. (1.24) (e.g., see [123]). In this case, the length of the Kuhn segment A of this chain is determined by the elasticity coefficient α according to the equation

$$A = 4\alpha/(3kT). \tag{5.115}$$

As in the elastic necklace model, in Hearst's molecular model it is assumed that longitudinal deformations along the chain having a fluctuation character are possible. However, in this case the condition is indispensable that the overall contour length L of the chain should remain constant when some of its parts undergo deformational changes in length. The requirement of constant length is mathematically supplied by the introduction of an indefinite Lagrange multiplier β_H into the dynamics equations and by the determination of β_H under the conditions L = const. These conditions lead to the following relationship between β_H and the mean-square end-to-end vector $\langle h^2 \rangle$ of the chain:

$$\beta_H = 3kTL/\langle h^2 \rangle. \tag{5.116}$$

Comparison of Eqs. (5.116) and (5.45) shows that β_H has the role of the coefficient of longitudinal chain elasticity or the force constant in the Rouse–Zimm bead-and-spring model.

Comparison of Eq. (5.116) with Porod's equation (1.24) for the force constant β_H yields

$$\beta_H = (3kT/A)[1 - (1 - e^{-x})/x]^{-1}. \tag{5.117}$$

Equation (5.117) for limiting conformations of a rod and a Gaussian coil gives $\beta_H = 3kT/L$ and $\beta_H = 3kT/A$ at $x \to 0$ and $x \to \infty$, respectively.

The hydrodynamic properties of the Hearst model in shear flow are described similarly to the Rouse–Zimm bead-and-spring model using the method of normal coordinates and mobility modes.

TABLE 5.1. Relative Value of Shear Optical Coefficient $\Delta \equiv ([n]/[\eta])/B\beta A$ and Value of the Coefficient $G = [\chi/g]RT/M[\eta]\eta_0$ at Various Values of Relative Chain Length $x = 2L/A$ in the Theory of FB of Kinetically Flexible Wormlike Chains [120]

					$x/2$					
	0	10^{-3}	1	10	49	100	400	10^3	10^4	∞
Δ	0	—	0.473	0.884	0.980	0.990	0.996	0.997	0.999	1
G	0.5	0.500	0.408	0.235	0.168	0.151	0.130	0.122	0.110	0.100

Furthermore, the maintenance of constant contour length in the shear field is ensured by the introduction of a certain dependence of the force constant β_H on the shear rate. Formally this procedure is equivalent to Peterlin's method [60, 61] in which the inverse Langevin's function $\mathscr{L}^*(h/L)$ is used [see Eq. (5.44)] to limit the extension of Gaussian chains in shear flow and to maintain a constant contour length. Hence, if the results of Noda–Hearst [120] and Peterlin [49] are applied to the dependence of intrinsic viscosity on shear rate for relatively long-chain molecules ($L/A \geq 10^2$), they virtually coincide: with increasing shear rate, $[\eta]$ decreases, and the shorter the chain (less than L/A), the greater is this decrease.

Since in the Noda–Hearst theory the conformation of the molecule is described by a wormlike chain, this theory using Eq. (5.117) allows the characterization of both the shear dependence of viscosity and FB at lower relative lengths L/A than those available in Peterlin's theory.

The optical model of the molecule in the Noda–Hearst theory is a kinetically flexible wormlike chain deformed in flow according to Rouse–Zimm dynamics.

The results obtained for the dependence of birefringence Δn and orientation angle φ_m on the hydrodynamic parameter $M[\eta]\eta_0 g$ are not presented in the form of analytical equations but rather by tabulated data obtained using a computer. For the shear optical coefficient $[n]/[\eta]$ and intrinsic orientation $[\chi/g]$ these data are listed in Table 5.1. As in all the other theories, the values of $[n]/[\eta]$ are independent of hydrodynamic interactions. The values of the coefficient $G = [\chi/g]RT/M[\eta]\eta_0$ given in Table 5.1 were obtained from the evaluation of hydrodynamic interactions.

The Rouse–Zimm hydrodynamic subchain model does not allow discussion of the dynamo-optical properties of chain molecules whose length is only a part of the Kuhn segment. Hence, Table 5.1 does not contain the data for the values of $L/A < 1$ with the exception of the case of a thin rigid rod ($L/A \to 0$).

The limiting values of $[n]/[\eta]$ in Table 5.1 are represented by relationships similar to those for a kinetically rigid wormlike chain: $([n]/[\eta])_{x \to 0} = B\beta L$ and $([n]/[\eta])_{x \to \infty} = B\beta A$. However, the value of factor B in the Noda–Hearst theory is greater by a factor of 5/3 than in Eq. (5.107). This may be due to differences in averaging when the optical properties of two systems are compared. The structural unit which defines optical anisotropy of the solution in the Zimm theory is the subchain, whose anisotropy in axes of end-to-end vector A [according to Eq. (5.36)] is $(3/5)\beta A$. In the Noda–Hearst theory the structural unit is an element of a wormlike chain whose anisotropy per length A is βA, which is 5/3 times larger than that of the Zimm model.

The dependence of $([n]/[\eta])/B\beta A$ on $x = 2L/A$ corresponding to the data given in Table 5.1 is represented by curve 3 in Fig. 5.23. In the chosen system of coordinates, the initial slopes and limiting values of curves 3 and 1 coincide. This fact implies that in both cases the use of the ratio of the limit to the initial slope for the determination of equilibrium rigidity from experimental data gives identical results. However, the general shape of curve 3 differs from that of curve 1: it is flatter and attains saturation at higher values than curve 1.

Curve 2 in Fig. 5.3 represents the dependence of the coefficient $G = [\chi/g]RT/M[\eta]\eta_0$ on x according to the data given in Table 5.1. In the range of $x \to 0$ the coefficient G is 1/2, the same as for a rodlike necklace in which hydrodynamic interaction is calculated according to Kirkwood, with the preaveraging of Oseen's tensor (see Chapter 2). However, in the range of $x > 2$ (i.e., $L > A$), G decreases sharply with increasing x, attaining the limiting value $G_{x \to \infty} = 0.1$ corresponding to the hydrodynamics of a kinetically flexible nondraining Gaussian coil. Hence, over a wide range of x values the dynamo-optical properties of the Hearst molecular model are virtually equivalent to Zimm's results for kinetically flexible Gaussian coils. This seems natural because the Noda–Hearst molecular model, just as that of Rouse–Zimm, does not contain the internal viscosity (kinetic rigidity) parameter. The decrease in G with increasing x shown by curve 2 in Fig. 5.3 can be regarded as the result of an increase in the parameter X of hydrodynamic interaction in the chain with increasing chain length.

4. APPARATUS

The instrument used for the investigation of FB consists of both a mechanical component and an optical component.

4.1. Mechanical Component

To achieve a state of laminar flow of the solution under investigation, a cylindrical apparatus (dynamo-optimeter) with an inner or outer rotor is employed.

The liquid to be studied is placed in the annular space between two concentric cylinders, one of which is rotated while the other remains stationary. The liquid layer is viewed along the direction parallel to the cylinder axes. The laminar flow in the annular gap can be considered to be virtually uniform in the small region observed within the viewing angle. The velocity gradient is a constant $g = (dU/dR)$ in the direction of the radius R and equal to

$$g = 2\pi R \nu / \Delta R, \tag{5.118}$$

where ν is the number of revolutions of the cylinder per second, $R = \frac{1}{2}(R_1 + R_2)$ is the mean radius, and $\Delta R = R_1 - R_2$ is the width of the annular gap.

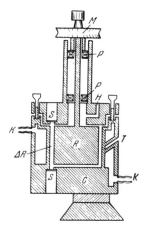

Fig. 5.25. Cylindrical apparatus with an inner rotor and bearings at the top of the stator.

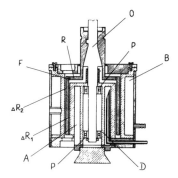

Fig. 5.26. Cylindrical apparatus with an outer rotor and bearings in the stator body.

A great number of different cylinder designs and different optical arrangements have been used in various investigations [1–11]. In a quantitative study of FB, one basic condition must be satisfied: the flow must remain laminar. A study of the conditions favoring laminar flow has revealed that a cylinder apparatus will produce vastly different results depending on whether the inner or outer cylinder is rotated [124].

When the inner cylinder is rotated, the critical velocity gradient g_k (maximum velocity at which laminar flow characteristics are still observed) is equal to

$$g_k = \frac{\pi^2}{\sqrt{0.057}} \frac{\eta}{\rho} \frac{\sqrt{R}}{(\Delta R)^{5/2}}, \tag{5.119}$$

where η is the viscosity and ρ is the density of the liquid. When the outer cylinder is rotated, the onset of turbulence occurs at a much higher speed of rotation than if the inner cylinder is rotated. For example, at $R/\Delta R = 60$ with an outer rotor the crit-

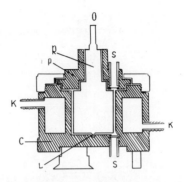

Fig. 5.27. Teflon cylindrical apparatus for operation with aggressive solvents.

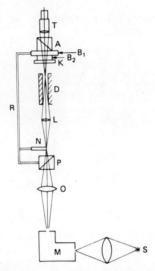

Fig. 5.28. Scheme for the optical part of the instrument for visual FB observation.

ical gradient is six times greater than if the inner cylinder of the same dimensions is rotated, and at $R/\Delta R = 10$ it is fifty times greater [124].

Figure 5.25 schematically describes the instrument with an inner rotor, widely used for the study of various polymers. Rotor R rotates on two ball bearings P mounted in the upper part of the top, which can be taken off with the rotor from stator C. Viewing windows S of thin quartz glass (without tension) allows the operation to be observed through gap ΔR in the direction of the axis of the instrument. A thermostating water jacket is placed in the stator (pipes K) and the temperature is controlled with a thermocouple (by opening T). H is the inlet opening for pouring the solution and M the pulley. The width of the rotor-to-stator gap is 0.2 to

0.5 mm; the rotor height is 30 to 50 mm; and the rotor diameter is 30 to 40 mm. All the parts of the instrument coming in contact with the solution are made of titanium and can therefore be used with any organic solvent and with aqueous polyelectrolyte solutions. The instrument is fed with only 35 ml of the polymer solution and may readily be used if several polymer fractions must be investigated and the amount of each fraction is limited.

In cylindrical instruments with an outer rotor, high-velocity gradients can be attained with relatively wide gaps without passing the limits of a laminar flow [124].

However, this can actually be used to advantage only in instruments with fixed windows [125–127]. One of this type of instrument [127] is described in Fig. 5.26. The rotor of the dynamo-optimeter is a hollow cylinder R open from below and rotating together with the O axis on ball bearings P mounted in the stator D. Observation through the gap ΔR_1 is accomplished with a system of perforations F in the upper base of the rotor. When the observation is made through the gap ΔR_2, the dynamo-optimeter can function as an instrument with an inner rotor. All parts in contact with the solutions are made of titanium. The solution is thermostated using an internal (A) and external system (B). The rotor height usually varies from 50 to 100 mm, the gap is 0.4 to 1.00 mm, and the rotor diameter 50 to 80 mm. The capacity of the dynamo-optimeter with the outer rotor is higher by one order of magnitude than that of the instrument with the inner rotor (Fig. 5.25). Consequently, these instruments are suitable for precise measurements of a weak FB in low-viscosity solvents.

For operations with aggressive solvents a Teflon dynamo-optimeter should be used (Fig. 5.27). This dynamo-optimeter is equipped with an inner rotor and differs from that described in Fig. 5.25 in that it contains no metal ball bearings. The step bearing L of Teflon rotor R is supported by stator C. The second bearing P is formed directly by the top of the stator. The only metal (titanium) part of the instrument is the axis O mounted inside the body of the rotor. Optimum dimensions of this instrument are as follows: rotor height 30 to 60 mm, rotor diameter 30 to 50 mm, and gap 0.4 to 0.8 mm.

The rotor of the cylinder dynamo-optimeter is activated by a motor through a pulley and a system of reduction gears.

The rotational velocity of the rotor is measured with the aid of a perforated disc that periodically interrupts a light beam during the rotation. The pulsing light beam is fed to a photodiode whose electric impulses are fed to a digital frequency meter. The frequency of impulses shown by the meter is proportional to the rotor speed. This scheme permits a reliable measurement of the rotation frequency of the rotor in the range from $2 \cdot 10^{-2}$ to 50 rps.

4.2. Optical Component

The optical method used in the investigation of FB may be visual or photoelectric.

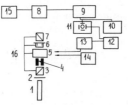

Fig. 5.29. Scheme of the photoelectric method for birefringence measurements. 1) He–Ne laser ($\lambda = 6300$ Å); 2) $\lambda/4$ plate ensuring circular polarization of the beams entering the polarizer (3); 4) solution under investigation; 5) modulator; 6) compensator; 7) analyzer; 8) photomultiplier; 9) selective (narrow-field) amplifier; 10) synchronous detector; 11) oscillograph; 12) phase shifter; 13) sonic generator; 14) modulator power supply; 15) photomultiplier power supply; 16) rotating lever fixing together the mechanical elements 3, 5, 6, and 7.

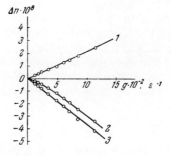

Fig. 5.30. Dependence of Δn on g for the solution of a fraction ($M = 4.75 \cdot 10^5$) of cellulose tribenzoate in dimethylphthalate. 1) Dependence of Δn on g for the solvent; 2) the same for the solution; 3) the same for the dissolved polymer.

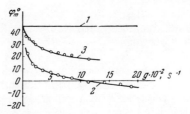

Fig. 5.31. Dependence of φ_m on g for the same polymer–solvent system as in Fig. 5.30. 1) Dependence of φ_m on g for the solvent; 2) the same for the solution; 3) the same for the dissolved polymer.

4.2.1. Visual Method. Figure 5.28 schematically describes the optical part of the instrument for the visual study of FB. A compensating device, according to Brace [128], is used in this scheme. It consists of a thin mica plate K

(elliptical compensator introducing phase difference δ_c corresponding to a few hundredths of a wavelength) rotating on a limb B_2 and a very thin half-shaded plate N (a few thousandths of a wavelength) obscuring half of the field of vision. Light source S (mercury lamp of superhigh pressure) is projected on the entrance slit of monochromator M. The image of the exit slit of the monochromator is projected by lens O on the edge of half-shaded plate N. Crossed polarizing prisms P and A rotate together with plate N and compensator K with the aid of lever R fixed on limb B_1. A thin lens L projects the image of plate N into gap D between the stator and the rotor of the cylindrical instrument. Telescope T is also focused on this gap.

The principle of the method using the elliptical compensator is based on Eq. (5.120) expressing the intensity $\mathscr{J}$ of the light passing through crossed polarizing prims P and A and two thin anisotropic plates placed between these prisms at azimuths η and η_c with respect to the polarizer P, and introducing the phase differences δ and δ_c, respectively [129]

$$\mathscr{J} = (\mathscr{J}_0/4)(\delta \sin 2\eta + \delta_c \sin 2\eta_c)^2, \qquad (5.120)$$

where $\mathscr{J}_0$ is the intensity of the light emerging from polarizer P.

Equation (5.120) is approximate and holds for low δ values if $\delta \approx \sin \delta$.

If the parameters δ and η in Fig. 5.28 refer to the anisotropic layer being investigated, δ_c and η_c refer to the elliptical compensator K, and if the half-shaded plate N is absent, then it follows that, according to Eq. (5.120), the darkening of the field of vision $(\mathscr{J} = 0)$ can occur at $\delta \sin 2\eta = -\delta_c \sin 2\eta_c$. In this case, and if the optical axis of the anisotropic layer is in the diagonal position ($\eta = \pi/4$), which occurs in the measurements of δ, the condition of darkening is expressed as $\delta = -\delta_c \sin 2\eta_c$ and can be used for the determination of δ from the known value of δ_c by rotating the compensator K (limb B_2) to the position of darkening η_c. Actually, the darkening is incomplete but a very low minimum of illumination of the field of vision is observed. The precision of mounting the compensator into the compensating position greatly increases if the half-shaded plate N is introduced. In this case, the desired phase difference δ is determined according to

$$\delta = -\delta_c(\sin 2\eta_c - \sin 2\eta_c), \qquad (5.121)$$

where η_c and η_0 are half-shaded azimuths of the compensation (i.e., azimuths corresponding to equal illumination of the two halves of the field of vision) registered from the main plane of the polarizer P in the presence and absence of birefringence in solution, respectively. The value of birefringence Δn is determined from that of δ according to $\Delta n = \delta \lambda/2\pi l$, where l is the thickness of the anisotropic layer in the path of light beams, and λ is the wavelength.

The main optical directions of the anisotropic layer (and, correspondingly, the extinction angle in FB) are found when the compensator is switched on [$\delta_c = 0$ or $\eta_c = 0$ in Eq. (5.120)] and the half-shaded position is established by rotating the whole optical system (limb B_1). The extinction angle is determined as the half-angle between two half-shaded positions, one of which corresponds to the clockwise rotation of the rotor of the dynamo-optimeter and the other to the counterclockwise rotation.

4.2.2. Photoelectric Method. The photoelectric method [130] successfully employed in the study of FB also uses the compensation principle based on Eq. (5.120). However, the anisotropy in solution and its compensation are recorded with the aid of the scheme shown in Fig. 5.29 rather than visually.

In contrast to the scheme in Fig. 5.28, in the optical part of this scheme, the half-shaded plate N is replaced by a harmonic modulator 5 of the ellipticity of the polarized light. The main detail of the modulator rigidly secured to lever 16 is a ferrite shaft in which an alternating magnetic field excites longitudinal mechanical vibrations at the resonance frequency ω of the shaft. Harmonic optical anisotropy (photoelastic effect) is established in a glass plate rigidly fixed to the end of the shaft $\delta_1 = \delta_{10} \sin \omega t$. The axis of this anisotropy forms an angle $\pi/4$ with the axis of analyzer 7.

In this scheme, the intensity $\mathscr{I}$ of light passing through crossed polarizer 3 and analyzer 7 is determined by the combined action of the anisotropic layer (introducing the phase difference $\delta_{10} \sin 2\eta$), the compensator ($\delta_c \sin 2\eta_c$) and the modulator ($\delta_1 = \delta_{10} \sin \omega t$). According to Eq. (5.120), $\mathscr{I}$ is given by

$$\mathscr{I} = B\,[\delta \sin 2\eta + \delta_c \sin 2\eta_c + \delta_{10} \sin \omega t]^2 = B\,[(\delta \sin 2\eta + \delta_c \sin 2\eta_c)^2$$
$$+ (\delta_{10}^2/2)(1 - \cos 2\omega t) + 2\delta_{10}(\delta \sin 2\eta + \delta_c \sin 2\eta_c) \sin \omega t], \qquad (5.122)$$

where B is constant.

Hence, the intensity of the light emerging from analyzer 7 and reaching photomultiplier 8 is expanded into three components, a constant component and two sinusoidal components with frequencies ω and 2ω. The corresponding electric signal at the output of the multiplier is received by a narrow-band resonance amplifier synchronous with modulator 5 and separating from the overall signal a harmonic component at frequency ω. As a result, according to Eq. (5.122) the amplitude V of the harmonic signal fed to oscillograph 11 or to synchronous detector 10 is given by

$$V = 2\delta_{10}(\delta \sin 2\eta + \delta_c \sin 2\eta_c), \qquad (5.123)$$

where V is linearly dependent on the anisotropy $\Delta n = \delta\lambda/2\pi l$ of the solution. This fact ensures a higher (not less than by an order of magnitude) sensitivity of the photoelectric method than the visual method in which, according to Eq. (5.120), the light flux affecting the eye is proportional to δ^2.

Equation (5.123) demonstrates that the value of the signal V recorded with an oscillograph or a detector becomes zero at $\delta \sin 2\eta = -\delta_c \sin 2\eta_c$, i.e., under the same conditions as in visual observation. Hence, the procedures for the determina-

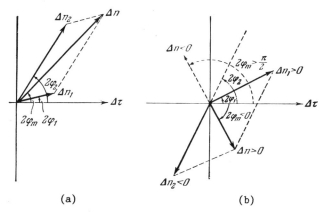

Fig. 5.32. Flow birefringence Δn when Δn_1 and Δn_2 are summed: a) when they are of the same sign; b) when they are opposite in sign.

tion of the anisotropy of the solution and the orientation angle in the photoelectric and visual methods are identical.

4.3. Determination of Excess Flow Birefringence of a Dissolved Polymer

In the investigation of dilute solutions in which the magnitude of the effect introduced by the dissolved polymer may prove to be comparable to that of pure solvent, the latter effect should be taken into account in order to be able to determine correctly the Δn value and the orientation φ_m of the excess flow birefringence introduced by the polymer.

For this purpose the solution should be considered as a two-component system in which the solvent is characterized by the values $\varphi_0 = \varphi_1 = \pi/4$ (low-molecular-weight liquid) and $\Delta n_1 = \Delta n_0$ and, correspondingly, the polymer induces flow birefringence $\Delta n_2 = \Delta n_p$ with the orientation angle $\varphi_2 = \varphi_p$. Then using Eqs. (5.70) and (5.71) we obtain

$$\tan 2\varphi_p = \tan 2\varphi_m - \Delta n_0/(\Delta n \cos 2\varphi_m), \tag{5.124}$$

$$\Delta n_p = \Delta n \cos 2\varphi_m/\cos 2\varphi_p. \tag{5.125}$$

Determining by independent measurements Δn_0 in a pure solvent and Δn and φ_m in solution, it is possible to calculate Δn_p and φ_p for the dissolved polymer.

Figures 5.30 and 5.31 presented as illustrations show the dependences $\Delta n = f(g)$ and $\varphi_m = f(g)$ for a sample of cellulose tribenzoate in dimethylphthalate.

At the selected concentration ($c = 1 \cdot 10^{-4}$ g/cm^3) flow birefringences in the solvent and in the dissolved polymer Δn_p are close in absolute values but opposite in sign. This leads to an "anomalous" shape for the dependences $\varphi_m(g)$ of the solution (curve 2 in Fig. 5.31) and markedly displaces the experimental curve of the dependence $\Delta n(g)$ for the solution (curve 2 in Fig. 5.30) from that of the curve $\Delta n_p(g)$ for the dissolved polymer (curve 3 in Fig. 5.30). The application of experimental values of Δn_0, Δn, and φ_m and the calculation of Δn_p and φ_p according to Eqs. (5.124) and (5.125) leads to "normal" curves 3 in Fig. 5.30 and 5.31 describing the dynamo-optical properties of molecules of the dissolved polymer.

Another example of the application of Eqs. (5.124) and (5.125) to a binary polymer–solvent system is provided by the investigation of FB in partially polymerized styrene [131].

Calculations carried out according to Eqs. (5.124) and (5.125) may be replaced by the plotting of the corresponding appropriate diagrams in which flow birefringence Δn_i of the component i is described by a vector-forming angle $2\varphi_i$ with the direction of the shear $\Delta\tau$ (Fig. 5.32a, b). The value of Δn and the orientation φ_m of the total effect are found as the vector sums of the components (Δn_i and φ_i). The angles φ_i reckoned from the direction $\Delta\tau$ in the counterclockwise direction are assumed to be positive, and those in the opposite direction to be negative. Birefringence is considered to be positive if the corresponding angle 2φ is taken from the direction $\Delta\tau$ to that of the vector Δn. If the corresponding double orientation angle is taken from $\Delta\tau$ to the direction opposite to Δn, then birefringence is considered to be negative with respect to the selected axis. Thus, for example in Fig. 5.32b, the total anisotropy is positive with respect to the axis with $2\varphi_m < 0$ and negative with respect to that with $2\varphi_m > \pi/2$.

REFERENCES

1. J. T. Edsall, Adv. Colloid Sci., **1**, 269 (1942).
2. A. Peterlin and H. A. Stuart, in: Hand-und Jahrbuch der Chemischen. Physik, A. Eucken and K. L. Wolf (eds.), Vol. 8, 1B, Akadem. Verlagsges., Leipzig (1942), p. 1.
3. W. Kuhn, H. Kuhn, and P. Buchner, Ergeb. Exakt. Naturwiss., **25**, 1 (1951).
4. R. Cerf and H. A. Scheraga, Chem. Rev., **51**, 185 (1952).
5. H. A. Stuart (ed.), Die Physik der Hochpolymeren, Vols. 1 and 2, Springer Verlag, Berlin (1952, 1953).
6. A. Peterlin, in: Rheology, F. R. Eirich (ed.), Vol. 1. Academic Press, New York (1956), p. 615.
7. R. Cerf, Adv. Polym. Sci., **1**, 383 (1959).
8. H. A. Scheraga and R. Signer, in: Physical Methods of Organic Chemistry, A. Weissberger (ed.), Vol. 1, New York (1960), p. 2388.
9. V. N. Tsvetkov, in: Newer Methods of Polymer Characterization, B. Ke (ed.), Interscience, New York (1964), p. 563.
10. H. Janeschitz-Kriegl, Adv. Polym. Sci., **6**, 170 (1969).
11. A. Peterlin and P. Munk, in: Physical Methods of Chemistry, A. Weissberger and B. Rossiter (eds.), Interscience, New York (1972), p. 271.

12. V. N. Tsvetkov and L. N. Andreeva, in: Polymer Handbook, J. Brandrup and E. H. Immergut (eds.), 2nd edn., Interscience, New York (1974).
13. V. N. Tsvetkov and L. N. Andreeva, Adv. Polym. Sci., 39, 98 (1981).
14. A. Peterlin and H. A. Stuart, Z. Phys., 112, 1 (1939).
15. P. Boeder, Z. Phys., 75, 258 (1932).
16. A. Peterlin, Z. Phys., 111, 232 (1938).
17. H. A. Scheraga, J. T. Edsall, and J. O. Gadd, J. Chem. Phys., 19, 1101 (1951).
18. D. Vorlander, Z. Phys. Chem., A152, 47 (1931); A178, 93 (1936).
19. V. N. Tsvetkov, V. E. Eskin, and S. Y. Frenkel, Structure of Macromolecules in Solution, Butterworths, London (1970).
20. V. N. Tsvetkov and E. V. Frisman, Dokl. Akad. Nauk SSSR, 67, 49 (1949).
21. E. V. Frisman and V. N. Tsvetkov, Zh. Fiz. Khim., 25, 682 (1951).
22. J. V. Champion and G. H. Meeten, Trans. Faraday Soc., 64, 238 (1968).
23. L. Silberstein, Philos. Mag., 33, 92, 215, 521 (1927).
24. F. Perrin, J. Phys. Rad., (7)5, 497 (1934).
25. C. V. Raman and K. J. Krishnan, Philos. Mag. (7), 769 (1928).
26. J. Maxwell, Treatise on Electricity and Magnetism, London (1873), p. 65.
27. W. Kuhn and F. Grün, Kolloid Z., 101, 248 (1942).
28. V. N. Tsvetkov, Dokl. Akad. Nauk SSSR, 165, 360 (1965).
29. W. Kuhn, Kolloid Z., 68, 2 (1934).
30. H. Kuhn, Experientia, 1, 28 (1945).
31. H. Kuhn, Helv. Chim. Acta, 31, 1677 (1948).
32. V. N. Tsvetkov and E. V. Frisman, Acta Physicochim. URSS, 20, 363 (1945).
33. V. N. Tsvetkov and E. V. Frisman, Dokl. Akad. Nauk SSSR, 97, 647 (1954).
34. V. N. Tsvetkov, Int. Symp. Macromol. Chem., Milan (1954).
35. V. N. Tsvetkov, J. Polym. Sci., 23, 151 (1957).
36. V. N. Tsvetkov, E. V. Frisman, O. B. Ptitsin, and S. Y. Kotljar, Zh. Tekh. Fiz., 28, 1428 (1958).
37. M. Copic, J. Polym. Sci., 20, 593 (1956).
38. M. Copic, J. Chem. Phys., 26, 1382 (1957).
39. R. Koyama, J. Phys. Soc. Jpn., 16, 1366 (1961).
40. J. G. Kirkwood, J. Chem. Phys., 4, 592 (1936).
41. W. Kuhn and H. Kuhn, Helv. Chim. Acta, 26, 1394 (1943).
42. V. N. Tsvetkov, Vysokomol. Soedin., 5, 740 (1963).
43. V. N. Tsvetkov and A. E. Grishchenko, J. Polym. Sci., C16, 3195 (1968).
44. V. N. Tsvetkov, A. E. Grishchenko, L. E. De-Millo, and E. N. Rostovskii, Vysokomol. Soedin., 6, 384 (1964).
45. A. E. Grishchenko and M. G. Vitovskaya, Vysokomol. Soedin., 8, 800 (1966); A9, 1280 (1967).
46. E. V. Frisman, A. K. Dadivanian, and G. A. Dujev, Dokl. Akad. Nauk SSSR, 153, 1062 (1963).
47. E. V. Frisman and A. K. Dadivanian, Vysokomol. Soedin., 8, 1359 (1966).
48. E. V. Frisman and A. K. Dadivanian, J. Polym. Sci., C16, 1001 (1967).
49. W. Kuhn and H. Kuhn, Helv. Chim. Acta, 28, 1533 (1945); 29, 71 (1946).
50. J. Hermans, Physica, 10, 777 (1943); Recl. Trav. Chim., 63, 219 (1944).
51. P. Rouse, J. Chem. Phys., 21, 1272 (1953).

52. B. Zimm, J. Chem. Phys., **24**, 269 (1956).
53. G. V. Thurston and J. L. Schrag, J. Chem. Phys., **45**, 3373 (1966).
54. G. V. Thurston and A. Peterlin, J. Chem. Phys., **46**, 4881 (1967).
55. J. W. Miller and J. L. Schrag, Macromolecules, **8**, 361 (1975).
56. T. P. Lodge and J. L. Schrag, Macromolecules, **15**, 1376 (1982).
57. B. Zimm, G. Roe, and L. Epstein, J. Chem. Phys., **24**, 279 (1956).
58. N. W. Tshoegl, J. Chem. Phys., **39**, 149 (1963); **40**, 473 (1964); **44**, 4615 (1966).
59. A. Peterlin, Pure Appl. Chem., **12**, 563 (1966).
60. A. Peterlin, Polymer, **2**, 257 (1961).
61. C. Reinhold and A. Peterlin, J. Chem. Phys., **44**, 4333 (1966).
62. A. Peterlin and R. Signer, Helv. Chim. Acta, **36**, 1575 (1953).
63. A. Peterlin, J. Polym. Sci., **12**, 45 (1954).
64. A. S. Lodge, Nature, **176**, 838 (1955); Trans. Faraday Soc., **52**, 127 (1956).
65. L. R. G. Treloar, The Physics of Rubber Elasticity, Clarendon, Oxford (1958).
66. V. N. Tsvetkov, V. E. Bichkova, S. M. Savvon, and I. N. Nekrasov, Vysokomol. Soedin., **1**, 1407 (1959).
67. V. N. Tsvetkov and E. V. Frisman, Acta Physicochim. URSS, **20**, 61 (1945).
68. V. N. Tsvetkov, E. V. Frisman, and L. N. Muhina, Zh. Eksp. Teor. Fiz., **30**, 649 (1956).
69. V. N. Tsvetkov and I. N. Shtennikova, Zh. Tekh. Fiz., **29**, 885 (1959).
70. V. N. Tsvetkov, N. N. Boitzova, and A. E. Grishchenko, Vestn. Leningr. Gos. Univ., Ser. Phys. Chem., **4**, 59 (1962).
71. V. N. Tsvetkov, E. V. Frisman, and N. N. Boitzova, Vysokomol. Soedin., **2**, 1001 (1960).
72. V. N. Tsvetkov and S. Ya. Ljubina, Vysokomol. Soedin., **1**, 857 (1959).
73. E. V. Frisman and V. N. Tsvetkov, J. Polym. Sci., **30**, 297 (1958).
74. V. N. Tsvetkov and V. E. Bychkova, Vysokomol. Soedin., **6**, 600 (1964).
75. V. N. Tsvetkov and A. E. Grishchenko, J. Polym. Sci., **C16**, 3195 (1968).
76. O. B. Ptitsyn, Vysokomol. Soedin., **3**, 390 (1961).
77. V. N. Tsvetkov and S. Ya. Ljubina, Vysokomol. Soedin., **2**, 75 (1960).
78. V. N. Tsvetkov, Usp. Fiz. Nauk, **81**, 51 (1963).
79. V. N. Tsvetkov, Vysokomol. Soedin., **5**, 747 (1963).
80. V. N. Tsvetkov and I. N. Shtennikova, Vysokomol. Soedin., **2**, 640 (1960).
81. V. N. Tsvetkov and I. N. Shtennikova, in: Carbochain Molecular Compounds, Moscow, pp. 118–127.
82. Ch. Sadron, J. Phys. Rad., **9**(7), 381 (1938).
83. R. Cerf, J. Polym. Sci., **12**, 15, 35 (1954).
84. R. Cerf, J. Phys. Rad., **19**, 122 (1958); J. Polym. Sci., **23**, 125 (1957).
85. M. C. Williams and R. D. Zimmerman, Trans. Soc. Rheol., **17**, 23 (1973).
86. A. Peterlin, J. Polym. Sci. Symp., **43**, 187 (1973).
87. R. Cerf, in: Advances in Chemical Physics, I. Prigogine and A. Rice (eds.), Vol. 33, Wiley, New York (1975), p. 73.
88. R. Cerf, J. Phys., **4**, 358 (1977).
89. Ch. Chaffey, J. Chim. Phys., **63**, 1385 (1966).
90. R. Cerf, C. R. Acad. Sci. (Paris), **230**, 81 (1950); J. Chim. Phys., **48**, 85 (1951).

91. J. Leray, J. Chim. Phys., **57**, 323 (1960).
92. V. N. Tsvetkov and V. P. Budtov, Vysokomol. Soedin., **6**, 16, 1209 (1964).
93. V. P. Budtov, Vestn. Leningr. Gos. Univ., **4**, 47, 60, 152 (1964).
94. V. P. Budtov and S. N. Penkov, Vysokomol. Soedin., **15**, 42 (1973).
95. W. H. Stockmayer, Pure Appl. Chem., Suppl. Macromol. Chem., **8**, 379 (1973).
96. Y. Y. Gotlib, Vysokomol. Soedin., **6**, 389 (1964).
97. V. N. Tsvetkov, Vysokomol. Soedin., **4**, 894 (1962).
98. V. N. Tsvetkov, S. Y. Magarik, T. Kadyrov, and G. A. Andreeva, Vysokomol. Soedin., **A10**, 943 (1968).
99. S. H. Aharoni, Polymer, **19**, 401 (1978).
100. K. Solc and W. H. Stockmayer, J. Chem. Phys., **54**, 2756 (1971).
101. K. Solc, J. Chem. Phys., **55**, 335 (1971); Macromolecules, **6**, 378 (1978).
102. W. Gobush, K. Solc, and W. H. Stockmayer, J. Chem. Phys., **60**, 12 (1974).
103. I. Mazur, C. Guttman, and F. McCrackin, Macromolecules, **6**, 872 (1973); **10**, 139 (1977).
104. D. E. Kranbuehl, P. H. Ferdier, and J. Spencer, J. Chem. Phys., **59**, 3861 (1973); **67**, 361 (1977).
105. P. H. Linden, J. Appl. Phys., **46**, 4235 (1975).
106. D. Y. Yoon and P. J. Flory, J. Chem. Phys., **61**, 5366 (1974).
107. W. L. Mattice, Macromolecules, **12**, 944 (1979); **13**, 506 (1980).
108. W. L. Mattice, J. Am. Chem. Soc., **102**, 2242 (1980).
109. M. Doi and H. Nakajama, Chem. Phys., **6**, 124 (1974).
110. T. Minato and A. Hatano, Macromolecules, **11**, 195 (1978).
111. J. E. Hearst, J. Chem. Phys., **38**, 1062 (1963).
112. W. L. Mattice, Macromolecules, **10**, 1177 (1977).
113. V. N. Tsvetkov, Vysokomol. Soedin., **A20**, 2066 (1978).
114. V. N. Tsvetkov, Dokl. Akad. Nauk SSSR, **192**, 380 (1970).
115. V. N. Tsvetkov, Vysokomol. Soedin., **7**, 1468 (1965).
116. V. N. Tsvetkov, Dokl. Akad. Nauk SSSR, **266**, 670 (1982).
117. V. N. Tsvetkov, Vysokomol. Soedin., **A25**, 1571 (1983).
118. Y. Y. Gotlib and Y. E. Svetlov, Dokl. Akad. Nauk SSSR, **168**, 621 (1966).
119. J. Shimada and H. Yamakawa, Macromolecules, **9**, 583 (1976).
120. I. Noda and J. E. Hearst, J. Chem. Phys., **54**, 2342 (1971).
121. J. E. Hearst and R. A. Harris, J. Chem. Phys., **44**, 2595 (1966); **45**, 3106 (1966); **46**, 398 (1967).
122. J. E. Hearst, E. Beals, and R. A. Harris, J. Chem. Phys., **48**, 537 (1968).
123. L. D. Landau and E. M. Lifshitz, Statistical Physics, Part 1, Nauka, Moscow (1976), pp. 431–435.
124. G. Taylor, Proc. R. Soc. London, **A157**, 546 (1936).
125. E. V. Frisman and V. N. Tsvetkov, Zh. Eksp. Teor. Fiz., **23**, 690 (1952).
126. E. V. Frisman and V. N. Tsvetkov, Zh. Tekh. Fiz., **25**, 447 (1955).
127. E. V. Frisman and Sjui-Mao, Vysokomol. Soedin., **3**, 276 (1961).
128. D. B. Brace, Phys. Rev., **18**, 70 (1904); **19**, 218 (1904).
129. G. Szivessy, in: Handbuch der Physik, H. Geiger and K. Scheel (eds.), Vol. 19, Springer Verlag, Berlin (1928), p. 918.
130. S. N. Pen'kov and B. Z. Stepanenko, Opt. Spektrosk., **14**, 156 (1963).
131. V. N. Tsvetkov and E. V. Frisman, Zh. Fiz. Khim., **21**, 261 (1947).

Chapter 6

FLOW BIREFRINGENCE.
EXPERIMENTAL DATA

1. GENERAL PROPERTIES

1.1. Shear Optical Coefficient and Form Effect

In considering experimental data on FB in solutions of polymers exhibiting high chain rigidity it is useful to compare the dynamo-optical properties of these polymers with those of typical flexible-chain polymers.

At low shear rates g the value of birefringence Δn in solutions of both flexible and rigid polymers is proportional to shear stress $\Delta \tau = g(\eta - \eta_0)$ and, correspondingly, the shear optical coefficient $\Delta n/\Delta \tau$ is independent of $\Delta \tau$.

For flexible-chain polymers in a matching solvent ($n_k = n_s$) the value of $\Delta n/\Delta \tau$ is independent of solution concentration c but in a solvent in which the form effect is manifested ($n_k \neq n_s$) the positive part of $\Delta n/\Delta \tau$ decreases with increasing concentration as a result of the weakening of the macroform effect (Fig. 5.11).

In solutions of a polymer exhibiting high equilibrium chain rigidity, the value of $\Delta n/\Delta \tau$ is virtually *independent of concentration* both in matching solvents and when the form effect exists. This is illustrated in Fig. 6.1, which shows the dependence of $\Delta n/\Delta \tau$ on the volume concentration of the polymer $[\eta]c$ for a series of samples of aromatic polyamide in sulfuric acid [1]. In the solutions considered here the refractive index of the polymer n_k is much higher than that of the solvent n_s (increment $dn/dc \approx 0.25$ cm^3/g), and hence the form effect constitutes over half of the measured positive value of $\Delta n/\Delta \tau$. However, the value of $\Delta n/\Delta \tau$ plotted in Fig. 6.1 remains invariable over the entire range of volume concentrations in which this value for flexible-chain molecules (Fig. 5.11) decreases twice. This fact implies that for a rigid-chain polymer the form effect is the microform effect virtually independent of solution concentrations.

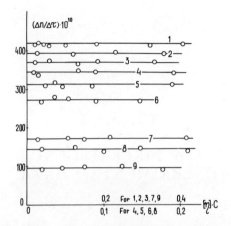

Fig. 6.1. Shear optical coefficient $\Delta n/\Delta \tau$ versus volume concentration $[\eta]c$ for solutions of several samples of a copolymer of aromatic amide and benzimidazole in sulfuric acid [1]. $M \cdot 10^3$ equal to: 1) 33; 2) 29; 3) 23; 4) 13; 5) 10; 6) 7; 7) 3.6; 8) 2.9; 9) 2.3.

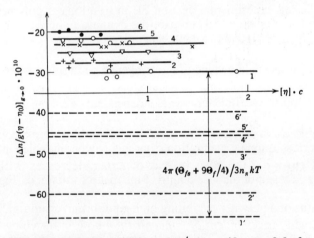

Fig. 6.2. Dependence $\Delta n/\Delta \tau = [\Delta n/g(\eta - \eta_0)]_{g \to 0}$ on $[\eta]c$ for cellulose nitrate fractions in butyl acetate. M equal to: 1) $1.03 \cdot 10^6$; 2) $8.14 = 10^5$; 3) $7.78 \cdot 10^5$; 4) $5.2 \cdot 10^5$; 5) $4.25 \cdot 10^5$; 6) $1.53 \cdot 10^5$ [2].

The higher the molecular weight of the polymer, the greater are the values of $\Delta n/\Delta \tau$ corresponding to different characteristics of rigid-chain polymers: the molecular-weight dependence of $\Delta n/\Delta \tau$ which will be discussed below.

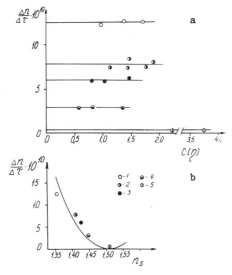

Fig. 6.3. a) Shear optical coefficient $\Delta n/\Delta \tau$ versus volume concentration $c[\eta]$ for cellulose acetate solutions (degree of substitution 2.3) in various solvents: 1) acetone; 2) dioxane; 3) dimethylformamide; 4) cyclohexanone; 5) pyridine. b) Dependence of $\Delta n/\Delta \tau$ on the refractive index of the solvent n_s for cellulose acetate.

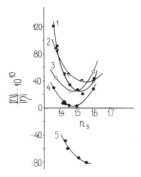

Fig. 6.4. Shear optical coefficient $[n]/[\eta]$ versus refractive index of the solvent n_s for cellulose esters and ethers [3]: 1) cyanoethyltritylcellulose; 2) cellulose monophenylacetate; 3) ethyl cellulose; 4) cyanoethylacetylcellulose; 5) cellulose benzoate.

Figure 6.2 shows a similar dependence of $(\Delta n/\Delta \tau)_{g \to 0}$ on volume concentration $c[\eta]$ for a series of cellulose nitrate fractions (curves 1–6) in butyl acetate ($n_k - n_s = 0.1$). The negative value of $\Delta n/\Delta \tau$ obtained here increases (in absolute value) with molecular weight, but remains independent of concentration for each fraction. Broken lines (1'–6') represent the same dependence for the same fractions in a solvent in which the form effect is absent (cyclohexanone). Hence, the distance along the ordinate between the intercepts of the solid curve and the corresponding broken curve characterizes the total value of the macro- and microform effects for a given fraction [see Eq. (5.64)]. Since this value is independent of the concentration $[\eta]c$, the contribution of the macroform effect contained in it is negligible.

The same conclusions may be drawn if the concentration dependence of $(\Delta n/\Delta\tau)_{g\to0}$ is studied for a single polymer (fraction) in solvents with different refractive indices n_s. The results of these investigations for cellulose acetate solutions are shown in Fig. 6.3. Here, as in the preceding cases, $\Delta n/\Delta\tau$ is independent of concentration (Fig. 6.3a), which demonstrates the absence of the form effect. However, $\Delta n/\Delta\tau$ is strongly dependent on the refractive index of the solvent, n_s, in accordance with the large contribution of the microform effect to FB. The shape of the parabolic dependence of $\Delta n/\Delta\tau$ on n_s in Fig. 6.3b is similar to that of curves 1 and 2 in Fig. 5.9 for flexible-chain polymers, but markedly differs from them in the fact that the latter curves characterize the macroform effect, whereas the dependence of $\Delta n/\Delta\tau$ on n_s in Fig. 6.3 is determined by the microform effect. Hence, the curves in Fig. 5.9 have been used to evaluate the form asymmetry of a Gaussian coil [2], whereas the parabolic dependence plotted in Fig. 6.3 may serve for the determination of the equilibrium rigidity of the cellulose acetate chain. With this purpose, Eqs. (5.43), (5.65), and (5.98) are compared to the experimental dependence of $\Delta n/\Delta\tau$ on n_s. In fact, the theoretical curve shown in Fig. 6.3 represents the theoretical dependence (5.65) corresponding to $e = 1/2$ and $S = 25$, or to the length of the Kuhn segment, $A = 130\cdot10^{-8}$ cm.

This method of determination of the equilibrium rigidity from the value of the microform effect in the FB phenomenon for a rigid-chain polymer has been applied to various cellulose derivatives. Some results are shown in Fig. 6.4. According to Eqs. (5.43) and (5.65), the experimental dependence of $[n]/[\eta]$ on n_s for each polymer is a parabola. The minimum of this parabola may be used to determine the intrinsic anisotropy of the segment $S\cdot\Delta a_i$, and its other part may be used to calculate the molecular weight of the segment M_0S and the number of monomer units in a segment S according to Eq. (5.65). The values of S and $A = \lambda\cdot S$ obtained by this method are given in Table 6.1. They are in satisfactory agreement with the values of A in Table 4.9, obtained from hydrodynamic data.

These experimental data confirm the theoretical prediction (Chapter 5) that no appreciable influence of the macroform effect on FB is observed in rigid-chain polymer solutions. Although the microform effect in the FB phenomenon for a rigid-chain polymer solution in a nonmatching solvent may be significant, it can change only the value and sign of the effective anisotropy of the segment and the monomer unit of the chain; however, it is independent of the molecular weight of the polymer.

This fact is of great importance in the consideration of the molecular-weight dependence of the shear optical coefficient $\Delta n/\Delta\tau$ in solutions of polymers exhibiting

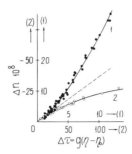

Fig. 6.5. Flow birefringence Δn versus shear stress $\Delta \tau = g(\eta - \eta_0)$ for polyisobutylene in benzene (1) ($M = 132{,}000$; concentration range $c = 0.84–0.28$ g/100 cm³) [4] and for cellulose nitrate in cyclohexanone (2) ($M = 500{,}000$; degree of substitution 2.8; concentration range $c = 0.15–0.03$ g/100 cm³) [5].

TABLE 6.1. Number of Monomer Units S in a Segment and Length of the Kuhn Segment A for Cellulose Esters and Ethers according to FB Data (microform effect) [3]

Polymer	Degree of substitution	S	$A \cdot 10^8$, cm
Cellulose nitrate	2.6	50	260
Ethyl cellulose	2.6	35	180
Cellulose benzoate	3.0	52	270
Cellulose carbanilate	1.9	35	180
Cellulose monophenylacetate	2.5	49	250
Cyanoethyl-acetyl cellulose	1.7	49	250
Cyanoethyl-trityl cellulose	2.0	54	280

high equilibrium rigidity in the chain. It shows that the character of this dependence should remain invariable for the polymer investigated, regardless of whether investigations are carried out in a matching solvent or when a considerable form effect exists. This feature sharply distinguishes the steady-state dynamo-optical properties of a rigid-chain polymer from those of a flexible-chain polymer for which the molecular-weight dependence $[n]/[\eta]$ undergoes a considerable change upon varying the increment dn/dc in a polymer–solvent system (Fig. 5.10).

1.2. Dependence of FB on Shear Stress and Chain Rigidity

The foregoing differences in the steady-state characteristics of FB for rigid- and flexible-chain polymers are related to the difference in the equilibrium rigidity (length of the Kuhn segment) of their chains. This difference is also detected and determined quantitatively by hydrodynamic methods (see Chapter 4).

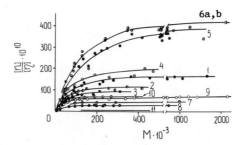

Fig. 6.6. Shear optical coefficient $[n]/[\eta]$ versus molecular weight $M = M_{sD}$ for solutions of fractions of various ladder polysiloxanes. 1–3) Ladder poly(phenyl siloxane)s in bromoform [8–11]; 4) ladder poly-(phenyl siloxane) in benzene [12]; 5) ladder poly(*m*-chlorophenyl siloxane) in tetrachloromethane [11–13]; 6) ladder poly(dichlorophenyl siloxane) in bromoform and tetrabromoethane [14, 15]; 7) ladder poly(3-methyl–1-butene siloxane) in benzene [11, 16, 17]; 8) ladder poly(3-methyl–1-butene siloxane) in butyl acetate [10]; 9) ladder poly-(phenylisobutyl siloxane) (1:1) in benzene [11]; 10) ladder poly-(phenylisohexyl siloxane) (1:1) in benzene [11]; 11) linear (nonladder) poly(methylphenyl siloxane) in benzene [10, 18].

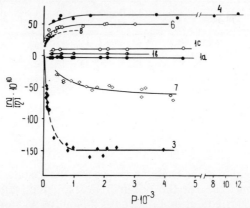

Fig. 6.7. Shear optical coefficient $[n]/[\eta]$ versus degree of polymerization P for solutions of cellulose esters and ethers. 1) Cellulose butyrate [19] in tetrachloroethane (a), in bromoform (b), and in methyl ethyl ketone (c); 2) cellulose carbanilate in dioxane [20, 21]; 3) cellulose monophenyl acetate in bromoform [22]; 4) cellulose diphenylphosphonocarbamate in dioxane [22]; 5) cellulose nitrate in cyclohexanone [23, 24]; 6) ethyl cellulose in dioxane [22].

The problem of whether the kinetic rigidities of these molecules also differ cannot be solved by the comparison of their equilibrium dynamo-optical characteristics. A certain information on this problem may be obtained by the study of FB of polymer solutions as a function of shear stress $\Delta\tau = (\eta - \eta_0)g$.

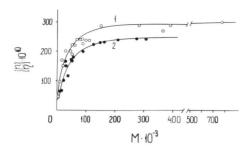

Fig. 6.8. Dependence of $[n]/[\eta]$ on M for solutions of poly(alkyl isocyanate)s. 1) Poly(butyl isocyanate) in tetrachloromethane [25]; 2) poly(chlorohexyl isocyanate) in tetrachloromethane [16].

Figure 6.5 shows some results of this study. Curve 1 represents the dependence of birefringence Δn on $\Delta\tau$ for solutions of polyisobutylene [4], a typical flexible-chain polymer with positive anisotropy (length of Kuhn segment $20\cdot10^{-8}$ cm) in benzene (matching solvent). In accordance with this, curve 1 exhibits a pronounced upward concavity (positive curvature) indicating that an important role in the observed birefringence is played by coil deformation due to the effect of shear stress in flow. This deformation effect, well known for flexible-chain polymers, corresponds to theoretical predictions [Eq. (5.63)] and has been discussed in connection with data presented in Figs. 5.12-5.14.

Curve 2 in Fig. 6.5 represents a similar dependence of Δn on $\Delta\tau$ for a solution of cellulose nitrate [3], a typical polymer with high-equilibrium chain rigidity. Cyclohexanone was used as the solvent, in which the form effect is absent $(n_k - n_s)$ and Δn is negative (in accordance with the negative intrinsic chain anisotropy).

The points fit a single curve for both polyisobutylene and cellulose nitrate (curves 1 and 2, respectively), which illustrates the invariance of the value of $\Delta n/\Delta\tau$ when the polymer concentration is varied. However, unlike curve 1, curve 2 exhibits a concavity toward the abscissa (negative curvature), which corresponds qualitatively to the dependence predicted theoretically for rigid (undeformed) asymmetric particles (Fig. 5.4). This fact implies that the molecular conformation of cellulose nitrate in a shear flow does not undergo such a profound change as that of polyisobutylene, and birefringence in solutions is, at least partially, due to the nonuniform rotation and the kinematic orientation of the asymmetric coil. A similar type of dependence of Δn on $\Delta\tau$ is observed for other rigid-chain polymers (see, e.g., [6]).

1.3. Dependence of Shear Optical Coefficient on Molecular Weight (Chain Length)

1.3.1. General Relationships. One of the most characteristic properties of polymers exhibiting high chain rigidity is the molecular-weight dependence of the shear optical coefficient. For typical flexible-chain polymers in matching sol-

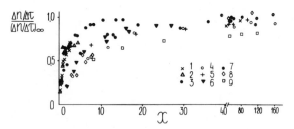

Fig. 6.9. Relative value of the shear optical coefficient $([n]/[\eta])/([n]/[\eta])_\infty$ versus $x = 2L/A$. Points represent experimental data for various polymers: 1) poly(γ-benzyl-L-glutamate) in dichloroethane [27]; 2) poly(butyl isocyanate) in tetrachloromethane [25]; 3) poly(chlorohexyl isocyanate) in tetrachloromethane [26]; 4) ladder poly(phenyl siloxane) in benzene [15]; 5) ladder poly(chlorophenyl siloxane) in benzene [15]; 6) ladder poly(dichlorophenyl siloxane) in bromoform [14]; 7) cellulose nitrate in cyclohexanone [23]; 8) cellulose carbanilate in dioxane [20]; 9) polycarbonate in bromoform [28].

vents $(n_k - n_s)$ the value of $\Delta n/\Delta\tau$ remains virtually constant over the entire molecular-weight range beyond the oligomeric range, whereas in solutions of rigid-chain polymers the shear optical coefficient drastically changes with molecular weight, regardless of the value of $n_k - n_s$ and the sign of FB.

This is demonstrated in Figs. 6.6-6.8, which show the dependence of $[n]/[\eta]$ on M for a number of samples of ladder polysiloxanes, cellulose esters and ethers, and poly(alkyl isocyanate)s.

For all these polymers the value of $[n]/[\eta]$ remains constant only at high molecular weights, but with decreasing M the shear optical coefficient sharply decreases, approaching zero at $M \to 0$. This type of dependence of $[n]/[\eta]$ on M, revealed experimentally for rigid-chain polymers, is qualitatively similar to the theoretical curves shown in Fig. 5.23, which demonstrates the fact that over the molecular-weight range considered, with decreasing M the conformational properties of the molecules of these polymers deviate from those of Gaussian chains and approach those of a straight rod, i.e., the molecules behave as "semirigid" chains.

For a quantitative comparison of experimental curves plotted in Figs. 6.6-6.8 with the theoretical curves in Fig. 5.23, the experimental data should be represented as a generalized dependence of the value of $([n]/[\eta])/([n]/[\eta])_\infty$ on the parameter $x = 2L/A$ using for each polymer the asymptotic limit $([n]/[\eta])_\infty$ of the curves in Figs. 6.6-6.8, the chain length $L = M/M_L$, and the length of the Kuhn segment if it is known from other experimental data. For all the polymers shown in Figs. 6.6-6.8, the values of A have been determined by sedimentation, diffusion, and viscometry, and are given in the table in Chapter 4. With the aid of these values, the experimental values of x and $([n]/[\eta])/([n]/[\eta])_\infty \equiv \Delta$ have been obtained for some polymers investigated. The values of Δ are shown by points in Fig. 6.9 as a function of x.

The experimental points for various polymers in Fig. 6.9 fill a wide range which cannot be even approximately represented by a single curve. This means that

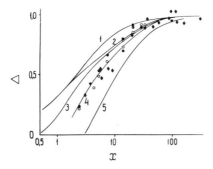

Fig. 6.10. Dependence of $\Delta = ([n]/[\eta])/([n]/[\eta])_\infty$ on $x = 2L/A$ at various values of the parameter A/d. Curve 1) According to Eq. (5.111); curves 2–5) according to the Noda-Hearst theory (Table 5.1) with an additional multiplier b_0 according to Eq. (5.102) at $p_x = \infty$ and at various values of d/A equal to 0 (2), 0.25 (3), 0.5 (4), and 1 (5). Points represent experimental data: cellulose carbanilate in dioxane (♦) [20], ladder poly(dichlorophenyl siloxane) in bromoform (○) and in tetrabromoethane (●) [14, 15].

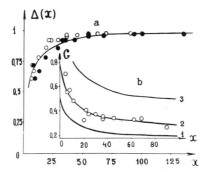

Fig. 6.11. a) Dependence $\Delta(x) = ([n]/[\eta])/([n]/[\eta])_\infty$ for cellulose nitrate fractions (degree of substitution 2.7) in cyclohexanone. Curve – according to the Noda-Hearst theory. Points represent experimental values of $\Delta(x)$ at $A = 200 \cdot 10^{-8}$ cm (open circles) and at $A = 300 \cdot 10^{-8}$ cm (filled circles). b) Dependence of the coefficient G on x. Curve 1 according to the Noda-Hearst theory. Curves 2 and 3 are corrected, taking into account the polydispersity for $M_w/M_n = 1.2$ and $M_w/M_n = 1.6$, respectively. Points represent experimental data for cellulose nitrate fractions in cyclohexanone [31].

the dependence of Δ on the chain length for the totality of the polymers investigated cannot be expressed in the form of a universal dependence of Δ on x using as the only parameter the persistence length a or the length of the Kuhn segment A of an equivalent wormlike model.

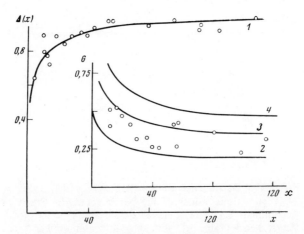

Fig. 6.12. Dependence $\Delta(x) = ([n]/[\eta])/(\Delta n)_\infty$ on x (where $x = 2L/A$): curve 1, according to the Noda-Hearst theory (Table 5.1); points near the curve represent experimental data for cellulose nitrate fractions in cyclohexanone (degree of substitution 2.1) corresponding to the value of $A = 200 \cdot 10^{-8}$ cm. Dependence of the coefficient G on x: curve 2, according to the Noda-Hearst theory; curves 3 and 4, the same as curve 1 but corrected for polydispersity at $M_w/M_n = 1.5$ and 2.1, respectively. Points represent experimental data for cellulose nitrate fractions in cyclohexanone [32].

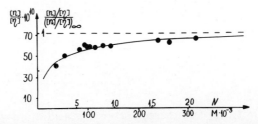

Fig. 6.13. Dependence of $[n]/[\eta]$ on M for cyanoethyl-cellulose fractions (degree of substitution 2.6) in cyclohexanone. Points represent experimental data. Solid curve represents the dependence $([n]/[\eta])/([n]/[\eta])_\infty$ on $N = x/2$ according to the Noda-Hearst theory. Experimental points fit the theoretical curve at the length of the Kuhn segment $A = 260 \cdot 10^{-8}$ cm [32].

However, it is possible to trace some regularities in the cloud of points plotted in Fig. 6.9.

Thus, the experimental curves (which could be drawn for the points of each polymer) differ for different polymers in the slope of the initial part at low x and in the rate at which they approach the limiting value at high x. If these properties of the curves are compared with the data presented in the table in Chapter 4, in which the lengths of the Kuhn segment A are given for the corresponding polymers, it can be easily seen that the slope of the curves $\Delta = \Delta(x)$ is much steeper for polymers with the highest equilibrium chain rigidity, such as poly(γ-benzyl-L-glutamate) (PBLG) or poly(alkyl isocyanate)s (PAIC) than for ladder polymers with moderate chain rigidity. This feature is particularly pronounced in the range of low x in which the experimental points for most samples investigated fall much lower than the theoretical curves in Fig. 5.23, which describe flow birefringence in the framework of the model of an infinitely thin wormlike chain. Curve 9 is the most gently-sloping among the curves in Fig. 6.9. This curve describes the dependence $\Delta(x)$ for oligomers of polycarbonate [28], a flexible-chain polymer whose equilibrium rigidity is lower by one order of magnitude than that of ladder polymers or cellulose esters and ethers.

1.3.2. Polymers with Moderately High Chain Rigidity. The deviation (toward decreasing values) of the initial slopes of experimental curves from the theoretical dependence shown in Fig. 5.23 for polymer molecules exhibiting low equilibrium rigidity is qualitatively understandable if we bear in mind the finite value of the diameter d of a real polymer chain. This point is not accounted for in the theories represented by the curves in Fig. 5.23, but may be taken into account if Eq. (5.102) is used. For polymer molecules with high equilibrium rigidity the d/A value is small ($d/A < 0.1$) and the second multiplier in Eq. (5.102) is virtually equal to unity over the entire range of x experimentally available; at low x, b_0 is correspondingly equal to ≈ 1.

In contrast, for flexible-chain polymers with a relatively high value of d/A, according to Eq. (5.102), at low x the value of b_0 can be considerably less than unity and, correspondingly, the value of $[n]/[\eta]$ can be lower than the theoretical value for an infinitely thin chain. This is shown in Fig. 6.10, in which curves 1–5 represent the theoretical dependence of $\Delta \equiv ([n]/[\eta])/([n]/[\eta])_\infty$ on x (x is plotted on the logarithmic scale), taking into account the finite value of d at various values of the parameter d/A in Eq. (5.102) [29].

The points in Fig. 6.10 represent the experimental values of Δ for ladder poly(dichlorophenyl siloxane) [14] and cellulose carbanilate [20]. The experimental data for both polymers are in agreement with theoretical curve 4, corresponding to the value of $d/A = 0.5$ for a kinetically flexible-chain polymer. This demonstrates qualitatively the fact that at low x the hydrodynamic properties of the molecules of these two polymers do not correspond to the infinitely thin wormlike model. For quantitative agreement between theoretical and experimental data, in accordance with curve 4 in Fig. 6.10, A/d should be taken equal to 2. This value may be reasonable in order of magnitude for many flexible-chain polymers, but is not realistic for such rigid-chain polymers as ladder polysiloxanes or cellulose esters and ethers.

In fact, this result implies that for polymers with moderate equilibrium rigidity and large chain diameter (ladder polysiloxanes and cellulose derivatives), the use of experimental values of the shear optical coefficient at low M (low x) and a comparison of these values with theoretical dependence (5.111) or Table 5.1 cannot lead to reasonable values for the chain-equilibrium rigidity parameter. Since the initial slopes of experimental curves $\Delta(x)$ are much less than those of the theoretical curves in Fig. (5.23), according to experimental data on FB the length of the Kuhn segment will greatly exceed that obtained from the hydrodynamic data presented in Chapter 4.

The situation becomes more favorable if the experimental values of Δ at relatively high M are used. For moderately rigid polymers these values are usually available to the experimenter. Thus, if the experimental values of Δ in the range $\Delta_{\infty} > \Delta > 0.5\Delta_{\infty}$ are discussed, then the application of the Noda-Hearst theory (Table 5.1) to both ladder polysiloxanes [15] and cellulose derivatives [30–33] leads to values of A in satisfactory agreement with the data from hydrodynamic methods. This agreement is shown in Figs. 6.11–6.13, in which the experimental dependences $\Delta(x)$ are compared with the results of the Noda-Hearst theory (Table 5.1).

In all these cases the values of A obtained by comparing the experimental points for FB with the theoretical curve $\Delta(x)$ are in satisfactory agreement with those determined for the same polymers by hydrodynamic methods (Chapter 4).

The fact that at relatively high M for moderately rigid-chain polymers the dependence $\Delta(x)$ is adequately described by the Noda-Hearst theory shows that FB in solutions of these polymers is related to the manifestation of intramolecular mobility and kinetic flexibility of their chains.

At low molecular weights, when the chain length decreases, the properties of molecules of typical flexible-chain polymers also deviate from Gaussian properties. In the FB phenomenon this deviation is revealed by a decrease in $\Delta n/\Delta\tau$ with decreasing molecular weight in the oligomeric range, i.e., at much lower M values (and much lower chain lengths L) than for rigid-chain polymers [28, 34–36]. If experimental data for flexible-chain polymers are represented as the dependence $\Delta(x)$ (points 9 in Fig. 6.9), then their comparison with the theoretical curves in Fig. 5.23 leads to much higher values of A than those obtained from hydrodynamic data. The treatment of experimental data [28, 35] using the theory of photoelasticity of swollen Langevin networks according to Kuhn [37] leads to more reasonable results for flexible-chain polymers. These results show that the wormlike chain cannot serve as a good model for an adequate description of hydrodynamic and dynamo-optical properties of molecules of flexible-chain polymers.

1.3.3. Polymers with High-Equilibrium Chain Rigidity. If the equilibrium rigidity of the chain (A/d) is high, then at low M its shape is actually close to that of a thin straight rod and, according to Eq. (5.112), it is possible to determine the anisotropy of the monomer unit Δa and that of the unit length of the chain β from the initial slope $B\beta/M_L = B\Delta a/M_0$ of the curve describing the dependence of $[n]/[\eta]$ on M (if M_L is known from the structural formula of the polymer). For this purpose several relatively low-molecular-weight fractions of the rigid-chain polymer investigated should be available.

If a sample of the same polymer with relatively high molecular weight and, correspondingly, a Gaussian coil conformation in solution is also available, then it is possible to determine the molecular weight of the segment M_s from the ratio of the limiting value $(\Delta n/\Delta \tau)_\infty$ to the initial slope according to Eq. (5.112):

$$(\Delta n/\Delta \tau)_\infty/[\partial (\Delta n/\Delta \tau)/\partial M]_{x \to 0} = M_s = SM_0. \qquad (6.1)$$

Hence, for a polymer with very stiff chains the parameter of equilibrium rigidity S and the optical anisotropy of the monomer unit Δa are unequivocally determined from the data on FB if these data are obtained for several low-molecular-weight fractions (with known values of M) and for at least one high-molecular-weight sample.

This favorable situation is observed for poly(alkyl isocyanate) samples for whose fractions the values of $\Delta n/\Delta \tau$ have been determined over a wide molecular-weight range [25, 26] (Fig. 6.8). Both curves in Fig. 6.8 cover fairly completely both the low- and high-molecular-weight regions of the dependence $[n]/[\eta] = f(M)$. This allows a reliable determination of the initial slopes and the limiting values of curves 1 and 2. Using these initial slopes and the limiting values of curves 1 and 2, it has been possible to determine the values of $S = 370$ and $\Delta a = 11 \cdot 10^{-25}$ cm^3 for poly(butyl isocyanate) [25], and $S = 280$ and $\Delta a = 11 \cdot 10^{-25}$ cm^3 for poly(chloro-hexyl isocyanate) [26].

However, since experimental curves 1 and 2 contain the values of $\Delta n/\Delta \tau$ over the entire range of M for which the conformation of the molecule changes from rodlike to a Gaussian coil, they may be compared with the theoretical dependences plotted in Fig. 5.23. In this case it was found that for poly(alkyl isocyanate)s the experimental data are in better agreement with the dependences (5.111) and (5.112) for kinetically rigid wormlike chains than with the Noda-Hearst dependence (Table 5.1), which is more shallow. This is illustrated in Fig. 6.14, where the points represent the experimental values of $[n]/[\eta]$ at various M for poly(chloro-hexyl isocyanate) [26] and the curve describes the dependence $([n]/[\eta])/B\beta A$ on curve x according to Eq. (5.111). The coincidence of the theoretical curve and the experimental points leads to the values of $\Delta a = 15 \cdot 10^{-25}$ cm^3 and $S = 235$. The latter value corresponds to the length of the Kuhn segment $A = S\lambda = 470 \cdot 10^{-8}$ cm, which is in reasonable agreement with the value of A evaluated by hydrodynamic methods (Table 4.10).

Figure 6.14 indicates that for the polymer fractions with the lowest M of those investigated, the chain length is less than half the Kuhn segment and, consequently, the molecule has the shape of a weakly bending rod.

Poly(γ-benzyl-L-glutamate) (PBLG) in a helixogenic solvent is another example of a polymer with high equilibrium chain rigidity for which the experimental dependence of $\Delta n/\Delta \tau$ on M is in agreement with Eq. (5.112). This is demonstrated in Fig. 6.15, where the points represent the experimental values of $[n]/[\eta]$ as a function of M for PBLG fractions in dichloroethane [27] and the curve describes the dependence of $([n]/[\eta])/B\beta A$ on x according to Eq. (5.111). Although the range

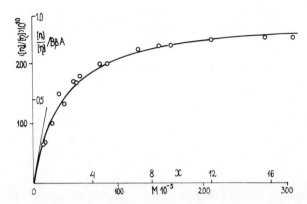

Fig. 6.14. Dependence of $[n]/[\eta]$ on M for poly(chlorohexyl isocyanate) fractions in tetrachloromethane (points) [26] and dependence of $([n]/[\eta])/B\beta A$ on x according to Eq. (5.111) (curve). Theoretical curve and experimental points coincide at $\Delta a = 15 \cdot 10^{-25}$ cm^3 and $A = 470 \cdot 10^{-8}$ cm.

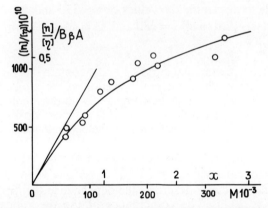

Fig. 6.15. Dependence of $[n]/[\eta]$ on M for poly(γ-benzyl-L-glutamate) fractions in dichloroethane (points) and dependence of $([n]/[\eta])/B\beta A$ on x according to Eq. (5.111) (curve). Theoretical curve and experimental points coincide at $\Delta a = 25 \cdot 10^{-25}$ cm^3 and $S = 1100$.

of investigated fractions for this polymer extends to $M = 3 \cdot 10^5$, as in the case of polyisocyanates (Fig. 6.14), in contrast to the latter the conformation of PBLG molecules at $M = 3 \cdot 10^5$ is still far from the Gaussian coil because of their high rigidity, and the value of $[n]/[\eta]$ is far from the limiting value $([n]/[\eta])_\infty$. Over the entire M range investigated the experimental points fit the theoretical curve at values of $\Delta a = 25 \cdot 10^{-25}$ cm^3 and $S = 1100$. This value of S corresponds to the length of the segment $A = (1670–2220) \cdot 10^{-8}$ cm, depending on the assumed value of the monomer unit $\lambda = (1.5–2) \cdot 10^{-8}$ cm. These values of A agree with the data of hydrodynamic methods (Table 4.10). For fractions with the lowest value of M among

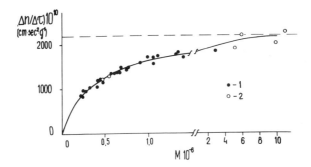

Fig. 6.16. Dependence of $\Delta n/\Delta\tau$ on M for DNA samples in solutions at neutral pH, 0.2 M NaCl. Points represented experimental data for samples obtained by enzymatic (1) or ultrasonic degradation (2) of native DNA (thymus) [37]. Curves represent theoretical data according to Eq. (5.111). Experimental points and theoretical curve coincide at $\Delta a = -67 \cdot 10^{-25}$ cm^3 and $M_s = 3 \cdot 10^5$.

those shown in Fig. 6.15 the chain length is only $0.2A$, and hence its conformation is close to a straight rod.

Double-helix chains of native DNA are a classical example of polymer molecules with high equilibrium rigidity. However, the molecular weights of the investigated DNA samples are usually so high that although their molecules are very rigid, their conformations are not so close to a rodlike conformation as in the case of polyisocyanates or helixogenic polypeptides. In contrast, the samples available to the experimenter are usually in the molecular-weight range in which, with increasing M, it is possible to follow in detail the changes in the conformation of the molecule up to its transformation into a Gaussian coil.

This can be clearly seen in Fig. 6.16, in which the points represent the dependence of the shear optical coefficient $\Delta n/\Delta\tau$ on molecular weight for DNA samples obtained by enzymatic and ultrasonic degradation of high-molecular-weight native DNA [37]. The curve representing the theoretical dependence (5.112) coincides with the experimental points shown in Fig. 6.16 at the values of $M_s = 300 \cdot 10^3$ and $\Delta a = -67 \cdot 10^{-25}$ cm^3. In this case, the chain length of the lowest-molecular-weight sample is about one segment A, whereas in the high-molecular-weight part of the curve $\Delta n/\Delta\tau$ remains virtually invariable with the change in M, which corresponds to the Gaussian range. This value of M_s corresponds to $S = 450$ (at $M_0 = 660$) and the segment length $A = 1500 \cdot 10^{-8}$ cm (at $\lambda = 3.4 \cdot 10^{-8}$ cm). This value of A is in the range of A values determined for DNA molecules using hydrodynamic and light-scattering methods [38–40].

The above examples show that the molecular-weight dependence of the shear optical coefficient observed experimentally for polymers with high-equilibrium chain rigidity may be adequately interpreted on the basis of Eqs. (5.111) and (5.112) characterizing the dynamo-optical properties of kinetically rigid wormlike chains. Hence, we may conclude that molecules with high-equilibrium rigidity also exhibit high kinetic rigidity in the FB phenomenon.

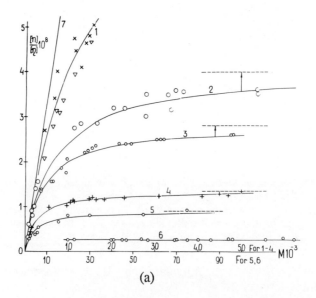

(a)

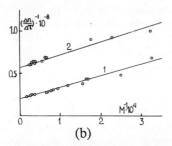

(b)

Fig. 6.17. a) Dependence of $[n]/[\eta]$ on M for solutions of samples and fractions of some aromatic polymers. 1) Poly(*para*-benzamide) in H_2SO_4. Points represent experimental data for samples: ∇ [42] and $\times$ [43]. Molecular weights M_w are according to viscosity data given in Table 4.12. The curve is according to Eq. (5.112) at $\beta_M = 0.87 \cdot 10^{-25}$ cm³/dalton, $M_s = 14{,}000$ ($A = 700 \cdot 10^{-8}$). 2) Poly(amido benzimidazole) in H_2SO_4. Points represent experimental data for fractions [44]. The curve is according to Eq. (5.112) at $\beta_M = 0.87 \cdot 10^{-25}$ cm³/dalton, $M_s = 6000$ ($A = 300 \cdot 10^{-8}$ cm). 3) Poly(amide hydrazide) in dimethylsulfoxide. Points represent experimental data for samples [45]. The curve is according to Eq. (5.112) at $\beta_M = 0.91 \cdot 10^{-25}$ cm³/dalton, $M_s = 4000$ ($A = 200 \cdot 10^{-8}$ cm). 4) Poly(*para*-phenylene oxadiazole) in H_2SO_4. Points represent experimental data for samples [46]. Molecular weights M, according to the data given in Table 4.13. The curve is according to Eq. (5.112) at $\beta_M = 0.7 \cdot 10^{-25}$ cm³/dalton, $M_s = 2500$ ($A = 120 \cdot 10^{-8}$ cm). 5) Aromatic copolyester (ACPE) in tetrachloroethane (see Table 4.12). Points represent experimental data for fractions [47]. The curve is according to Eq. (5.112) at $\beta_M = 0.28 \cdot 10^{-25}$ cm³ dalton, $M_s = 4000$; at $M = 26 \cdot 10^{-8}$ dalton cm⁻¹, $A = 155 \cdot 10^{-8}$ cm. 6) Poly(*meta*-phenylene isophthalamide) in H_2SO_4. Points represent experimental data for fractions [48, 49]. 7) Straight line corresponding to the initial slope of curves 1 and 2. b) Dependence of $(\Delta n/\Delta\tau)^{-1}$ on M^{-1} for solutions of aromatic polyamides: PpHPhBOTPhA (1) and PmHPhBOTPhA (2) in sulfuric acid.

Consequently, the study of the dependence of $\Delta n/\Delta \tau$ on M as applied to these polymers may serve as an independent method for the quantitative determination of equilibrium rigidity and optical anisotropy of their chains.

When the parameters of chain rigidity, determined from hydrodynamic and dynamo-optical investigations, are compared, it should be taken into account that the initial structural parameters obtained by these two methods are not quite identical. Hydrodynamic methods (Chapters 2 and 4) give the results in the form of a combination of the values of A and λ from which A is determined if λ is known. Apart from the value of Δa, the molecular-weight dependence of FB yields *the molecular weight* M_s of the Kuhn segment or the number of monomer units S in a segment (because M_0 is usually known). The transition from these values to A also requires knowledge of the value of λ.

Another peculiarity of the FB method as compared to hydrodynamic methods and light scattering is the fact that the value of $\Delta n/\Delta \tau$ is not sensitive to excluded-volume effects.

Although in solutions of rigid-chain polymers these effects are slight and very often may be neglected, they can be manifested in the study of samples of very high molecular weight, such as DNA samples. Hence, when the methods in which the geometric dimensions of molecules are measured (hydrodynamic methods and light scattering) are used for the investigation of high-molecular-weight polymers, caution should be exercised and the possible expansion of molecules in thermodynamically good solvents should be evaluated (Chapter 2, Section 7).

In the FB phenomenon the situation is quite different. As already mentioned (Chapter 5, Section 2.3.2.2), when a chain molecule expands in thermodynamically good solvents, the intrinsic viscosity $[\eta]$ and the intrinsic FB (in those parts which depend on the segmental anisotropy of the chain, i.e., in parts $[n]_i$ and $[n]_{fs}$) increase equally. Hence, for a flexible-chain polymer in matching solvents ($n_k = n_s$) and for a rigid chain polymer in any solvent $[n]/[\eta]$ is independent of the excluded-volume effects {41}. This point is of considerable practical importance because it allows the application of Eq. (5.111) to solutions of a rigid-chain polymer of any molecular weight in any solvents, regardless of their thermodynamic strength and the excluded-volume effects. This possibility is an indisputable advantage of the FB method over hydrodynamic methods.

1.3.4. Aromatic Polymers. Although the equilibrium rigidity of many synthetic molecules containing para-aromatic rings in the main chain is not as high as for polymers considered in the preceding section, their solutions usually exhibit a marked dependence of $[n]/[\eta]$ on M.

However, the quantitative comparison of this dependence with theoretical equations (5.111) or (5.112) should be carried out using the experimental data obtained for fractions or samples over as wide a range of reliably determined molecular weights as possible. This can be accomplished more easily if the polymer is soluble in common organic solvents because, in this case, it is possible to carry out its fractionation by standard methods.

Poly(amide benzimidazole) (PABI) is a polymer of this type. For its fractions the experimental dependence of $[n]/[\eta]$ on M has been obtained over a wide molecular-weight range, M_{sD} [44]. For solutions in H_2SO_4 this dependence is represented by points near curve 2 in Fig. 6.17. The experimental points make it possible to determine not only the initial slope of the dependence of $[n]/[\eta]$ on M (line 7 in Fig. 6.17) and the limiting value $([n]/[\eta])_\infty$, but also to compare over the entire M range the experimental data with theoretical curve 2, plotted according to Eq. (5.112) at the values of $M_s = 6000$ and $\beta_M = \beta/M_L = 0.87 \cdot 10^{-25}$ cm^3/dalton. At these chain-rigidity parameter and anisotropy values the experimental points fit the theoretical dependence (5.112) for a kinetically rigid chain.

Another polymer for samples of which the molecular weights M_w have been determined by light scattering is poly(amide hydrazide). For these samples the dependence of the experimental values of $[n]/[\eta]$ on M_w in dimethylsulfoxide is shown by points near curve 3 in Fig. 6.14 [45]. Here, just as for PABI, the experimental points are in good agreement with the theoretical dependence [Eq. (5.112)] (curve 3) at values of $\beta_M = 0.91 \cdot 10^{-25}$ cm^3/dalton and $M_s = 4000$.

Similar results have been obtained for poly(*para*-phenylene oxadiazole) samples in sulfuric acid [46]. The molecular weights $M_{D\eta}$ of these samples have been determined by the diffusion–viscosity method (Table 4.13). The experimental values of $[n]/[\eta]$ for poly(phenylene oxadiazole) (points near curve 4 in Fig. 6.17) are in agreement with the theoretical dependence [Eq. (5.112)] represented by curve 4 at values of $\beta_M = 0.7 \cdot 10^{-25}$ cm^3/dalton and $M_s = 2500$.

For these aromatic polymers it may be assumed that M_L is $20 \cdot 10^8$ dalton/cm, which permits the calculation of the length of the Kuhn segment from the relation $A = M_s/M_L$. The values of A are given in the caption to Fig. 6.17.

The points near curve 1 in Fig. 6.17 represent the dependence of the experimental values of $[n]/[\eta]$ on M for poly(*para*-benzamide) (PPBA) samples in sulfuric acid [42, 43] (see Table 4.12). The investigated samples cover a relatively narrow molecular-weight range, and this prevents the complete analysis of the entire dependence of $[n]/[\eta]$ on M carried out in the preceding cases. However, the initial slope of this dependence is quite distinct and may be roughly assumed to coincide with the initial slope of curve 2 for PABI (curve 7). Thus, the chain anisotropy per unit molecular weight of PPBA is assumed to be equal to $\beta_M = 0.87 \cdot 10^{-25}$ cm^3/dalton. Under these conditions, the equilibrium rigidity of PPBA molecules can be evaluated using the deviation of experimental values of $[n]/[\eta]$ from straight line 7 with increasing M, assuming that this deviation is described by Eq. (5.112). This evaluation yields $M_s = 14,000$, which corresponds to the value $A = 700 \cdot 10^{-8}$ cm.

The points close to curve 5 in Fig. 6.17 represent the molecular weight M_{sD} dependences of the experimental values of $[n]/[\eta]$ for fractions of an aromatic copolyester (ACPE) in tetrachloroethane [47]. The structural formula of this polymer is given in Table 4.12. The experimental points are in agreement with theoretical curve 5, plotted according to Eq. (5.112) at values of $\beta_M = 0.28 \cdot 10^{-25}$

cm^3/dalton and $M_s = 4000$, which corresponds (at $M_L = 26 \cdot 10^8$ dalton/cm) to the value $A = 155 \cdot 10^{-8}$ cm.

Curve 6 and its corresponding points represent the values of $[n]/[\eta]$ for poly(*meta*-phenylene isophthalamide) (PMPhIPhA) fractions in H_2SO_4 [48, 49]. Over the entire molecular-weight range investigated the shear optical coefficient for this polyamide is independent of M. This means that in this range of M the PM-PhIPhA molecules have the conformation of a Gaussian coil in accordance with their considerable flexibility exceeding that of other aromatic polyamides.

Hence, for all aromatic polymers shown in Fig. 6.17 the experimental dependence of $[n]/[\eta]$ on M is adequately described by Eq. (5.112) for kinetically rigid wormlike chains. Moreover, the values of the rigidity parameter A determined for these polymers by the FB method are in quite reasonable agreement with those determined by hydrodynamic methods (Tables 4.12 and 4.13). The observed differences are actually within the experimental errors of hydrodynamic and dynamo-optical methods.

Relationship (5.112) is equivalent to a linear dependence between the reciprocal values $(\Delta n/\Delta \tau)^{-1}$ and M^{-1}. Such dependence obviously must be satisfied for all such rigid-chain polymers which obey relationship (5.112). For example, Fig. 6.17a demonstrates the data for solutions of two rigid-chain aromatic polyamides – poly(*para*-hydroxyphenylbenzoxazole terephthalamide) (PpHPhBOTPhA) and poly(*meta*-hydroxyphenylbenzoxazole terephthalamide) (PmHPhBOTPhA) – in sulfuric acid.

Experimental points representing the dependence $(\Delta n/\Delta \tau)^{-1}$ on M^{-1} for each polymer fit a straight line with slope equal to $M_L/(B\beta)$ and with ordinate intercept equal to $(B\beta A)^{-1}$. The parameters of anisotropy β and rigidity A, calculated from the data of Fig. 6.17a for these two polymers, are given in Table 6.9 (Nos. 7 and 8).

Although the observed limiting values $([n]/[\eta])_\infty$ for various polymers shown in Fig. 6.17 may differ by a factor of several tens or higher in accordance with their different rigidities, for all aromatic polymers dissolved in the same solvent (H_2SO_4) the specific anisotropies β_M have similar values. This is a natural consequence of the similarity of the chemical structure and optical properties of their chains (refractive indices of the polymer and the optical anisotropy of chain units).

However, attention should be directed to the much lower value of β_M for ACPE molecules (No. 5) in tetrachloroethane than for other polymers with similar structures dissolved in sulfuric acid. This is due to different refractive index increments, *dn/dc*, in these polymer–solvent systems (*dn/dc* is 0.15 in tetrachloroethane and 0.25–0.30 in sulfuric acid) and the corresponding difference in the microform effect in these two solvents. The fact that β_M for solutions of aromatic polymers in sulfuric acid is three times as high as for their solutions in tetrachloroethane implies that FB in sulfuric acid solutions of these polymers is due to a considerable extent to the microform effect.

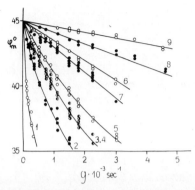

Fig. 6.18. Extinction angle φ_m versus rate gradient g for solutions of ladder poly(dichlorophenyl siloxane) fractions in tetrabromoethane [14]. Curves correspond to the following molecular weights M_{sD}: 1) $2.16 \cdot 10^6$; 2) $0.59 \cdot 10^6$; 3) $0.515 \cdot 10^6$; 4) $0.37 \cdot 10^6$; 5) $0.35 \cdot 10^6$; 6, 7) $0.25 \cdot 10^6$; 8) $0.13 \cdot 10^6$; 9) $0.09 \cdot 10^6$. Points near each curve correspond to concentrations in the range shown in Fig. 6.19.

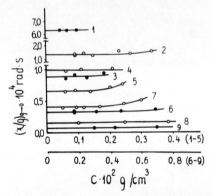

Fig. 6.19. Initial slope $(\chi/g)_{g \to 0}$ of orientation angles versus concentration c for solutions of poly(dichlorophenyl siloxane) fractions in tetrabromoethane [14]. Numbers at curves are fraction numbers according to Fig. 6.18.

The values of β_M and M_s for polymers, shown in Fig. 6.17, have been calculated by comparing the experimental data with Eq. (5.112). If Table 5.1 is used for the same purpose instead of Eq. (5.112) and only the initial slopes and limiting values of curves $[n]/[\eta] = f(M)$ are used for comparison, this will not change the results with respect to M_s and will only decrease the β_M values by a factor of 5/3 because the coefficient B in the Noda–Hearst theory is 5/3 times greater than in Eq. (5.112). However, if the *whole curve* is used for the comparison of the experimental and theoretical dependences $[n]/[\eta] = f(M)$, then it becomes evident that the experimental points are in better agreement with Eq. (5.112) than with the data listed in Table 5.1. This result may be regarded as an indication of the kinetic rigidity of aromatic polymers shown in Fig. 6.17.

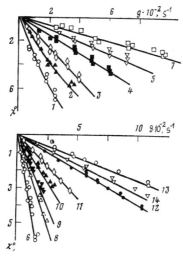

Fig. 6.20. Orientation angle χ versus rate gradient g for solutions of cellulose nitrate fractions in cyclohexanone [23]. Curves correspond go the following molecular weights M_{sD}: 1) $7.7 \cdot 10^5$; 2) $5.9 \cdot 10^5$; 3) $5.7 \cdot 10^5$; 4) $3.8 \cdot 10^5$; 5, 6) $3.0 \cdot 10^5$; 7) $2.5 \cdot 10^5$; 8) $1.9 \cdot 10^5$; 9) $1.8 \cdot 10^5$; 10) $1.6 \cdot 10^5$; 11) $1.0 \cdot 10^5$; 12) $0.7 \cdot 10^5$; 13) $0.5 \cdot 10^5$; 14) $0.4 \cdot 10^5$. Points near each curve correspond to concentrations in the range shown in Fig. 6.21.

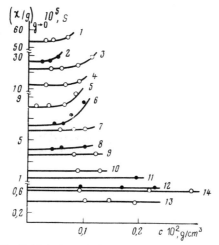

Fig. 6.21. Initial slope $(\chi/g)_{g \to 0}$ of orientation angles versus concentration c for solutions of cellulose nitrate fractions in cyclohexanone [23]. Numbers at the curves are fraction numbers according to Fig. 6.20.

In consideration of the experimental dependence of the shear optical coefficient on molecular weight, it should be borne in mind that the polydispersity of the polymer may affect the character of this dependence. At high M at which the curve

$[n]/[\eta] = f(M)$ virtually tends to the limit (saturation effect), $[n]/[\eta]$ is independent of M and hence is insensitive to polydispersity. In contrast, in the initial range of the curve $x \to 0$, $[n]/[\eta]$ increases proportionally to M and for a polydisperse sample is determined by the ratio $[\overline{W}(\gamma_1 - \gamma_2)/\overline{M}]/(\overline{W}/\overline{M})$. For a rodlike molecule this ratio is proportional to $\overline{M}^3/\overline{M}^2 = \overline{M}_Z$. Hence, if the values of $\overline{M}_w$, $\overline{M}_{sD}$, or $\overline{M}_{D\eta}$ are used to plot the experimental dependence of $[n]/[\eta]$ on M, for a polymer with a wide molecular-weight distribution this may lead to an undue increase in the initial slope of the curve and, correspondingly, to a decrease in M_s and A determined from the ratio of the limit of the curve to its initial slope.

It is possible to avoid the errors due to molecular-weight polydispersity if we compare the relative rigidities of two polymers for which the values of β_M are known to be close to each other. In this case the ratio of the limiting values $([n]/[\eta])_\infty$ for the two polymers yields the ratio of molecular weights M_s of their segments, regardless of the polydispersity of the samples being compared.

1.4. Intrinsic Orientation

The measurement of extinction angles φ_m is a more difficult experimental problem than the determination of birefringence and, in many cases, requires the use of a highly sensitive photoelectric procedure (Chapter 5, Section 4.2.2). This is particularly true for the determination of intrinsic orientation $[\chi/g]$ when rigorous extrapolation of the measured values to both zero shear rates g and zero concentrations c is required.

1.4.1. Cellulose Esters and Ethers and Ladder Polymers. Effect of Polydispersity. Figures 6.18 and 6.20 show as examples the experimental data for the dependence of the extinction angle φ_m for $\chi = \pi/4 - \varphi_m$ on the rate gradient g for fractions of ladder polysiloxane and cellulose nitrate, respectively. The experimental installation permits the measurement of the angles χ for these polymers when these angles are about one degree and larger. This ensures the possibility of carrying out measurements in the range of linear portions of the curves $\chi = \chi(g)$ and of determining reliably the initial slopes $(\chi/g)_{g \to 0}$ of these curves (see comment to Fig. 5.2 in Chapter 5, Section 1.2). Figures 5.19 and 6.21 show the concentration dependence of the initial slopes $(\chi/g)_{g \to 0}$ for the same fractions. The extrapolation of curves in Figs. 6.19 and 6.21 to $c \to 0$ allows the determination of the values of intrinsic orientation $[\chi/g]$ for the fractions investigated.

Comparing the values of $[\chi/g]$ with the corresponding values of the parameter $\eta_0[\eta]M$ it is possible to calculate the coefficient G in Eq. (5.6) for each sample or fraction investigated:

$$G = (RT/\eta_0)\,([\chi/g]/[\eta])\,M^{-1}. \tag{6.2}$$

The values of G as a function of chain length ($x = 2L/A$) for cellulose nitrate fractions in cyclohexanone are shown in Figs. 6.11b and 6.12. The theoretical curves plotted according to Noda–Hearst and corresponding to the data listed in Table 5.1 are also shown in these figures.

The general character of the experimental dependence corresponds to the theoretical curve: with increasing chain length the value of G decreases. However, in absolute magnitude the experimental values of G greatly exceed theoretical values. One of the reasons for this may be the molecular-weight polydispersity of the fractions investigated, which is not taken into account in the theory under consideration. A high-molecular-weight polymer is oriented and deformed in flow at lower shear rates than a low-molecular-weight polymer. Hence, the presence of a high-molecular-weight component in a system of real polymer molecules should lead to an increase in the initial slopes of χ/g and, correspondingly, in the values of G. In a similar manner, the polydispersity of "frozen" conformations in an assembly of kinetically rigid chains leads to the same result.

The sensitivity of orientation angles to the molecular-weight polydispersity of polymers has already been repeatedly discussed [50–54]. In particular, it has been shown for flexible-chain molecules that for a polymolecular sample, Eq. (6.2) remains valid if the coefficient G contained in it is replaced by $G_w = \gamma G$, where the correction factor for polydispersity γ is given by

$$\gamma = [\overline{(M^{a+1})^2} / (\overline{M^{a+1}})^2]/U. \tag{6.3}$$

Here, $U = M_w/M_n$ is the generally used polydispersity parameter of the polymer and a is the exponent for viscosity in the Mark–Kuhn equation. It is assumed in Eq. (6.3) that in Eq. (6.2) the weight-average value of molecular weight M_w is used as the experimental value.

If the type of the molecular-weight distribution of the polymer is known, then, using Eq. (6.3), it is possible to express γ as a function of only two parameters: U and a. Thus, for the Schultz–Zimm distribution [55, 56] Eq. (6.3) yields the following equation [53]:

$$\gamma = [z!\,(z+1+2a)!]/\{(z+1)[(z+a)!]^2\}, \quad z = 1/(U-1). \tag{6.4}$$

For the logarithmic distribution [57] we obtain [53]

$$\gamma = U \exp[(a+1)^2 - 1]. \tag{6.5}$$

For cellulose derivatives over the experimentally investigated molecular-weight range, a is unity. Assuming that the fractions investigated exhibit the Schultz–Zimm distribution, the values of γ and, correspondingly, those of $G_w = \gamma G$, have been calculated according to Eq. (6.4) for various values of M_w/M_n. They are shown by curves in Fig. 6.11 and 6.12. These curves indicate that the difference between the experimental values of G and those predicted by the Noda–Hearst theory may be explained, at least qualitatively, by the polydispersity of the cellulose nitrate fractions investigated. Similar results have been obtained for other cellulose esters and ethers and ladder polysiloxanes [58].

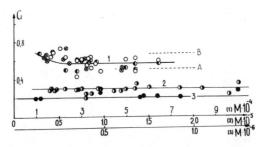

Fig. 6.22. Coefficient $G = [\chi/g]RT/(M[\eta]\eta_0)$ versus molecular weight M for some polymers. 1) Poly(amide benzimidazole) in H_2SO_4 (○) and dimethyl-acetamide (◑), $(M = M_{sD})$ [62, 63]; poly(amide hydrazide) in dimethylsulfoxide (⊗, $M = M_w$) [45]. 2) Poly(*meta*-phenylene isophthalamide) in H_2SO_4 (◐, $M = M_{sD}$). 3) Poly(α-methylstyrene) in tetra-bromoethane (●, $M = M_{sD}$) [60]. Broken curves represent the theoretical values of G for kinetically rigid Gaussian coils with strong (A) and weak (B) hydrodynamic interactions.

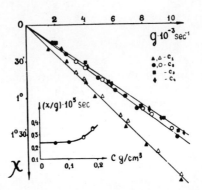

Fig. 6.23. Orientation angle χ versus shear rate g and χ/g versus concentration c for a copolymer of para-aromatic amides ($M = 5 \cdot 10^3$) in H_2SO_4 [64]. Open and filled circles correspond to visual and photoelectric recording, respectively, at solution concentrations $c_1 = 0.186$, $c_2 = 0.154$, $c_3 = 0.101$, $c_4 = 0.056 \cdot 10^{-2}$ g/cm³.

However, to be able to compare quantitatively the predictions of the Zimm or the Noda–Hearst theories based on the model of a kinetically flexible elastic necklace with experimental data for real chain molecules, these data should be obtained for polymers with a very narrow molecular-weight distribution and with relatively flexible molecules.

Early experimental data on extinction angles in flexible-chain polymer solutions [2] cannot be used for this purpose since they have not been obtained for sharp fractions.

The few later investigations have been carried out with sharper fractions [53, 59]. However, their results did not contain a detailed analysis of the concentration dependence of the *initial slopes* of curves $\chi = \chi(g)$. Consequently, these data have been interpreted [54] on the basis of a general equation (5.54) in which the parameter β_0 was replaced by its reduced value $\beta_r = g(\eta - \eta_0)M/cRT$ determined at finite values of concentration c and shear rate g. For the above-mentioned reasons (Chapter 5, Section 2.2.2.2) this method seems to be less reliable that that in which only the initial (linear) portions of the curves $\chi/(g)$ are used with subsequent extrapolation to $c \to 0$ (Figs. 6.18–6.21).

In a recent paper, the measurements of intrinsic orientation have been carried out for very sharp fractions of anionic poly(α-methylstyrene) in tetrabromoethane [60]. The molecular weights of these fractions have been in the range $1 \cdot 10^5 \leq M_{sD} \leq 3 \cdot 10^6$ with their polydispersity parameters $U = M_w/M_n \leq 1.07$. At this value of U the correction factor γ in Eq. (6.4) or (6.5) can differ from unity by not more than 0.1.

Figure 6.22 shows by points on curve 3 the values obtained for these fractions using plots similar to the curves in Figs. 6.18–6.21. Over the entire molecular-weight range the coefficient G remains constant, as expected for coil-shaped molecules whose conformation does not undergo changes upon variation in M. Although in the investigated range of M the molecules of poly(α-methylstyrene) according to their hydrodynamic properties are nondraining Gaussian coils, the experimental points on curve 3 correspond to the value of $G = 0.2$, which is two times greater than the value predicted for these molecules by the Zimm theory. This result implies that the increased experimental values of G as compared to theoretical values cannot be uniquely interpreted by the effect of molecular polydispersity of the polymer investigated because the observed discrepancies may be due to other reasons (for example, the inadequacy of the theoretical values of G or a manifestation of the kinetic rigidity of chains).

The latter observation should be borne in mind, in particular, in the investigation of aromatic polymers with FB affected by the kinetic rigidity of molecules.

1.4.2. Aromatic Polymers. A high positive anisotropy of aromatic polymers and the corresponding high FB value in their solutions ensure favorable conditions for measuring extinction angles at low shear rates and concentrations. Moreover, the relatively low molecular weights of polycondensation polymers, which is of practical importance, makes it possible (in particular if a photoelectric procedure is used) to reliably determine orientation angles at values of χ less than $1°$. This is demonstrated in Figs. 6.23 and 6.24, which show the corresponding data for some aromatic polyamides. The linear character of the $\chi(g)$ curves allows a reliable determination of the initial slopes $(\chi/g)_{g \to 0}$ and a reliable extrapolation of the data to zero concentration (Fig. 6.25) for obtaining the intrinsic values of these orientations $[\chi/g]$ in solutions of these polymers.

In those cases when the molecular weights M of the samples of fractions investigated are known from measurements by an independent method, the values of $[\chi/g]$ may be correlated with those of M and $[\eta]$ and the coefficients G may be calculated according to Eq. (6.2).

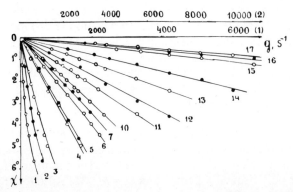

Fig. 6.24. Orientation angle χ versus shear rate g at minimum concentrations used for samples of poly(*para*-hydroxyphenylbenzoxazole terephthalamide) in H_2SO_4 [65]. Molecular weights M of samples range from $5 \cdot 10^4$ (1) to $6 \cdot 10^3$ (17). Minimum concentrations range from $1 \cdot 10^{-5}$ g/cm^3 (1) to $3 \cdot 10^{-3}$ g/cm^3 (17).

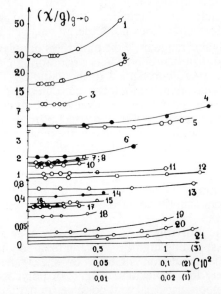

Fig. 6.25. Initial slopes $(\chi/g)_{g \to 0}$ of orientation angles versus concentration c of solutions for the same samples as in Fig. 6.24. Molecular weights of samples range from $5 \cdot 10^4$ (1) to $1.5 \cdot 10^3$ (21). Scale 1 on the abscissa – for samples 1–5; scale 2 for samples 6–17; scale 3 for samples 18–21 [65].

The dependence of the values of G obtained in this manner on molecular weight ($\overline{M}_w$ or $\overline{M}_{sD}$) for fractions and samples of some aromatic polyamides is shown in Fig. 6.22.

Consideration of the curves plotted in Fig. 6.22 (and of others reported in the literature [58]) makes it possible to draw some general conclusions.

If the fractions (or samples) of the polymer being investigated are in the range of molecular weights M in which the shear optical coefficient $\Delta n/\Delta \tau$ remains constant, then if M is varied, the value of G also remains constant. For example, this occurs for poly(m-phenylene isophthalamide) in sulfuric acid (curve 2 in Fig. 6.22). This result seems natural because, in this case, the constancy of both $\Delta n/\Delta \tau$ and G is ensured by the fact that in this range of M polymer molecules exhibit the Gaussian coil conformation.

In the low-molecular-weight range where for aromatic polymers $\Delta n/\Delta \tau$ sharply increases with molecular weight (curves in Fig. 6.17), the coefficient G for these polymers somewhat decreases with increasing M (although much less drastically than $\Delta n/\Delta \tau$), attaining values which remain virtually constant in the higher-molecular-weight range ($M \geq 10^4$). This character of the dependence of G on M is shown by curve 1 in Fig. 6.22 plotted according to experimental points for poly(amide benzimidazole) fractions and poly(amide hydrazide) samples.

It is noteworthy that the values of G obtained (at $\overline{M}_w > 10^4$) for rigid-chain para-aromatic polymers are much higher than those for flexible-chain molecules [poly(α-methylstyrene), Fig. 6.22, curve 3] and are twice as high as that for a less-rigid meta-aromatic polymer: poly($meta$-phenylene isophthalamide) (curve 2). These high values of G do not change when the solvent viscosity η_0 is varied by an order of magnitude or higher, and, as the analysis of molecular-weight homogeneity shows, cannot be attributed to the effect of sample polydispersity [62, 63]. If weight-average molecular weights $\overline{M}_w$ are used (or $\overline{M}_{sD}$, which are close to them), then for the para-aromatic polymers investigated [58] at $\overline{M}_w > 10^4$ the coefficients G are found to be close to 0.6. This value is within the range of $G = 0.5$–0.67 predicted by the theory for kinetically rigid Gaussian coils with various extents of draining (Fig. 5.17).

The character of the dependence $G(M)$ corresponding to curve 1 in Fig. 6.22 is specific not only for the two polymers represented by this curve. It is general for all other investigated para-aromatic condensation polymers with high equilibrium chain rigidity [58].

Taking into account the foregoing considerations, it may be assumed that the high values of $G \approx 0.6$ obtained experimentally for this class of polymers reflect the specific features of their molecular structure and are mainly due to the high kinetic rigidity of their chains (at least in the molecular-weight range which includes the experimental data discussed here).

According to the considerations presented in Chapter 5 (Section 3.4), when the length of kinetically rigid chains decreases and their conformations vary correspondingly from a Gaussian coil to a rodlike conformation, the theoretical value of G remains virtually constant and close to 0.5–0.6. However, for rigid-chain para-aromatic polymers in the range of low molecular weights the experimental values of G increase with decreasing M. This type of the dependence $G(M)$ may be explained, at least qualitatively, by the molecular-weight polydispersity of the samples under investigation.

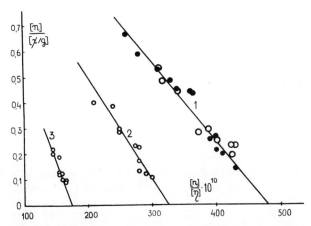

Fig. 6.26. Dependence of $[n]/[\chi/g]$ on $[n]/[\eta]$ for some aromatic polyamides. 1) Poly(*para*-hydroxyphenylbenzoxazole terephthalamide) in H_2SO_4 (●) [65] and copolymer of *para*-phenylene terphthalamide and benzimidazole in H_2SO_4 (○) [1]. 2) Copolymer of *para*-phenylene terephthalamide and *para*-benzamide in H_2SO_4 [64]. 3) Poly(*meta*-hydroxyphenylbenzoxazole terephthalamide) in H_2SO_4 [65].

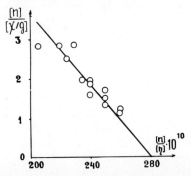

Fig. 6.27. Dependence of $[n]/[\chi/g]$ on $[n]/[\eta]$ for poly-(amide hydrazide) in dimethylsulfoxide [45].

According to Eq.(6.2), for a sample polydisperse with respect to molecular weight the value of G is determined by the average value $(\overline{W/[\eta]})/M$, which has different dependences on M in the range of the Gaussian coil and in that of rodlike conformations. For a draining Gaussian coil W changes proportionally to M^2 and $[\eta]$ changes proportionally to M. Consequently, the average value $\overline{W/[\eta]}$ is proportional to $\overline{M^2}/\overline{M} = \overline{M}_w$. Hence, it might be expected that in this range of molecular conformations the value of $(\overline{W/[\eta]})/M$ and, therefore, that of G, will remain invariable with the change in M if the values of $\overline{M}_w$ are substituted for M in Eq. (6.2). This conclusion is confirmed by the above experimental data (curve 1 in Fig. 6.22 in the range of $M > 10^4$).

TABLE 6.2. Abscissa Intercepts $([n]/[\eta])_\infty$ and Slopes of Straight Lines Representing the Dependences of $[n]/[\chi/g]$ on $[n]/[\eta]$ for Solutions of Various Aromatic Polyamides and Their Anisotropy β_M and Equilibrium Rigidity (M_s and A) Parameters

No.	Polymer	$([n]/[\eta])_\infty \cdot 10^{10}$, cm sec^2 g^{-1}	Slope, g^{-1}cm$^3 \cdot$ sec^{-1}	$\beta \cdot 10^{25}$, cm^3	M_s	$M_L \cdot 10^{-8}$, cm^{-1}	$A \cdot 10^8$, cm	References
1	Poly(para-hydroxy-phenylbenzoxazole terephthalamide) in H$_2$SO$_4$	480	$3.1 \cdot 10^7$	0.97	6500	20	330	[65]
2	Poly(meta-hydroxy-phenylbenzoxazole terephthalamide) in H$_2$SO$_4$	175	$7.1 \cdot 10^7$	0.81	2820	23	120	[65]
3	Copolymer of para-phenylene terephthal-amide and benzimidazole in H$_2$SO$_4$	480	$3.1 \cdot 10^7$	0.97	6500	20	330	[1]
4	Copolymer of paraphenyl-ene terephthalamide and parabenzamide in H$_2$SO$_4$	325	$4 \cdot 10^7$	0.85	5000	20	250	[64]
5	Poly(amide hydrazide) in dimethylsulfoxide	280	$4.6 \cdot 10^8$	0.85	4300	20	220	[45]

In the range of rodlike conformations W is proportional to M^3 and $[\eta]$ is proportional to M^2. Hence, $\overline{W/[\eta]}$ is proportional to $\bar{M}^3/\bar{M}^2 = \bar{M}_z$. Consequently, in order to maintain the constant value of G in this range, the value of $M = \bar{M}_z$ should be substituted into Eq. (6.2). In the calculation of G over the entire molecular-weight range investigated, the values of M obtained by the same method (e.g., M_w or M_{sD}) are usually employed. Therefore, with decreasing M (i.e., x for a wormlike chain), an increase in G should be expected (since for a polydisperse polymer $\bar{M}_z > \bar{M}_w \approx \bar{M}_{sD}$), which is actually observed experimentally. Therefore, in the molecular-weight range in which the conformational and hydrodynamic properties of rigid-chain aromatic polymers are close to those of draining random coils, the coefficient G [calculated according to Eq.(6.2) using the values of $\bar{M}_w$] is virtually constant and close to the theoretical value 0.6. This fact is of great practical importance because it is the basis for using Eq. (6.2) in the determination of the molecular weight of the polymer $\bar{M}_w$ from the experimental values of $[\chi/g]$, $[\eta]$, and η_0 and the theoretical value of G (≈ 0.6), as has been done in a number of papers [42, 64, 67]. For many rigid-chain aromatic polymers this possibility is of considerable importance because the application of other methods of determination of M involves experimental difficulties which have already been repeatedly mentioned.

1.5. Intrinsic Values of FB and Orientation

It has been shown in Chapter 5 that for kinetically rigid wormlike chains in the molecular-weight range in which the coefficient G is invariable, the dependence of $[n]/[\chi/g]$ on $[n]/[\eta]$ should be linear. For rigid-chain aromatic polymers in the

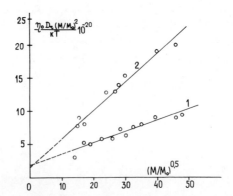

Fig. 6.28. Dependence of $\eta_0 D_r (M/M_0)^2/kT$ on $(M/M_0)^{0.5}$ for solutions of cellulose nitrate fractions in cyclohexanone. Nitrogen content N is equal to: 13.4% (1) [23]; 12.1% (2) [31].

range of constant G this relationship is actually observed. This is demonstrated in Figs. 6.26 and 6.27, where the experimental values of $[n]/[\chi/g]$ are plotted against $[n]/[\eta]$ for homologous series of some aromatic polyamides differing in molecular weights.

In accordance with Eq. (5.114), the experimental points fit straight lines forming the abscissa intercepts

$$([n]/[\eta])_\infty = B\beta A = B\beta_M M_s \qquad (6.6)$$

and having slopes to the abscissa

$$\partial\,([n]/[\chi/g])/\partial\,[n]/[\eta]) = -(RT/\eta_0 G)(AM_L)^{-1}. \qquad (6.7)$$

For each polymer the intercept and the slope of the straight lines determine the specific chain anisotropy β_M and the molecular weight of the segment M_s from the relations

$$\beta_M = (\text{slope} \times \text{intercept})/(BRT/\eta_0 G), \qquad (6.8)$$

$$M_s = (\text{intercept})/B\beta_M. \qquad (6.9)$$

Table 6.2 lists the slopes and the intercepts of the straight lines for all polymers represented by plots in Figs. 6.26 and 6.27, and also the corresponding values of β_M and M_s. It is assumed in the calculations that $G = 0.63$. If the values of M_L are evaluated according to the structural formulas of the molecules investigated, it is possible to determine the lengths of the Kuhn segment $A = M_s/M_L$ presented in the last column of Table 6.2.

The parameters of anisotropy and rigidity of the polymers investigated obtained from the dependence of $[n]/[\chi/g]$ on $[n]/[\eta]$ agree within experimental error with those determined from the dependence of $[n]/[\eta]$ on M (see Fig. 6.17).

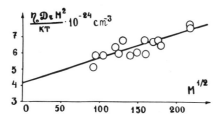

Fig. 6.29. Dependence of $\eta_0 D_r M_w{}^2/kT$ on $M_w{}^{1/2}$ for solutions of poly(amide hydrazide) samples in dimethylsulfoxide [45].

1.6. Intrinsic Orientation and Rotational Friction of Molecules

Intrinsic orientation $[\chi/g]$ of FB in a solution of kinetically rigid molecules is directly related by Eq. (6.6) to the rotational diffusion coefficient D_r or the rotational friction coefficient W of the molecule in its rotation about the short (transverse) axis.

The theory of rotational friction of rigid wormlike chains relates using Eqs. (2.118) and (2.119) the rotational friction coefficients W of the molecule with its length or molecular weight M. The coefficient W contained in these equations corresponds to the rotation of the wormlike chain about one of the axes X or Y normal to Z, the symmetry axis of the molecule (Fig. 2.8). As we have seen [Eq. (2.125) and Fig. 5.21, curve 1], in this case at low $x = 2L/A$, when the chain conformation is close to rodlike, the axis X or Y (Fig. 2.8) is actually the minor axis of the molecule and hence the coefficient W contained in Eq. (2.118) corresponds to W in Eq. (5.5). However, at high x in a system of XYZ coordinates in Fig. 2.8 the wormlike chain is spherically symmetrical and W in Eq. (2.119) is the rotational friction coefficient of a wormlike coil averaged over all its spatial orientations. Hence, the values of W in Eqs. (5.5) and (2.119) are not quite identical.

However, the model experiments carried out by Kuhn, who has studied the rotational friction of Gaussian coils in their rotation about the short molecular axis, have shown that in this case also the dependence of W on M [Eq. (2.57)] remains identical in form with the dependence (2.119), differing from it only in the numerical coefficients. This point is the basis for applying Eq. (2.119) and using in this equation the friction coefficients W determined from the intrinsic orientation of FB according to Eq. (5.5).

Equation (2.119) describes the rotational friction of a wormlike chain which exhibits a marked draining effect, although its conformation is close to the Gaussian coil ($L/A \gg 1$). Hence, to be able to apply Eq. (2.119) to real rigid-chain polymers, the molecular weights of the samples or fractions investigated should be relatively high. For this purpose it is convenient to use ladder polysiloxanes or cellulose esters and ethers for which experimental data on intrinsic orientations $[\chi/g]$ [and, hence on the values of D_r in Eq. (5.5)] are available in fractions over a wide molecular-weight range M_{sD}.

Figure 6.28 shows the dependence of $\eta_0 D_r (M/M_0)^2/kT$ on $(M/M_0)^{1/2}$ for a series of fractions of two cellulose nitrate samples (differing in the degree of nitration) in cyclohexanone [23, 31]. In accordance with the theoretical prediction [Eq. (2.119)], the points for the fractions of each sample fit a straight line with slope $0.72(M_L/A)^{3/2}$ and intercept with the ordinate $0.64M_L^2 A^{-1}[\ln (A/d) - 1.43]$.

The slopes of straight lines 1 and 2 (at the values of $M_L = M_0/5.15 \cdot 10^{-8} = 54.7 \cdot 10^8$ cm^{-1} and $51.6 \cdot 10^8$ cm^{-1}, respectively) make it possible to determine the values $A = 240 \cdot 10^{-8}$ cm and $A = 130 \cdot 10^{-8}$ cm for cellulose nitrates with a nitrogen content of 13.4 and 12.1%, respectively. The hydrodynamic diameters d of the chain, determined from the intercepts of straight lines 1 and 2, are $d = 10 \cdot 10^{-8}$ cm and $d = 12 \cdot 10^{-8}$ cm, respectively. These values of A and d are in reasonable agreement with those determined by diffusion and viscometry (Table 4.9).

A plot similar to Fig. 6.28 is shown in Fig. 6.29 for poly(amide hydrazide) samples in dimethylsulfoxide [45]. Maximum molecular weights of the samples are about $50 \cdot 10^3$, which poorly agrees with the condition $L/A \gg 1$, which is indispensable for a reliable application of Eq. (2.119). The experimental points in Fig. 6.28 correspond to a straight line (plotted by the least-squares method), the slope of which gives $A = (240 \pm 50) \cdot 10^{-8}$ cm. Although this value corresponds in order of magnitude to the data of diffusion and viscometry (Table 4.12), the error in its determination is relatively great because no high-molecular-weight samples of this poly(amide hydrazide) are available. The result is general for all condensation para-aromatic polymers used in practice whose molecular weights generally do not exceed a few ten thousand [68].

2. MOLECULAR STRUCTURE, RIGIDITY, AND OPTICAL ANISOTROPY OF THE POLYMER CHAIN

2.1. Cellulose Esters and Ethers

Table 6.3 shows the limiting values of shear optical coefficients ($[n]/[\eta]$)$_\infty$ and the corresponding values of segmental anisotropy $\alpha_1 - \alpha_2$ calculated according to Eq. (5.60) for various cellulose esters and ethers in different solvents. For most of the polymers the equilibrium rigidity (number of monomer units S in a segment) is known from hydrodynamic data (Table 4.9), investigation of the microform effect (Table 6.1), or the dependence of $[n]/[\eta]$ on molecular weight [23, 30–32] (Figs. 6.7 and Figs. 6.11–6.13). The anisotropies of the monomer unit $a_{\cdot \parallel} - a_\perp = \Delta a = (\alpha_1 - \alpha_2)/S$ calculated using these data and the values of $\alpha_1 - \alpha_2$ are also listed in Table 6.3. The values of $\alpha_1 - \alpha_2$ and Δa in Table 6.3 contain not only the intrinsic anisotropy, but also the anisotropy of microform. Hence, for the same polymer in solvents with different refractive indices they may differ not only in value, but also in sign. However, even in matching solvents ($n_k - n_s$) for different cellulose

TABLE 6.3. Limiting Values $\left([n]/[\eta]\right)_\infty$ and Anisotropy of the Segment $\alpha_1 - \alpha_2$ and the Monomer unit $\Delta a = a_\parallel - a_\perp$ of Cellulose Esters and Ethers

No.	Cellulose esters and ethers	Substituting radical	Degree of substitution	Solvent	$\left(\dfrac{[n]}{[\eta]}\right)_\infty$ cm s² g⁻¹	$(\alpha_1-\alpha_2)\cdot10^{25}$ cm³	$\Delta a \cdot 10^{25}$ cm³	References
1	Cellulose butyrate	$-O-CO-C_3H_7$	3.0	Tetrachloroethane	-2.8	-34	-2.6	[19]
				Bromoform	$+3.3$	$+35$	$+2.7$	
				Methyl ethyl ketone	$+11.0$	$+144$	$+5.6$	
2	Cellulose benzoate	$-O-CO-C_6H_5$	3.0	Dimethylformamide	-48.4	-617	-10.4	[69]
				Chloroform	-60.5	-763	-16.4	
				Bromobenzene	-79.3	-914	-19.6	
				Dimethylphthalate	-73.0	-830	-17.8	
3	Cellulose carbanilate	$-O-CO-NH-C_6H_5$	2.2	Dioxane	-150	-1880	-47.0	[20]
			3.0	Dioxane 21°	-144	-1830	-45.8	[21]
				Dioxane 65°	-68	-872	—	[21]
				Ethyl acetate	-42	-560	-15.2	[22]
4	Cellulose monophenyl acetate	$-O-CO-CH_2C_6H_5$	2.8	Bromobenzene	$+52$	$+600$	$+13.0$	[22]
				Bromoform	$+42$	$+478$	$+10.3$	
5	Cellulose diphenyl acetate	$-O-CO-CH(C_6H_5)_2$	—	Dioxane	$+82$	$+1030$	—	[71]
6	Cellulose diphenylphosphonocarbamate	$-O-CO-NH-P(OC_6H_5)_2$	2.2	Dioxane	$+50$	$+640$	$+13.7$	[22]
7	Ethylcellulose	$-O-C_2H_5$	2.5	Dioxane	$+40$	$+512$	$+14.6$	[22]
8	Benzylcellulose	$-O-CH_2-C_6H_5$	2.5	Dioxane	$+25$	$+291$	—	[71]
9	Cyanoethylcellulose	$-O-CH_2-CH_2-CN$	2.6	Cyclohexane	$+72$	$+900$	$+17.8$	[32]
10	Cyanoethylacetylcellulose	$-O-CH_2-CH_2-CN,$ $-O-CO-CH_3$	—	Acetone	$+29$	$+390$	$+17.0$	[70]
				Dimethylformamide	$+7.5$	$+90$	$+3.9$	
11	Cellulose nitrate	$-O-NO_2$	2.8	Cyclohexanone	-65	-820	-14	[24]
			2.7	Cyclohexanone	-43	-540	-9.0	[23]
			2.7	Amyl acetate	-25	-320	-5.4	[23]
			2.7	Butyl acetate	-23	-300	-5.0	[23]
			2.7	Ethyl acetate	-10	-140	-2.3	[23]
			2.7	Acetone	-8.4	-115	-1.9	[23]
			2.3	Cyclohexanone	-27	-330	-8.4	[31]
			1.9	Cyclohexanone	-5	-62	-2.0	[5]
			1.9	Dioxane	$+11.7$	$+149$	$+4.8$	[5]

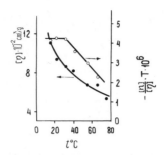

Fig. 6.30. Temperature dependence of $[\eta]$ and $[n]/[\eta]$ for solutions of cellulose carbanilate in dioxane [21].

TABLE 6.4. Hydrodynamic and Optical Characteristics and Equilibrium Rigidity of Molecules of a Sodium Salt of Cellulose Sulfoester at Various Ionic Strengths of Solution, I [3, 73]

I, mole/liter	$[\eta]$, cm³/g	$([n]/[\eta]) \cdot 10^{10}$, cm s² g⁻¹	$(\alpha_1 - \alpha_2) \cdot 10^{25}$, cm³	$(\alpha_1 - \alpha_2)/[\eta] \cdot 10^{25}$, g	$A \cdot 10^{8}$, cm
0.200	290	46.6	634	2.19	195
0.150	290	47.4	645	2.22	197
0.100	310	50.0	680	2.19	216
0.010	500	72.0	980	1.96	330
0.005	610	96.7	1320	2.16	410
0.001	870	127	1730	2.00	575

derivatives the values of $[n]/[\eta]$ and, hence, those of $\alpha_1 - \alpha_2$ and Δa can differ in sign and be very different in absolute values, although the difference in the equilibrium rigidities of molecules of different cellulose esters and ethers is not very great. This fact implies that the optical anisotropy of chains of these esters and ethers is determined to a great extent by the structure and anisotropy of substituent side groups and the degree of substitution. This is demonstrated, for example, by the data for cellulose nitrates differing in nitro-group content. Another example of a profound effect of the structure of substituents on the chain anisotropy is provided by a comparison of Δa for samples 9 and 10 in Table 6.3: the replacement of some of the cyanoethyl groups by cyanoacetyl groups decreases Δa more than fourfold. This high sensitivity of the optical anisotropy of cellulose derivatives to side substituent structure is due to the fact that the anisotropy of the cellulose chain itself is not high [72].

Chapter 4 was concerned with the problem of the well-known capacity of cellulose ester and ether molecules to undergo a considerable change in equilibrium rigidity upon a variation in the solvent or temperature (Table 4.2). This feature is interpreted as an indication of the fact that the chain rigidity of cellulose esters and ethers is profoundly affected by intramolecular hydrogen bonds. These bonds can weaken either when the temperature is increased or under the effect of a solvent capable of breaking these bonds. This feature is not less distinctly manifested in the FB phenomenon of cellulose derivatives than in their hydrodynamic properties. This can be seen both from the data in Table 6.3 and the temperature dependence of the shear optical coefficient [21, 73]. Figure 6.30 shows, as an example, the temperature dependence of $[n]/[\eta]$ for a solution of cellulose carbanilate [21]. When the temperature is increased by $50°K$, the value of $[n]/[\eta]$ proportional to βA decreases by two times. Since in this case the chain anisotropy per unit length β does not undergo any considerable changes, it may be assumed that the observed temperature effect corresponds to a twofold decrease in chain rigidity (A). In this case, intrinsic viscosity also decreases by two times, as expected for a hydrodynamically draining Gaussian coil.

· The skeletal rigidity of molecules of cellulose derivatives containing ionogenic groups may be further increased by the electrostatic interaction between these groups. When the ionic strength of the solvent decreases and the polyion chain expands correspondingly, both the intrinsic viscosity $[\eta]$ and the $[n]/[\eta]$ ratio increase. This increase implies that chain expansion is accompanied by an increase in segmental anisotropy $\alpha_1 - \alpha_2$ and, correspondingly, in the length of the segment A. These properties are demonstrated in Table 6.4, which gives the corresponding data for aqueous solutions of a sodium salt of cellulose sulfoether $(M = 10^5$, the side substituent is SO_3Na) [74]. The data listed in the table show that the $(\alpha_1 - \alpha_2)/[\eta]$ ratio remains virtually constant when the ionic strength decreases. This fact implies that (since for a Gaussian chain of constant length L the value of $\alpha_1 - \alpha_2$ changes proportionally to $\langle h^2 \rangle$) when the coil undergoes polyelectrolytic uncoiling, intrinsic viscosity $[\eta]$ increases proportionally to $\langle h^2 \rangle$, which is characteristic of hydrodynamically draining Gaussian coils.

TABLE 6.5. Limiting Values $([n]/[\eta])_\infty$ and Anisotropy of the Segment $\alpha_1 - \alpha_2$ and of the Monomer Unit $\Delta a = a_\parallel \neq a_\perp$ of Ladder Polysiloxanes*

No.	Polymer	Solvent	$-([n]/[\eta])_\infty \cdot 10^{10}$, cm s^2 g^{-1}	$-(\alpha_1 - \alpha_2) \times 10^{25}$, cm^3	$-\Delta a \cdot 10^{25}$, cm^3	References
1	.Ladder poly(phenyl siloxane)	Bromoform	160	1800	25	[8, 10, 11]
2	The same	Bromoform	110	1230	23	[9—11]
3	The same	Bromoform	95	1050	31	[9, 10]
4	The same	Benzene	240	—	—	[12]
5	Ladder poly(m-chloro- phenyl siloxane)	Tetrachloro- methane	380	4700	40	[11]
6	Ladder poly(dichloro- phenyl siloxane)	Tetrabromoethane Bromoform	425 425	4700 4450	53 50	[14]
7	Ladder poly(3-methyl- butene siloxane)	Benzene	47	570	6.5	[16]
8	The same	Butyl acetate	30	400	4.2	[11, 16, 17]
9	Ladder poly(phenyl iso- butyl siloxane) 1:1	Benzene	69	830	21	[11]
10	Ladder poly(phenyl iso- hexyl siloxane) 1:1	Benzene	81	980	19.5	[11]
11	Linear poly(methyl phenyl siloxane)	Benzene	6.1	85.5	17	[10, 18]

*Structure of side groups in Table 4.6.

2.2. Ladder Polysiloxanes

Table 6.5 gives the limiting values $([n]/[\eta])_\infty$ for a number of ladder polysiloxanes according to the data plotted in Fig. 6.6 and to the values of $\alpha_1 - \alpha_2$ calculated according to Eq. (5.60) and the values of $([n]/[\eta])_\infty$. The last column contains the values of the anisotropy of a monomer unit $\Delta a = a_\| - a_\perp = (\alpha_1 - \alpha_2)/S$ calculated from the values of $\alpha_1 - \alpha_2$ and S. The numbers of monomer units in a segment S are taken from Tables 4.6 and 4.7 using the equality $S = A/\lambda$ and $\lambda = 2.5 \cdot 10^{-8}$ cm (see Fig. 1.4). For all ladder polymers $([n]/[\eta])_\infty$ is high in absolute value and negative in sign. Consequently, the main part of the optical anisotropy of the molecule is contributed by the side groups. This conclusion agrees with the well-known fact that the optical anisotropy of the SiO bonds forming the backbone of the macromolecule is very slight [75].

The segmental anisotropy of various ladder polymers is higher by one order of magnitude, or even more than that of linear poly(methylphenyl siloxane), because the equilibrium rigidity of the ladder structure is more than ten times higher than that of a single-strain chain. However, the anisotropy Δa of a monomer unit of ladder poly(phenyl siloxane)s exceeds that of a linear poly(methylphenyl siloxane) only by a factor of 1.5–2. This difference is due to the presence of two phenyl rings in the monomer unit of ladder poly(phenyl siloxane)s and one ring in that of a single-strain poly(methylphenyl siloxane).

This fact implies that the contributions provided to the anisotropy Δa of the monomer unit by one phenyl ring in ladder and single-strain poly(phenyl siloxane)s virtually coincide. Hence, the double-chain structure has no great effect on the rotational mobility of phenyl side groups. The differences in the values of Δa for ladder polymers differing in the structure of the side radicals are mainly due to different anisotropies of these radicals. The replacement of the phenyl side group with an aliphatic group decreases the absolute value of Δa, whereas the introduction of chlorine atoms as substituents of hydrogen atoms in the phenyl ring increases the negative optical anisotropy of the monomer unit.

Moreover, the differences in the segmental anisotropy of ladder polymers may be due to different degrees of imperfection of their double-strain structure whose presence has been indicated above (Chapter 4). This may be demonstrated by comparing the anisotropies of different samples of ladder poly(phenyl siloxane)s for which the chemical structure of monomer units is the same, but the values of $([n]/[\eta])_\infty$ and, correspondingly, those of $\alpha_1 - \alpha_2$ may differ by a factor of 2.5. In fact, the introduction of single bonds into the cyclic chain leading to a partial ladder structure drastically decreases the equilibrium chain rigidity as, for example, in the case of semiladder molecules of poly-N-maleimide and polyacenaphthylene.

2.3. Stepladder Polymers

The investigation of the hydrodynamic properties of stepladder molecules of N-substituted poly(isobutyl maleimide)s and polyacenaphthylenes (Chapter 4, Section 4) has shown that their equilibrium rigidity twice and even more than twice ex-

TABLE 6.6. Shear Optical Coefficients $[n]/[\eta]$ and Anisotropy of the Segment $\alpha_1 - \alpha_2$ and of the Monomer Unit Δa of Stepladder (Nos. 1–4) and Nonladder (Nos. 5 and 6) Polymers with Similar Structures of Cyclic Side Groups

No.	Polymer	Monomer	Solvent	$([n]/[\eta]) \cdot 10^{10}$, cm s^2 g^{-1}	$(\alpha_1 - \alpha_2) \cdot 10^{25}$, cm^3	$\Delta a \cdot 10^{25}$, cm^3	References
1	Poly(isobutyl maleimide)		Chlorobenzene	$+14$	$+160$	$+10$	[78]
2	Poly(tolyl maleimide)		Bromoform	-14	-160	-10	[79]
3	Poly(2,4-dimethyl phenyl-maleimide)		Bromoform	-19	-200	-12.5	[80]
4	Polyacenaphthylene		Bromoform	-18	-200	-12	[81]
5	Poly(vinyl pyrrolidone)		Benzyl alcohol	-6.5	-75	-10	[84]
6	Poly-β-vinylnaphthalene		Bromoform	-40	-440	-30	[81]

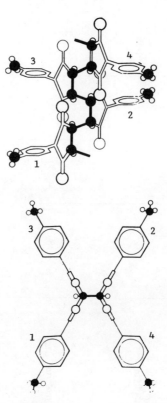

Fig. 6.31. Possible structure of poly(N-tolylmaleimide) chain in the extended conformation. Top: view transverse to the direction of the main chain; bottom: view along the direction of the main chain. Numbers 1–4 are numbers of phenyl rings [79].

ceeds that of common flexible-chain polymers. In this case, however, in the molecular-weight range of tens and hundreds of thousands the molecules of these polymers in solutions are random coils and in θ-solvents adopt the conformation of Gaussian coils (Table 4.5).

This general conclusion is confirmed by the investigations of FB in solutions of these polymers [76–81] showing that in matching solvents their shear optical coefficient $[n]/[\eta]$ remains constant over the entire molecular-weight range $M \geq 10^5$. Accordingly, for polymaleimide and polyacenaphthylene molecules the segmental anisotropy $\alpha_1 - \alpha_2$ has been calculated from the values of $[n]/[\eta]$ according to Eq. (5.60), and the anisotropy of the monomer unit $\Delta a = (\alpha_1 - \alpha_2)\lambda/A$ has been determined using segment lengths A (from hydrodynamic data, Table 4.5) and $\lambda = 2.5 \cdot 10^{-8}$ cm (for an extended trans-chain). The data obtained are listed in Table 6.6.

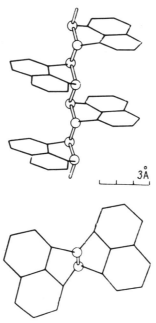

Fig. 6.32. Possible structure of polyacenaphthylene chain in extended trans-conformation. Top: view transverse to the direction of the main chain; bottom: view along the direction of the main chain.

Although the equilibrium rigidities of chains of stepladder polymers 1–4 in Table 6.6 and, correspondingly, their hydrodynamic properties (Table 4.5) are close to each other, the dynamo-optical properties differ greatly and hence may yield additional information about the structure of their molecules.

Thus, in contrast to all the other polymers listed in Table 6.6, the FB and, consequently, the anisotropy of the monomer unit of poly(isobutyl maleimide) (PBMI) is positive in sign. To elucidate this fact it is useful to compare the anisotropy of PBMI with that of polyvinylpyrrolidone (PVP) the data for which are also given in Table 6.6 (No. 5) for this purpose. In the PVP molecule the chain anisotropy mainly depends on the difference in the polarizabilities of the pyrrolidone ring. However, this ring, which is similar in structure and anisotropy to the imide ring, differs from it in that it induces negative anisotropy in the axes of the main chain of the polymer molecule. This means that the plane of the pyrrolidone ring forms a considerable angle (close to 90°) with the direction of the main chain. This position of the ring bonded to the main chain by a single bond evidently corresponds to a minimum of steric hindrance between rings and is usual for chain molecules with cyclic side groups [84] [polystyrene, poly(phenyl siloxane), etc.]. If this is taken into account, it becomes evident that the positive anisotropy introduced into the monomer unit by the imide ring (Table 6.6, No. 1) implies that the plane of this ring forms only a small angle with the direction of the main chain.

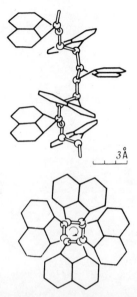

Fig. 6.33. Possible structure of the polyacenaphthylene chain (helix 4₁) in the extended conformation. Top: view transverse to the direction of the main chain; bottom: view along the direction of the main chain.

This position of the imide ring corresponds to the structure of polyimides whose rings, in contrast to those of PVP, form a part of the main chain.

The replacement of an isobutyl substituent (which is virtually isotropic) by a *para*-tolyl substituent in the molecule of N-substituted polymaleimide changes the sign of its optical anisotropy from positive to negative (Table 6.6, No. 2). In this case, the absolute value of the anisotropy of the segment and the monomer unit Δa remains virtually invariable. This drastic change in the optical properties of the chain can be easily explained by the effect of anisotropic phenyl substituents with planes that form large angles with the chain direction as in all other molecules with cyclic side groups.

A possible structure of the extended chain of poly(N-tolyl maleimide) corresponding to the optical properties of this polymer is shown in Fig. 6.31.

If a 2,4-dimethylphenyl substituent is introduced into the polymaleimide chain instead of a tolyl substituent (Table 6.6, No. 3), this virtually does not change chain rigidity but increases the negative anisotropy of the monomer unit by 25%. Since the CH_3 group inserted in the phenyl ring in position 2 is almost isotropic, this increase in optical anisotropy should be attributed to a decrease in the rotational mobility of aromatic side rings. The change in their mobility is probably due to steric interactions of these CH_3 groups with the neighboring atomic groups.

The optical anisotropy of the monomer unit Δa in the molecule of a stepladder polymer polyacenaphthylene (Table 6.6, No. 4) is determined by the difference in the polarizabilities of the anisotropic naphthene ring. Although, owing to the stepladder structure of the molecule, this ring is rigidly bonded to the main chain, just as the imide ring in the poly(isobutyl maleimide) molecule (Table 6.6, No. 1), in contrast to the latter polymer the anisotropy of polyacenaphthylene in chlorobenzene is negative. If the macro- and microform effects are taken into account, the negative intrinsic anisotropy of polyacenaphthylene is found to be one-and-a-half times higher than the values shown in the table, and hence is comparable to the anisotropy of poly(β-vinylnaphthalene) (Table 6.6, No. 6). This polymer is a flexible-chain polymer of the nonladder type in which the naphthene side group is bonded to the main chain by a single bond and the plane of this group is normal to the chain direction (as a result of steric hindrances). This is the structure that ensures very high negative anisotropy of the monomer unit of polyvinylnaphthalene. If the foregoing considerations are taken into account, it becomes evident that the high negative anisotropy introduced into the monomer unit of polyacenaphthylene by the naphthene ring is due to the fact that the plane of this ring is inclined to the chain direction by an angle much larger than 45°. The structure of the repeat unit of the molecule satisfying this condition can be formed when the valence angles are greatly deformed and the structure of the five-membered ring linking the naphthene group to the main chain is not planar [81].

Two possible structures of polyacenaphthylene corresponding to these conditions are shown in Figs. 6.32 and 6.33.

The first of them, with an extended conformation (Fig. 6.32), is a planar trans-chain in which the neighboring naphthene rings are located on different sides of the chain plane. However, in this structure all the bonds of the main chain about which rotation is possible are parallel and the chain should have the "crankshaft" conformation [82] in which the equilibrium rigidity and anisotropy of the molecule should greatly exceed the experimental values obtained for polyacenaphthylene.

The second possible structure of the polyacenaphthylene molecule (Fig. 6.33) leads to a chain conformation of the type of 4_1 helix with nonparallel rotation axes in the chain. Hence, in this case the expected equilibrium rigidity of the chain is lower than for the first type of structure and is in better agreement with the experimental data for polyacenaphthylene.

It is possible that a real polyacenaphthylene molecule is a statistical array of structures of the first and second types. In this case the polymer chain consists of alternating more rigid and more flexible fragments, which may favor the appearance of lyotropic mesomorphism revealed in concentrated polyacenaphthylene solutions [83].

2.4. Polypeptides and Polyisocyanates

2.4.1. Intramolecular Orientational Order. The high equilibrium rigidity of polypeptide molecules (in a helical conformation) and poly(alkyl isocyanate)s implies a very high degree of intramolecular orientational order of their structure [3, 85].

In fact, the degree of uniaxial orientational order Q in a chain molecule may be determined by the well-known equation [86]

$$Q = \left(\overline{3\cos^2\vartheta} - 1\right)\big/2, \tag{6.10}$$

where ϑ is the angle formed by a chain element (monomer unit or segment) and the direction chosen in the molecule as its axis.

A comparison of Eqs. (6.10) and (5.30) or (5.82) gives

$$Q = (\gamma_1 - \gamma_2)/\beta L = (\gamma_1 - \gamma_2)/\beta_M M, \tag{6.11}$$

It follows from Eq. (6.11) that Q is the ratio of the anisotropy of the molecule $\gamma_1 - \gamma_2$ in the conformation investigated to its anisotropy βL in the extended (rodlike) conformation. To put it another way, Q is the ratio of the specific anisotropy of the entire molecule $(\gamma_1 - \gamma_2)/M$ to the specific anisotropy of a chain element β_M.

If the properties of a chain molecule are described by the wormlike chain model, the axis of which is the vector $\mathbf{h}$, then, according to Eq. (5.86) or (5.97), the degree of order may be represented by one of the following ratios:

$$R = (6/5)\,f(x)/x \tag{6.12}$$

or

$$Q = (6/5)(1 - e^{-5x/6})/x, \tag{6.13}$$

where $f(x)$ is defined by Eq. (5.87).

It follows from Eqs. (6.12) and (6.13) that the degree of orientational order in the wormlike chain at low x (in the rodlike conformation) is equal to unity and decreases with increasing chain length (x) tending in the Gaussian range to the limiting value of $Q = 0$ at $x \to \infty$. The higher the chain flexibility, the greater is the rate of decrease in Q with increasing L. Hence, in the investigations of real chain molecules with high equilibrium chain rigidity, which are carried out almost entirely at low x (Figs. 6.14 and 6.15), the value of Q is close to unity. For the fractions of PBLG investigated exhibiting the highest molecular weight the value of Q is not lower than 0.5, and for poly(alkyl isocyanate)s its lowest value is several tenths. For ladder polysiloxanes and cellulose esters and ethers for which the high-molecular-weight range is much more comprehensively represented, the lowest Q values are of the order of 10^{-2}. In real flexible-chain polymers Q usually ranges from 10^{-2} to 10^{-3}.

Hence, for such molecules as polypeptides and poly(alkyl isocyanate)s in the molecular-weight range actually investigated, the degree of order is close to the Q

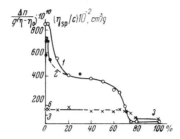

Fig. 6.34. 1, 2) Shear optical coefficient $\Delta n/\Delta \tau$ and 3) reduced viscosity $\eta_{sp}/c \equiv (\eta - \eta_0)/\eta_0 c$ of PBLG solutions ($M = 1.8 \cdot 10^5$) in a mixed dichloroethane–dichloroacetic acid (DCA) solvent at various component concentrations. Curve 2 refers to solutions with the addition of 2% dimethylformamide [88]. The abscissa gives the percentage of DCA in the solvent.

TABLE 6.7. Optical Characteristics of Polypeptide and Polyisocyanate Molecules in Various Conformations

No.	Polymer and solvent	$M \cdot 10^{-5}$	$[\eta] \cdot 10^{-2}$, cm³/g	$(\Delta n/\Delta \tau)_\infty \cdot 10^{10}$, cm s² g⁻¹	$(\alpha_1 - \alpha_2) \cdot 10^{25}$, cm³	$\Delta a \cdot 10^{25}$, cm³	$A \cdot 10^8$, cm	References
1	Poly(γ-benzyl-L-glutamate) (PBLG) in dichloroethane helix	1.8	4.8	+850	+25000	+25	2000	[88]
2	PBLG in dichloroacetic acid (coil)	1.8	1.7	+8.0	+230	+25	18	[88]
3	Poly(butyl isocyanate) (PBI) in tetrachloromethane	2.3	24	+240	+4000	+11	740	[7]
4	PBI in tetrachloromethane + 10% pentafluorophenol mixture	2.3	3.6	+46	+800	+11	150	[90]
5	Poly(tolyl isocyanate) in bromoform	—	0.4	−3.5	−39	−5	15	[91]

values for low-molecular-weight thermotropic liquid crystals [85]. Hence, it is possible to call these molecules "crystal-like" [3, 87].

It is noteworthy that the anisotropy of PBLG molecules and poly(alkyl isocyanate)s is positive, although the monomer units of these molecules contain side radicals in the form of relatively long chains. The positive sign of anisotropy shows that the degree of orientational order in the main chain of these molecules is much higher than that in the side chains and, hence, the contributions of these chains to the anisotropy of the molecule is not very large.

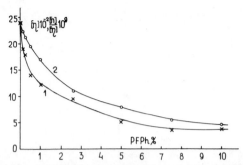

Fig. 6.35. 1) Intrinsic viscosity [η] and 2) shear optical coefficient $[n]/[\eta]$ of poly(butyl isocyanate) solutions ($M = 2.25 \cdot 10^5$) in mixed tetrachloromethane–pentafluorophenol (PFPh) solvents versus amount (in volume %) of PFPh in the solvent [90].

The anisotropy of DNA molecules and monomer units is negative and very high (Fig. 6.16). This is due to the fact that the main contribution to this anisotropy is provided by purine and pyrimidine bases rigidly attached by hydrogen bonds to the double helical backbone of the molecule. In this case the planes of the bases are oriented at angles close to 90° to the axes of helices. Thus, these bases form a system of aromatic rings with a high degree of orientational order.

Since according to Eqs. (6.10)–(6.13) the degree of orientational order is directly related to the optical anisotropy of the molecule, both its value and its change due to the variation in molecular structure are determined by the FB method with the greatest sensitivity. The changes in Q with chain length are evidently determined by the dependence of $[n]/[\eta]$ on M considered in detail in the preceding sections.

2.4.2. Conformational Transformations. The fundamental changes in the structure and conformation of polypeptide molecules after the breaking of intramolecular hydrogen bonds and the helix–coil transition are usually determined by the viscometric method or from the chirality of the polymer, but are much more apparent in the FB phenomenon. The high sensitivity of FB to the destruction of the regular helical structure of the molecule is demonstrated in Fig. 6.34. This figure shows the reduced viscosity and the shear optical coefficient $[n]/[\eta]$ in PBLG solutions as a function of the composition of the mixed solvent, the change in which leads to the helix–coil conformational transition. When the helical structure of the polypeptide molecules is destroyed and the molecule adopts a random-coil conformation, the reduced viscosity decreases by three times, which is an indication of the corresponding decrease in the rotational friction coefficient of the molecule. In this case, the value of $[n]/[\eta]$ remains positive but decreases by two orders of magnitude (from $8.5 \cdot 10^{-8}$ to $8 \cdot 10^{-10}$). According to the FB theory, this very high decrease in $[n]/[\eta]$ (at a constant molecular weight) is the result of an equal decrease in the optical anisotropy $\gamma_1 - \gamma_2$ of the molecule and [according to Eq. (6.11)] of the degree of order Q.

TABLE 6.8. Shear Optical Coefficient $(\Delta n/\Delta\tau)_\infty$, Length of the Kuhn Segment A, and Hindrance Parameter σ for Some Aromatic Polymers

Polymer	Solvent	$(\Delta n/\Delta\tau)_\infty \cdot 10^{10}$, cm s^2 g^{-1}	$A \cdot 10^8$, cm	$A_f \cdot 10^8$, cm	σ	References
Poly(phenylene hydroxydiazole)	Sulfuric acid	135	120	71	1.3	[46]
Poly(phenyl quinoxaline)	Tetrachloroethane, chloroform					[93]
I		140	120	58	1.4	
II		85	70	27	1.6	
III		50	40	21	1.4	
Poly(tetraphenylmethane terephthalamide)	Sulfuric acid	36	130	50	1.6	[94]
Poly(amide benzimidazole)	Sulfuric acid	400	290	170	1.3	[44]
Poly(amide benzoxazole)	Sulfuric acid	460	330	170	1.4	[109]
Poly(para-hydroxyphenyl-benzoxazole terephthalamide)	Sulfuric acid	460	330	170	1.4	[65]
Poly(meta-hydroxyphenylbenzoxazole terephthalamide)	Sulfuric acid	180	130	45	1.7	[65]

If the molecule is represented by a wormlike chain, then the change in $[n]/[\eta]$ accompanying the helix–coil transition may be interpreted using Eq. (5.111), which indicates that despiralization decreases the anisotropy of the segment βA more than one hundred times (Table 6.7). If the value of β (or the anisotropy of the monomer unit Δa) remains invariable, the segment length A for PBLG molecules in dichloroacetic acid (DCAA) is found to be $10 \cdot 10^{-8}$ cm, which corresponds to the rigidity of the molecules of flexible-chain polymers.

This result is confirmed by FB investigations carried out in solutions of a series of PBLG fractions in DCAA. The concentration dependence of FB makes it possible to reveal the macroform effect in these solutions [88], just as is observed for typical flexible-chain polymers (Fig. 5.11).

Similar relationships have been found not only in PBLG solutions but also for other polypeptides investigated by the FB method [88].

All these results show that the helix–coil transition fundamentally changes the structure and conformation of the polypeptide chain, transforming a "crystal-like" molecule into the molecule of a common flexible-chain polymer.

In their rigidity, the hydrodynamic and dynamo-optical properties of poly-(alkyl isocyanate) molecules are close to helical polypeptides and, at least in the low-molecular-weight range, may be regarded as crystal-like molecules. However, as already mentioned (Chapter 4), the nature of their rigidity greatly differs from that of polypeptides since the rigidity of isocyanates is caused by the quasi-conjugation of valence bonds of the amide groups rather than by the secondary structure (hydrogen bonds). Hence, it might be expected that the flexibility of polyisocyanate molecules increases in those cases when the resonance interaction in their amide groups is weakened for some reason.

These changes have been observed in the study of poly(butyl isocyanate) (PBIC) solutions in the mixed solvent system tetrachloromethane–pentafluorophenol (PFPh). The viscosity of solutions decreased several times with an increase in relative concentration of the second component. This has been interpreted as the conformational "phase" transition similar to the helix–coil transition in polypeptides [89].

The investigation of this phenomenon by the FB method has led to the results presented in Fig. 6.35 and Table 6.7.

When the PFPh concentration in the mixed solvent increases, not only the intrinsic viscosity of the solution but also the shear optical coefficient $[n]/[\eta]$ decrease. When the PFPh content is large (25 volume % and higher), these changes are irreversible because of the degradation of the polymer. In solutions with PFPh concentration less than 25%, the molecular weight of PBIC does not vary for a long time and the changes in $[\eta]$ and $[n]/[\eta]$ are reversible.

It should be noted that the curves of the dependence of $[\eta]$ and $[n]/[\eta]$ on solvent composition (Fig. 6.35) are similar, and their limiting values (in the absence of polymer degradation) correspond to a fivefold decrease in both $[\eta]$ and $[n]/[\eta]$ as compared to that in a tetrachloromethane solution. At these limiting values of $[\eta]$ and $[n]/[\eta]$ the anisotropy of the PBIC molecules (and, correspondingly, the degree of intramolecular order) is relatively high and the segment length is $150 \cdot 10^{-8}$ cm, i.e., one order of magnitude higher than that of common flexible-chain polymers.

Hence, the conformational changes which poly(alkyl isocyanate) molecules undergo under the influence of PFPh are not nearly as profound and their consequences are not nearly as catastrophic as in the helix–coil transition in polypeptide molecules. The effect of the addition of moderate amounts of PFPh to the solvent probably results in some weakening of resonance interaction in the amide groups of the chain. This weakening leads to a decrease in the degree of coplanarity of these groups and, correspondingly, to a certain increase in chain flexibility (degree of coiling). In this case the PBIC molecule retains all the properties characteristic of a rigid-chain polymer, in contrast to a polypeptide molecule for which the helix–coil transition means the transformation from a crystal-like molecule into a common flexible-chain polymers.

The introduction of an aromatic ring into the side group of the chain leads to much more profound changes in the optical properties of polyisocyanate molecules. As shown by hydrodynamic data (Chapter 4), the replacement of an alkyl side radical with an aromatic radical in the poly(tolyl isocyanate) (PTI) molecule drastically decreases the rigidity of the main chain to the value corresponding to a flexible-chain molecule (Table 4.10).

These structural changes are even more distinctly revealed in the FB of PTI solutions (Table 6.7). For PTI in bromoform the shear optical coefficient and the segmental chain anisotropy are lower in absolute value by two orders of magnitude than for PBI in tetrachloromethane. Moreover, both $[n]/[\eta]$ and $\alpha_1 - \alpha_2$ *are nega-*

tive in sign. The negative intrinsic anisotropy of the monomer unit Δa for PTI completely corresponds to the optical properties of flexible-chain polymers with phenyl side radicals [84]. In these molecules the main contribution to the negative anisotropy of the molecule is provided by anisotropic side rings whose orientational order does not differ from that of the main chain and is much lower than in the chains of crystal-like molecules.

The profound changes induced by the aromatic side ring in the structure of polyisocyanate molecules are evidently due to a sharp decrease in resonance interaction and a virtually complete disappearance of the coplanar structure of the amide groups of the chain.

2.5. Aromatic Polymers

Data on the equilibrium rigidity of aromatic polymers obtained by FB may be used to characterize their molecular structure and conformations.

2.5.1. Hindrance to Intramolecular Rotation. The structure of the repeat unit of aromatic polymers can be much more complex than that of the monomer units of common flexible-chain polymers (e.g., carbon-chain polymers). However, in many cases it is possible to distinguish "virtual" rotating bonds in their chains, using these bonds for the calculation of the statistical chain size, as has been done, for example, for poly(*meta*-phenylene isophthalamide) (PMPhIPhA) in Chapter 4 (Section 9.1.4).

The simplest structure is that of poly(*para*-phenylene–1,3,4-oxadiazole) (PPhOD) for which the data obtained by FB are shown in Fig. 6.17 and listed in Table 6.8. The repeat unit of this polymer contains one "virtual" bond $\Delta \approx 7 \cdot 10^{-8}$ cm, which, together with that of the neighboring unit, makes the angle $\vartheta \approx 36°$. In this case, in order to calculate the number of monomer units S_f in a segment with unhindered rotation, Eq. (4.25) may be used with δ taken equal to zero. At $\vartheta = 36°$ we obtain $S_f = 10.4$ and, correspondingly, $A_f = 75 \cdot 10^{-8}$ cm. Comparison of this value with the experimental value of $A = 120 \cdot 10^{-8}$ cm in Table 6.8 leads to the value of the hindrance parameter $\sigma = (A/A_f) = 1.3$.

The structure of another aromatic polymer, poly(phenyl quinoxaline) (PPhQ), can be much more complex and varied. This structure depends on both the chemical structure of atomic groups bonding neighboring quinoxaline rings in the chain (structures I, II, and III) and on the stereochemical configuration of the quinoxaline nucleus. The latter is determined by the way in which the quinoxaline ring is attached to the neighboring bonds of the main chain (*trans–trans, cis–cis,* or *trans–cis* bonds). The FB method has been used [93] to investigate the solutions of PPhQ samples I, II, and III whose structures may be represented by the general formula

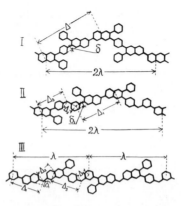

Fig. 6.36. Extended conformations of poly(phenyl quinoxaline) I, II, and III molecules with the *trans*-structure of the quinoxaline nucleus. $\Delta_1, \Delta_2, \Delta_3, ..., \Delta_v$ are the virtual rotating bonds in the repeat unit of the chain. λ is the length of the repeat unit in the chain direction.

where R is a single bond (I, II) or O (III); Ar is $-\langle\bigcirc\rangle-$ (I) or $-\langle\bigcirc\rangle-O-$ $\langle\bigcirc\rangle-$ (II, III).

Extended conformations of the molecules of these polymers with the *trans–cis* stereoisomeric structure of the quinoxaline nucleus are shown in Fig. 6.36. It is clear that for each polymer the repeat unit may be represented by an equivalent chain of linear parts (virtual bonds), $\Delta_1, \Delta_2, ..., \Delta_v$, joined at an angle $\vartheta = 60°$ about which rotation is possible and a certain number of bonds normal to them, δ_i, about which rotation is impossible. Thus, for sample I the number v of bonds Δ is one, for sample II it is three, and for sample III it is four. For other stereoisomeric structures of the quinoxaline heterocycle (*trans–trans* or *cis–cis*) the number of virtual bonds may be different [93].

For molecular structure I in Fig. 6.36, v is one and Eq. (4.25) may be used for calculating S_f and A_f.

For structures II and III, in which v is greater than unity, Eq. (1.16) should be used (assuming δ_i to be approximately zero). The values of A_f calculated in this manner for the three samples investigated (with a mixed trans–cis structure of the quinoxaline nucleus as the most probable structure) are given in Table 6.8.

The experimental values of the equilibrium rigidity of molecules of samples I, II, and III may be obtained using the data on the FB of their solutions in organic solvents [93]. The values of $(\Delta n/\Delta \tau)_\infty = B\beta A$ for these polymers are listed in Table 6.8. Bearing in mind that the anisotropy parameter β for PPhQ should be close to

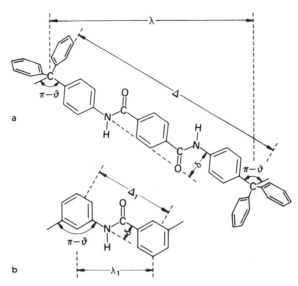

Fig. 6.37. Structure of the repeat unit of poly(tetra-phenylmethane terephthalamide) (a) and poly(*meta*-phenylene isophthalamide) (b).

its values for other aromatic polymers (Fig. 6.17), the value of A may be evaluated by comparing the values of $(\Delta n/\Delta \tau)_\infty$ for PPhQ and for these polymers for which the values of A are known. The values of A obtained by this method and the corresponding values of σ are given in Table 6.8.

It should be noted that the conformational properties of sample I of PPhQ are profoundly affected by the isomeric form of the quinoxaline nucleus. If this is the *trans–trans* form, it can be clearly seen that at any chain conformation all its valence bonds about which rotation is possible are mutually parallel ("crankshaft" conformation [82]) and the equilibrium rigidity should be much higher than the experimental value. This fact implies that even if the *trans–trans* form of the quinoxaline nucleus is present in the molecules of PPhQ I investigated, it is not the main component.

Still et al. [94] have obtained much more regular polymers of the PPhQ I type in the *trans–trans* form. Although the authors do not report data on the conformational properties and rigidity of the molecules of these polymers in dilute solutions, the high values of viscosity and the lyotropic mesomorphism found by the authors in solutions of these polymers doubtlessly indicate that these molecules have an extended shape characteristic of the "crankshaft" conformation.

For the molecules of poly(tetraphenylmethane terephthalamide) (PTPhMTA, Fig. 6.37) [95], the length of the Kuhn segment may also be determined from the value of $(\Delta n/\Delta \tau)_\infty$ (Table 6.8) by comparing this value with that of $(\Delta n/\Delta \tau)_\infty$ and the rigidity of the molecules of PMPhIPhA. In this case it is necessary to take into account the fact that the anisotropy β of the unit length of the PTPhMTA chain is more

TABLE 6.9. Values of Shear Optical Coefficient $(\Delta n/\Delta \tau)_\infty$, Anisotropy of Unit Length β, and Length of the Kuhn Segment A for Some Rigid-Chain Polyamides and Polyesters according to FB Data for Their Solutions

No.	Polymer, solvent	$(\Delta n/\Delta\tau)_\infty \cdot 10^{10},$ cm s^2 g^{-1}	$\beta \cdot 10^{17},$ cm^2	$A \cdot 10^8,$ cm	References
1	Poly(para-benzamide) (PPBA) in sulfuric acid	800 *	17.5	600	[42, 43]
2	Poly(para-phenylene terephthalamide) (PPPhTPhA) in sulfuric acid	400	17.5	300	[64, 110]
3	Poly(meta-phenylene isophthalamide) (PMPhIPhA) in sulfuric acid	28	8	45	[48, 49]
4	Poly(amide hydrazide) (PAH) in dimethylsulfoxide	280	14.5	250	[45]
5	Poly(amide benzimidazole) (PABI) in sulfuric acid	400	18	290	[44]
6	Poly(amide benzoxazole) (PABO) in sulfuric acid	460	18	330	[109]
7	Poly(para-hydroxyphenyl benzoxazole terephthalamide) (PPHPhBOTPhA) in sulfuric acid	460	18	330	[65]
8	Poly(meta-hydroxyphenyl benzoxazole terephthalamide) (PMHPhBOTPhA) in sulfuric acid	180	18	130	[65]
9	Copolymer of PPPhTPhA and PPBA in sulfuric acid	330	17.5	250	[64]
10	Poly(cyclohexane amide) in sulfuric acid	32	2.3	170	[98]
11	Polycaprolactam (nylon 6) (PCL) in sulfuric acid	5.1	3.7	17	[98]
12	Aromatic copolyester I–II ($Z = 0.3$) in tetrachloroethane	90	7.2	154	[47]
13	Aromatic polyester II in tetrachloroethane	19	4.25	55	[47]
14	Para-aromatic polyester I in tetrachloroethane	490 †	8.5	700	[47]

*Obtained by extrapolation to $M \to \infty$ according to Eq. (5.112).
†Obtained by extrapolation of the straight line in Fig. 6.44 to $Z \to 0$.

than twice as low as the corresponding value for PMPhIPhA since the former contains the optically isotropic tetraphenylmethane groups (Fig. 6.37). Hence, although the experimental value of $(\Delta n/\Delta \tau)_\infty$ for PTPhMTA only slightly exceeds that for PMPhIPhA (Fig. 6.17), the length of the segment A for the former is 2.5 times greater than the latter.

The value of A_f corresponding to free rotation of PTPhMTA chains is calculated using Eqs. (4.22)–(4.25) since the repeat unit of its chain consists of one virtual bond. On passing to the neighboring unit this bond is displaced in the perpendicular direction by δ and rotates by the tetrahedral angle $\vartheta = 71°$ (Fig. 6.37). Comparison of the value of A_f obtained here with A permits the determination of the hindrance parameter σ (Table 6.8). Table 6.8 also lists the values of σ calculated from the data in Table 6.9 for molecules of some other aromatic polyamides.

For all the samples listed in Table 6.8, the hindrance parameters are similar in value and, just as for other aromatic polymers (Chapter 4, Section 9), these values lie below those characteristic of typical flexible-chain polymers. This fact implies that, in contrast to cellulose derivatives, in this case the cause of high equilibrium

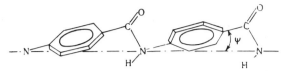

Fig. 6.38. Extended conformation of the poly(*para*-benzamide) chain. Broken line indicates the main direction of the chain; ψ is the angle between the chain direction and the Δ bond about which rotation takes place.

chain rigidity of the aromatic polymers considered is the geometric features of their molecular structure rather than hindrance to intramolecular rotation.

2.5.2.   Chain Anisotropy for Para- and Meta-Structures. The optical anisotropy (and, correspondingly, the sign and value of FB) of aromatic polymers including aromatic polyamides is mainly determined by the type of arrangement of anisotropic aromatic rings providing the major contribution to the anisotropy of the molecule.

Thus, the phenyl ring bonded to the chain as a side radical always provides a negative contribution to its anisotropy, as for example for PTPhMTA (Fig. 6.37 and Table 6.8). However, even in those cases where the aromatic ring is rigidly bonded to the main chain, it can lead to negative anisotropy of the molecule if, on the average, the normal to the ring plane does not form too large an angle with the chain direction. Polyacenaphthylene may serve as an example (Figs. 6.32 and 6.33 and Table 6.6).

The polarizability of the benzene ring is symmetric with respect to the axis normal to the ring plane, being a minimum in the direction of the axis of symmetry, and the difference between the main polarizabilities of benzene is $\Delta b_{\mathrm{B}} = -60 \cdot 10^{-25}$ cm^3. In accordance with this, the average contribution provided by the aromatic ring to the anisotropy of the molecule is given by

$$\overline{\Delta b_{\mathrm{B}}} = \Delta b_{\mathrm{B}} \left[(3 \,\overline{\cos^2 \varphi} - 1)/2 \right] \left[(3 \,\overline{\cos^2 \psi} - 1)/2 \right], \tag{6.14}$$

where φ is the angle formed by the normal of the ring and the virtual bond Δ of the chain which contains the ring, and ψ is the angle between this bond and the main direction of the chain.

In aromatic polyamides the aromatic rings (or hetero-rings) are separated along the chain by amide groups, and hence the arrangement and orientation of these rings in the molecule depend to a great extent on the structure and conformation of the amide groups. The analysis of experimental data on FB provides conclusive evidence of the fact that the main structural form of the amide groups in the chains of aromatic polyamides is a planar *trans*-form, while the adoption of cis-conformations in noticeable amounts is ruled out [82]. Accordingly, the main conformation of the chain of para-aromatic polyamide is the "crankshaft" conformation in which all the rotation axes in the chain are approximately parallel to each other (Fig. 6.38).

In both para- and meta-aromatic polyamides (Fig. 4.53) the virtual bond Δ lies in the plane of the corresponding phenyl ring. Hence, φ is equal to $\pi/2$ and Eq. (6.14) becomes

$$\overline{\Delta b_B} = -\Delta b_B \left(3 \overline{\cos^2 \psi} - 1\right)/4. \tag{6.15}$$

The angle ψ formed by the axis Δ and the direction of the extended chain of a para-aromatic polyamide (Fig. 6.38) is close to the value $\psi \approx 12°$ and, correspondingly, according to Eq. (6.15) for this polymer we have $\Delta b_B = -0.47 \Delta b_B$.

In the case of PMPhIPhA, an aromatic polyamide in which all the phenyl rings in the chain are in the meta-position (Fig. 4.53), we have $\psi = \vartheta/2 = 30°$ and, in accordance with Eq. (6.1), for this polymer the contribution of the phenyl ring to the anisotropy of the monomer unit Δb_B is $-0.31 \Delta b_B$. Hence, the phenyl ring inserted in both the para- and the meta-position introduces positive anisotropy into the polyamide chain, but it is greater by a factor of 1.5 in the former case than in the latter.

As already mentioned (e.g., Section 1.3.4) in such solvents as sulfuric acid the microform effect provides an important positive contribution to the anisotropy of an aromatic polymer. It might be expected that the contribution of this effect will also be greater for para- than for meta-aromatic units because of the more extended shape of the former.

Hence, it may be expected that the specific positive anisotropy β_M of para-aromatic polyamides should be higher than β_M for those polyamides in which the phenyl rings (or most of them) are incorporated into the chain in the meta-position. This expectation is confirmed by the data presented in Fig. 6.17 and Table 6.9. These data show that for all para-aromatic polyamides the values of $(\Delta n/\Delta \tau)_\infty$ are approximately proportional to the lengths of the Kuhn segment A, whereas for PMPhIPhA the value of $(\Delta n/\Delta \tau)_\infty$ is lower than for PPPhTPhA by one order of magnitude, but the rigidities of these polymers differ fivefold.

2.6. Ring-Containing and Linear Aliphatic Polyamides

The extended shape of the molecules typical of para-aromatic polyamides can also be retained if the aromatic rings are replaced by other groups, provided these groups ensure an approximate parallelism of rotation axes in the chain. The dynamo-optical properties of molecules of polycyclohexaneamide (PCHA), a ring-containing aliphatic polyamide, may serve as an example confirming this statement.

This polymer can be obtained as a hydrated analog of poly(*para*-benzamide) (PPBA) ($-NHC_6H_{10}CO-$) or poly(*para*-phenylene terephthalamide) (PPPhTPhA) ($-NHC_6H_{10}NHCOC_6H_{10}CO-$). For this purpose, anionic polymerization is used in the former case [96] and high-temperature polycondensation is employed in the latter [97].

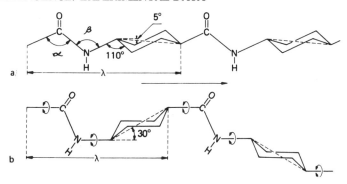

Fig. 6.39. Conformation of the PCHA chain. a) Extended "crankshaft"; b) coiled "crankshaft."

Comparative investigations [98] of FB for PCHA solutions in sulfuric acid and solutions of a linear aliphatic polyamide, polycaprolactam (PCL, nylon 6), in the same solvent have shown that the shear optical coefficient $(\Delta n/\Delta\tau)_\infty$ for the former is six times higher than for the latter, although the anisotropy per unit length of the chain β is lower by a factor of 1.5 for PCHA than for PCL (Table 6.9). The use of these data and the equation $(\Delta n/\Delta\tau)_\infty = B\beta A$ leads to a value of A for the PCHA molecules which exceeds by one order of magnitude that for PCL (Table 6.9), and is typical of a rigid-chain polymer.

In contrast, PCL is a typical flexible-chain polymer in its dynamo-optical characteristics, which corresponds to the data obtained by light scattering [99].

The results obtained for PCHA correspond to the structure of its molecules shown in Fig. 6.39. Although in the PCHA molecules the phenyl rings are replaced by cyclohexane rings, if the bond angles α and β in the amide group are taken to be equal, it can be clearly seen that the rotating bonds in the chain remain parallel, just as in the chain of a para-aromatic polyamide, regardless of whether the "chair" or "boat" shape is assumed by the cyclohexane ring. Hence, the PCHA molecule exhibits the "crankshaft" structure in which the chain dimensions in the direction parallel to the rotation axes increase proportionally to molecular weight regardless of whether the chain adopts the "extended" or "coiled" conformation. Both conformations correspond to a high degree of intramolecular orientational order revealed by the high values of $\Delta n/\Delta\tau$ and A.

2.7. Rigid-Chain Copolymers

In the copolymerization of flexible-chain molecules the chain rigidity of the copolymers is close in order of magnitude to that of the components undergoing copolymerization since these rigidities usually do not differ by more than a few tens of a percent.

For rigid-chain copolymers the situation is quite different because, in contrast to flexible-chain molecules, rigid-chain molecules may differ in their equilibrium flexibility by one order of magnitude or even more, depending on their structure

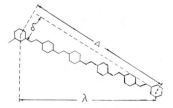

Fig. 6.40. Model for a repeat unit of the chain of the copolymer of meta- and para-aromatic polyamide at $Z = 0.2$. Δ and δ are the lengths of virtual bonds and λ is the unit length in the chain direction.

(see, e.g., Fig. 6.17). Hence, it may be expected that the flexibility of the copolymer containing rigid-chain components will be intermediate between the values for the components and may differ greatly depending on their structure and content in the copolymer.

The character of these variations may be clearly demonstrated using the copolymers of para- and meta-aromatic polyamides.

2.7.1. Para-Meta-Aromatic Copolyamides. Additivity of Flexibilities [100]. If the structures of all the repeat units of the chain of a para–meta-aromatic copolyamide are assumed to be identical, then the structure of this chain may be represented in the form shown in Fig. 6.40. In this scheme all the amide groups are in the planar conformation and the bond angles at the carbon and nitrogen atoms of the amide group, α and β, are assumed equal. The composition of the copolymer is determined by the ratio $Z = n/(m + n)$, where n and m are the numbers of phenyl rings in its chain in the meta- and para-positions, respectively. The assumed approximations ($\alpha = \beta$ and the rigid *trans*-structure of the amide groups) are equivalent to the assumption that copolymer chain flexibility is determined only by the meta-aromatic component, whereas the para-aromatic component is inserted into the chain as ideally rigid sequences of a straight crankshaft.

Each repeat unit of the chain shown in Fig. 6.40 consisting of $1/Z$ monomer units of polyamide (phenyl rings) can evidently be replaced by two virtual bonds $\Delta = \Delta_m Z$ and $\delta = \delta_m/Z$, where Δ_m and δ_m are the Δ and δ virtual bonds in the PMPhIPhA chain in Fig. 4.53. Hence, the number of "monomer units" (repeated structural units) S in the Kuhn segment for a chain of para–meta-copolyamide with free rotation can evidently be calculated according to Eq. (4.25) where, just as for PMPhIPhA, the angle ϑ is $60°$ and $\delta/\Delta = 0.2$.

Hence, it is clear that the number S obtained for the polymer will be the same as for its flexible component PMPhIPhA. However, the length of the repeat unit λ and that of the virtual bond Δ in the copolymer are $1/Z$ times greater than in PMPhIPhA. Similarly, according to Eq. (4.24) the length A_Z of the Kuhn segment for the copolymer with free rotation is given by

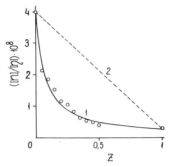

Fig. 6.41. Dependence of $[n]/[\eta]$ on Z $[Z = n/(n + m)$ is the relative content of *meta*-phenyl rings in the chain] for solutions of para–meta-aromatic copolymers in sulfuric acid [100]. Points represent experimental data. Straight line 1 corresponds to the additivity of rigidities of the components, and curve 2 corresponds to the additivity of their flexibilities.

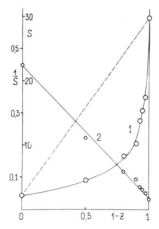

Fig. 6.42. Number of monomer units S in the Kuhn segment (curve 1) and the value of $1/S$ (curve 2) for the copolymers of caprolactam (CL) and cyclohexaneamide versus molar fraction of Z of the flexible component (CL) in the copolymer.

$$A_Z = \Delta[(\delta/\Delta)^2 + (1 + \cos\vartheta)/(1 - \cos\vartheta)]/[\cos(\vartheta/2) + (\delta/\Delta)\sin(\vartheta/2)], \quad (6.16)$$

where δ/Δ and ϑ for the copolymer coincide with their values for PMPhIPhA, whereas Δ and, correspondingly, the length of the Kuhn segment for the copolymer, are $1/Z$ times greater than for PMPhIPhA.

The result may be represented in the following form:

$$1/A_Z = Z(1/A_1), \quad (6.17)$$

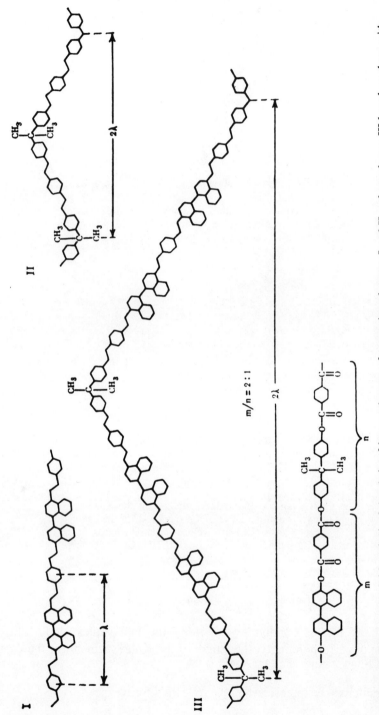

Fig. 6.43. Repeat units (of length λ) of the chain of homopolymers of aromatic polyesters I and II and copolymer III based on them with the composition (molar fraction of the flexible component) $Z = n/(n + m) = 1/3$. n and m are the molar fractions of I (rigid) and II (flexible) components in the copolymer, respectively.

where A_Z and A_1 are the lengths of the Kuhn segment for the copolymer chain and its flexible component (PMPhIPhA), respectively.

Equation (6.17) shows that the meta-aromatic component in the copolymer introduces flexibility (expressed by the value of $1/A$) into the copolymer chain additively, i.e., proportionally to its flexibility $1/A_1$ and its molar fraction Z in the copolymer.

Equation (6.17) is approximate because it does not take into account the chain flexibility of a real para-aromatic component which, although being small compared to $1/A_1$, is finite and characterized by the value of $1/A_0$, where A_0 is the length of the Kuhn segment for the chain of para-aromatic homopolyamide (at $Z = 0$). Equation (6.17) may be corrected if the flexibility of para-aromatic chain sequences is taken into account and, according to the additivity rule of flexibilities, if the second term proportional to $1/A_0$ and the molar content $(1 - Z)$ of the second component is introduced into the right side of Eq. (6.17). Then Eq. (6.17) becomes

$$1/A_z = Z\,(1/A_1) + (1 - Z)\,(1/A_0). \tag{6.18}$$

The validity of Eq. (6.18) has been checked in the investigation of FB in solutions of the PPPhTPhA and PMPhIPhA copolymers differing in the composition of the components [100]. The results are shown in Fig. 6.41 as the dependence of the shear optical coefficient $[n]/[\eta]$ on the molar fraction Z of the flexible-chain component (PMPhIPhA) in the copolymer. The experimental points illustrate a sharp decrease in $[n]/[\eta]$ with increasing Z, which agrees with theoretical curve 2 plotted according to Eq. (6.18) with the simplifying assumption that $[n]/[\eta]$ is proportional to copolymer chain rigidity. Straight line 2 represents the dependence of $[n]/[\eta]$ on Z assuming that the shear optical coefficients (and correspondingly the rigidities) of components in the copolymer are additive. This assumption sharply contradicts the experimental data shown in Fig. 6.41, which indicate, at least qualitatively, the validity of Eq. (6.18) and demonstrate the possibility of drastically decreasing the rigidity of a para-aromatic polymer by adding small amounts of the more-flexible-chain components. This possibility is of great practical importance for the modification of physicomechanical properties of rigid-chain polymers. In particular, this method is widely used in the synthesis of polymer with thermo- and lyotropic–mesogenic properties [101].

2.7.2. Copolymers of Cyclohexaneamide and Caprolactam. The relationships analogous to those described in the preceding section have been established in the investigation of FB in sulfuric acid solutions of copolymers of cyclohexaneamide and caprolactam [98]. As has been shown in Section 2.6 and according to the data listed in Table 6.9, the rigidity of molecules of the PCHA homopolymer [and the value of $(\Delta n/\Delta \tau)_\infty$ in its solutions] is much higher than that for the PCL homopolymer. In accordance with this, the values of $([n]/[\eta])_z$ for copolymers are intermediate between those for homopolymers and are profoundly affected by the composition of the copolymer, determined as the molar fraction Z of the flexible component. In order to relate quantitatively the experimental values of $([n]/[\eta])_\infty$ to the rigidity of the copolymer, it should be considered that with variation in composition Z not only does the flexibility of the copolymer chain change, but also its optical anisotropy (determined by the anisotropy of unit length β or by that of the monomer unit Δa). For this purpose, the following equations should be used:

$$([n]/[\eta])_1 = B\beta_1 A_1 = B\Delta a_1 S_1, \quad ([n]/[\eta])_0 = B\beta_0 A_0 = B\Delta a_0 S_0,$$
$$([n]/[\eta])_Z = B\beta_Z A_Z = B\Delta a_Z S_Z, \tag{6.19}$$

where the subscripts show that the corresponding value refers to a more flexible ($Z = 1$) or more rigid ($Z = 0$) homopolymer, or to a copolymer (of the composition Z). Moreover, the shear optical coefficients in Eq. (6.19) are the limiting values $(\Delta n/\Delta\tau)_\infty$.

The values of β_Z or Δa_Z for the copolymers investigated are unknown. However, they may be calculated from the values of β_1 and β_0 for PCL and PCHA (listed in Table 6.9) if Eqs. (6.20) or (6.21) assuming the additivity of component anisotropies are used:

$$\beta_Z = \beta_1 Z + (1 - Z)\beta_0, \tag{6.20}$$
$$\Delta a_Z = \Delta a_1 Z + (1 - Z)\Delta a_0. \tag{6.21}$$

The values of S_Z of the monomer units in the Kuhn segment have been determined according to Eq. (6.19) for copolymers with various compositions Z using the values of Δa_Z obtained in this manner and the experimental values of $([n]/[\eta])_Z$ for the copolymers. These values of S_Z are shown as a function of Z by points on curve 1 in Fig. 6.42. The experimental points and curve 1 show that the equilibrium rigidity of the copolymer undergoes a sharp decrease when small amounts of the flexible component are introduced. The deviation of the curve from the broken straight line shows that the rigidities of components are not additive.

The points near curve 2 in Fig. 6.42 representing the experimental values of $1/S_Z$ are in agreement with the theoretical straight line corresponding to the dependence

$$1/S_Z = Z(1/S_1) + (1 - Z)(1/S_0), \tag{6.22}$$

equivalent to Eq. (6.18).

Hence, the experimental data on FB in solutions of PCL and PCHA copolymers confirm quantitatively the additivity rule of flexibilities of the components in the chains of these copolymers.

2.7.3. Copolymers of Aromatic Esters. Hydrodynamic and dynamo-optical properties of an aromatic copolymeric ester (ACPE) have been dis-

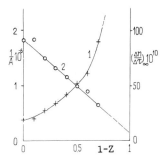

Fig. 6.44. 1) Shear optical coefficient $(\Delta n/\Delta\tau)_\infty$ and 2) parameter of chain flexibility $1/A$ for copolymers of aromatic esters in tetrachloroethane versus copolymer composition Z.

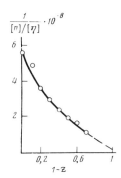

Fig. 6.45. Dependence of $1/([\eta]/[\eta])_\infty$ on $1 - Z$ for solutions of aromatic polyesters in tetrachloroethane. Points represent experimental data, and solid curve represents the theoretical dependence according to Eq. (6.23) at the values of the rigidity and anisotropy parameters of the components given in Table 6.9.

cussed in Chapter 4, Section 10 and Chapter 6, Section 1.34. The corresponding data are presented in Tables 4.12 and 6.9 and in Fig. 6.17. The structures of the repeat units of components I and II are shown in Fig. 6.43. Component I is a totally para-aromatic polyester (containing phenyl and naphthene rings) and component II contains not only the phenylene ester groups but also the diphenylpropane groups, ensuring its greater flexibility over component I.

If a simplifying assumption is made that the structures of all the repeat units in the chain of the ACPE copolymer are homogeneous, a fragment of its chain with the molar fraction of the flexible component (with the composition) $Z = n/(n + m) = 1/3$ can be represented as III in Fig. 6.43. The composition of ACPE I–II, the data for which are given in Tables 4.12 and 6.9 and shown in Fig. 6.17, is $Z = 0.3$.

Apart from this copolymer, FB has also been investigated in solutions of a series of similar copolymers differing in composition [47]. The values of $(\Delta n/\Delta\tau)_\infty$

obtained in tetrachloroethane for the series investigated are shown in Fig. 6.44 by curve 1 as a function of $1 - Z$. As for other rigid-chain copolymers, in ACPE solutions $(\Delta n/\Delta \tau)_\infty$ increases drastically with decreasing molar fraction Z of the flexible-chain component, tending to the limiting high value (at $Z = 0$) for the para-aromatic polyester (curve 1 in Fig. 6.44). However, it is evidently impossible to determine this limit by extrapolating curve 1 to $Z = 0$.

The procedure of extrapolating the value of $1/([n]/[\eta])_\infty$ to $Z \to 0$ seems more suitable to the determination of $(\Delta n/\Delta \tau)_\infty \equiv ([n]/[\eta])_\infty$ of a para-aromatic polyester since the value $1/([n]/[\eta])_\infty$ in the first approximation characterizes the flexibility of the copolymer, varying linearly with Z according to the additivity rule. Figure 6.45 shows the corresponding plot for the copolymers considered here. The extrapolation of experimental data (points) in Fig. 6.45 to $Z \to 0$ allows the evaluation of the limiting value $([n]/[\eta])_{Z \to 0}^\infty$ but the precision of this evaluation is not high because the dependence of $1/([n]/[\eta])_\infty$ on Z is nonlinear.

The nonlinearity of this dependence in Fig. 6.45 may be associated with the fact that when Z is varied, not only the rigidity of the copolymer (A) but also the anisotropy β of unit length of its chain undergoes a change. This seems natural if it is taken into account that in structure III in Fig. 6.43 the average number of aromatic rings per unit length increases with decreasing Z.

A combination of Eqs. (6.18), (6.19), and (6.20) gives for the copolymer the equation

$$([n]/[\eta])_Z^{-1} = (1/B\beta_0 A_0)\,\{1 - Z\,[1 - (A_0/A_1)]\}/\{1 - Z\,[1 - (\beta_1/\beta_0)]\}. \qquad (6.23)$$

It follows from Eq. (6.23) that the linear dependence of $([n]/[\eta])_Z^{-1}$ on Z might be expected only if the anisotropies β_1 and β_0 of the components are equal. It also follows from Eq. (6.23) that for positively anisotropic components at $\beta_1 > \beta_0$ (flexible component is more anisotropic than rigid component) the curve of the dependence of $([n]/[\eta])_Z^{-1}$ on Z has a downward curvature, whereas at $\beta_1 < \beta_0$ it has an upward curvature. The former case has already been mentioned in the discussion of PCL and PCHA copolymers in the preceding section. In contrast to this, the type of curve in Fig. 6.45 corresponds to the condition $\beta_0 > \beta_1$.

If a certain value of $\beta_1/\beta_0 < 1$ is chosen, and experimental values of $\Delta n/\Delta \tau$, shown by points on curve 1 in Fig. 6.44, as well as values of β and A for the copolymer for which $Z = 0.3$ (No. 12 in Table 6.9) are used, it is possible to determine for all the polymers investigated β_Z according to the additivity rule of anisotropies [Eq. (6.20)] and the values of A_Z according to Eq. (6.19).

If the β_1/β_0 ratio is chosen correctly, the values of $1/A_Z$ will have a linear dependence on Z according to the additivity rule of flexibilities {Eq. (6.18)].

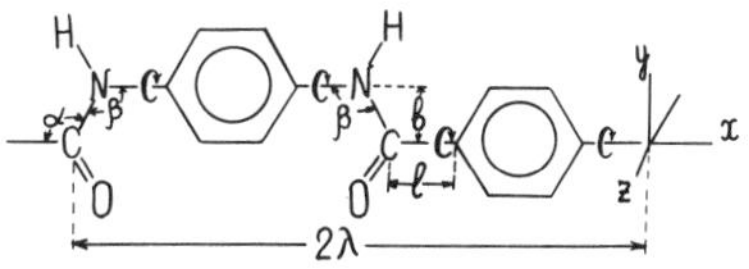

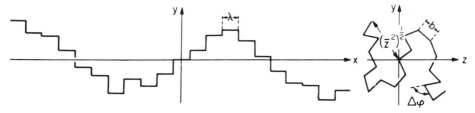

Fig. 6.46. Idealized structure of the repeat unit (length λ) of the poly(*para*-phenylene terephthalamide) (PPPhTPhA) chain.

Fig. 6.47. "Crankshaft" chain conformation of a para-aromatic polyamide.

This dependence is adequately obeyed at $\beta_0/\beta_1 = 2$, as is demonstrated by straight line 2 in Fig. 6.44. The extrapolation of this line to $Z \to 0$ makes it possible to determine the length of the Kuhn segment for para-aromatic polyester I whose data are given in Table 6.9 (No. 14). The value of A obtained here is greater by one order of magnitude than the segment length for homopolyester II containing a flexible propane fragment (No. 13 in Table 6.9). Comparison of No. 14 and No. 1 in Table 6.9 shows that the chain rigidity of para-aromatic polyester is close to that of PPBA, the most rigid of all known para-aromatic polyamides.

When the values of β_1, A_1, β_0, and A_0 are substituted into Eq. (6.23), according to this equation the theoretical dependence of $([n]/[\eta])^{-1}$ on Z is described by the solid curve in Fig. 6.45. The coincidence of experimental points and the theoretical curve demonstrates that the flexibilities and anisotropies of components in the molecules of rigid-chain copolymers are additive.

2.8. Flexibility of Molecules of Aromatic Polyamides

2.8.1. Polyamides with *para*-Phenylene Rings in the Chain. The chain of the para-aromatic component of copolymers of para–meta-aromatic amides is shown for simplicity in Fig. 6.40 as an ideally rigid linear sequence of monomer units. This conformation is possible in the case of an idealized molecular structure in which all the amide groups of the chain are in the planar *trans*-conformation and the α and β angles at the carbon and nitrogen atoms of these groups are equal (Fig. 6.46).

In this structure all the axes C_{ar}–N and C_{ar}–C about which rotation in the chain is possible are parallel, although they may not lie on the same straight line. The chain exhibits the "crankshaft" conformation [82], characterized by an ideal orientational order in the molecule in the direction of the rotation axes (axis x in

Figs. 6.46 and 6.47). In this chain both the geometrical dimensions in the direction of the x axis and the optical anisotropy of the molecule should increase proportionally to the degree of polymerization as in an ideal "rodlike" structure.

In directions y and z (Fig. 6.47) normal to the x axis linear order is absent because the chain rotations about para-aromatic axes are random. Hence, the chain projection on the yz plane may be represented by a chain of freely jointed links b – projections of the C–N amide bonds on the yz plane. The transverse dimensions of this model characterized by the value of $(r^2)^{1/2}$ in Fig. 6.47 are evidently proportional to the square root of the degree of polymerization. Hence, the "degree of asymmetry" of this model proportional to $L/(r^2)^{1/2}$ (L is the chain length in the x direction) at relatively high L is high, just as for a "rodlike" molecule.

It can be clearly seen that the molecules retaining this structure should exhibit exceptionally high equilibrium rigidity, which is observed to a certain extent for para-aromatic polyamides. However, both the hydrodynamic and optical properties of these molecules determined experimentally quite definitely demonstrate the existence of flexibility in their chains, quantitatively expressed in the finite values of the Kuhn segments of para-aromatic polyamides.

The main mechanism for the flexibility of molecules of para-aromatic polyamides is due to the difference in the values of the bond angles α and β at the carbon and nitrogen atoms of the amide group. If β differs from α, then when we move along the chain by one monomer unit, not only is the axis of internal motion displaced by the value of $\delta = b$, but it also changes its direction by the angle $\beta - \alpha$, which leads to "bending" of the chain (Fig. 6.48). The rigidity of this chain may be characterized by the number of monomer units S_β in the Kuhn segment using [100] Eq. (6.24), which is analogous to Eqs. (4.25) and (6.16),

$$S_\beta = \sigma^2 \{(\delta/\Delta)^2 + [1 + \cos(\beta - \alpha)]/[1 - \cos(\beta - \alpha)]\}/\{\cos[(\beta - \alpha)/2] + (\delta/\Delta)\sin[(\beta - \alpha)/2]\}^2, \quad (6.24)$$

where, as previously, $\sigma^2 = S_\beta/S_f$ characterizes the hindrance to rotation about the bonds C_{ar}–C and C_{ar}–N in the polyamide chain, and δ/Δ is the ratio of lengths of virtual bonds for the chain of an aromatic polyamide equal to 0.2.

The possible difference between the bond angles, $\beta - \alpha$, in the amide group lies in the range 6–12° [102]. In combination with Eq. (6.24), this leads to the possible values of S_β in the range of $S_\beta/\sigma^2 = 370 - 90$ or (at $\lambda = 6.5 \cdot 10^{-8}$ cm) to the length of the Kuhn segment A in the range of $A/\sigma^2 = (2400 - 600) \cdot 10^{-8}$ cm.

The values of A obtained here are too high because even with complete freedom of rotation or complete symmetry of hindrance potentials ($\sigma = 1$) the minimum possible value of $A = 600 \cdot 10^{-8}$ cm is twice as high as the experimental value of A (see Table 6.9) for PPPhIPhA and other para-aromatic polyamides.

Hence, the flexibility mechanism arising from the inequality of angles α and β cannot be considered to be the only mechanism for polymers of this class.

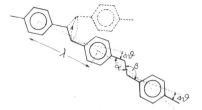

Fig. 6.48. Chain bending of a para-aromatic polyamide with unequal α and β angles at the carbon and nitrogen atoms of the amide group.

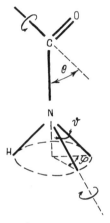

Fig. 6.49. Deformation of the amide group distorting its coplanar structure.

Another mechanism of flexibility whose appearance may be expected in polyamide chains can be provided by the deviation of the amide group configuration from the planar *trans*-structure. These deviations may arise in the process of thermal chain motion leading to random deformations of bond angles and bonds of the amide groups. Without the existence of this "deformational" mechanism it would be difficult to explain the finite flexibility of poly(alkyl isocyanate) chains whose molecules consist completely of amide groups.

The deviation of the amide group *trans*-structure from coplanarity may be the result of the rotation of N–phenyl or C–phenyl bonds by the azimuth φ with the retention of the invariable bond angle $\pi - \theta$ in the main chain (Fig. 6.49). As a consequence of this amide group deformation, two neighboring para-aromatic axes about which rotation in the chains of aromatic polyamide is possible (shown by arrows) form an angle ϑ with each other with a value determined by the equation

$$\sin(\vartheta/2) = \sin\theta\sin(\varphi/2). \qquad (6.25)$$

If, as a result of thermal motion of the chain, the angle of deviation from coplanarity of the amide groups attains the average value of φ, the chain acquires a flexibility characterized by the number of monomer units S_d in a segment and determined by Eq. (6.24) with the substitution of ϑ from Eq. (6.25) for $\beta - \alpha$. At low values of φ and correspondingly of ϑ, the combination of Eqs. (6.24) and (6.25) makes it possible to write approximately

$$S_d = \sigma^2/\sin^2(\varphi/2)\sin^2\theta. \tag{6.26}$$

The order of the average value of the deviation angle φ of the amide group from the *trans*-conformation in the process of thermal rotational vibrations may be evaluated from the equation

$$\langle\cos\varphi\rangle = \left\{\int_{-\pi/2}^{\pi/2}\exp[-U_0(1-\cos\varphi)/2RT]\cos\varphi\,d\varphi\right\}\Big/\left\{\int_{-\pi/2}^{\pi/2}\exp[-U_0(1-\cos\varphi)/2RT]\,d\varphi\right\},$$

$$\tag{6.27}$$

where $U_0 = 21$ kcal/mole is the energy of conjugation of the amide group maintaining its planar *trans*-structure [103]. At $T = 300°K$, we obtain from Eq. (6.27) the value $\langle\cos\varphi\rangle = 0.9711$ and, correspondingly, $\varphi \approx 14°$. The substitution of this value of φ into Eq. (6.26) at $\theta = 60°$ gives $S_d = 90\sigma^2$, which corresponds to the length of the Kuhn segment of a para-aromatic polyamide A_d equal to $590 \cdot 10^{-8}\sigma^2$ cm.

In real chains of para-aromatic polyamides both of the above-mentioned flexibility mechanisms exist: the structural mechanism arising from the inequality of angles α and β in the amide group and the deformational mechanism due to deformation of this group during thermal chain motion. Assuming the additivity of these mechanisms, we obtain, by analogy with Eqs. (6.18) or (6.22), the expression for the number of monomer units S in the chain segment of a para-aromatic polyamide

$$1/S = 1/S_\beta + 1/S_d, \tag{6.28}$$

where S_β and S_d are determined from Eqs. (6.24) and (6.26).

If it is assumed that $\beta - \alpha \approx 10°$, then according to Eq. (6.24) $S_\beta = \sigma^2 \cdot 130$ and $A_\beta = \sigma^2 = 845 \cdot 10^{-8}$ cm. Under these conditions, Eq. (6.28) gives $S = 53$ for σ^2 and $A = 346 \cdot 10^{-8}\sigma^2$ cm.

If these theoretical values for the rigidity parameters are compared with experimental data for the most rigid para-aromatic polyamides and polyesters with $A \approx (600–700) \cdot 10^{-8}$ cm, according to Table 6.9 an agreement exists between them at reasonable value of the hindrance parameter $\sigma = 1.4$. A more detailed comparison of experimental and theoretical values of A is meaningless because a reliable value of $\beta - \alpha$ is unknown. Hence, it may be assumed that for these polymers the experimentally observed chain flexibility may be explained, at least qualitatively, by the combined action of two mechanisms, the structural and deformational mechanisms represented by Eq. (6.28). In this case, the contributions provided by these two mechanisms to chain flexibility are approximately equal.

For another para-aromatic polyamide, PPPhTPhA, the experimental values of A (Table 6.9) are lower than the minimum values of A (at $\sigma = 1$) predicted by Eq. (6.28) and, therefore, the flexibility of its chains cannot be due exclusively to the action of the two mechanisms considered above. It is possible that a considerable influence on the conformational properties of PPPhTPhA molecules is exerted by certain "defects" in the molecular structure determined mainly by the presence of amide groups in *cis*-conformations. Even a negligible quantity of these "defective" groups should drastically decrease the rigidity of the para-aromatic polyamide chain. It is possible that the conditions of chemical synthesis (polycondensation) play an important part in the formation of defective structures. The existence of these structures may lead to different conformational properties of the chains with the same "average" chemical structure and composition.

It also seems likely that the decrease in the regularity of structure of a para-aromatic polymer may lead to decreasing equilibrium rigidity of the chain. This possibility is illustrated by the data reported for the copolymer of PPPhTPhA and PPBA (Table 6.9, No. 9), whose rigidity is slightly lower than that of PPPhTPhA molecules, although the copolymer has been obtained by the addition to PPPhTPhA of a more rigid component 2, PPBA. In this connection it might be of importance that, in the absence of any defects, the structure of the PPPhTPhA chain is less regular than that of the PPBA chain. However, when the conformational characteristics of PPBA and PPPhTPhA molecules are compared, it should be borne in mind that although the stereochemical structure of monomer units of these molecules is very similar, they differ greatly in their dipolar properties: the PPBA molecule exhibits a considerable component of the dipole moment parallel to the chain and in PPPhTPhA molecules this component is absent.

The differences in the effect of the solvent (sulfuric acid) on the conformational properties of the PPBA and PPPhTPhA molecules can also be of great importance. According to the viscometric data obtained by Millaud and Strazielle [108], the conformation of the PPPhTPhA molecules in chlorosulfonic and methanesulfonic acids is more extended than in sulfuric acid and is characterized by a length of the Kuhn segment close to that for PPBA in sulfuric acid. Hence, the comparison of the conformational characteristics of PPPhTPhA and PPBA presented here refers only to those conditions when both these polymers are dissolved in the same solvent: sulfuric acid.

Flory et al. [104] have calculated the rigidity parameters of the chains of para-aromatic polyamide and polyester using a system of virtual bonds differing somewhat from that used here and without taking into account the deformability of the amide group ($S_d = \infty$). Under this condition, chain rigidity is determined only by the value of S_β, and at $\sigma = 1$ the values of S and A from [104] coincide with those calculated here according to Eq. (6.24).

In these papers [104, 105], the energy of dipole interaction between the neighboring amide groups of the chain is considered to be a potential hindering rotation in this chain. The results of calculations yield the values of $\sigma = 1.15$–1.30 which are in agreement with the above value of $\sigma = 1.4$ based on the experimental data on the rigidity of the molecules considered.

It should be recalled that the low value of σ obtained for rigid-chain polyamides from experimental data on hydrodynamic properties and FB of their solutions does not necessarily imply that their type of intramolecular rotation is close to "truly free rotation." In principle, a low value of σ can correspond to the existence of high potential barriers hindering rotation if the potential hindrance curve is relatively symmetrical. However, even in those cases when these rotations are "truly free," in para-aromatic polymers they cannot lead to a considerable deviation of the shape of the molecule from "rodlike" because the axes of these rotations are approximately parallel. The duration of the extended conformation of the molecule is mainly determined by the stability of bond angles and bonds of the chain rather than by the type of intramolecular motions. Hence, regardless of the degree of hindrance to intramolecular motions, the molecule of para-aromatic polyamide of polyester maintaining its extended shape behaves as a kinetically rigid chain.

On the other hand, the kinetics of intramolecular rotations and the character of potential hindrance curves in the molecules of various para-aromatic polyamides and polyesters may be of great importance for determining other properties of these polymers, such as their capacity to generate lyotropic mesomorphism. These problems have been considered by Lauprêtre and Monnerie [106, 107].

2.8.2. Polyamides with Para-, Meta-, and Heteroaromatic Rings in the Chain. The mechanisms of flexibility described by Eqs. (6.24) and (6.26) are characteristic not only of para- but also of meta-aromatic polyamides since both contain amide groups. However, the part played by these mechanisms in the flexibility of meta-aromatic chains is much less important and, in many cases, may be neglected in practice as we did in the case of PMPhIPhA: we restricted ourselves to Eq. (4.25) in the calculation of S for this polymer.

In the general case, if all the mechanisms determining chain flexibility are taken into account, the number S of repeat units in the Kuhn segment can be calculated from the equation

$$1/S = \sum_{i}^{\nu} (1/S_i), \tag{6.29}$$

in which all the parameters of flexibility $1/S_i$ introduced by various elements of the repeat unit are summed.

Thus, for PMPhIPhA, the value of ν is three. In this case we have $S_1 = \sigma^2 \cdot S_f$, where S_f is calculated according to Eq. (4.25), and $S_2 = S_\beta$ and $S_3 = S_d$ are determined from Eqs. (6.24) and (6.26), respectively. In accordance with this, $S_1 = 3.3\sigma^2$ (Chapter 4, Section 9.1.4), $S_2 = 130\sigma^2$ (at $\beta - \alpha = 10°$), and $S_3 = 90\sigma^2$. Hence, the total contribution of the two latter mechanisms to overall chain flexibility is less than 5%.

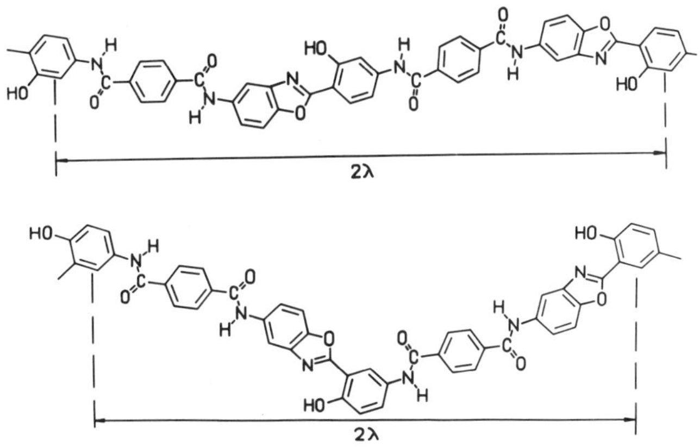

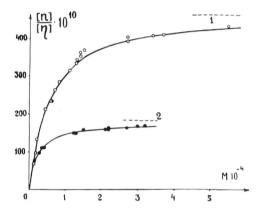

Fig. 6.50. Structure of the repeat unit of the chains of PPHPhBOTPhA (a) and PMHPhBOTYPhA (b) in the extended conformation.

Fig. 6.51. Shear optical coefficient $[n]/[\eta]$ versus molecular weight M for polymer solutions in sulfuric acid: 1) PPHPhBOTPhA; 2) PMHPhBOTPhA. Points represent experimental data [65]. Curves are plotted according to Eq. (5.111) at the values of the parameter β and A given in Table 6.9.

It is also possible to evaluate chain flexibility using Eq. (6.29) for polyamides with a more complex structure, such as poly(*para*- and *meta*-hydroxyphenyl-benzoxazole terephthalamide) (PPHPhBOTPhA and PMHPhBOTPhA). The structures of the repeat units of these two polymers are shown in Fig. 6.50. They differ in the position of the hydroxyphenyl ring, inserted in the para-position in the former polymer and in the meta-position in the latter.

The repeat unit of the former polymer contains two amide groups, each of which introduces two flexibility mechanisms: the structural mechanism (flexibility

$1/S_\beta$) and the deformational mechanism (flexibility $1/S_d$). Moreover, the unit of the para-polymer contains one benzoxazole heterocycle which changes the direction of the rotation axis in the chain by an angle $\vartheta \approx 30°$ without displacement of the axis. The flexibility parameter $1/S_\vartheta$ corresponding to this mechanism is evidently determined, just as in the case of PMPhIPhA, using Eq. (4.25) at $\vartheta = 30°$ and $\delta/\Delta = 0$. According to Eq. (6.29), the total action of all the mechanisms leads to the flexibility of the PPHPhBOTPhA chain $1/S = 2/S_\beta + 2/S_d + 1/S_\vartheta \approx (9\sigma^2)^{-1}$.

For this polymer the summed contribution $2/S_\beta + 2/S_d$ to the flexibility of the molecule by the structural and deformational mechanisms of the amide groups is very marked: it is 0.40 of the total chain flexibility.

Apart from the above-mentioned mechanisms, the repeat unit of the PMHPhBOTPhA chain introduced another mechanism related to the presence of the meta-phenyl ring changing the direction of the rotation axis (without displacement) by an angle $\vartheta = 60°$. If this mechanism is taken into account [according to Eq. (4.25)] in Eq. (6.29), this leads to a value $1/S = (2.5\sigma^2)^{-1}$ for the meta-polymer. For this isomer the total contribution provided to the flexibility of the molecule by the structural and deformational mechanisms of the amide groups is 0.1 of the overall chain flexibility. This relative contribution is four times less than in the case of the para-isomer. If it is assumed that the hindrance parameters in the molecules of *para*- and *meta*-oxazoles coincide, then the rigidity of the former should exceed that of the latter by more than a factor of three: $S_{para}/S_{meta} = 9/2.5 = 3.6$.

The experimental data on the equilibrium rigidity of PPHPhBOTPhA and PMHPhBOTPhA have been obtained by methods of hydrodynamics (Table 4.13) and FB (Table 6.2) of their solutions in sulfuric acid.

Figure 6.51 shows the dependence of $[n]/[\eta]$ on molecular weight for the solutions of these polymers. The curves differ greatly in their limiting values ($[n]/[\eta])_\infty$ in accordance with different rigidities of the two samples investigated. The treatment of experimental data using Eqs. (5.111) and (5.112) gives the values of the chain anisotropy β and the length of the Kuhn segment A listed in Table 6.9 (Nos. 7 and 8). These data agree within experimental error with those given in Tables 6.2 and 4.13 and also correspond to the relative evaluation of the rigidity of *para*- and *meta*-oxazoles carried out using Eq. (6.29). Comparison of experimental and theoretical values of rigidity parameters yields the values of the hindrance parameters $\sigma = 1.5$ and 1.7 for para- and meta-isomers, respectively [65].

The investigation of FB in solutions of aromatic poly(amide benzoxazole) (PABO) [109], whose structure differs from that of PPHPhBOTPhA (Fig. 4.50) only by the absence of the hydroxyl substituent in the phenyl ring, has shown complete coincidence between the dependences $[n]/[\eta] = f(M)$ for these two polymers. Correspondingly, both the data given in Table 6.9 and the values of σ for these polymers are identical.

TABLE 6.10. Shear Optical Coefficients $\Delta n/\Delta\tau$ and Equilibrium Rigidity of Molecules of Some Amide and Ring-Containing Polymers in Solution

No.	Polymer	Repeat unit	Solvent	$(\Delta n/\Delta\tau)\cdot10^{10}$, cm s^2 g^{-1}	$A\cdot10^8$, cm
1	Heterocyclic polyamide		Sulfuric acid	390	280
2	Polyamic acid		Dimethylacetamide	70	50
3	Polyamidoimide		Sulfuric acid	270	190
4	Polyamic acid		Dimethylacetamide	60	45
5	Polypyrromellitimide		Sulfuric acid	200—350	140—250

The structure of the poly(amide benzimidazole) chain (PABI) differs from that

of PABO only in the presence of the benzimidazole heterocycle $-\overset{\text{N}}{\text{C}}$ in

PABI instead of .the benzoxazole heterocycle $-\overset{\text{N}}{\text{C}}$. Taking into account

the similarity of stereochemical structures of these two heterocycles, we will obtain for PABI, according to Eq. (6.29), the value of $1/S \approx (9\sigma^2)^{-1}$, just as for PPHPhBOTPhA and PABO.

The experimental data on the FB of PABI solutions in sulfuric acid (Fig. 6.17 and Table 6.9) lead to the values of the rigidity parameter A for this polymer close to those for PABO. The slightly lower value of $([n]/[\eta])_\infty$ and A for PABI may be ascribed to the slight differences in bond angles and bonds of the benzimidazole and benzoxazole rings. Comparison of the experimental value of $S = A/\lambda = 290/17.8 = 16.3$ with the theoretical value $S = 9\sigma^2$ for PABI yields for the hindrance parameter $\sigma = 1.3$, which corresponds to the values of σ for this class of polymers.

Just as for PPHPhBOTPhA, the contribution of chain flexibility provided by the structural and deformational mechanisms of the amide groups to the total flexibility of the PABI chain is 40% and the neglect of this contribution would lead to the theoretical value of $S = 15\sigma^2$ and, correspondingly, to the degree of hindrance $\sigma \approx 1$. This is the explanation for the similar result obtained in the calculation of σ for PABI molecules using the experimental data on their translational diffusion in solution (Chapter 4, Section 9.2). In these calculations the amide group has been assumed to have a rigid *trans*-structure with equal bond angles at the carbon and nitrogen atoms. As a result, neither the deformational nor the structural mechanism of flexibility of the amide groups have been taken into account.

The above examples show that, among various sources of flexibility of aromatic polyamides, the introduction of the phenyl ring into the chain in the meta-position is of major importance because, in this case, the direction of the rotation axis in the chain changes drastically – by 60°. Hence, if chemical reactions in the polymer chain are used to form additional bonds excluding the possibility of rotation about the meta-aromatic axes of the phenyl rings, then the equilibrium rigidity of the chain modified in this manner should increase sharply.

This kind of change in the chemical structure of the chain occurs in imidization reactions for which examples are given in Table 6.10. The phenyl rings in the meta-position in the chains of polyamic acids (Nos. 2 and 4) during imidization are inserted into the heterocycles of the polyimides formed (Nos. 3 and 5). As a result, in all the chains of both polyamidoimide (No. 3) and poly(pyrromellitimide benzidine) (No. 4) all the bonds about which internal rotation is possible are mutually parallel; this is accompanied by a drastic increase in chain rigidity.

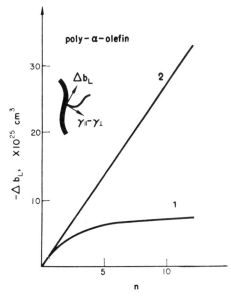

Fig. 6.52. Contribution of Δb_L of the side radical to the anisotropy of the monomer unit of the poly-α-olefin molecule versus number of carbon atoms n in the side chain [3]. 1) Theory of noninteracting polyethylene chains [Eq. (6.30)]; 2) experimental data [112] corresponding to the dependence of Δb_L = $-2.73 \cdot 10^{-25} n$ cm³.

FB is a sensitive indicator of the increase in intramolecular order and equilibrium rigidity of the chain during its imidization, as shown by the data given in Table 6.10. The same data also show that although the rigidity of the molecules of polyimides (in particular, that of No. 5) should be higher than that of the molecules of heterocyclic polyamide (No. 1), the opposite case is actually observed. This fact indicates that in the examples illustrated in Table 6.9 imidization is incomplete.

These data show that FB may be used as a sensitive method for studying imidization processes.

2.9. Comb-Shaped Molecules

The hydrodynamic properties of dilute solutions of comb-shaped polymers have been considered in Chapter 4, Section 3. It was shown that the hydrodynamic characteristics of comb-shaped molecules may be used to determine the equilibrium rigidity of their main chains. Moreover, it was found that the flexibility of the main chain decreases somewhat with increasing side-radical length (Fig. 4.15), which can be ascribed to the effect of increasing the interaction between side chains.

The study of FB of solutions of comb-shaped molecules makes it possible to obtain additional important information on the conformational characteristics not

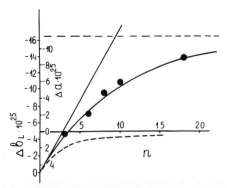

Fig. 6.53. Anisotropy of the monomer unit Δa and contribution Δb_L of the side radical to this anisotropy versus number n of valence bonds in the side chain for alkyl esters of polymethacrylic acid in benzene [113]. Points represent experimental data; solid curve describes the theoretical dependence according to Eq. (6.30) at $\Delta a^* = 3.3 \cdot 10^{-25}$ cm^3 and $v^* = 60$. Broken curve – according to Eq. (6.30) at $\Delta a^* = 3.3 \cdot 10^{-25}$ cm^3 and $v^* = 16$.

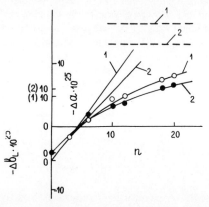

Fig. 6.54. Anisotropy of the monomer unit Δa and contribution Δb_L of the side radical to this anisotropy versus number n of valence bonds in the side chain for poly(alkyl acrylate)s in: 1) toluene; 2) decalin [114]. Points – experimental data. Solid curves describe the theoretical dependence according to Eq. (6.30) at $\bar{\omega}^* = 80$ and values of Δa equal to: 1) $2.6 \cdot 10^{-25}$ cm^3 in toluene; 2) $2.1 \cdot 10^{-25}$ cm^3 in decalin. Solid straight lines show the initial slopes of the curves, and broken lines show their limits.

only of the main chain but also of the side chains, since the value and sign of the anisotropy of the entire chain molecule are determined to a considerable extent by the structure and anisotropy of its side radicals [2, 111].

The side radical of a comb-shaped molecule may be regarded as a wormlike chain whose initial element is rigidly bonded to the main chain of the molecule and forms with it an angle of 90° (Fig. 6.52). Then the optical anisotropy of the side radical in a system of coordinates of its first element, $\gamma_{\parallel} - \gamma_{\perp}$, is determined by Eq. (5.94). In accordance with this, the contribution Δb_L of the side radical to the anisotropy of the monomer unit of the comb-shaped molecule is given by

$$\Delta b_L = -(\gamma_{\parallel} - \gamma_{\perp})/2 = -\Delta a^* v^* (1 - e^{-6n/v^*})/12. \tag{6.30}$$

where n and v^* are the numbers of valence bonds in the side chain and in its Kuhn segment, respectively, and Δa^* is the anisotropy of the side chain per valence bond.

Equation (6.30) can be applied to the analysis of conformational properties of comb-shaped molecules using experimental data on their optical anisotropy.

2.9.1. Poly-α-olefins −RCH−CH$_2$− (R = C$_n$H$_{2n+1}$). Systematic experimental data concerning the dependence of optical anisotropy of the comb-shaped molecule on the number n of methylene groups in its side chains have been obtained for a series of poly-α-olefins [112]. It has been shown that in this series an increase in the length of the side group is accompanied by a virtually linear increase in negative segmental anisotropy $\alpha_1 - \alpha_2$ of the molecule. In principle, this increase may be caused by both an increase in the rigidity of the main chain (increase in A or S) and an increase in the negative contribution to the anisotropy of the monomer unit of the molecule by its side radical C_nH_{2n+1} when its length increases.

Since no quantitative data are available on the dependence of the rigidity of the main chain of poly-α-olefins on side chain length, we must restrict ourselves to an approximate assumption that this dependence is not strong (which is actually observed only for the lower homologs of the series). On this assumption the experimental dependence [112] may be represented in the form $\Delta b_L = -2.73 \cdot 10^{-25} n$ cm^3. On the other hand, if the flexibility of the hydrocarbon side chain of the poly-α-olefin is taken to be equal to that of the polyethylene chain, then assuming v^* to be 16 [111], Eq. (6.30) yields the theoretical dependence $\Delta b_L = -7.3 \cdot 10^{-25}(1 - e^{-3n/8})$. Both dependences are shown in Fig. 6.52.

The theoretical curve shows that, as a result of the flexibility of the side chain, when its length (i.e., n) increases, the anisotropy Δb_L should attain the asymptotic limit determined by the degree of flexibility of the side radical.

However, in the investigated range of n values the experimental dependence of Δb_L on n does not exhibit this tendency. This fact implies that the rigidity and orientational order of the side chains of poly-α-olefins greatly exceed the values obtained from experimental investigations of the conformational and optical properties of single hydrocarbon chains in a dilute solution.

2.9.2. Poly(alkyl acrylate)s and Poly(alkyl methacrylate)s. The study of the optical anisotropy of poly(alkyl acrylate) (PAA) [114] and poly(alkyl methacrylate) (PAMA) [113] molecules as a function of alkyl side chain length leads to similar conclusions to those reached in the preceding section.

As in the case of poly-α-olefins, the value of $(\Delta n/\Delta \tau)_{\infty}$ obtained from the FB of PAA and PAMA solutions makes it possible to determine the segmental optical anisotropy $\alpha_1 - \alpha_2 = \beta A = \Delta a S$ of the main chain. However, in contrast to poly-olefins, for these polymers the hydrodynamic data yield quantitative information on the flexibility of the main chain (values of S and A) discussed in Chapter 4, Section 3.1 (Table 4.2 and Fig. 4.15). Hence, the consideration of the conformational properties of PAA and PAMA molecules can be more quantitative than in the case of poly-α-olefins: comparison of the values of S and $\alpha_1 - \alpha_2$ makes it possible to determine the anisotropy of $\Delta a = a_{\parallel} - a_{\perp}$ of the monomer unit containing the contribution Δb_L provided by the alkyl side group.

The points in Fig. 6.53 represent experimental values of the anisotropy of the monomer unit Δa of the PAMA series in benzene as a function of the number of valence bonds in the side chain. The solid curve describes the theoretical dependence (6.30) obtained at the values of $\Delta a^* = 3.3 \cdot 10^{-25}$ cm^3 and $v^* = 60$, ensuring good agreement between the theoretical curve and the experimental data. The value of Δa^* is in good agreement with the anisotropy of the polymethylene chain per CH$_2$ group determined from FB in polyethylene solutions [111]. The value of $v^* = 60$ obtained in this case is four times greater than that of $v = 16$ corresponding to the rigidity of both the polyethylene molecule and many other flexible-chain molecules in solutions. The theoretical dependence (6.30) corresponding to the value of $v^* = 16$ is shown by the broken curve in Fig. 6.53.

Analogous data are shown in Fig. 6.54 for a series of poly(alkyl acrylate)s differing in side chain length. For these polymers, just as for polyolefins and PAMA, with increasing side chain length the optical anisotropy of the monomer unit Δa reverses its sign from positive to negative as a result of the negative contribution Δb_L by the side chain. The experimental points obtained in two solvents for each of these solvents fit the curve representing the dependence [Eq. (6.30)] for a wormlike chain. The deviation of these curves from their initial slopes characterizes side chain flexibility and leads to the value of $v^* = 80$ for the number of valence bonds of the side chain in the Kuhn segment. This value exceeds by a factor of 5 the rigidity parameter of polyethylene molecules, which corresponds within experimental and calculation errors to the results obtained for poly(alkyl methacrylate)s.

The high rigidity of the side chains of comb-shaped molecules indicates that their intramolecular orientational order greatly exceeds that of flexible polymer chains. This might be caused by the interaction between densely placed side chains in comb-shaped molecules. The longer the alkyl side groups, the stronger is this interaction which leads to both a certain decrease in the flexibility of the main chain and to a more pronounced increase in the rigidity and optical anisotropy of the side chains.

TABLE 6.11. Shear Optical Coefficient $[n]/[\eta]$, Segmental Anisotropy $a_1 - a_2$, and Anisotropy of the Monomer Unit Δa of Molecules of Polymethacryloyl Derivatives $\left(\begin{array}{c} \text{H}_3\text{C}-\overset{|}{\text{C}}-\text{C}-\text{O}-\text{R} \\ \overset{|}{\text{CH}_2}\ \overset{\|}{\text{O}} \\ | \end{array} \right)$ with Mesogenic Side Chains

No.	Radical R	Solvent	$([n]/[\eta]) \cdot 10^{10}$, $cm\, s^2\, g^{-1}$	$(a_1 - a_2) \cdot 10^{25}$, cm^3	$\Delta a \cdot 10^{25}$ cm^3	References
1	—(ring)—C(=O)—O—C$_{16}$H$_{33}$	Tetrachloro-methane	−35	−445	−18	[122]
2	—(ring)—C(=O)—O—(ring)—CN	Dimethyl-formamide	−20	−240	—	[123]
3	—(ring)—C(=O)—O—(ring)—O—CH$_3$	Tetrachloro-ethane	−40	−500	—	
4	—(ring)—O—C(=O)—(ring)—O—C$_3$H$_7$	Dimethyl-formamide–toluene (1:1)	−25	−320	—	[124]
5	—(ring)—O—C(=O)—(ring)—O—C$_6$H$_{13}$	Benzene	−30	−370	—	[125]
6	—(ring)—O—C(=O)—(ring, NO$_2$)—O—C$_6$H$_{13}$	Dioxane	−95	−1200	—	[123]
7	—(ring)—O—C(=O)—(ring)—O—C$_9$H$_{19}$	Tetrachloro-methane	−50	−600	−40	[126]
8	—(ring, ·)—C(=O)—O—(ring)—O—C$_9$H$_{19}$	Tetrachloro-methane	−220	−2700	−110	[127]
9	—(ring)—C(=O)—O—(ring)—O—C$_{12}$H$_{25}$	Tetrachloro-methane	−190	−2350	—	
10	—(ring)—O—C(=O)—(ring)—O—C$_{16}$H$_{33}$	Tetrachloro-methane	−220	−2700	−110	[124]
11	Copolymer No. 10 with cetyl methacrylate (7 : 3)	Tetrachloro-methane	−85	−1050	−44	[128]
12	The same (1 : 1)		−55	−680	−28	[123]
13	—(ring)—C(=O)—O—(ring)—O—C(=O)—(ring)—O—C$_{16}$H$_{33}$	Chloroform Benzene	−390 −245	−4900 −3000	−100 −150	[129]
14	—(CH$_2$)$_2$—O—(ring)—C(=O)—O—(ring)—O—CH$_3$	Tetrachloro-methane	−10	—	—	[127]
15	—(CH$_2$)$_{10}$—C(=O)—O—(ring)—C(=O)—O—(ring)—O—C$_4$H$_9$	Dioxane	−6.5	—	—	[127]

TABLE 6.12. Shear Optical Coefficient $[n]/[\eta]$, Segmental Anisotropy $\alpha_1 - \alpha_2$, and Anisotropy of the Monomer Unit Δa of Molecules of Polyacryloyl Derivatives

$$\left(\begin{array}{c} H-\overset{|}{C}-\overset{\|}{C}-O-R \\ \overset{|}{C}H_2 \;\; O \\ | \end{array} \right)$$

with Mesogenic Side Chains and Poly(nonyloxybenzamide styrene) $\left(-CH_2-\overset{R}{\underset{H}{\overset{|}{C}}}-, \;\; № \; 9 \right)$

No.	Radical R	Solvent	$([n]/[\eta]) \cdot 10^{10}$ cm s^2 g^{-1}	$(\alpha_1-\alpha_2) \cdot 10^{25}$, cm^3	$\Delta a \cdot 10^{25}$, cm^3	References
1		Tetrachloro-ethane	−43	−520	—	
2		Tetrachloro-ethane	−43	−520	—	
3		Tetrachloro-ethane	−52	−630	—	
4		Tetrachloro-ethane	−67	−810	—	
5		Tetrachloro-ethane	−65	−800	—	
6		Tetrachloro-ethane	−36	−450	−19	[130]
7		Tetrachloro-ethane	−45	−510	−21	[130]
8		Benzene	−25	−300	−15	[132]
9		Benzene	−210	−2500	−100	[131]

The fact that the specific conformational and optical properties of comb-shaped molecules are due to the interaction between their side chains may be confirmed by the conformational properties of comb-shaped molecules bearing mesogenic side groups.

2.9.3. Molecules with Mesogenic Side Groups. The hydrodynamic properties of molecules whose side groups contain fragments capable of forming the mesophase have been considered in Chapter 4, Section 3.2. The equilibrium rigidity of the main chain of these molecules, determined from their translational friction and viscometry in dilute solutions, is relatively low; it is close to the rigidity of other comb-shaped molecules and exceeds that of typical flexible-chain polymers by only three or four times.

In accordance with this, the hydrodynamic behavior of these molecules is typical of that of a flexible-chain polymer, which is revealed by the FB relationships in their solutions. Thus, the shear optical coefficient $\Delta n/\Delta\tau$ in solutions of comb-shaped mesogenic molecules is virtually independent of their molecular weight, in contrast to rigid-chain polymers in which this dependence is always observed. Hence, the experimental value of $\Delta n/\Delta\tau$ for these polymers is the limiting value $(\Delta n/\Delta\tau)_\infty$ characteristic of Gaussian chains. The value of $(\Delta n/\Delta\tau)_\infty$ may be used to determine the anisotropy $\alpha_1 - \alpha_2 = \beta A$ of the Kuhn segment characterizing the rigidity of the main chain. Moreover, if the length of the segment A is also known from hydrodynamic data (Table 4.4), the anisotropy of the monomer unit of the molecule can also be determined. The corresponding experimental data are given in Tables 6.11 and 6.12 for polymethacryloyl and polyacryloyl derivatives, respectively.

The mesogenic character of the side radicals of these molecules is determined by the presence of *para*-phenyl rings separated by rigid ester, amide, or other groups with conjugated bonds, and by the presence of relatively long and flexible alkyl chains at the end of the side radical.

A characteristic feature of all the polymers listed in Tables 6.11 and 6.12 is the high negative FB observed in their solutions and, correspondingly, their high negative segmental anisotropy $\alpha_1 - \alpha_2$. At relatively low main chain rigidity this feature is a direct consequence of the high negative anisotropy Δa of the monomer unit of the molecule. Thus, for polymer No. 10 (Table 6.11) the value of Δa exceeds by more than one order of magnitude that of poly(cetyl methacrylate) (Fig. 6.53) at the same alkyl group length ($C_{16}H_{33}$). The negative anisotropy of the monomer unit of poly(nonyloxybenzamide styrene) (PNOBS, No. 9 in Table 6.12) is also higher by one order of magnitude than that of polystyrene, although the number of anisotropic phenyl rings in the PNOBS side radical is only twice as large as polystyrene.

The analysis of data given in Tables 6.11 and 6.12 suggests that the high negative values of Δa determined by the contribution of side radicals to the anisotropy can be interpreted only by the fact that the orientational order in the side groups (determining their optical anisotropy) greatly exceeds that of polymer chains with a similar structure, but existing in the "free" state in a dilute solution.

Hence, the interaction between the side groups of comb-shaped molecules leading to a decrease in the flexibility of their main chains greatly increases the orientational order and rigidity of the side chains if these chains contain mesogenic fragments.

These properties are increasingly pronounced with increasing "mesogenic character" of the side groups. Thus, an increase in the number of phenyl rings in the side group from one to two, the alkyl end group remaining invariable, leads to a sixfold increase in Δa and, correspondingly, in the degree of order in the side chain (samples 1 and 10 in Table 6.11). An increase in the length of the alkyl sequence with the number of phenyl rings remaining invariable also results in the ordering of the structure of this radical and a great increase in Δa (samples 2–5 and 6–13 in Table 6.11).

On the other hand, an increase in the distance between mesogenic side radicals by the copolymerization of the mesogenic component (No. 10 in Table 6.11) with cetyl methacrylate weakens the anisotropic interaction between the side chains decreasing the orientational order in the molecule and the optical anisotropy of its monomer units (Nos. 11 and 12 in Table 6.11).

The above factors show that the interaction between the mesogenic side chains does not reduce to simple steric hindrances. This interaction is presumably of a more complex "anisotropic" type in which a tendency toward mutual parallel orientation of the side chains is shown. This tendency is characteristic of both nematic and smectic order in liquid crystals.

For this "mesomorphic" intramolecular order in the orientation of the side chains to become evident in the FB of the solution, a certain correlation should exist between the orientations of the side and main chains since the FB phenomenon is eventually related to the orientation of macromolecules as a whole, i.e., to that of their main chains. The increase in the distance between the mesogenic part of the side chain and the main chain, achieved by introducing a flexible "spacer" between them, should evidently lead to a weaker correlation between the orientations of the main chain and the mesogenic groups of the side chain. This weakening is distinctly observed through FB and is manifested in a decrease in the absolute values of both $[n]/[\eta]$ and the anisotropy of the monomer unit Δa. This effect may be demonstrated if we compare the data for polymers Nos. 3 and 14 in Table 6.11 and those for polymers No. 5 in Table 6.12 and No. 15 in Table 6.11. It can be seen that the introduction of the spacer group $(CH_2)_2$ in the former case decreases by four times the values of $[n]/[\eta]$ and Δa, whereas the introduction of a longer spacer $(CH_2)_{10}$ in the latter case leads to a tenfold decrease in the same parameters.

Hence, the increase in the distance between the mesogenic groups and the main chain results in a decrease in intramolecular orientational order which, according to Eq. (6.10), is determined with respect to the chosen molecular axis (e.g., vector $\mathbf{h}$) taking into account the orientation angles ϑ of *all* chain elements (including the side groups).

On the other hand, the weakening of the orientational correlation between the main chain and the mesogenic side segments of the comb-shaped molecule due to the introduction of spacer groups facilitates the generation of orientational order between the side groups of different molecules, thus favoring the formation of the mesomorphic structure in the polymer bulk [115].

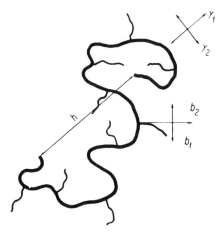

Fig. 6.55. Configuration of graft copolymer molecule. γ_1 and γ_2 are the main polarizabilities of the molecule, and b_2 and b_1 are those of the grafted chain.

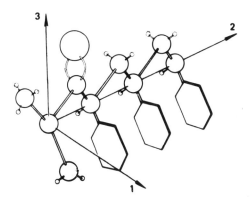

Fig. 6.56. Poly(methyl methacrylate) (PMMA) monomer unit with grafted polystyrene (PS) chain. Axis 1 is parallel to the extended *trans*-chain of PMMA, the plane of which is plane 1, 3. Axis 2 is parallel to the extended *trans*-chain of grafted PS.

2.9.4. Graft Copolymers. The hydrodynamic properties of molecules of graft copolymers of polystyrene (PS) and poly(methyl methacrylate) (PMMA) have been considered in Chapter 4, Section 3.3, and it has been shown that at appropriate ratios of the length of the main chain (L_A) to that of the grafted polystyrene (L_B) the structure of the copolymer molecules may be regarded as comb-shaped.

The study of FB in solutions of these copolymers [116–120] provides important information on the structural, conformational, and optical characteristics

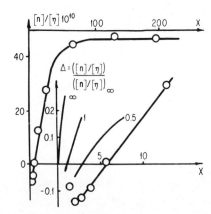

Fig. 6.57. Shear optical coefficient $[n]/[\eta]$ and its relative value $\Delta \equiv ([n]/[\eta])/([n]/[\eta])_\infty$ versus parameter $x = 2L/A$ for solutions of fractions of a graft copolymer of methyl methacrylate and styrene in bromoform. Numbers at the theoretical curves $\Delta(x)$ correspond to various values of the $A/2d$ ratio [121]; points represent experimental data [118, 119].

of these molecules. It was the FB method that made it possible to prove unequivocally for the first time the "graft" structure of these copolymers [116].

A unique feature of the optical properties of PMMA-PS graft copolymers is the fact that the segmental anisotropy of their molecules (determined by FB) is *positive* in sign, i.e., it is opposite in sign to the *negative* segmental anisotropy of the grafted PS, although the content of the latter in the copolymer exceeds 90%. Moreover, the segmental anisotropy of molecules of the graft copolymer $\alpha_1 - \alpha_2$ exceeds in numerical value many times that of grafted homopolymers.

These properties can be understood and interpreted quantitatively for the molecules exhibiting the comb-shaped structure, i.e., for the molecules for which the degree of polymerization P_A of the main chain (PMMA) is much higher than that, P_B, of the grafted PS chains.

The positive sign of the anisotropy $\gamma_1 - \gamma_2$ of the molecule of a graft copolymer and of its segmental anisotropy $\alpha_1 - \alpha_2$ implies that the polarizability γ_1 of the molecule in the direction of its greatest geometrical length (parallel to h) is higher than in that normal to h (γ_2, Fig. 6.55). It is impossible to explain the high positive value of $\gamma_1 - \gamma_2$ found experimentally by the presence of the PMMA chains (main chain) exhibiting positive anisotropy because its anisotropy is very low and its content in the copolymer is small ($\approx 10\%$). The main significance in the observed anisotropy should evidently be ascribed to PS whose polarizability b_2 (Fig. 6.55) along the PS chain (axis 2 in Fig. 6.56) is much lower than the average polarizability b_1 (Fig. 6.55) in the perpendicular direction. Hence, the PS chain with negative anisotropy bonded to the main chain at the right angle provides the *positive* contribution Δb_L to the anisotropy of the main chain.

According to Eq. (6.30), this contribution, as for any side chain represented by a wormlike chain, is determined by the equation

$$\Delta b_L = -\Delta a_B S_B \left(1 - e^{-6P_B/S_B}\right)/12,\qquad(6.31)$$

where Δa_B is the anisotropy of the monomer unit of grafted PS, S_B is the number of monomer units in the Kuhn segment of the grafted PS chain, and P_B is the degree of polymerization of this chain. Since for PS we have $\Delta a_B < 0$, it follows from Eq. (6.31) that $\Delta b_L > 0$.

The segmental anisotropy of a comb-shaped molecule, $\alpha_1 - \alpha_2$, is a sum of the segmental anisotropy of the main PMMA chain $(\alpha_1 - \alpha_2)_A = \Delta a_A S_A$ and the total anisotropy, $\Sigma \Delta b_L$, introduced by all the side chains grafted onto a part of the main chain of one segment in length. Taking this into account and applying Eq. (6.31), we obtain

$$\alpha_1 - \alpha_2 = S_A \left\{ \Delta a_A - \Delta a_B \left[x_p/(1 - x_p) \right] (S_B/P_B) \left(1 - e^{-6P_B/S_B}\right)/12 \right\},\qquad(6.32)$$

where Δa_A is the anisotropy of the monomer unit of the main chain (PMMA), S_A is the number of monomer units in the Kuhn segment of the main chain, and x_p is the molar fraction of the grafted component (PS) in a comb-shaped polymer (Chapter 4, Section 3.3.1).

For a comb-shaped polymer $(P_A \gg P_B)$ of relatively high molecular weight the segmental anisotropy $\alpha_1 - \alpha_2$ can be determined according to experimental data on FB by applying Eq. (5.60), just as for a Gaussian chain. The value of S_A is determined from the experimental hydrodynamic data (Chapter 4, Section 3.3.2). Hence, for a graft polymer with a known composition and known length of the grafted chain (x_p, P_B) the number S_B characterizing the rigidity of the side chains of the grafted PS can be determined using the hydrodynamic and dynamo-optical experimental data and Eq. (6.32).

These investigations have shown that for comb-shaped graft copolymers $(P_A \gg P_B)$ with a relatively high grafting frequency $(x_p \approx 0.9)$ the equilibrium rigidity of the side chains is close to that of the main chain and the length of the Kuhn segment is $A \approx (100-200)\cdot10^{-8}$ cm. Hence, the conformation of the relatively short PS side chains $(P_B \approx 50-100)$ is close to the weakly bending rod perpendicular to the main chain.

These properties are increasingly pronounced with increasing grafting frequency and disappear when this frequency is low, i.e., when the distance between the grafted chains increases. This point shows, just as in the case of other comb-shaped molecules, that the interaction between the side groups of graft copolymers profoundly affects the specific properties of their molecules.

As for other rigid-chain polymers, with decreasing main chain length of the graft copolymer, the decrease in the shear optical coefficient, $[n]/[\eta]$, can be observed in the FB phenomenon in its solutions (Fig. 6.57).

However, the observed dependence of $[n]/[\eta]$ on $x = 2L/A$ cannot be described by Eq. (5.111) derived for a thin wormlike chain because the effective chain diameter d of the graft copolymer is large and the condition $d/L \ll 1$ is not obeyed at low L. As can be shown by Eq. (5.102) for a wormlike chain of large diameter, at a relatively small length L (and, correspondingly, x) the parameter of the shape asymmetry b_0 can become zero and even negative, which should be accompanied by a change in the FB sign. In graft copolymer solutions this phenomenon is actually observed with decreasing length of the main chain (Fig. 6.57). This change in the FB sign corresponds to a change in the molecular structure from a comb-shaped to a starlike structure.

Figure 6.57 shows, apart from the values of $[n]/[\eta]$, the initial parts of curves representing the dependence of $([n]/[\eta])/([n]/[\eta])_{\infty}$ on x corresponding to different values of $A/2d$ for a wormlike chain if Eq. (5.102) is applied. The comparison of these curves with experimental data allows the estimation of chain diameter, thus confirming the high value of this diameter $(30–100) \cdot 10^{-8}$ cm for the molecules of the graft copolymers investigated.

All the above data on the structure and conformation of graft copolymer molecules obtained by the FB method agree with the results of investigations of the same samples by hydrodynamic methods (Chapter 4, Section 3.3.2).

REFERENCES

1. N. V. Pogodina, I. N. Bogatova, and V. N. Tsvetkov, Vysokomol. Soedin., **A27**, 1405 (1985).
2. V. N. Tsvetkov, in: Newer Methods of Polymer Characterization, Chapter 14, B. Ke (ed.), Wiley–Interscience, New York (1964).
3. V. N. Tsvetkov, Usp. Khim., **38**, 1674 (1969).
4. V. N. Tsvetkov and E. V. Frisman, Acta Physicochim. URSS, **20**, 61 (1945).
5. N. V. Pogodina, Y. B. Tarabukina, L. V. Starchenko, G. N. Marchenko, and V. N. Tsvetkov, Vysokomol. Soedin., **A22**, 2219 (1980).
6. J. Yang, J. Am. Chem. Soc., **80**, 5138 (1958).
7. I. N. Shtennikova, Thesis, Inst. Macromolec. Compounds, Acad. Sci. USSR, Leningrad (1973).
8. V. N. Tsvetkov, K. A. Andrianov, I. N. Shtennikova, G. I. Okhrimenko, L. N. Andreeva, G. A. Fomin, and V. I. Pakhomov, Vysokomol. Soedin., **A10**, 547 (1968).
9. V. N. Tsvetkov, K. A. Andrianov, G. I. Okhrimenko, and M. G. Vitovskaya, Eur. Polym. J., **7**, 1215 (1971).
10. V. N. Tsvetkov, Makromol. Chem., **160**, 1 (1972).
11. V. N. Tsvetkov, K. A. Andrianov, N. N. Makarova, and M. G. Vitovskaya, Vysokomol. Soedin., **9**, 27 (1973).
12. K. A. Andrianov, S. V. Bushin, M. G. Vitovskaya, V. N. Yemelyanov, P. N. Lavrenko, N. N. Makarova, A. M. Muzafarov, V. Y. Nikolaev, G. F. Kolbina, I. N. Shtennikova, and V. N. Tsvetkov, Vysokomol. Soedin., **A19**, 469 (1977).

13. V. N. Tsvetkov, K. A. Andrianov, E. I. Rjumtsev, I. N. Shtennikova, N. V. Pogodina, G. F. Kolbina, and N. N. Makarova, Vysokomol. Soedin., **A17**, 2493 (1975).

14. L. N. Andreeva, P. N. Lavrenko, E. V. Belyaeva, A. A. Boikov, A. M. Muzafarov, V. N. Emelianov, K. A. Andrianov, and V. N. Tsvetkov, Vysokomol. Soedin., **A21**, 362 (1979).

15. L. N. Andreeva, S. V. Bushin, A. I. Mashoshin, Y. B. Lysenko, K. P. Smirnov, V. N. Yemelyanov, and V. N. Tsvetkov, Vysokomol. Soedin., **A24**, 2101 (1982).

16. V. N. Tsvetkov, K. A. Andrianov, M. G. Vitovskaya, N. N. Makarova, E. N. Zakharova, S. V. Bushin, and P. N. Lavrenko, Vysokomol. Soedin., **A14**, 369 (1972).

17. V. N. Tsvetkov, K. A. Andrianov, M. G. Vitovskaya, N. N. Makarova, S. V. Bushin, E. N. Zakharova, A. A. Gorbunov, and P. N. Lavrenko, Vysokomol. Soedin., **A15**, 872 (1973).

18. V. N. Tsvetkov, K. A. Andrianov, E. L. Vinogradov, I. N. Shtennikova, S. E. Yakushkina, and V. I. Pakhomov, J. Polym. Sci., **C23**, 385 (1968).

19. S. Y. Ljubina, S. I. Klenin, I. A. Strelina, A. V. Troitskaya, V. I. Kurliankina, and V. N. Tsvetkov, Vysokomol. Soedin., **A15**, 691 (1973).

20. L. N. Andreeva and B. Y. Elohovski, Vysokomol. Soedin., **19B**, 111 (1977).

21. P. N. Lavrenko, E. U. Urinov, L. N. Andreeva, K. J. Linow, H. Dautzenberg, and B. Philipp, Vysokomol. Soedin., **A18**, 2579 (1976).

22. L. N. Andreeva, T. I. Garmonova, I. N. Shtennikova, and V. N. Tsvetkov, The Fifth Meeting on Cellulose, Vladimir (1976).

23. N. V. Pogodina, K. S. Pozhivilko, A. B. Melnikov, S. A. Didenko, G. N. Martchenko, and V. N. Tsvetkov, Vysokomol. Soedin., **A23**, 2454 (1981).

24. V. N. Tsvetkov, I. N. Shtennikova, N. A. Medjeritskaya, and L. S. Bolotnikova, High-Molecular-Weight Compounds. Cellulose and Its Derivatives (1963), pp. 74–85.

25. I. N. Shtennikova, V. N. Tsvetkov, T. V. Peker, E. I. Rjumtzev, and Y. P. Getmanchuk, Vysokomol. Soedin., **A16**, 1086 (1974).

26. I. N. Shtennikova, T. V. Peker, Y. P. Getmanchuk, and V. A. Kudrenko, Vysokomol. Soedin., **A20**, 1246 (1978).

27. V. N. Tsvetkov, I. N. Shtennikova, E. I. Rjumtsev, and G. I. Okhrimenko, Vysokomol. Soedin., **7**, 1104 (1965).

28. T. I. Garmonova, M. G. Vitovskaya, P. N. Lavrenko, V. N. Tsvetkov, and E. V. Korovina, Vysokomol. Soedin., **A13**, 884 (1971).

29. V. N. Tsvetkov, Dokl. Akad. Nauk SSSR, **192**, 380 (1970).

30. N. V. Pogodina, K. S. Pozhivilko, N. P. Yevlampieva, A. B. Melnikov, S. V. Bushin, S. A. Didenko, G. N. Marchenko, and V. N. Tsvetkov, Vysokomol. Soedin., **A23**, 1252 (1981).

31. N. V. Pogodina, P. N. Lavrenko, K. S. Pozhivilko, A. B. Melnikov, T. A. Kolobova, G. N. Marchenko, and V. N. Tsvetkov, Vysokomol. Soedin., **A24**, 332 (1982).

32. V. N. Tsvetkov, P. N. Lavrenko, L. N. Andreeva, A. I. Mashoshin, O. V. Okatova, O. I. Mikrjukova, and L. I. Kutzenko, Eur. Polym. J., **20**, 823 (1984).

33. E. V. Frisman and M. A. Sibileva, Vysokomol. Soedin., **7**, 674 (1965).

34. V. N. Tsvetkov, T. I. Garmonova, and R. P. Stankevich, Vysokomol. Soedin., **8**, 980 (1966).

35. G. B. Thurston and J. L. Schrag, J. Polym. Sci., **A2**, 1331 (1968).

36. W. Kuhn, R. Pasternak, and H. Kuhn, Helv. Chim. Acta, **30**, 1705 (1948).

37. V. N. Tsvetkov, L. N. Andreeva, and L. N. Kvitchenko, Vysokomol. Soedin., **7**, 2001 (1965).

38. H. Eisenberg, in: Basic Principles in Nucleic Acid Chemistry, Vol. 2, Academic Press, New York (1974), p. 171.

39. D. Jolly and H. Eisenberg, Biopolymers, **15**, 61 (1976).

40. J. E. Goldfrey and H. Eisenberg, Biophys. Chem., **5**, 301 (1976).

41. V. N. Tsvetkov, V. E. Bichkova, S. M. Savvon, and I. N. Nekrasov, Vysokomol. Soedin., **1**, 1407 (1959).

42. V. N. Tsvetkov, G. I. Kudryavtsev, I. N. Shtennikova, T. V. Beker, E. N. Zakharova, V. D. Kalmykova, and A. V. Volokhina, Eur. Polym. J., **12**, 517 (1976).

43. N. V. Pogodina, I. N. Bogatova, and V. N. Tsvetkov, Vysokomol. Soedin., **A27**, 1405 (1985).

44. I. N. Shtennikova, T. V. Peker, T. I. Garmonova, G. F. Kolbina, L. V. Avrorova, A. B. Tokarev, G. I. Kudryavtsev, and V. N. Tsvetkov, Vysokomol. Soedin., **A23**, 2510 (1981).

45. V. N. Tsvetkov, N. V. Pogodina, and L. V. Starchenko, Eur. Polym. J., **19**, 837 (1983).

46. V. N. Tsvetkov, N. A. Mikhailova, V. B. Novakovski, A. V. Volokhina, and A. B. Raskina, Vysokomol. Soedin., **A22**, 1028 (1980).

47. V. N. Tsvetkov, L. N. Andreeva, S. V. Bushin, A. I. Mashoshin, V. A. Chercasov, Z. Edlinski, and D. Sek, Eur. Polym. J., **20**, 373 (1984).

48. V. N. Tsvetkov, E. N. Zakharova, and N. A. Mikhailova, Dokl. Akad. Nauk SSSR, **224**, 1365 (1975).

49. N. A. Mikhailova, Thesis, University of Leningrad (1982).

50. C. J. Sadron, Phys. Rad., **9**, 381, 384 (1938).

51. M. Goldstein, J. Chem. Phys., **20**, 677 (1952).

52. A. Peterlin, J. Chem. Phys., **39**, 224 (1963).

53. U. Daum, J. Polym. Sci., **A26**, 141 (1968).

54. H. Janeschitz-Kriegle, Adv. Polym. Sci., **6**, 170 (1969).

55. G. V. Schulz, Z. Phys. Chem., **B43**, 25 (1939).

56. B. H. Zimm, J. Chem. Phys., **16**, 1099 (1948).

57. H. Wesslau, Makromol. Chem., **20**, 111 (1956).

58. V. N. Tsvetkov and L. N. Andreeva, Adv. Polym. Sci., **39**, 95 (1981).

59. H. Janeschitz-Kriegle, Kolloid Z., **203**, 119 (1965).

60. L. N. Andreeva, E. V. Belyaeva, A. A. Boikov, P. N. Lavrenko, and V. N. Tsvetkov, Vysokomol. Soedin., **A25**, 1631 (1983).

61. P. N. Lavrenko, A. A. Boikov, L. N. Andreeva, E. V. Belyaeva, and A. F. Podolski, Vysokomol. Soedin., **A23**, 1937 (1981).

62. I. N. Shtennikova and T. I. Garmonova, Vysokomol. Soedin., **A25**, 1643 (1983).

63. I. N. Shtennikova, T. V. Peker, T. I. Garmonova, and N. A. Mikhailova, Eur. Polym. J., **20**, 1003 (1984).

64. N. V. Pogodina, L. V. Startchenko, K. S. Pozhivilko, V. D. Kalmykova, T. A. Kulitchikhina, A. V. Volokhina, G. V. Kudryavtsev, and V. N. Tsvetkov, Vysokomol. Soedin., **A23**, 2185 (1981).
65. N. V. Pogodina, L. V. Startchenko, and V. N. Tsvetkov, Eur. Polym. J., **19**, 841 (1984).
66. V. N. Tsvetkov, I. N. Shtennikova, and T. V. Peker, Eur. Polym. J., **13**, 455 (1977).
67. V. N. Tsvetkov, N. V. Pogodina, and L. V. Startchenko, Eur. Polym. J., **17**, 397 (1981).
68. J. Preston, in: Liquid-Crystalline Order in Polymers, A. Blumstein (ed.), Academic Press, New York (1978).
69. V. N. Tsvetkov and I. N. Shtennikova, Vysokomol. Soedin., **6**, 1041 (1964).
70. V. N. Tsvetkov, A. E. Grischenko, and P. A. Slavetskaya, Vysokomol. Soedin., **5**, 856 (1964).
71. V. N. Tsvetkov, E. I. Rjumtsev, I. N. Shtennikova, T. V. Peker, and N. V. Tsvetkova, Eur. Polym. J., **9**, 1 (1973).
72. S. Y. Lyubina, S. I. Klenin, I. A. Strelina, A. V. Troitskaya, A. K. Khripunov, and E. U. Urinov, Vysokomol. Soedin., **A19**, 244 (1977).
73. J. W. Nordermeer, R. Daryanani, and H. Janeschitz-Kriegel, Polymer, **16**, 359 (1975).
74. V. N. Tsvetkov, E. N. Zakharova, and M. M. Krunchak, Vysokomol. Soedin., **A10**, 685 (1968).
75. V. N. Tsvetkov, E. V. Frisman, and N. N. Boitzova, Vysokomol. Soedin., **2**, 1001 (1960).
76. M. G. Vitovskaya, V. N. Tsvetkov, L. I. Godunova, and T. V. Sheremeteva, Vysokomol. Soedin., **A9**, 1682 (1967).
77. T. V. Sheremeteva, G. N. Larina, V. N. Tsvetkov, and I. N. Shtennikova, J. Polym. Sci., **C22**, 185 (1968).
78. V. N. Tsvetkov, G. A. Fomin, P. N. Lavrenko, I. N. Shtennikova, T. V. Sheremeteva, and L. I. Godunova, Vysokomol. Soedin., **A10**, 903 (1968).
79. V. N. Tsvetkov, N. N. Kupriyanova, G. V. Tarasova, P. N. Lavrenko, and I. I. Migunova, Vysokomol. Soedin., **A12**, 1974 (1970).
80. V. N. Tsvetkov, G. V. Tarasova, E. L. Vinogradov, N. N. Kupriyanova, V. M. Yamshikov, V. S. Skazka, V. S. Ivanov, V. K. Smirnova, and I. I. Migunova, Vysokomol. Soedin., **A13**, 620 (1971).
81. V. N. Tsvetkov, M. G. Vitovskaya, P. N. Lavrenko, E. N. Zakharova, I. F. Gavrilenko, and N. N. Stefanovskaya, Vysokomol. Soedin., **A13**, 2532 (1971).
82. V. N. Tsvetkov, Eur. Polym. J., **12**, 867 (1976).
83. S. M. Aharoni, J. Macromol. Sci. Phys., **21**, 105 (1982).
84. V. N. Tsvetkov and L. N. Andreeva, in: Polymer Handbook, J. Brandrup and E. H. Immergut (eds.), Wiley, New York (1975).
85. V. N. Tsvetkov, E. I. Rjumtsev, and I. N. Shtennikova, in: Liquid-Crystalline Order in Polymers, A. Blumstein (ed.), Academic Press, New York (1978).
86. V. N. Tsvetkov, Acta Physicochim. URSS, **16**, 132 (1942).
87. V. N. Tsvetkov, E. I. Rjumtsev, and I. N. Shtennikova, Vysokomol. Soedin., **A11**, 132 (1969); **A13**, 506 (1971).

88.   V. N. Tsvetkov, I. N. Shtennikova, E. I. Rjumtsev, and G. F. Pirogova, Vysokomol. Soedin., **A9**, 1575, 1583 (1967); J. Polym. Sci., **C16**, 3205 (1968).

89.   L. J. Fetters, Polym. Lett., **10**, 577 (1972).

90.   E. I. Rjumtsev, I. N. Shtennikova, N. V. Pogodina, and T. V. Peker, Vysokomol. Soedin., **A18**, 743 (1976).

91.   V. N. Tsvetkov, I. N. Shtennikova, E. I. Rjumtsev, L. N. Andreeva, Y. P. Getmanchuk, Y. L. Spirin, and R. I. Dryagileva, Vysokomol. Soedin., **A10**, 2132 (1968).

92.   V. N. Tsvetkov, Vysokomol. Soedin., **5**, 740 (1963).

93.   V. N. Tsvetkov, V. V. Korshak, I. N. Shtennikova, H. Raubach, E. S. Krongauz, G. M. Pavlov, G. F. Kolbina, and S. O. Tsepelevich, Vysokomol. Soedin., **A21**, 83 (1979); Macromolecules, **12**, 645 (1979).

94.   P. D. Sybert, W. H. Beever, and J. K. Stille, Macromolecules, **14**, 493 (1981).

95.   V. N. Tsvetkov, I. N. Shtennikova, P. N. Lavrenko, G. F. Kolbina, O. V. Okatova, G. Rafler, and G. Reinisch, Acta Polym., **31**, 434 (1980).

96.   O. P. Rokachevskaya, A. V. Volokhina, and G. I. Kudryavtsev, Vysokomol. Soedin., **7**, 1092 (1965).

97.   V. D. Kalmykova, M. N. Bogdanov, N. P. Okromchelidze, I. B. Zhmaeva, and V. Ya. Efremov, Vysokomol. Soedin., **8**, 1586 (1966).

98.   N. V. Pogodina, N. V. Starchenko, and V. N. Tsvetkov, Eur. Polym. J., **16**, 387 (1980).

99.   P. R. Saunders, J. Polym. Sci., **A2**, 3755 (1964).

100. V. N. Tsvetkov, N. V. Pogodina, and L. V. Starchenko, Eur. Polym. J., **17**, 397 (1981).

101. Polymeric Liquid Crystals, Prepr. Sympos. A.C.S., Washington, August (1983).

102. U. Shmueli, W. Traub, and K. Rosenheck, J. Polym. Sci., **A2**(7), 515 (1969).

103. L. Pauling, Nature of the Chemical Bond, Cornell Univ. Press, Ithaca, New York (1960).

104. B. Erman, P. J. Flory, and J. P. Hummel, Macromolecules, **13**, 484 (1980).

105. J. P. Hummel and P. J. Flory, Macromolecules, **13**, 479 (1980).

106. F. Laupretre and L. Monnerie, Eur. Polym. J., **14**, 415 (1978).

107. F. Laupretre, L. Monnerie, and B. Fayolle, J. Polym. Sci., Polym. Phys. Ed., **18**, 2243 (1980).

108. B. Millaud and C. Strazielle, Makromol. Chem., **179**, 1261 (1978).

109. N. A. Mikhailova, V. N. Tsvetkov, and V. B. Novakovski, Vysokomol. Soedin., **B24**, 770 (1982).

110. V. N. Tsvetkov, I. N. Shtennikova, and T. V. Peker, Eur. Polym. J., **13**, 455 (1977).

111. J. Brandrup and E. H. Immergut (eds.), Polymer Handbook, Wiley, New York (1975),Chapter IV, p. 377.

112. W. Philippoff and E. G. M. Tornqvist, J. Polym. Sci., **C23**, 881 (1968).

113. V. N. Tsvetkov, D. Hardy, I. N. Shtennikova, E. V. Korneeva, G. F. Pirogova, and K. Nitray, Vysokomol. Soedin., **A11**, 349 (1969).

114. V. N. Tsvetkov, L. N. Andreeva, E. V. Korneeva, and P. N. Lavrenko, Dokl. Akad. Nauk SSSR, **205**, 895 (1972).

115. H. Finkelmann, M. Happ, M. Portugall, and H. Ringsdorf, Makromol. Chem., Rapid Commun., **179**, 2541 (1978).
116. V. N. Tsvetkov, S. Ya. Magarik, S. I. Klenin, and V. E. Eskin, Vysokomol. Soedin., **5**, 3 (1963).
117. V. N. Tsvetkov, S. Ya. Magarik, T. Kadyrov, and G. A. Andreeva, Vysokomol. Soedin., **A10**, 943 (1968).
118. V. N. Tsvetkov, L. N. Andreeva, S. Ya. Magarik, P. N. Lavrenko, and G. A. Fomin, Vysokomol. Soedin., **A13**, 2011 (1971).
119. G. M. Pavlov, S. Ya. Magarik, G. A. Fomin, V. M. Yamschikov, G. A. Andreeva, V. S. Skaska, I. G. Kirillova, and V. N. Tsvetkov, Vysokomol. Soedin., **A15**, 1696 (1973).
120. S. Ya. Magarik, G. M. Pavlov, and G. A. Fomin, Macromolecules, **11**, 294 (1978).
121. V. N. Tsvetkov, Dokl. Akad. Nauk SSSR, **192**, 380 (1970).
122. V. N. Tsvetkov, I. N. Shtennikova, E. I. Rjumtsev, N. V. Pogodina, G. F. Kolbina, E. V. Korneeva, P. N. Lavrenko, O. V. Okatova, Y. B. Amerik, and A. A. Baturin, Vysokomol. Soedin., **A18**, 2016 (1976).
123. G. F. Kolbina, Thesis, Inst. Macromolecular Compounds, Acad. Sci. USSR, Leningrad (1981).
124. V. N. Tsvetkov, I. N. Shtennikova, E. I. Rjumtsev, G. F. Kolbina, I. I. Konstantinov, Y. B. Amerik, and B. A. Krenzel, Vysokomol. Soedin., **A11**, 2528 (1969).
125. V. N. Tsvetkov, E. I. Rjumtsev, I. I. Konstantinov, Y. B. Amerik, and B. A. Krenzel, Vysokomol. Soedin., **A14**, 67 (1972).
126. V. N. Tsvetkov, I. N. Shtennikova, E. I. Rjumtsev, G. F. Kolbina, E. V. Korneeva, B. A. Krenzel, Y. B. Amerik, and I. I. Konstantinov, Vysokomol. Soedin., **A15**, 2158 (1973).
127. I. N. Shtennikova, T. V. Peker, G. F. Kolbina, V. R. Petrov, V. S. Greneva, I. I. Konstantinov, and Y. B. Amerik, Vysokomol. Soedin., **A24**, 2047 (1982).
128. V. N. Tsvetkov, E. I. Rjumtsev, I. N. Shtennikova, E. V. Korneeva, G. I. Okhrimenko, N. A. Mikhailova, A. A. Baturin, Y. B. Amerik, and B. A. Krenzel, Vysokomol. Soedin., **A15**, 2570 (1973).
129. V. N. Tsvetkov, I. N. Shtennikova, G. F. Kolbina, A. I. Mashoshin, P. N. Lavrenko, S. V. Bushin, A. A. Baturin, and Y. B. Amerik, Vysokomol. Soedin., **A27**, 319 (1985).
130. G. M. Pavlov, E. V. Korneeva, T. I. Garmonova, D. Hardy, and K. Nitrai, Vysokomol. Soedin., **A22**, 1634 (1978).
131. E. I. Rjumtsev, I. N. Shtennikova, N. V. Pogodina, G. F. Kolbina, I. I. Konstantinov, and Y. B. Amerik, Vysokomol. Soedin., **A18**, 439 (1976).
132. V. N. Tsvetkov, E. V. Korneeva, I. N. Shtennikova, P. N. Lavrenko, G. F. Kolbina, D. Hardy, and K. Nitrai, Vysokomol. Soedin., **A14**, 427 (1972).

Chapter 7

ELECTRIC BIREFRINGENCE

1. MAIN RELATIONSHIPS FOR LOW-MOLECULAR-WEIGHT SUBSTANCES

Electric birefringence (EB) – the Kerr effect – is a phenomenon widely used in molecular optics as a method for investigating the molecular structure of low-molecular-weight substances. The study of this effect in the gas phase or in solutions in combination with other methods, such as refraction, light scattering, dielectric measurements, etc., may reveal the spatial arrangement of atoms in the molecule and thus can be used to calculate the main values of the polarizability tensor of the molecule and obtain information on the value and direction of its dipole moment [2–11].

The optical anisotropy appearing in the substance due to the effect of a weak steady-state electric field is proportional to the square of the field strength. This relationship, established by Kerr [12], is expressed by the equation

$$\Delta n/n \equiv (n_p - n_s)/n = K_e E^2, \qquad (7.1)$$

where $\Delta n \equiv n_p - n_s$ is the difference between extraordinary n_p and ordinary n_s refractive indices for two light beams with directions of the electric vector parallel (n_p) and perpendicular (n_s) to the electric field. E is the strength of the electric field whose direction normal to light beams n is a mean value of the refractive index of the substance. The coefficient K_e is usually called the Kerr constant.

If a solution is investigated, the characteristic value of EB introduced by the solute can be expressed by the constant K determined by the equation

$$K = \lim_{\substack{E \to 0, \\ c \to 0}} (\Delta n/cE^2), \qquad (7.2)$$

399

where $\Delta n = n_p - n_s$ is the excess difference (i.e., the difference introduced by the solute) between the extraordinary and ordinary refractive indices, and c is the weight concentration (g/cm^3) of the solute. We will call the coefficient K the Kerr constant of the solute.

The molar Kerr constant mK is often used to characterize the molecular properties of the liquid. According to Debye [5], it is determined by the equation

$$mK = 6K_e (M/\rho) n^2 / [(n^2 + 2)(\varepsilon + 2)]^2, \qquad (7.3)$$

where ε is the dielectric permittivity of the liquid, M and ρ are its molecular weight and density, respectively, and K_e is determined according to Eq. (7.1).

In solution the molar constant of the solvent mK_1 and the solute mK_2 are additive, and the molar constant of the solution mK_{12} is determined by the equation [8]

$$mK_{12} = 6K_{e12} n_{12}^2 [(n_{12}^2 + 2)(\varepsilon_{12} + 2)]^{-2} (M_1 f_1 + M_2 f_2)/\rho_{12}$$
$$= 6K_{e1} n_1^2 [(n_1^2 + 2)(\varepsilon_1 + 2)]^{-2} M_1 f_1 / \rho_1 + 6K_{e2} n_2^2 [(n_2^2 + 2)(\varepsilon_2 + 2)]^{-2} M_2 f_2 / \rho_2, \quad (7.4)$$

where the subscripts 1 and 2 refer to the solvent and solute, respectively; f_1 and f_2 are the molar fractions of the components in solution.

It follows from Eq. (7.4) that the molar constant of the solute extrapolated to zero concentration, $(mK)_{c \to 0}$, is given by

$$(mK)_{c \to 0} = 6KnM/[(n^2 + 2)(\varepsilon + 2)]^2. \qquad (7.5)$$

where the Kerr constant K is determined according to Eq. (7.2), M is the molecular weight of the solute, and n and ε are the refractive index and dielectric permittivity of the solvent, respectively.

It should be borne in mind that both the factor $(\varepsilon + 2)$ and the value of the numerical coefficient in Eq. (7.5) are determined by the choice of the internal field according to Lorentz. If the evaluation of the internal field is different, the factor containing ε and, correspondingly, the numerical coefficient, can change.

The molecular theory of the Kerr effect in gases and liquids [2, 3] explains the EB phenomenon by the orientation of optically anisotropic molecules in the electric field. The causes of the orientation are the anisotropy of dielectric polarizability of the molecules [2] and their dipole moments [3].

The Langevin–Born theory relates the molar constant mK of the substance to the main components of the tensor of optical (γ_1, γ_2, and γ_3) and dielectric (δ_1, δ_2, and δ_3) polarizabilities of the molecule and to the corresponding components (μ_1, μ_2, and μ_3) of its dipole moments by the equation

$$mK = \frac{2\pi N_A}{405kT} \left[\sum_{i,\,j=1,\,2,\,3} (\gamma_i - \gamma_j)(\delta_i - \delta_j) + \frac{1}{kT} \sum_{i,\,j=1,\,2,\,3} (\gamma_i - \gamma_j)(\mu_i^2 - \mu_j^2) \right]. \quad (7.6)$$

For molecules with axial symmetry of optical and dielectric polarizabilities ($\gamma_2 = \gamma_3$ and $\delta_2 = \delta_3$) Eq. (7.3) is simplified to

$$mK = \frac{4\pi N_A}{405kT}(\gamma_1 - \gamma_2)[\delta_1 - \delta_2 + (\mu^2/kT)(3\cos^2\theta - 1)/2],\qquad(7.7)$$

where θ is the angle formed by the dipole of the molecule μ and the axis of symmetry of its optical and dielectric polarizabilities (γ_1 and δ_1).

For a substance investigated in solution and consisting of molecules with axial symmetry of optical and dielectric properties according to Eqs. (7.5) and (7.7), the Kerr constant K is given by

$$K = \frac{2\pi N_A}{135kT}\frac{(n^2+2)^2}{p}\left(\frac{\varepsilon+2}{3}\right)^2\frac{\gamma_1-\gamma_2}{M}\left[\delta_1 - \delta_2 + \frac{\mu^2}{kT}(3\cos^2\theta - 1)/2\right].\qquad(7.8)$$

Equation (7.8) shows that the sign of EB depends on the value of μ and the direction (angle θ) of the dipole moment of the molecule. Thus, for nondipole molecules ($\mu = 0$), K is always positive since the signs of $\gamma_1 - \gamma_2$ and $\delta_1 - \delta_2$ coincide. For strongly polar molecules for which the term $\delta_1 - \delta_2$ in Eq. (7.8) may be neglected, the Kerr effect is positive at $\theta < 54.72°$ and negative at $\theta > 54.72°$. In any case, the sign of K is determined by the combination of values of $\delta_1 - \delta_2$, μ, and θ.

Equation (7.8) is widely used in the experimental study of optical and dipole structures of molecules by the EB method in dilute solutions [13–19]. According to Eq. (7.8), the Kerr effect in combination with the data on equilibrium dielectric properties of the same solutions allows the determination of the angle characterizing the direction of the dipole moment in the molecule. As could be seen above, this structural parameter is of major importance to both the value and the sign of the Kerr effect in liquid. For mesogenic substances the angle θ essentially determines the electrooptical and dielectric properties of the mesophase. These mesogenic substances are widely used in modern technology for the visual presentation of information (displays, etc.) [17-19].

2. SATURATION EFFECT IN STRONG FIELDS

The Langevin–Born theory [2, 3] leading to Eqs. (7.6) and (7.7) is based on the concept that the potential energy of molecules due to their rotation in the electric field is small compared to the energy kT of their thermal motion. This condition is expressed by the equations

$$\mu\mathcal{E}/(kT)\equiv\alpha\ll1,\quad(\delta_1-\delta_2)\mathcal{E}^2/(2kT)\equiv\beta\ll1,\qquad(7.9)$$

where $\mathcal{E}$ is the field acting on the molecule proportional to the applied field E [in Eqs. (7.1) and (7.2)] and differing from it in the factor of the internal field. Thus, according to Lorentz, $\mathcal{E} = (\varepsilon + 2)E/3$, but according to Onsager [20], another relationship exists.

If a strong field E is used, or rigid macromolecules (particles) with high μ values and a high anisotropy of polarizability $\delta_1 - \delta_2$ are studied, the simplifying assumptions in Eq. (7.9) cannot be used and the EB theory requires a more complete analysis of the phenomenon.

The general Lorentz–Lorenz equation relating the refraction of liquid to the polarizability of its molecules leads to Eq. (7.10), analogous to Eq. (5.10):

$$(n_p - n_s)/c \equiv \Delta n/c = 2\pi N_A (n^2 + 2)^2 (\overline{\gamma_p - \gamma_s})/(9Mn), \qquad (7.10)$$

where $\overline{\gamma_p - \gamma_s}$ is the average difference between the optical polarizabilities introduced by one molecule of the solute in the two main directions.

For molecules with axial symmetry of optical properties, the value $\overline{\gamma_p - \gamma_s}$ is related to the anisotropy of the molecule $\gamma_1 - \gamma_2$ by the general equation

$$\overline{\gamma_p - \gamma_s} = (\gamma_1 - \gamma_2) Q_\gamma = (\gamma_1 - \gamma_2) (3 \overline{\cos^2 \vartheta_\gamma} - 1)/2, \qquad (7.11)$$

where Q_γ is the degree of uniaxial orientational order of solute molecules determined by Eq. (6.10), and the angle ϑ_γ characterizes the orientation of the *optical* axis of the molecule with respect to the field direction E.

2.1. Dipole Orientation of Particles

In the case of a purely dipolar orientation of molecules in the electric field, when the anisotropy parameter β in Eq. (7.9) may be neglected, the average value of the cosine of the orientation angle ϑ_μ of molecular dipole moments with respect to the field is given by [4]

$$\overline{\cos \vartheta_\mu} = \mathscr{L} (\alpha), \qquad (7.12)$$

where $\mathscr{L}(\alpha)$ is the known Langevin function [21] mentioned previously [see Eq. (5.26)].

Under the same conditions, the following expression has been obtained for the mean-square cosine of the orientation angle of dipoles [22, 23]:

$$\overline{\cos^2 \vartheta_\mu} = 1 - 2\mathscr{L} (\alpha)/\alpha \qquad (7.13)$$

and, correspondingly,

$$Q_\mu \equiv (3 \overline{\cos^2 \vartheta_\mu} - 1)/2 = 1 - 3\mathscr{L} (\alpha)/\alpha = Q (\alpha). \qquad (7.14)$$

The combination of Eqs. (7.10), (7.11), and (7.14) leads to Eq. (7.15), representing the specific EB, $\Delta n/c$, of the solute as a function of the parameter of dipole orientation α over the entire range of values $0 \leq \alpha \leq \infty$:

$$\Delta n/c = [2\pi N_A (n^2 + 2)^2/9Mn] [(3 \cos^2 \theta - 1)/2] (\gamma_1 - \gamma_2) Q (\alpha), \qquad (7.15)$$

where θ is the angle formed by the optical axis (γ_1) of the molecule and its dipole moment μ, and the orientation parameter $Q(\alpha)$ is defined by Eq. (7.14).

According to Eq. (7.15), the sign of EB is determined by the value of the angle θ and the dependence of Δn on E^2 is determined by that of $Q(\alpha)$ on α^2.

At low α the function $Q(\alpha)$ is approximated by the dependence

$$Q(\alpha)_{\alpha \ll 1} = (\alpha^2/15)(1 - 2\alpha^2/21 + \ldots),\tag{7.16}$$

where the first term corresponds to the Kerr law (7.1). According to Eqs. (7.1) and (7.9), Eq. (7.16) leads to the Kerr constant determined by the second term on the right side of Eq. (7.8).

With increasing field strength E (and correspondingly increasing α) the function $Q(\alpha)$ and, correspondingly, the function $\Delta n(E)$, reflect the trend toward saturation. At high α the function $Q(\alpha)$ may be represented by the equation

$$Q(\alpha) = 1 - 3/\alpha + 3/\alpha^2 - \ldots\tag{7.17}$$

The limiting value of $Q(\alpha) = 1$ corresponds to complete dipole orientation parallel to the field E and to the saturation of the curve representing the dependence of Δn on E. If in this case the angle θ is zero, then according to Eq. (7.15) the optical axes of the molecules are also oriented parallel to the field.

2.2.   Orientation Induced by the Dielectric Anisotropy of Particles

If a molecule (particle) has no rigid dipole ($\mu = 0$ and $\alpha = 0$), but is dielectrically anisotropic ($\delta_1 - \delta_2 \neq 0$), then in the electric field a dipole is induced in this molecule as a result of the anisotropy of its polarizability. In this case the orientation of molecules in the field is caused by the orientation of dipoles induced in this molecule and the degree of orientation is determined by the parameter β in Eq. (7.9). For particles with axial symmetry of dielectric (δ) and optical (γ) polarizabilities with coinciding axes δ and γ, according to Eqs. (7.11) and (7.10) the degree of orientational order in the system $Q(\beta) = Q_\delta = Q_\gamma$ determines the value of Δn of the observed EB.

In this case the ordering parameter $Q(\beta)$ is determined by the equation [24, 25]

$$Q(\beta) = 3e^\beta \left/ \left(4\sqrt{\beta} \int_0^{\sqrt{\beta}} e^{-x^2} dx \right) \right. - 3(4\beta) - 1/2,\tag{7.18}$$

which is tabulated as a function of β [25].

In the range of weak fields E ($\beta \ll 1$) it follows from Eq. (7.18) that

$$Q(\beta)_{\beta \ll 1} = 2\beta/15, \tag{7.19}$$

which corresponds to the Kerr law. The substitution of this value into Eqs. (7.11) and (7.10) and the application of Eq. (7.9) lead to the Kerr constant determined by the first term on the right side of Eq. (7.8).

In the range of a weak field E the combination of Eqs. (7.16) and (7.19) leads to a general expression of the Kerr constant [Eq. (7.8)].

At a high field strength at $\beta > 10$, according to Eq.(7.18) the value of $Q(\beta)$ may be represented to a good approximation by the equation [25]

$$Q(\beta)_{\beta > 10} = 1 - 3/(2\beta - 1), \tag{7.20}$$

describing the phenomenon of saturation of the dependences $Q \equiv Q(\beta)$ and $\Delta n = \Delta n \cdot (E^2)$. According to Eq. (7.11), the value of $Q(\beta) = Q_\gamma = 1$ corresponds to the orientation of optical axes of the molecules parallel to the field.

Some researchers [25–27] have considered more general cases of particles with finite values of parameters α and β at various values of the $\alpha^2/(2\beta)$ ratio as well as cases of particles without axial symmetry of optical and dielectric properties. The results are presented in the form of graphs and tables [28].

There are numerous investigations of EB of solutions in which saturation effects have been observed. They have been carried out using colloidal systems, suspensions with macroscopic particles, virus solutions, rigid protein molecules, polypeptides, and nucleic acids, as well as some other polyelectrolyte systems. The results were interpreted using the theories considered above, and were used to characterize the optical, electrical, and dipole properties of the particles investigated [28, 29, 11]. It should be noted that this interpretation, applied to polyelectrolyte molecules (to which many of the investigated samples belong), can yield qualitative rather than quantitative results since for polyelectrolyte systems the separation of the dipole mechanism of orientation from the anisotropic mechanism presents a difficult problem [29–32].

3. NONSTEADY-STATE EB: RELAXATION PHENOMENA

The orientation of a particle (or molecule) due to the effect of the electric field does not occur instantaneously, but requires a certain length of time depending on the rotational mobility of particles determined by the coefficients of its rotational diffusion D_r.

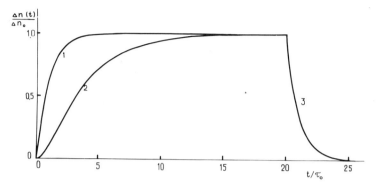

Fig. 7.1. Relative value of EB, $\Delta n(t)/\Delta n_0$, versus time t when the electric field is instantaneously switched on (1, 2) and off (3). 1, 3) Nondipolar ($\alpha = 0$) electrically anisotropic particles; 2, 3) dipolar electrically isotropic particles ($\beta = 0$).

The investigations of dielectric relaxation and NMR data have shown that the time of molecular reorientation in low-molecular-weight liquids ranges from 10^{-9} to 10^{-11} s. However, in solutions containing supermolecular particles or rigid macro-molecules, the reorientation times should be much longer and hence can be determined by electro-optical methods using fields variable with time in the radiofrequency range.

Hence, the study of the kinetics of the Kerr effect in these liquids or solutions with the application of an electric field varying with time can provide important information on the dynamics of the rotational motion of the particles or macro-molecules investigated.

The principal methods used for this purpose are the studies of EB in the pulse field and in the periodical (sinusoidal) field.

3.1. Transient EB in the Pulse Field

The electric pulse used may have quite varied shapes [33]. The rectangular pulse obtained by the instantaneous switching of the electric field on and off is used most frequently.

The process of the disappearance of EB when the field is instantaneously switched off is the most simple and universal process because it is determined only by the free disorientation of particles (molecules) under the influence of their thermal rotational motion and is independent of the electrical properties of particles. In this case, for a monodisperse system of rigid particles the decrease in birefringence Δn with time t (after the field has been switched off) proceeds by the simple exponential law [34, 35] characterized by the relaxation time τ_0:

$$\Delta n = \Delta n_0 e^{-t/\tau_0}, \tag{7.21}$$

where Δn_0 is the initial EB of the system in the steady-state field.

For particles with axial symmetry of geometrical shape and optical properties, the relaxation time τ_0 is related to the diffusion coefficient D_r of the particle in its rotation about the axis normal to the axis of symmetry:

$$\tau_0 = 1/6D_r. \tag{7.22}$$

Equations (7.21) and (7.22) are universal equations describing the process of free relaxation of the anisotropy of solution regardless of the type of external forces inducing the orientation of particles. In particular, Eq. (7.22) may also be applied to the relaxation of FB in a system of rigid particles [Eq. (5.50a) and the following equations]. The dependence (7.21) is described by curve 3 in Fig. 7.1.

The process of the appearance of EB when the electric field is switched on is more complex. In this case, the rate of the process is determined not only by the thermal motion of particles but also by the type of effects induced by the electric field. Hence, under these conditions the dependence of Δn on t is different for dipolar particles and for dielectrically anisotropic particles.

In the general case, for a system of particles with axial symmetry of shape and dielectric polarizability ($\delta_1 - \delta_2$ is the anisotropy of polarizability) and with the dipole moment μ parallel to the axis of symmetry δ_1, the dependence of Δn on t after the switching on of a weak field (in the range in which the Kerr law is obeyed) is determined by the equation [34, 35]

$$\Delta n / \Delta n_0 = 1 - 3\,(\alpha^2/\beta)\,(2\alpha^2/\beta + 4)^{-1}\,e^{-t/\tau} + (\alpha^2/\beta - 4)\,(2\alpha^2/\beta + 4)^{-1}\,e^{-t/\tau_0}, \tag{7.23}$$

where

$$\tau = 1/2D_r; \tag{7.24}$$

τ_0 is determined from Eq. (7.22) and α and β are determined from Eq. (7.9).

The relaxation time τ is three times as long as τ_0 and coincides with that of the polarization of the liquid containing dipolar molecules [4].

If the particle has no dipole moment ($\alpha = 0$) and EB is determined only by the anisotropy of its polarizability $\delta_1 - \delta_2$ (induction effect), then Eq. (7.23) becomes

$$\Delta n / \Delta n_0 = 1 - e^{-t/\tau_0}. \tag{7.25}$$

In this case, according to Eq. (7.25) the curve of the appearance of Δn after the field has been switched on (curve 1 in Fig. 7.1) is identical with that of its disappearance [Eq. (7.21)].

If the anisotropy of polarizability of the particle, $\delta_1 - \delta_2$, is small and the orientation in the field is virtually due only to the dipole moment μ, then $\alpha^2/\beta \gg 1$, and

Eq. (7.23) reduces to

$$\Delta n/\Delta n_0 = 1 - (3/2)\, e^{-t/\tau} + (1/2)\, e^{-t/\tau_0}. \tag{7.26}$$

In this case the increase in Δn proceeds much more slowly than in the case of induction orientation and the curve $\Delta n(t)$ is tangential to the axis t at $t = 0$ (curve 2 in Fig. 7.1).

In general, when the signs of α^2 and β coincide according to Eq. (7.23), the process of establishing the equilibrium value Δn_0 becomes increasingly slow with increasing α^2/β ratio, i.e., with an increasing contribution of the permanent dipole of the particle to EB.

If the signs of α^2 and β are different (which occurs if the dipole is inclined to the axis of maximum polarizability of the particle at a large angle), then after the field has been switched on the increase in EB can occur with the reversal of the sign of Δn.

Hence, the study of the type of dependence of $\Delta n(t)$ when a rectangular pulse is switched on in principle provides information about the relative contribution of the permanent and induced moments of the particle to the EB phenomenon.

The method of reversing pulses in which the field direction is instantaneously reversed [36, 37] can also provide useful information on this problem. For particles with axial symmetry of optical properties and the dipole parallel to the axis of symmetry, the reversal of the field is accompanied by a change in EB according to the dependence [36]

$$\Delta n/\Delta n_0 = 1 + 3\,(\alpha^2/\beta)\,(\alpha^2/\beta + 2)^{-1}(e^{-t/\tau_0} - e^{-t/\tau}). \tag{7.27}$$

It follows from Eq. (7.27) that for nonpolar particles ($\alpha = 0$) $\Delta n/\Delta n_0$ is unity, i.e., EB does not vary upon field reversal but remains equal to the equilibrium value Δn_0. In contrast, when a permanent dipole exists, the curve $\Delta n(t)$ should be expected to exhibit a minimum Δn_m whose depth is determined by the equation

$$\alpha^2/\beta = 2\,(1 - \Delta n_m/\Delta n_0)/(0.1547 + \Delta n_m/\Delta n_0), \tag{7.28}$$

and the time required for the attainment of the minimum is given by

$$t_{\min} = 0.2747/D_r. \tag{7.29}$$

The study of this curve can evidently provide information on the values of α^2/β and D_r.

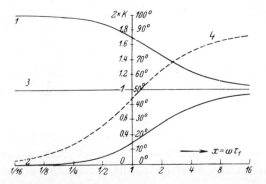

Fig. 7.2. EB of solution of dielectrically anisotropic particles versus cyclic frequency ω of the sinusoidal electric field:
$\tau_1 = 2\tau_0 = 1/3D_r = 2\tau/3$. 1, 2) Maximum (1) and minimum (2)
values of the instantaneous Kerr constant $K(t)$ attained during
the field period; 3) average value of the Kerr constant K independent of the field frequency; 4) phase shift φ between the
field and $K(t)$.

Pulse methods are widely applied to the study of the dynamics of EB in colloidal systems and solutions of viruses, nucleic acids, proteins, and many other
biogenic particles and molecules [30, 33]. In the investigation of electro-optical
properties of these samples the application of the pulse procedure is virtually un-
avoidable because of the considerable electric conductivity of these systems.

3.2. Dispersion of EB in the Sinusoidal Field

In the sinusoidal electric field the strength $E = E_0 \cos \omega t$ is a function of time t,
and hence the distribution function of the orientations of dielectrically anisotropic
and polar particles should also be dependent on time. Correspondingly, the degree
of order Q and the observed excess birefringence Δn introduced by the solute also
vary with time.

The EB theory in a weak sinusoidal field (in the range of strengths in which
the Kerr law is obeyed) has been developed by Peterlin and Stuart [38] for rigid
dipole particles with axial symmetry of geometrical shape, optical and dielectric
polarizabilities, and a dipole which coincides with the axis of symmetry of the particle. The cases of dielectrically anisotropic particles (induced dipole) and particles
with a permanent dipole are considered separately.

3.2.1. Dielectrically Anisotropic Particles.
If the orientation of
particles in an electric field is induced by their dielectric anisotropy, then the instantaneous value of the excess FB of solution, $\Delta n(t)$, and the corresponding instantaneous Kerr constant, $K(t)$, according to theory [38], are defined by the equation

$$K(t) \equiv \frac{\Delta n(t)}{cE_{\text{eff}}^2} = \frac{2\Delta n(t)}{cE_0^2} = K\left[1 + \frac{\cos(2\omega t - \varphi)}{(1 + 4\omega^2\tau_0^2)^{1/2}}\right], \quad \tan\varphi = 2\omega\tau_0, \qquad (7.30)$$

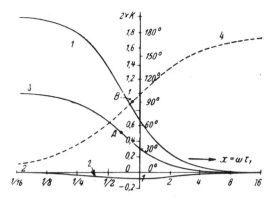

Fig. 7.3. EB of a solution of dipolar particles versus cyclic frequency ω of the sinusoidal electric field: $\tau_1 = 2\tau_0 = 2\tau/3 = 1/3D_r$. 1, 2) Maximum (1) and minimum (2) values of the instantaneous Kerr constant $K(t)$ attained during the field period; 3) the average value of the Kerr constant K_v depending on the field frequency $\nu = \omega/2\pi$ according to Eq. (7.32); 4) the phase shift between the field and $K(t)$.

where ω and E_0 are the cyclic frequency and the amplitude value of the field strength, and τ_0 is the relaxation time of anisotropy determined from Eq. (7.22). The constant K is identical with the value of K determined from Eq. (7.2).

It follows from Eq. (7.30) that EB is the sum of the constant component (curve 3 in Fig. 7.2) representing the EB averaged over time and the variable component

$$E_{\text{eff}} = E_0/\sqrt{2}.$$

The latter component is the harmonic function with double frequency 2ω and amplitude which at low frequencies ($4\omega^2\tau_0^2 \ll 1$) is equal to the constant part K, and with increasing frequency decreases to zero (Fig. 7.2, curves 1 and 2). The phase angle φ formed by the field and $\widetilde{K}$ increases with increasing frequency ω tending to the limit $\pi/2$ at $\omega \to \infty$ (Fig. 7.2, curve 4).

Hence, in a sinusoidal field EB and, correspondingly, $K(t)$ oscillate at a double frequency symmetrically with respect to the average value of K, and its sign coincides with that of K at any moment in time. With increasing frequency ω the amplitude of these oscillations decreases to zero. Hence, when a method recording the values of EB averaged over time (e.g., the visual method) is used, the measured EB value is independent of the field frequency and remains equal to the equilibrium value in a constant field with strength equal to $E_{\text{eff}} = E_0/\sqrt{2}$.

These relationships arise from the fact that the torque acting upon a nondipolar dielectrically anisotropic particle is proportional to $(\delta_1 - \delta_2)E^2$ and, hence, to the square of the field strength. Hence, the rotation of this particle depends on the field strength and not on its sign, and the particle does not react to the change in the field direction. In a sinusoidal field the particle is subjected to rotational pulses of only one sign; its rotation is not reversible and reduces to rotational vibrations abut the angular equilibrium position. In an assembly of particles in the equilibrium position the axes of their maximum polarizability δ_1 are not randomly oriented in space but exhibit a predominant orientation in the direction of the field E.

With increasing field frequency the particles having limited rotational mobility cease to react to field variations, which leads to a decrease in the amplitude of oscillations of the variable EB component ($\widetilde{K}$) down to its complete disappearance. In this case, however, the particles retain the predominant orientation from the field, and EB remains invariable (constant).

3.2.2. Particles with a Permanent Dipole. The behavior of EB is quite different from that for dielectrically anisotropic particles if an assembly of particles with permanent dipoles is present in the sinusoidal field.

In this case, the torque acting upon the particle reverses its sign twice during one period. At each reversal of the field direction the particles are forced to reverse in such a manner that their dipoles follow the field. At the moment of this reversal the dipoles and the optical axes of particles are oriented predominantly normal to the field and hence at these moments the sign of EB is opposite to that in the equilibrium state (Fig. 7.3, curve 2).

Rotational friction results in a retardation of particles following the field in their rotation and leads to the phase shift φ between the sinusoidal field and the sinusoidal anisotropy $\Delta n(t)$ of the medium. With increasing field frequency ω the angle φ increases, approaching π at $\omega \to \infty$ (Fig. 7.3, curve 4).

Moreover, with increasing frequency the rotational friction increasingly restricts the reversal of polar particles in accordance with the field. At a relatively high frequency the particles cease to respond to reversing of the field, their orientations become random, and the anisotropy of solution disappears completely.

Taking into account the above mechanisms of particle motion, the theory [38] leads to the following frequency and time dependence of the instantaneous EB, $\Delta n(t)$, and instantaneous Kerr constant $K(t)$:

$$K(t) = \frac{\Delta n(t)}{cE_{\text{eff}}^2} = \frac{2\Delta n(t)}{cE_0^2} = K\left[\frac{1}{1+\omega^2\tau^2} + \frac{\cos(2\omega t - \varphi)}{(1+4\omega^2\tau_0^2)^{1/2}(1+\omega^2\tau^2)^{1/2}}\right],$$

$$\tan\varphi = \frac{5\omega\tau}{3 - 2\omega^2\tau^2}, \qquad (7.31)$$

where K as before is the equilibrium value of the Kerr constant determined from Eq. (7.2) and τ is the relaxation time of dipole orientation determined from Eq. (7.24).

Equation (7.31) shows that, as in the case of dielectrically anisotropic particles, in a solution of dipole particles at an invariable field frequency the birefringence is the sum of the constant component [first term on the right side of Eq. (7.31)] and the component which is the harmonic function of time [second term in Eq. (7.31)]. The amplitude values of EB are shown by curves 1 and 2 in Fig. 7.3.

At the time moments when $2\omega t - \varphi = \pi, 3\pi, 5\pi, \ldots$ (moments of particle reversal), the second term in brackets in Eq. (7.31) is negative in sign and greater in numerical value than the first, which is positive (since $\tau_0 = \tau/3$). Hence, at these moments the sign of EB is opposite to that of K, which is shown by curve 2 in Fig. 7.3.

The first term in Eq. (7.31), K_v, which is independent of time,

$$K_v = K/(1 + \omega^2\tau^2),\qquad(7.32)$$

is the value of the Kerr constant averaged over time (Fig. 7.3, curve 3). This average value K decreases with increasing field frequency and disappears at high frequencies as a result of the relaxation of the rotational motion of dipole particles.

Hence, EB in a solution of dipole particles in a sinusoidal field exhibits dispersion, in contrast to the Kerr effect in a solution of electrically anisotropic nondipolar particles. Thus, the study of the frequency dependence of the Kerr effect can serve as the most sensitive method of separating the anisotropic and dipole effects in the total EB usually observed in an assembly of real dielectrically anisotropic and polar particles. The type of frequency dependence of EB for these particles is shown in Fig. 7.4 for the cases in which the signs of the dipole and the anisotropic components coincide and for when they are opposite.

4. KERR EFFECT IN NONPOLYMERIC SUBSTANCES WITH CHAIN MOLECULES

The investigations of the Kerr effect in low-molecular-weight substances exhibiting a linear chain structure have already been carried out in early works. Thus, Stuart et al. [8] have measured EB in a series of normal alkanes C_nH_{2n+2} in the liquid phase.

The experimental values of the Kerr constant K obtained from Eq. (7.2) are shown in Fig. 7.5 (curve 1) as a function of the number $Z (= n)$ of carbon atoms in the chain. Over the entire range of Z values investigated (from 5 to 30) the value of K is positive and monotonically increases with chain length. Moreover, in the range $Z < 15$ the value of K may be considered to be proportional to Z within experimental error.

For nonpolar molecules of normal hydrocarbons ($\mu = 0$) the proportionality of K to the chain length (or to M) according to Eq. (7.8) implies an increase in the anisotropy of the chain $\gamma_1 - \gamma_2$ and $\delta_1 - \delta_2$ proportional to its length, as for a "rodlike" structure. This relationship is consistent with the concept that the con-

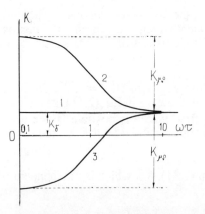

Fig. 7.4. Frequency dependence of the average Kerr constant K of a solution of dielectrically anisotropic dipolar particles. 1) K_δ, the anisotropic part of the Kerr constant; 2) the overall effect $K = K_\delta + K_\mu$ at coinciding signs of K_δ and K_μ (dipolar part of the Kerr constant); 3) overall effect $K = K_\delta + K_\mu$ at different signs of K_δ and K_μ; $K_{\mu,0}$ is the value of K_μ under equilibrium conditions ($\omega \to 0$).

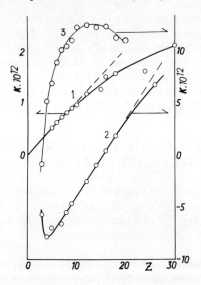

Fig. 7.5. Kerr constant (in $g^{-1} cm^5 statvolts^{-2}$) for normal alkanes and their derivatives versus the number Z of carbon atoms in the chain: 1) n-alkanes in bulk; 2) n-alcohols in benzene; 3) dibromoalkanes in cyclohexane.

formation of a linear hydrocarbon chain containing less than 10–15 carbon atoms may be approximated by a planar *trans*-chain which is oriented in the electric field "as a whole" under the influence of the moment induced in it. The value of $Z = 10$–15 is in good agreement with the number of monomer units $S = 5$–8 in a Kuhn segment for typical flexible carbon chain polymers.

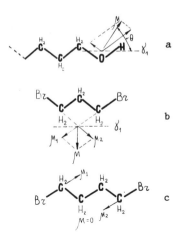

Fig. 7.6. Dipole moment μ of molecules of *n*-alkane derivatives in the *trans*-conformation. a) Normal alcohol: μ forms an angle $\theta = 62°$ with the axis of maximum polarizability γ_1 of the molecule; b) dibromopropane: μ is perpendicular to the axis of maximum polarizability γ_1 of the molecule; c) dibromobutane, $\mu = 0$.

As Z increases further, curve 1 shows a trend toward saturation, which can be ascribed to the effect of natural chain flexibility with chain length.

The initial slope of curve 1 in Fig. 7.5 and the application of Eq. (7.8) (if it is assumed that $\gamma_1 - \gamma_2 = \delta_1 - \delta_2$) make it possible to evaluate the average anisotropy of the hydrocarbon chain per CH_2 group $\Delta\gamma = 3.2\cdot10^{-25}$ cm³, which is equivalent to the specific anisotropy of the chain $\beta_M = 0.23\cdot10^{-25}$ cm³/dalton. The latter value corresponds in order of magnitude to that obtained by the FB method for polymer chains in matching solvents (e.g., see Fig. 6.17, curve 5). The value of $\Delta\gamma = 3.2\cdot10^{-25}$ cm³ is less by a factor of 1.5–2 than that for normal hydrocarbons in the gaseous state [39, 40].

Another example of EB investigation in a homologous series of low-molecular-weight chain molecules is found in [41], in which solutions of normal alcohols ($C_nH_{2n+1}OH$) in benzene have been studied. The experimental dependence of the Kerr constant K on the number Z of carbon atoms in the chain is represented by points on curve 2 in Fig. 7.5. Just as for *n*-alkanes, the points for *n*-alcohols in the range $4 \leq Z \leq 18$ fit a straight line of which a considerable part lies, however, in the range of negative K values. A characteristic feature of the dependence $K(Z)$ for *n*-alcohols is the change in K from negative to positive when the chain length increases.

In the interpretation of this dependence with the aid of Eq. (7.8) it should be taken into account that the molecules of *n*-alcohols are polar and the dipole moment $\mu = 1.7\cdot10^{-18}$ cgse $= 1.7$ D is located on one chain end and does not vary with chain length [41]. Correspondingly, the dipole component of K remains invariable with increasing Z, whereas the positive anisotropic part, as in the case of n-alkanes, increases proportionally to the length of the "rodlike" *trans*-chain. Naturally, lower homologs ($Z < 4$) should be excluded from this consideration because they do not correspond to the model under discussion.

Just as for n-alkanes, the slope of the straight part of curve 2 makes it possible to evaluate the contribution $\Delta\gamma$ of each CH_2 group to the anisotropy of the chain in the *trans*-conformation. Using the first term in Eq. (7.8), and assuming that $\gamma_1 - \gamma_2 = \delta_1 - \delta_2$ we obtain $\Delta\gamma = 7.8 \cdot 10^{-25}$ cm^3 for a hydrocarbon chain in benzene. This value is slightly higher than that of $\Delta\gamma$ in the gas phase [39, 40] and exceeds more than twice that obtained for n-alkanes in bulk.

This result demonstrates again the fact that the optical (and dielectric) anisotropy of a chain molecule in solution depends on the properties of the medium and cannot be uniquely calculated from the "valence optical scheme" using the anisotropy of bonds and groups in the gaseous state without taking into account the specific solvent. In particular, the high value of $\Delta\gamma$ for alkanes in benzene is in good agreement with the well-known fact that the positive optical anisotropy of chain molecules in solvents with anisotropic molecules is always higher than in isotropic solvents (Chapter 5, Section 2.1.2.3).

The negative sign of the Kerr constant for lower homologs of n-alcohols and the reversal of sign with increasing chain length imply that the dipole term in Eq. (7.8) is negative and hence the dipole moment μ of the C–O–H group is inclined at a considerable angle θ in the direction of the *trans*-chain. The value of the angle θ may be determined using the fact that, according to curve 2 in Fig. 7.5, the value of K becomes zero in the range of $Z_0 = 15.6$ in which the anisotropy of the *trans*-chain is $(\delta_1 - \delta_2)_0 = (\gamma_1 - \gamma_2)_0 = Z_0 \cdot \Delta\gamma = (15.6)(7.8) \cdot 10^{-25} = 122 \cdot 10^{-25}$ cm^3. The substitution of these values into Eq. (7.8) gives $\theta = 62°$. This value reasonably agrees with that proceeding from the dipole moments of C–O and OH bonds forming the dipole of the molecule (Fig. 7.6a). This value only slightly exceeds the critical value $\theta = 54.7°$ at which the dipole term does not provide any contribution to EB. Hence, the contribution to the Kerr effect provided by the dipole term for n-alkanes is not high and, therefore, it is possible to evaluate relatively reliably the significance of the anisotropic term leading to the linear dependence of K on Z.

In the range of $Z > 20$, curve 2 in Fig. 7.5 exhibits a deviation from linearity which, just as for n-alkanes, should be attributed to the effect of chain flexibility.

In a more recent paper, similar investigations of EB have been carried out in a homologous series of dibromoalkanes ($BrC_nH_{2n}Br$) dissolved in cyclohexane [42].

As in the two preceding cases, here the predominant component of the molecule is a nondipole n-alkane chain containing from 3 to 20 CH_2 groups. However, in contrast to n-alcohols, the molecule of dibromoalkane contains a strongly polar C–Br group on both ends, which leads to a specific dependence of K on Z represented by the points and curve 3 in Fig. 7.5. In this case, the dipole

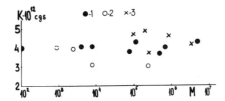

Fig. 7.7. Kerr constant K versus molecular weight of polystyrene in tetra-chloromethane: 1 cgse = 1 g^{-1} cm^5 statvolt^{-2}. The points correspond to the literature data: 1) [58]; 2) [59]; 3) [60].

moment of the molecule "as a whole" is determined by the vector sum of two dipoles located at chain ends and hence is strongly dependent on a mutual correlation of orientations of these dipoles. Thus, for the first homolog investigated ($n = Z = 3$) for the predominant *trans*-structure for the chain the dipole is normal to the axis of maximum polarizability γ_1 of the molecule (Fig. 7.6b). This leads to the negative value $K = -0.7 \cdot 10^{-12}$ cm^5 g^{-1} statvolt^{-2}. However, for the next homolog of the series ($Z = 4$) for a molecule with the *trans*-structure the dipole moment should already be equal to zero (Fig. 7.6c), which corresponds to a drastic change in the sign and value of the Kerr constant (Fig. 7.5, curve 3). The subsequent shape of the dependence $K(Z)$ shown by this curve can be qualitatively explained by a rapidly decreasing correlation between the orientations of polar groups on chain ends (i.e., by an increase in the array of conformational states [42]) with increasing chain length and decreasing contribution of the dipole [Eq. (7.8)] to the Kerr effect. Curve 3 virtually attains saturation at values $K \approx 10^{-11}$ cm^5 g^{-1} statvolt^{-2}, which is close to those of the highest homologs of the series of *n*-alcohols (Fig. 7.5, curve 2).

These examples show that when a dielectrically anisotropic chain molecule is subjected to the action of an electric field, mutual correlation in the orientations of its units separated along the chain by approximately the Kuhn segment length or less is distinctly manifested. For anisotropic molecules whose length does not exceed this limit, the process of orientation in the field may occur by the mechanism of rotation of the entire chain "as a whole" using Eq. (7.8) for the description of EB. The validity of these considerations is shown by the resemblance between the EB and flow birefringence (FB) phenomena in *n*-alkanes and *n*-alcohols. This resemblance is expressed both in the coinciding signs of the anisotropy in electric and mechanical fields and in the type of dependence of EB and FB on the chain length (compare Figs. 7.5 and 5.5).

This conclusion remains valid when only *one* polar group is present at the chain end, even if its presence leads to negative EB for the lowest homologs of the series.

When two (or more) polar groups separated along the chain are present in the molecule, the situation is more complex, since the orientation of the chain as a whole requires a joint orientation of all its polar units in the field. The data obtained for dibromoalkanes show, however, that the correlation in the orientations of polar groups decreases rapidly with increasing length of the chain segment between them.

Hence, it may be expected that the orientation of polar chain molecules in the electric field can occur by the mechanism of small-scale motions when the kinetic unit oriented in the field is much shorter than the Kuhn segment whose length is determined for this chain by other methods (e.g., by hydrodynamic methods).

5. SOLUTIONS OF FLEXIBLE-CHAIN POLYMERS

5.1.   Theoretical Aspects

The first theory of the Kerr effect in a constant electric field for flexible polymer molecules represented by a chain of freely jointed Kuhn segments has been proposed by Peterlin and Stuart [43, 44]. In accordance with the assumed model for the molecule, the theory is based on the supposition that the orientations of segments in the electric field are mutually independent. Hence, the result of this theory coincides completely with Eqs. (7.6)–(7.8) of the Langevin–Born theory for low-molecular-weight gases and liquids if the values of the dipole moment μ and optical (γ) and dielectric (δ) polarizabilities in these equations are assumed to refer to one segment of the polymer molecule. This assumption implies that in the Peterlin–Stuart theory birefringence is proportional to the overall segment concentration in solution regardless of whether the segments are distributed in long or short chain molecules. Hence, the main sequence of the theory is the conclusion that the Kerr constant K is independent of the molecular weight of the polymer if this molecular weight is sufficiently high for the application of conformational statistics of Gaussian chains to the molecules under investigation. This general conclusion also remains valid in all the subsequent theories in which the freely jointed chain model is not used but K is directly related to such structural parameters of the molecule as polarizability and dipole moments of the chemical bonds and the type of rotation of these bonds in the chain [46–51].

The methods developed by Flory and his school [48] in principle permit the calculations of the Kerr constant averaged over configurations for a flexible polymer chain of any length and conformation from a Gaussian coil to the monomer. These calculations are based on the rotational isomeric state model and the principles of vector additivity of dipole moments and tensor additivity of polarizabilities of chain valence bonds (valence–optical scheme).

The rotational isomeric state model already mentioned in Chapter 1, Section 2 is an approximate scheme in which the real rotation of chain units is replaced by a set of discrete rotational–isomeric states with different energetic levels. These states and the energies related to their changes are selected by analogy with known low-molecular-weight compounds. This method of describing the conformational properties of polymer chains has been used in recent papers and leads to the determination of the geometrical dimensions of molecules (e.g., second moments $\langle h^2 \rangle$) and their equilibrium rigidity parameters which are in reasonable agreement with experimental data [48].

The principle of vector additivity of dipole moments of chemical bonds has long been widely and reliably used for low-molecular-weight compounds [4–8], and the possibility of its application to the calculation of dipole moments of polymer molecules is beyond doubt.

The situation is more complex for the principle of tensor additivity of polarizabilities of valence bonds. Extensive experimental data obtained in various laboratories show that the optical anisotropy of a single valence bond can differ depending on the type of compound and the atomic group which contains this bond. A clear example of this difference is offered by a polyisobutylene molecule which in a matching solvent exhibits considerable positive anisotropy in spite of the tetrahedral symmetry of its structure [52]. The anisotropy of a polymer molecule in solution can also vary depending on the type of solvent.

In a number of papers the deviations from the additivity rule of bond polarizabilities have been ascribed to induction interactions [53] between these bonds, and methods of quantitative evaluation of the effect of these interactions have been proposed [51, 54, 55]. In some papers, corrections have been made for anisotropy due to collisions [56], and in other papers resonance interaction in the molecule has been taken into account [57].

In some cases, using the values of anisotropy of bond polarizabilities corrected in this manner, it has been possible to obtain agreement between theoretical and experimental values of the Kerr constants in the range of oligomers with a known dependence of K on the chain length [50, 51]. Even in the simplest case of n-alkanes the anisotropies of bond polarizabilities in these works could differ by a factor of two or three [50] from published values [40]. Hence, although these theoretical results are doubtless a success of the theory, they show that the valence–optical scheme in its "classical" form cannot be used with standard tabulated values of the anisotropy of polarizability of valence bonds.

However, although the theory does not give standard schemes for calculating the absolute Kerr constant, the dependence of K on the chain length can provide information on structural properties and the type of orientation of the molecule in the electric field, as has been shown above with n-alkane derivatives (Fig. 7.5).

5.2. Some Experimental Data

5.2.1. Molecular-Weight Dependence. Marinin et al. [58] have measured the Kerr constant in solutions of a series of polystyrene fractions in tetrachloromethane in the molecular-weight range $4 \cdot 10^3$–$5 \cdot 10^6$ and have shown that K is positive in sign and does not change with molecular weight within experimental error but remains close to the value of $K = 4 \cdot 10^{-12}$ g^{-1} cm^5 statvolt^{-2}. Their experimental data are shown by points (1) in Fig. 7.7. According to these data, the value of K for styrene in this solvent is virtually the same as that for polystyrene.

More recently these results for polystyrene have been confirmed by Champion et al. [58]. Their data are shown by points (2) in Fig. 7.7. The scattering of points for individual fractions is within experimental error. According to the data of LeFebre et al., shown by points (3) in Fig. 7.7, a certain decrease in K is observed with increasing molecular weight, but this decrease is illusory because it is within experimental error.

TABLE 7.1. Kerr Constants, K, and Shear Optical Coefficients, $\Delta n/\Delta\tau$, for Solutions of Some Polymers and Anisotropy of the Monomer Unit, Δa, of Their Molecules

No.	Polymer	$M \cdot 10^{-3}$	Electro-optical data		Dynamo-optical data		$\Delta a \cdot 10^{25}$, cm³
			Solvent	$K \cdot 10^{12}$, g⁻¹ cm⁵ statvolt⁻²	Solvent	$(\Delta n/\Delta\tau) \cdot 10^{10}$, cm s² g⁻¹	
1	Styrene (monomer)	0.104	—	4.0 [58]—7.0 [59]	—	+2.2 [75]	—
2	Polystyrene	4—500	Tetrachloromethane	4.0 [58, 59, 73]	Bromoform	—13.6 [74]	—18
3	Poly(2,5-dichlorostyrene)	900	Dioxane	4.0 *	Bromoform	—24 [74]	—30
4	Poly(para-chlorostyrene)	100	Dioxane	3.5 [73]	Bromoform	—25 [74]	—30
5	Polyvinylchloride	42	Dioxane	—7.7 [70]	Tetrahydrofuran	+3.3 [76]	+4.6
6	Poly(oxyethylene glycol)	6—14	Tetrachloromethane	—0.9 [64]	Cyclohexanone	+1.1 [77]	+1.5
7	Poly(methyl methacrylate)	420	Benzene	9 *	Benzene	+0.2 [78, 79]	+0.3
		740	Benzene	10 *	Benzene	+0.2	+0.3
		1100	Benzene	9 *	Benzene	+0.2	+0.3
		740	Dioxane	3.2 *	—	—	—
8	Poly(butyl methacrylate)	500	Dioxane	9.0 *	Benzene	—1.2 [80]	—2.1
			Tetrachloromethane	1.3 *			
9	Poly(octyl-cetyl methacrylate) copolymer (1:1)	120—450	Tetrachloromethane	1.9 *	Heptane	—5.5 [81]	—5.6
10	Poly(cetyl methacrylate) (PMA-16)	500	Tetrachloromethane	2.6 *	Benzene, tetrachloromethane	—12 [79]	—8.9
11	Poly(tert-butylphenyl methacrylate)	570	Tetrachloromethane	—4.0 *	Tetrachloromethane Bromobenzene	+0.64 [82] —8.1 [82]	+0.7 —8.8
12	Poly(methyl acrylate) (PA-1)	70—2900	Toluene	8 *	Toluene	+1.3 [83]	+1.9
13	Poly(butyl acrylate) (PA-4)	40—900	Toluene	9 *	Toluene	—0.82 [83]	—1.1
			Decalin	3.3 *	Decalin	—1.45 [83]	—1.94
14	Poly(decyl acrylate) (PA-10)	63—1100	Decalin	3.0 *	Decalin	—6.0 [83]	—3.7
15	PA-16	47—13000	Decalin	3.0 *	Decalin	—11.5 [83]	—6.4
16	PA-18	180—3600	Decalin	3.2 *	Decalin	—15.5 [83]	—6.6
17	PA-22	500	Decalin	3.1 *	Decalin	—16.5 *	—6.8

*Data were obtained in the author's laboratory.

The constancy of the K value with changing molecular weight M of a flexible-chain polymer has always been confirmed by experimental data obtained in those investigations of molecular solutions and using solvents ensuring the required precision of measurements [10, 61–64].

The strong molecular-weight dependence of the Kerr constant observed in polyvinylchloride [65] and polyvinylbromide [66] solutions was doubtless due to the fact that these solutions are nonmolecular. The presence of aggregates in the solutions of these two polymers has repeatedly been proven by various experimental methods [67, 70]. Incomplete solubility of poly(methyl methacrylate) (PMMA) fractions in benzene due to their different stereoregularity (tacticity) can also lead to the molecular-weight dependence of the value and sign of EB of their solutions [71]. However, in well-purified molecular solutions of PMMA in benzene the value of K remains constant upon variation in molecular weights (Table 7.1).

The main difficulty in the experimental investigations of EB in solutions of flexible-chain polymers is due to the low value of the observed effect. Even for polar macromolecules the Kerr constants K of noncharged flexible-chain polymers are $K \approx 10^{-12}$ g^{-1} cm^5 statvolt^{-2}. These values are often close to those for isotropic nonpolar solvents such as tetrachloromethane ($K = 0.3 \cdot 10^{-12}$ g^{-1} cm^5 statvolt^{-2}) or cyclohexane ($K = 0.4 \cdot 10^{-12}$ g^{-1} cm^5 statvolt^{-2}) and sometimes can be lower than that for benzene ($K = 2.6 \cdot 10^{-12}$ g^{-1} cm^5 statvolt^{-2}), also a nonpolar but anisotropic solvent. Hence, the determination of excess EB (as compared to the solvent) of a polymer, Δn, in a dilute solution is a difficult problem, and the error in this determination (and, correspondingly, in K) is usually great. This fact can probably be demonstrated if we take as an example the data on the EB of polyhydroxyethylene glycols in benzene [72] according to which the Kerr constant for this polymer at high molecular weights sharply decreases (with sign reversal) with increasing chain length. However, for solutions of the same polymer in tetrachloromethane the dependence of K on M is quite different, and in the molecular-weight range of 5000 and higher the Kerr constant remains invariable with change in M at $K = -0.9 \cdot 10^{-12}$ g^{-1} cm^5 statvolt^{-2} [64]. This limiting negative K value is only three times higher (in absolute value) than that for tetrachloromethane and twice as low as that for benzene.

These examples demonstrate the fact that the quantitative interpretation of experimental data on the Kerr effect in solutions of flexible-chain polymers using modern theories [48] requires great caution, in particular if the polymers under investigation are poorly soluble in solvents with isotropic nonpolar molecules. This remark refers, for example, to polyvinylchloride solutions in tetrachloromethane containing molecular aggregates [70].

5.2.2. Comparison with FB Data. Table 7.1 lists the values of Kerr constants for some flexible-chain polymers in comparison with shear optical coefficients $\Delta n / \Delta \tau$ in FB phenomenon for their solutions in identical or similar solvents.

These data refer to the molecular-weight range in which the Kerr constant is independent of the chain length, which for flexible-chain polymers usually occurs even at as low a degree of polymerization as approximately ten and higher. This limiting value of K does not differ greatly in order of magnitude from that for the corresponding monomer in the same solvent, although it can differ from it in sign (polyhydroxyethylene glycol [64], data in Fig. 7.5). However, the values of K can vary for different solvents and are usually higher in those with anisotropic molecules (K values for No. 7 in Table 7.1 in dioxane and benzene).

It is instructive to compare the Kerr effect with FB data in the same solvent as has been done above (Section 4) for n-alcohols.

As for any low-molecular-weight liquid (Chapter 5, Section 1.4), FB in styrene is positive because the monomer molecule exhibits the greatest polarizability in the direction of its length. In contrast, in the polystyrene molecule the planes of phenyl rings are oriented normally to the direction of the main chain, i.e., to that of the maximum length of the molecule. As a result, this direction corresponds to the minimum polarizability of the molecule γ_1 (axis 2 in Fig. 6.56). Hence, negative FB in polystyrene solutions implies that the orientation of any phenyl ring of the molecule in the hydrodynamic field is not independent but exhibits correlation with those of the main chain and of other phenyl groups (large-scale rotation in the chain).

EB in polystyrene solutions, just as in monostyrene, is positive in sign and its K value is close to that for monostyrene. Hence, the type of orientation of phenyl units in the electric field is the same for both poly- and monostyrene. In both cases the orientation of the phenyl ring takes place by rotation together with the dipole of the monomer unit rigidly linked to this ring. As a result, both the dipole and the ring plane are oriented predominantly in the field direction, which leads to positive EB.

The close values of K for the monomer and the polystyrene solution indicate that correlation between the rotation of various monomer units of the polymer chain in the electric field is very weak or absent (small-scale rotation in the chain).

Hence, the types of motion and orientation of the polystyrene chain are quite different in the FB and EB phenomena. In the former the motion of the molecule occurs with close correlation between the orientations of the main chain and phenyl side rings. In contrast, in EB this correlation is absent and each monomer unit is oriented in the electric field virtually independently of others.

A similar conclusion may be drawn from a comparison of EB and FB in solutions of other vinyl polymers with polar side groups (polymethacrylates, polyacrylates, and polychlorovinyl) listed in Table 7.1. For all these polymers there is no general interdependence between the signs of EB and FB – they can either coincide or be opposite – which shows that the types of orientation of their molecules in electric and mechanical fields are quite different.

The intrinsic optical anisotropy of the monomer unit of poly(*tert*-butylphenyl methacrylate) (Table 7.1, No. 11) in a system of coordinates of the main chain of the molecule is negative (phenyl ring in the side chain), which leads to negative FB for solutions of this polymer in bromobenzene (matching solvent). In solutions of this polymer in tetrachloromethane FB is positive as a result of the great contribution of the macroform effect to the anisotropy of the macromolecule, whereas EB in the same solutions is negative. This is explained by the fact that EB is determined by the orientation of the phenyl ring in the side group due to the action of torque to which the dipole of this group is subjected in the electric field. Since the angle θ between the dipole of the group and the plane of the phenyl ring is relatively large, according to Eq. (7.8) this orientation leads to negative anisotropy of solution. In this case the positive macroform effect due to the orientation of the macromolecule as a whole does not appear because there is no correlation between the dipole orientation of the polar side group and the orientation of the whole macrochain.

The differences in the types of orientation of a flexible-chain molecule in electric and mechanical fields can be seen if EB and FB are compared for the homologs of polymethacrylates and polyacrylates with different lengths of the alkyl side chains (Table 7.1, Nos. 7–10 and 12–17). As can be seen (Figs. 6.53 and 6.54), an increase in the length of the side chain increases the negative anisotropy of the monomer unit of a comb-shaped molecule, which leads to a systematic increase in the negative FB of solutions in the polymethacrylate and polyacrylate series. In contrast, the EB of the same solutions is positive in sign and does not reveal any systematic changes in these series. These differences are due to the fact that in shear flow the orientation of the side chains responsible for negative FB is correlated with that of the main chain, whereas the orientations of polar groups of the molecule in the electric field virtually do not lead to the orientation of any extensive parts of the main or the side chains.

These differences in the types of orientation of a flexible-chain molecule in the electric and the shear field are also observed for polymers in which the polar bonds constitute a part of the main chain (Table 7.1, No. 6). In this case the mutual correlation in the orientations of polar units rapidly decreases with increasing distance between them along the chain, and the orientation of a chain molecule in the electric field occurs (in the Peterlin–Stuart theory) by the mechanism of small-scale rotation of its segments whose length is close to that of the monomer unit.

Both the data listed in Table 7.1 and the foregoing discussion show that any attempts to interpret EB in solutions of flexible-chain polymers using the optical anisotropy of molecules (or their segments) according to FB data would be erroneous in principle and would not lead to reasonable results.

The length of "kinetic units" of a flexible polymer chain which are oriented in the electric field more or less mutually independently could be evaluated by studying the nonequilibrium (relaxation) electro-optical characteristics of polymer solutions. However, no reliable data on this problem are available. On the basis of the few experimental data available it is only possible to assert that in the range of frequencies up to 10^7 Hz no dispersion of EB in solutions of flexible-chain polymers without charges was observed (as in the solutions of low-molecular-weight substances and monomers). This is also an indication of the mutually independent orientations of individual monomer units in the electric field.

5.2.3. Polyelectrolytes. The situation is more favorable for the study of EB in solutions of flexible-chain polyelectrolytes with K value higher by several orders of magnitude than for molecules without charge {25, 30–32, 34, 85].

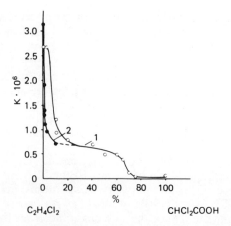

Fig. 7.8. Kerr constant K versus content of dichloroacetic acid (DCAA) in solutions of the following systems: 1) PPLG–dichloroethane (DCE)–DCAA–dimethylformamide; 2) PPLG–DCE–DCAA [94]. K is expressed in g^{-1} cm^5 statvolt^{-2}.

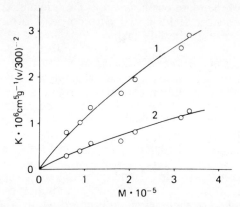

Fig. 7.9. Dependence of K on M for PBLG solutions in dichloroethane (1) and m-cresol (2) [93].

This seems plausible because the uncoiling of a flexible-chain polyion by electrostatic repulsion of ionogenic groups increases the persistent length of the chain, and the optical and hydrodynamic properties of the molecule approach those of a rigid-chain polymer [52, 86] (Table 6.4). This permits the interpretation of experimental data on the electro-optical properties of flexible-chain polyions in terms of a wormlike chain model [87–89]. However, EB in polyelectrolyte solutions is of a complex nature. The high value of the observed effect is caused by the polarization of the ionic atmosphere surrounding the ionized macromolecule rather than by the dipolar and dielectric structure of the polymer chain. This polarization induced by the electric field depends on the ionic state of the solution and the ionogenic properties of the chain, whereas its dependence on the chain structure and conformation is slight. Hence, information on the optical, dipole, and conformational properties of macromolecules obtained by using EB data in solutions of flexible-chain polyelectrolytes is usually only qualitative. Studies of the kinetics of the Kerr effect in polyelectrolytes (carried out by pulsed technique) are more useful, since in these systems relaxation phenomena are observed; they can be utilized to obtain information on the rotational mobility of the polymer chain [25, 87–89].

Numerous data are available on the Kerr effect in solutions of ionogenic chain molecules with secondary structures [28–31, 33–37], particularly polypeptides displaying a helical conformation, and other biological molecules. A high EB value in these solutions is ensured by both the polarization of ionic atmospheres and the character of the rigid secondary structure of the molecules. It is not always possible to distinguish between the effect of the structural order and the ionic state of the medium on the electro-optical properties of rigid polyions in solutions. This fact makes difficult a quantitative determination of the molecular structure of these polymers by the EB method.

6. NONIONOGENIC CHAIN MOLECULES WITH RIGID SECONDARY STRUCTURES

It is possible to obtain more reliable quantitative information on the conformational and structural characteristics of polymer chains by studying EB in solutions of polypeptides that do not contain ionogenic groups and using nonelectrolytic spiralizing systems as solvents [90–97].

Poly(γ-benzyl-L-glutamate) (PBLG) is a typical representative of these polymers. Its solutions in spiralizing solvents exhibit very high positive birefringence both in mechanical (Table 6.7 and Fig. 6.15) and electric fields.

The very high equilibrium rigidity of PBLG in the helical conformation (Table 6.7), resulting in high values of $\Delta n/\Delta \tau$, and of the Kerr constant K in solutions drastically decreases during the helix–coil conformational transition. This can be seen in Fig. 7.8, which shows the dependence of K on the percentage of a despiralizing component (dichloroacetic acid) in mixed solvents [94].

The curve in Fig. 7.8 is similar to that in Fig. 6.4, representing the conformational transition according to the change in the shear optical coefficient $\Delta n/\Delta \tau$. The only difference is that the decrease in K is even more drastic than that in $\Delta n/\Delta \tau$: during the helix–coil transition the value of K for PBLG with $M = 3.3 \cdot 10^5$ changes from $2.7 \cdot 10^{-6}$ to $3 \cdot 10^{-9}$ cm^5 g^{-1} statvolt^{-2}.

Another important difference in the electro-optical properties of rigid- and flexible-chain polymers is shown by the molecular-weight dependence of the Kerr constant. As already shown, for flexible-chain molecules K is virtually independent of M. In contrast, for rigid-chain polymers K usually increases with M. This is demonstrated in Fig. 7.9, which shows the dependence of K on M for PBLG solutions in dichloroethane and *meta*-cresol [93]. Although in both solvents the conformation of PBLG molecules is helical and equally rigid, the value of K in dichloroethane solutions is more than twice as large as that in *m*-cresol. This is caused by the high refractive index increment dn/dc in the former solvent and the correspondingly greater contribution of the microform effect [93, 95] to the optical anisotropy of the PBLG molecules in this solvent.

Finally, the third significant property of rigid-chain polypeptides distinguishing them from flexible-chain polymers is the dispersion of the Kerr effect observed in their solutions (in spiralizing solvents) in the range of radiofrequencies of the sinusoidal field. This is demonstrated in Fig. 7.10 [92], which shows the relative value of EB, $\Delta n_\nu/\Delta n_0$ (Δn_0 is EB at $\nu = 0$), for PBLG fractions in dichloroethane as a function of the frequency ν of the sinusoidal field. With increasing ν, EB decreases almost to zero. The range of dispersion is sharply shifted toward high ν values with decreasing molecular weight of the polymer. This frequency dependence of EB reveals that relaxation phenomena exist in the process of orientation of PBLG molecules and that this orientation is of dipole character (compare with Fig. 7.4).

All these features of EB in solutions of rigid helical molecules (high value of K, dependence of K on M, and the frequency dependence of EB) are more or less typical of other rigid-chain polymers without secondary structure: polyisocyanates, rigid-chain polyamides, cellulose derivatives, various polymer molecules with mesogenic structure, etc. However, in contrast to polypeptides, the electro-optical properties of these polymers are determined by the chemical structure of their molecules rather than by the specificity of their secondary structure. Hence, the study of the Kerr effect in the solutions of these polymers permits the establishment of a direct correlation between the EB of these solutions and the conformational and structural characteristics of their macromolecules.

7. RIGID-CHAIN POLYMERS: EQUILIBRIUM EB AND THE KERR CONSTANT

As already mentioned, the difficulty of obtaining reliable experimental values of K for flexible-chain polymers is due to the low value of EB in their solutions. The range of solvents that can be used for the corresponding measurements is very narrow because only nonconducting liquids with nonpolar optically and dielectrically isotropic molecules can be used. Many flexible-chain polymers are insoluble in these liquids and the application of EB to these polymers is virtually ruled out.

For rigid-chain polymers the situation is much more favorable. The considerable (and sometimes even very high) value of the Kerr effect in their solutions makes it possible to employ for its measurements various solvents, including those with anisotropic and polar molecules, and even those exhibiting high electric conductivity. Naturally, in the latter case, the pulsed technique should be used.

If all the required experimental conditions are obeyed (use of electric fields of moderate strength, exclusion of the electric conductivity effect, etc.), it is always possible to observe, in solutions of a rigid-chain polymer, an EB value consistent with the Kerr law; i.e., Δn is proportional to E^2. Some examples of this agreement are given below.

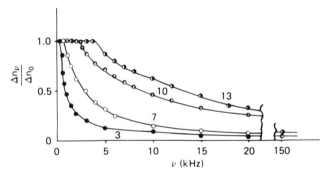

Fig. 7.10. Relative value of EB, $\Delta n_v/\Delta n_0$, for PBLG fractions in dichloroethane versus frequency v of the applied sinusoidal field [92]. Curves 3, 7, 10, and 13 refer to polymers of different molecular weights: 3) $M = 3.2 \cdot 10^5$; 7) $1.4 \cdot 10^5$; 10) $0.9 \cdot 10^5$; 13) $0.6 \cdot 10^5$.

Figure 7.11 shows Δn for solutions of cellulose carbanilate (CC) fractions in dioxane as a fraction of the square of the potential difference V^2 on the electrodes of the Kerr cell. Different curves correspond to different concentrations. At all concentrations the negative EB of solutions increases proportionally to V^2 in agreement with the Kerr law. The EB of a pure solvent (broken straight line) is positive and lower by two orders of magnitude in absolute value than that for a solution of the lowest concentration (curve 7).

Under these conditions the Kerr constant K for a dissolved polymer, determined by Eq. (7.2), can be calculated avoiding Eqs. (7.3)–(7.5) by the substitution of the following value of Δn:

$$\Delta n = \Delta n_c - \Delta n_{c=0} \qquad (7.33)$$

into Eq. (7.2).

The values of Δn_c and $\Delta n_{c=0}$ in Eq. (7.33) are the FB of the solution (at a concentration c) and the solvent, respectively, measured at the same field strength E.

Figure 7.11 illustrates the data of measurements carried out under the most favorable conditions because the polymer being investigated is soluble in a nonpolar isotropic solvent and its Kerr constant exceeds that for the solvent by more than four orders of magnitude.

In accordance with this, measurements were carried out in a sinusoidal field of low frequency using a visual method and the composition technique for the determination of Δn (see Section 12).

A similar situation is observed for cellulose nitrate (CN) solutions in dioxane whose data are shown in Fig. 7.12.

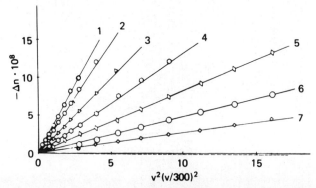

Fig. 7.11. Dependence of EB, Δn, for cellulose carbanilate solution in dioxane on the square of potential difference V^2 on the electrodes of the Kerr cell [98].
Sinusoidal field of frequency $\nu = 1$ kHz. Solution concentrations, $c \cdot 10^2$: 1)
0.885; 2) 0.853; 3) 0.407; 4) 0.298; 5) 0.180; 6) 0.107; 7) 0.057 g/cm^3. Broken straight line – pure solvent. K of the polymer is equal to $-27 \cdot 10^{-10}$ g^{-1} cm^5.
statvolt^{-2}. K of the solvent is equal to $0.4 \cdot 10^{-12}$ g^{-1} cm^5 statvolt^{-2}.

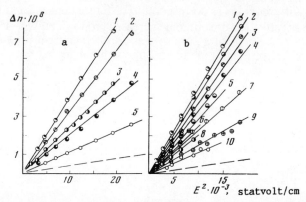

Fig. 7.12. Dependence of EB, Δn, on the square of field strength E^2 for solutions of cellulose nitrate with a low degree of substitution (CN, $M = 73 \cdot 10^3$) in
dioxane [99]. a) Measurements at the frequency $\nu = 0.15$ kHz of the sinusoidal
field and solution concentrations $c \cdot 10^2$ equal to: 1) 0.54; 2) 0.44; 3) 0.34; 4)
0.24; 5) 0.084 g/cm^3. Broken line – pure solvent. K of the polymer is equal to
$7.1 \cdot 10^{-10}$ g^{-1} cm^5 statvolt^{-2}. b) Measurements at concentration $c = 0.5 \cdot 10^{-2}$
g/cm^3 and field frequencies ν equal to: 1) 0.15; 2) 0.3; 3) 0.5; 4) 1; 5) 3; 6) 5; 7)
10; 8) 20; 9) 49; 10) 700 kHz.

In this case the EB of solutions is positive in sign as a result of high refractive
index increment in the polymer–solvent system and the correspondingly high positive contribution of the form effect of the chain (see Chapter 5, Section 2.1.2.2) to
the anisotropy of solution. This fact in itself is very important because it indicates
that a considerable difference exists in the EB mechanisms for rigid-chain (CN) and
flexible-chain polymers. As can be seen (Table 7.1, No. 11 and Section 5.2.3 in

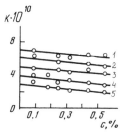

Fig. 7.13. Dependence of $K_{c,v} = \lim_{E \to 0}(\Delta n/cE)^2$ on concentration c for solutions of CN ($M = 73 \cdot 10^3$) with a low degree of substitution in dioxane at various frequencies v of the sinusoidal field [99] equal to: 1) 0.15; 2) 1; 3) 4; 4) 15; 5) 44 kHz.

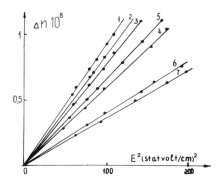

Fig. 7.14. Dependence of EB, Δn, on E^2 for solutions of CN fractions with low degree of substitution in cyclohexanone under the steady-state conditions of the rectangular pulsed field [100]. Different curves correspond to solutions of different concentrations c of fractions I and II: I) $M = 5.9 \cdot 10^5$, $K = -7.3 \cdot 10^{-8}$ g^{-1} cm^5 statvolt^{-2}, $c \cdot 10^2 = 0.012$ (2), 0.027 (3), 0.55 (4), 0.10 g/cm^3 (6). II) $M = 2.45 \cdot 10^5$, $K = -5.8 \cdot 10^{-8}$ g^{-1} cm^5. statvolt^{-2}, $c \cdot 10^2 = 0.06$ (5), 0.13 (7). 1) Pure cyclohexanone. K of cyclohexanone is equal to $9.5 \cdot 10^{-11}$ g^{-1} cm^5 statvolt^{-2}.

this chapter) the form effect is not pronounced in the EB phenomenon for flexible-chain polymers because of the small-scale character of chain motion in the electric field. In contrast, in the EB of solutions of rigid-chain polymers the form effect is just as distinct as in FB. This is an indication of the resemblance between the mechanisms for molecular motion in these two phenomena (large-scale motion).

The great difference (by more than three orders of magnitude) in the Kerr constants for the dissolved polymer and the solvent, as well as the low electric conductivity in the CN–dioxane system, make it possible to use the sinusoidal field in measurements and to calculate K for the polymer according to Eqs. (7.33) and (7.2). The data in Fig. 7.12b indicate that a dispersion of EB exists and, in this case, the effect introduced by the dissolved polymer decreases with increasing field frequency.

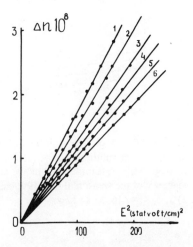

Fig. 7.15. Dependence of EB, Δn, on E^2 for solutions of cyanoethylcellulose (CEC, $M = 3.2 \cdot 10^5$, $K = 2.15 \cdot 10^{-8}$ g^{-1}· cm^5 statvolts^{-2}) in cyclohexanone under steady-state conditions of rectangular pulsed field. Solution concentration $c \cdot 10^2$ = 0.32 (1), 0.23 (2), 0.14 (3), 0.13 (4), 0.06 g/cm^3 (5); 6) pure solvent.

The slopes of the straight lines in Fig. 7.12a, b determine the values of $K_{c,\nu}$

$$K_{c,\nu} = \lim_{E \to 0} (\Delta n/cE^2),\qquad (7.34)$$

corresponding to the solution concentration c and the frequency of the applied field ν. Here Δn is the value determined according to Eq. (7.33) at predetermined values of c and ν.

Figure 7.13 shows the concentration dependence of the values of $K_{c,\nu}$ at various field frequencies ν. In the CN-dioxane system this dependence is linear and it is possible to reliably determine the Kerr constant K_ν of the dissolved polymer at a frequency ν by extrapolation to $c \to 0$.

When liquids with polar molecules are used as solvents, the experimental procedure becomes more complex because of the high electric conductivity of these solvents. In these cases the pulsed technique should be used. Some results obtained under these conditions for CN solutions in cyclohexanone are shown in Fig. 7.14. The form effect in the CN–cyclohexanone system is virtually absent, and hence the anisotropy of the polymer in solution is negative.

Similar data are shown in Fig. 7.15 for cyanoethylcellulose (CEC) solutions in cyclohexanone. In this system the Kerr effect is positive, in accordance with the positive optical anisotropy of CEC molecules. Although the Kerr constants for cyclohexanone and acetone are 200 times as high as those for nonpolar dioxane, the differences between the values of Δn for CN and CEC solutions on the one hand,

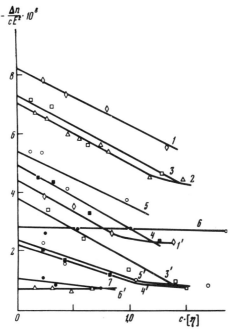

Fig. 7.16. Dependence of $\Delta n/cE^2$ [where Δn is determined according to Eq. (7.33)] on volume concentration $c[\eta]$ of the polymer in solution for CN fractions in cyclohexanone (nonprimed number) and in acetone (primed number) [100]. Molecular weights of fractions $M \cdot 10^{-5} = 7.7$ (1), 5.9 (2), 5.7 (3), 3.0 (4), 2.5 (5), 1.8 (6), 1.0 (7).

and those for these solvents on the other, can be distinguished fairly well, and thus it is possible to carry out the extrapolation of the Kerr effect to zero concentration (Fig. 7.16).

In solutions with considerable electric conductivity the concentration dependence of the Kerr effect is usually negative, high, and often nonlinear. Hence, in order to be able to interpret the experimental data on the Kerr effect in relationship to the molecular characteristics of the polymer, thorough extrapolation of these data to infinite dilution is needed.

Figure 7.17 shows the dependence of EB on E^2 for aromatic poly(amide hydrazide) (PAH) in dimethylsulfoxide. For this solvent the Kerr constant is 400 times higher than that for dioxane and other nonpolar solvents. However, even in a solution of negligibly low concentration (Fig. 7.17, curve 2) one-half of the measured value of Δn consists of EB introduced by the dissolved polymer. This results from the very high value of the Kerr constant for PAH which, for the lowest-molecular-weight samples, exceeds that for dimethylsulfoxide by two orders of magnitude, and for high-molecular-weight samples, by four orders of magnitude.

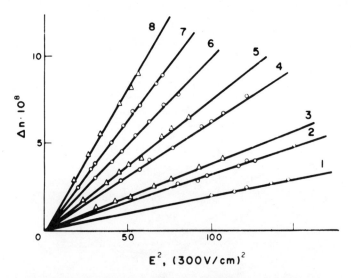

Fig. 7.17. Dependence of EB, Δn, on E^2 for solutions of two poly(amide hydrazide) samples in dimethylsulfoxide [101] under steady-state conditions of the rectangular pulsed field. Sample with $M = 3.4 \cdot 10^4$, $K = 1.7 \cdot 10^{-6}$ g^{-1} cm^5 statvolt^{-2} (curves 2, 4, 6, 7) and sample with $M = 2.4 \cdot 10^4$, $K = 1.4 \cdot 10^{-6}$ g^{-1} cm^5 statvolt^{-2} (curves 3, 5, 8). Solution concentrations $c \cdot 10^2 = 0.007$ (2), 0.103 (3), 0.03 (4), 0.22 (5), 0.052 (6), 0.07 (7), 0.7 g/cm^3 (8). 1) Pure solvent, $K = 18 \cdot 10^{-11}$ g^{-1}. $cm^5 \cdot$statvolt^{-2}.

Figure 7.18 shows the concentration dependence of $K = \Delta n/cE^2$ [where Δn is determined from Eq. (7.33)] for PAH solutions in dimethylsulfoxide. As for all polar rigid-chain polymers in strongly polar solvents, the concentration dependence of K for high-molecular-weight samples is strong. However, the available experimental points permit reliable extrapolation of K to the conditions of infinite dilution for the determination of characteristic Kerr constants of the samples.

These examples show that the Kerr effect in solutions of various rigid-chain polymers can differ greatly in value and may be measured using both sinusoidal electric fields and pulsed techniques depending on the electric conductivity of the solution. If an appropriate solvent (ensuring the molecular dispersion of solution) and appropriate measuring method are selected, in practice it is always possible to carry out measurements in that range of field strength in which the Kerr law is obeyed. The data extrapolated to infinite dilution ensure reliable determination of the Kerr constant for the polymer under investigation.

For rigid-chain high-molecular-weight polymers in the range of strong fields (nonconducting solutions) the effects of EB saturation are observed and the Kerr law is not obeyed (compare Section 2 of this chapter).

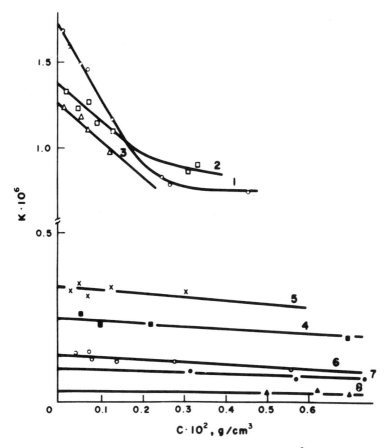

Fig. 7.18. Concentration dependence of $K = \Delta n / cE^2$ [where Δn is determined according to Eq. (7.33)] for solutions of PAH samples in dimethylsulfoxide. Molecular weights of samples $M \cdot 10^{-3} = 34$ (1), 24 (2), 19 (3), 12 (4), 11 (5), 9 (6), 8.6 (7), 5.2 (8) [101].

Poly(butyl isocyanate) (PBIC) is an example of these polymers. The electro-optical data for two PBIC samples are shown in Fig. 6.19 in the form of the dependence of Δn on E^2 for solutions in tetrachloromethane. The Kerr constant for this polymer is very high and, therefore, it is not necessary to take into account the contribution of the solvent to the FB of solutions, and the curves $\Delta n = \Delta n(E^2)$ in Fig. 7.19 may be compared with the theoretical dependences reported in Section 2 of this chapter.

It can be shown (as will be done below) that the Kerr effect observed in rigid-chain polymer solutions is virtually the effect of the dipole orientation of their molecules, whereas the part of EB induced by the dielectric anisotropy of molecules (induction effect) is negligible. Hence, the experimental dependence $\Delta n = \Delta n(E^2)$ reflecting the purely dipole-orientational nature of this effect at low E values should

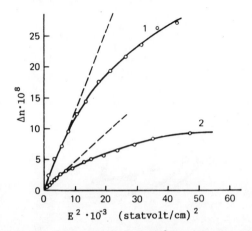

Fig. 7.19. Dependence of Δn on E^2 for solutions of two poly(butyl isocyanate) samples in tetrachloroethane [102]: 1) $M = 1.6 \cdot 10^5, c = 0.6 \cdot 10^{-3}$ g/cm^3; 2) $M = 0.66 \cdot 10^5$, $c = 0.7 \cdot 10^{-3}$ g/cm^3.

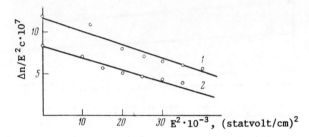

Fig. 7.20. Dependence of $\Delta n/cE^2$ on E^2 for solutions of two PBIC samples (1 and 2 in Fig. 7.19) in tetrachloromethane.

be described by Eqs. (7.15) and (7.16). With the application of Eq. (7.9) these equations give

$$\Delta n/cE^2 = (\Delta n/cE^2)_{E \to 0} [1 - (2/189) (\mu/kT)^2 (\varepsilon + 2)^2 E^2]. \qquad (7.35)$$

Equation (7.35) shows that the dependence of $\Delta n/cE^2$ on E^2 is linear, as is confirmed by the experimental straight lines shown in Fig. 7.20. The dipole moments μ of the molecules of these polymers can be determined from the slopes of these straight lines. For two PBIC samples represented by straight lines 1 and 2 in Fig. 7.20, the values of μ obtained in this manner are 500 and 370 D, respectively.

These very high dipole moments can be interpreted only by the assumption that the orientation of PBIC molecules in the electric field occurs by the mechanism of large-scale motion in which the kinetic unit responsible for the Kerr effect is a considerable part of the polymer chain rotating as a whole under the influence of the

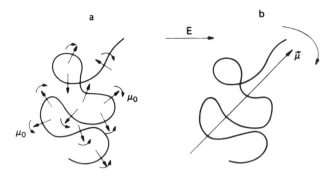

Fig. 7.21. Possible mechanisms for dipole orientation of a chain molecule in the electric field. a) Kinetically flexible chain (deformational mechanism); b) kinetically rigid chain (orientational mechanism).

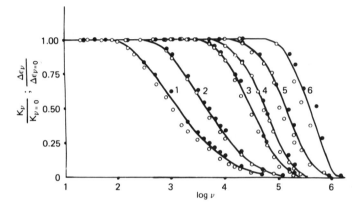

Fig. 7.22. Relative Kerr constant $K_\nu/K_{\nu=0}$ (open circles) and relative dielectric increment $\Delta\varepsilon_\nu/\Delta\varepsilon_{\nu=0}$ (filled circles) versus frequency of the sinusoidal electric field ν for poly(chlorohexyl isocyanate) fractions in dioxane: $M\cdot10^{-4} = 27.4$ (1), 12.5 (2), 5.0 (3), 3.7 (4), 2.3 (5), 1.0 (6) [103–106].

field. More detailed quantitative information on this problem can be obtained from investigations of the kinetics of the Kerr effect in solutions of rigid-chain polymers.

8. KINETICS OF THE KERR EFFECT IN SOLUTIONS OF RIGID-CHAIN POLYMERS

The study of the kinetics of EB in a solution of a rigid-chain polymer is of great importance because it helps to elucidate the physical nature of this phenomenon and provides information about the motion of the molecule in the electric field.

If the polymer chain contains polar groups (dipole μ_0), then in the electric field E each of these group experiences the torque orienting the dipole μ_0 in the field direction (Fig. 7.21a). Moreover, if the rotations of individual polar groups are in relatively weak correlation, their orientations in the field occur more or less mutually independently. As a result of these intramolecular rotations (small-scale motion) in the electric field the conformation of the molecule undergoes a change, i.e., the molecule undergoes deformation.

At the same time, in any conformation the chain as a whole exhibits a dipole moment μ which is the vector sum of dipole moments μ_0 of all its polar groups (Fig. 7.21b). The torque experienced by the dipole μ in the electric field can lead to dipole orientation in the field direction by the rotation of the chain molecule as a whole (large-scale motion). In this case the mechanism for the generation of polarization and macroscopic anisotropy of solution may be called orientational.

The discussion in Section 5.2.2 of this chapter suggests that the deformational mechanism is characteristic of kinetically flexible molecules with orientation time τ_{or} longer than the time of their deformation τ_{def}, i.e., the time during which the molecule in solution retains, on the average, a fortuitous conformation (relaxation time of conformation).

In contrast, for kinetically rigid molecules with $\tau_{or} < \tau_{def}$ the motion in the electric field should occur by the mechanism of orientation of the molecule as a whole and the rate of this process should be determined by the rotational fricton coefficient W or the rotational diffusion coefficient D_r (Section 3 in this chapter). Hence, the study of the orientation kinetics of kinetically rigid macromolecules in the electric field allows the determination of their rotational mobility (i.e., W and D_r).

Of the two principal methods for the study of the kinetics of the Kerr effect considered in Section 3 – the pulsed method and the method of continuous sinusoidal field – the latter is applicable only in those cases when the electric conductivity of the solution is low. However, for many reasons this method is more informative and its results are more precise. Hence, if the polymer being investigated exhibits solubility in a nonpolar solvent, the method of using sinusoidal fields of variable frequency should be preferred. Nevertheless, even if the electric conductivity of the solution is high, the sinusoidal method may be used in combination with the pulsed method (method of sinusoidal pulses).

8.1. EB Dispersion

As already mentioned, the dispersion of the Kerr effect in the range of radiofrequencies is a characteristic property of rigid-chain polymer solutions. This can be seen in Figs. 7.22–7.26, which show the frequency dependences of EB for some poly(alkyl isocyanate)s and cellulose esters and ethers.

As frequency increases, all dispersion curves descend virtually to zero. This indicates that EB is produced by the dipole–orientational mechanism, whereas the an-

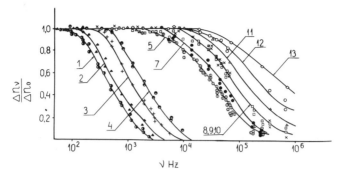

Fig. 7.23. Dependence of $\Delta n_\nu / \Delta n_0$ on frequency of the sinusoidal field for cellulose carbanilate fractions in dioxane. Δn_ν: EB at frequency ν; Δn_0: EB at a frequency $\nu = 0$. Curves correspond to the following molecular weights $M \cdot 10^{-3} = 870$ (1); 680 (2); 440 (3); 280 (4); 84 (5); 66 (7); 54 (8); 46 (9); 39 (10); 34 (11); 28 (12); 24 (13) [107–109].

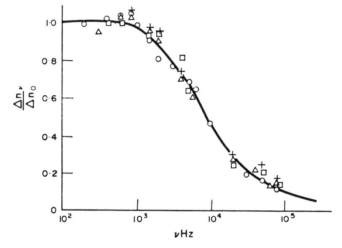

Fig. 7.24. Relative EB $\Delta n_\nu / \Delta n_0$ versus frequency ν of the sinusoidal field for ethyl cellulose solutions in dioxane. Points with different symbols refer to solutions of different concentrations in the range of $c \cdot 10^2 = (0.17\text{-}1.3)$ g/cm³ [110].

isotropy of the dielectric polarizability of the macromolecules is virtually impercep-tible (compare with Fig. 7.4).

This feature of the electro-optical properties of chain molecules can be under-stood on the basis of general concepts of the orientation of their segments in the electric field in the spirit of the Peterlin–Stuart theory [43, 44]. In fact, in terms of this theory the contribution introduced into the Kerr effect by the dielectric anisotropy of the segment is proportional to the difference between its principal po-larizabilities $\delta_1 - \delta_2$ [the first term on the right side of Eq. (7.8)], whereas the con-tribution of its dipole is proportional to μ^2, the square of its dipole moment μ [the second term in Eq. (7.8)] with increasing equilibrium chain rigidity; i.e., with increasing number S of monomer units in a segment, the anisotropy of the segment should increase proportionally to S, whereas the dipole moment of the segment increases proportionally to S^2. Hence, for rigid-chain polymers (characterized by high S) the contribution of the anisotropic term should be small compared to that of the dipole term.

The higher the molecular weight of the polymer, the lower is the range of frequencies in which the dispersion curves in Figs. 7.22–7.25 are located. This displacement of curves implies that relaxation times τ depend on M increasing with it. These properties are a direct proof of the fact that the mechanism for molecular motion responsible for EB in solutions of the above polymers is due to the rotation of macromolecules (or their considerable parts) as a whole. This may be interpreted as a manifestation of the kinetic rigidity of these macromolecules.

In the same frequency range the dispersion of the dielectric increment $\Delta \varepsilon$ of the solution can be observed, the curves of dielectric dispersions virtually coinciding with the dispersion curves of EB (Fig. 7.22). This indicates that the mechanisms for molecular motion in these two phenomena are identical and correspond to dispersion equation (7.32) for kinetically rigid molecules.

Figure 7.26 shows the frequency dependence of the Kerr effect in poly(butyl isocyanate) (PBIC) solutions in tetrafluoromethane containing various amounts of pentafluorophenol (PFPh). It is known (Chapter 6, Section 2.4.2) that the addition of PFPh into the solution decreases the equilibrium rigidity of PPIC chains, thus leading to their coiling and a decrease in their hydrodynamic dimensions without a change in molecular weight. As can be seen in Fig. 7.26, these conformational changes are accompanied by a displacement of the dispersion curves toward higher frequencies, illustrating the fact that the relaxation time τ of EB decreases in accordance with increasing rotational mobility of the molecule as a whole. This conclusion is confirmed by quantitative comparison of relaxation times τ with the rotational friction coefficients determined from the data of hydrodynamic investigations (viscometry) [112].

8.2.   Transient EB

The method of transient EB can also be used to study the kinetics and relaxation of the Kerr effect in solutions of rigid-chain polymers.

Thus, the method of reverse pulse can be used to show that EB in these solutions is induced by the dipole orientation of polar macromolecules without any appreciable effect on their dielectric anisotropy. An example of this can be seen in Fig. 7.27, which shows the oscillogram of the electric pulse with variable polarity and that of the corresponding EB pulse in the poly(amide anhydride) solution in dimethylsulfoxide [101]. The oscillogram shows that upon rapid reversion of polarity (i.e., a change in the direction of the electric field) an abrupt decrease in the EB pulse is observed. This corresponds to a marked decay in the anisotropy of solution Δn and its subsequent rise with time. According to Eqs. (7.27) and (7.28), this

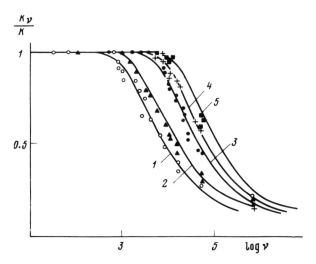

Fig. 7.25. Relative value of the Kerr constant K_v/K_0 versus frequency v of the sinusoidal field for solutions of CN fractions in dioxane. Curves correspond to molecular weights: $M \cdot 10^{-3} = 77$ (1); 70 (2); 51 (3); 46 (4), 34 (5) [111].

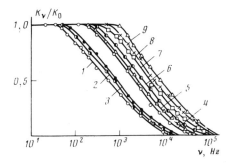

Fig. 7.26. Relative Kerr constant K_v/K_0 versus frequency of the sinusoidal field for poly(butyl isocyanate) ($M = 2.3 \cdot 10^5$) solutions in a mixed solvent (tetra-chloromethane–pentafluorophenol) at various solvent compositions. Curves correspond to solvents containing the following volume fraction of pentafluoro-phenol (%): 1) 0; 2) 0.07; 3) 0.12; 4) 0.26; 5) 0.5; 6) 1.0; 7) 2.6; 8) 5.0; 9) 7.5 [112].

curve of the dependence $\Delta n(t)$ with a deep minimum corresponds to the predominant part played by the dipole contribution to the observed Kerr effect.

For the quantitative determination of relaxation time of the Kerr effect and the related rotational mobility of macromolecules, it is advisable to study the curve of a decrease in EB in the rectangular pulsed field. This curve for a cellulose nitrate solution in cyclohexanone is shown in Fig. 7.28. The type of curve is determined by the total contribution of the nonrelaxing positive EB of the solvent and the relaxing negative part of EB introduced by the dissolved polymer. In accordance with this, when the field is switched on (moment $t = 0$), positive EB appears instanta-

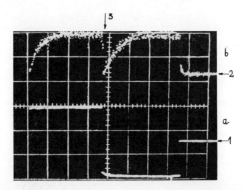

Fig. 7.27. Oscillogram of the reversed electric pulse (a) and the corresponding EB pulse for a poly(amide hydrazide) solution in dimethylsulfoxide (b). 1, 2) Zero levels of electric and optical pulses, respectively; 3) moment of reversion of the electric pulse [101].

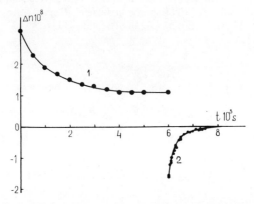

Fig. 7.28. Curve of increase and decay of transient EB for a rectangular pulsed field in a cellulose nitrate solution in cyclohexanone. $t = 0$ – moment when the field is switched on; $t = 6 \cdot 10^{-3}$ s – moment when the field is switched off [100].

neously as a result of the effect induced by the solvent. Subsequent decrease in positive Δn with time (curve 1) indicates the decrease in the negative contribution induced by the polymer. When the field is switched off ($t = 6 \cdot 10^{-3}$ s), the positive effect introduced by the solvent disappears instantaneously and the EB of solution reverses its sign. The subsequent decrease in negative EB with time (curve 2) characterizes free relaxation of the Kerr effect for the dissolved polymer. The shape of curve 2 of this relaxation allows the determination of relaxation time, τ_0, related to the rotational diffusion coefficient of the molecule D_r by Eq. (7.22).

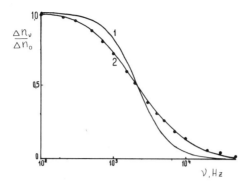

Fig. 7.29. Dependence of $\Delta n_v/\Delta n_0$ on the frequency v of the sinusoidal field. 1) Theoretical curve according to Eq. (7.32) with relaxation time $\tau = 10.6 \cdot 10^{-5}$ s. "Average" relaxation time determined from the half-height of the experimental curve coincides with τ for theoretical dependence 1; 2) experimental curve for a poly(butyl isocyanate) fraction ($M_w = 10^5$, $M_w/M_n = 1.2$) in tetrachloromethane [113].

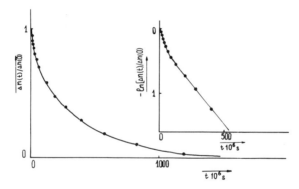

Fig. 7.30. Decay curves on linear (1) and logarithmic scales (2) obtained when the rectangular pulsed field is switched off for cellulose nitrate fraction ($M_w = 3 \cdot 10^5$, $M_w/M_n = 1.13$) in cyclohexanone [113].

8.3. Relaxation Time, Molecular Weight, and Viscosity

If, in an assembly of the molecules investigated, the dispersion of the Kerr effect in the sinusoidal field is characterized by the relaxation time τ, then the dispersion curves describing the dependence of $\Delta n_v/\Delta n_0$ or K_v/K_0 on field frequency should correspond to Eq. (7.32). In this case relaxation time τ is determined from the condition $\tau = 1/\omega_0$, where ω_0 is the cyclic frequency at which the dispersion curve falls to one-half of the initial value ($\Delta n_v/\Delta n_0 = K_v/K_0 = 1/2$).

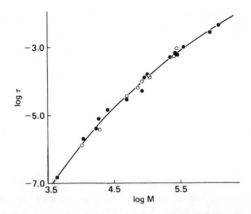

Fig. 7.31. Relaxation time τ in the Kerr effect (open circles) and in dielectric polarization (filled circles) for solutions of poly(butyl isocyanate) fractions in tetrachloromethane [103].

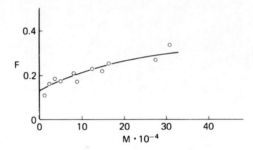

Fig. 7.32. Values of the coefficient F in Eq. (2.27) at various M according to the data of EB dispersion in solutions of poly(chlorohexyl isocyanate) fractions in tetrachloromethane [104].

Actually, the breadth of the experimental curves shown in Figs. 7.22–7.25 is often larger than that of the corresponding theoretical curve [Eq. (7.32)] (as is shown in Fig. 7.29) and hence should be described by a spectrum of relaxation times. In this case the value determined at the half-height of the dispersion curve can characterize only the "average" relaxation time τ.

Similarly, the relaxation curve of decay at transient EB in solution can differ from the simple exponential dependence (7.21) described by the relaxation time τ_0. An example of this can be seen in Fig. 7.30, which shows curve 1 of EB decay after switching off the rectangular pulse for a cellulose nitrate fraction in cyclohexanone. The deviation from dependence (7.21) is demonstrated by curve 2 in the same figure. According to Eq. (7.21), the change in the value of $\ln[\Delta n(t)/\Delta n(0)]$ with time t should be expressed by a straight line with slope $1/\tau_0$. However, dependence 2 in Fig. 7.30 is evidently nonlinear, which indicates that a spectrum of relaxation times exists. In this case the EB of the polymer may be approximately charac-

TABLE 7.2. Coefficients F Determined from the Equations $F = M[\eta]\eta_0/2RT\tau$ or $F = M[\eta]\eta_0 \cdot (6RT\tau_0)^{-1}$ according to Experimental Values of M, $[\eta]$, and τ (or τ_0)

No.	Polymer	Solvent	$M \cdot 10^{-4}$	$\tau \cdot 10^5$, s	$\tau_0 \cdot 10^5$, s	F	References
1	Poly(butyl isocyanate) fractions	Tetrachloromethane	5—140	3—630	—	0.13—0.40	[114]
2	Poly(chlorohexyl isocyanate) fractions	Tetrachloromethane	1.4—30.6	0.19—20	—	0.11—0.34	[104, 115]
3	Poly(butyl isocyanate)	Tetrachloromethane–pentafluorophenol mixture (vol. fraction 0–10%)	23	2.6—0.2	—	0.39—0.83	[112]
4	Cellulose carbanilate fractions	Dioxane	2.8—8.4	0.08—0.45	—	0.25—0.33	[107]
5	Cellulose carbanilate fractions*	Dioxane	2.0—6.8	0.05—0.64	—	0.19—0.28	[116]
6	Ethyl cellulose fractions*	Dioxane	0.9—6.3	0.08—2.0	—	0.14—0.25	[117]
7	Cellulose nitrate fractions	Dioxane	3.4—7.7	0.09—0.66	—	1.0—0.6	[111]
8	Ethyl cellulose	Dioxane	8.0	1.4	—	0.29	[110]
9	Benzyl cellulose	Dioxane	3.0	0.1	—	0.38	[110]
10	Cellulose butyrate	Dioxane	78	23.6	—	0.43	[110]
11	Cellulose benzoate	Dioxane	47	15.3	—	0.35	[110]
12	Cellulose diphenyl acetate	Dioxane	23	2.9	—	0.41	[110]
13	Cellulose diphenylphosphonocarbamate	Dioxane	100	11.8	—	0.83	[110]
14	Cellulose nitrate fractions	Cyclohexanone	77	—	28	0.68	[100]
			59	—	25	0.52	
			30	—	12	0.32	
			18	—	10	0.14	
15	Cyanoethyl cellulose fractions	Cyclohexanone	26	—	2.3	0.9	[118]
			18	—	1.6	0.9	
			15	—	1.3	0.7	
			10	—	1.1	0.2	
16	Poly(amide hydrazide)	Dimethylsulfoxide	3.4	—	0.78	0.42	[101]
			2.4	—	0.46	0.37	
			1.9	—	0.28	0.12	

*Values of τ are determined from the dispersion of dielectric permittivity of solutions.

terized by the "average" relaxation time τ_0 determined by the area Q under relaxation curve 1 in Fig. 7.30.

The average relaxation time determined by the above methods for some rigid-chain polymers are given in Table 2.

The existence of a spectrum of relaxation times in the EB of rigid-chain polymers may be due to both a set of mechanisms for intramolecular motion and the polydispersity of the polymer on the masses and conformations of molecules (Chapter 5, Section 3.3.2).

The dependence of relaxation time τ of EB on the molecular weight of the polymer is of fundamental importance for the determination of the character of motion of macromolecules in the electric field. The molecular-weight dependence of τ (average time determined from dispersion curves) for poly(butyl isocyanate) is shown in Fig. 7.31. The experimental points fit a curve whose slope decreases with increasing M, remaining in the 2.7–1.5 range. Taking into account Eq. (7.29), this can be expressed by

$$D_r = K_{D_r} M^{-b}, \tag{7.36}$$

where b increases from 2.7 to 1.5 with increasing M. This relationship corresponds to the concept that the kinetic unit rotating in the electric field is the polar molecule as a whole, and that with increasing M its hydrodynamic properties change from those of a straight rod to those of the nondraining Gaussian coil (Chapter 2, Sections 1.2 and 3.4).

Comparison of relaxation times τ with intrinsic viscosities $[\eta]$ of the same samples allows us to draw similar conclusions. As we can see, the theoretical relationship between the rotational friction coefficient for kinetically rigid particles $W = kT/D_r = 2kT\tau$ and the product $M[\eta]\eta_0$ is given by Eq. (2.27). Here coefficient F changes from $2/15 = 0.13$ to $5/12 = 0.42$, when the shape of the particle varies from rodlike to spherical. For the Gaussian coil F has an intermediate value of 0.25 [see Eq. (2.96)]. Figure 7.32 shows the values of F calculated according to Eq. (2.27) using the experimental values of τ, M, $[\eta]$, and η_0 for fractions of poly(chlorohexyl isocyanate) in tetrachloromethane [104]. The values of F are not only within the range predicted by the theory for kinetically rigid particles, but also clearly demonstrate the decrease in F with M. This is in accordance with the simultaneous change in the shape of the molecule from the random coil to the rod. Hence, the character of the molecular motion responsible for EB is similar to that of the motion accounting for the viscosity of the solution. This is the rotation of the molecule as a whole in the electric field or in the field of mechanical shear forces. Comparison of τ and $[\eta]$ for other rigid-chain polymers leads to a similar result [107]. The values of coefficients F for some rigid-chain molecules are listed in Table 7.2.

Depending on the experimental conditions, the coefficients F in Table 7.2 have been calculated using the experimental values of τ (determined from the dispersion of EB) or the values of τ_0 (determined from the curve of EB decay after the field is switched off). The values of F obtained by these two methods for the same solution virtually coincide in accordance with the equation $\tau = 3\tau_0$.

Table 7.2 also lists the values of τ and, correspondingly, those of F obtained from the dispersion of dielectric permittivity of solutions (Nos. 5 and 6). These

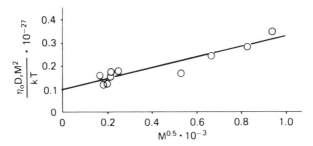

Fig. 7.33. Dependence of $\eta_0 D_r M^2/kT$ on $M^{1/2}$ for cellulose carbanilate fractions in dioxane [107, 108].

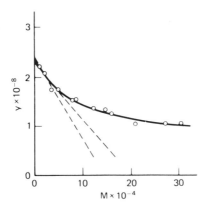

Fig. 7.34. Parameter $y = \{(3kT/\pi\eta_0 D_r)[\ln(2\lambda M/M_0 d) - \gamma]\}^{1/3} M_0 M$ versus M for poly(chlorohexyl isocyanate) fractions in tetrachloromethane [104].

data do not differ from those obtained by the method of EB dispersion in the same solutions. This fact demonstrates the identity of the mechanisms for molecular motion responsible for the relaxation of dipole polarization and the Kerr effect in solutions of rigid-chain polymers.

Although the values of τ or τ_0 for various polymers (or their fractions) given in Table 7.2 can differ by a factor of several hundreds, all these polymers are characterized by F values close to those predicted theoretically [Eq. (2.27)] for rigid particles of various shapes.

However, in the homologous series of fractions of the same polymer in Table 7.2 a tendency toward increasing F with increasing molecular weight can be observed similar to the data shown in Fig. 7.32, and in accordance with the simultaneous increase in chain coiling.

The transition to coiled molecular conformations with invariable molecular weight is also accompanied by an increase in the coefficient F (No. 3) according to the theoretical predictions for kinetically rigid molecules.

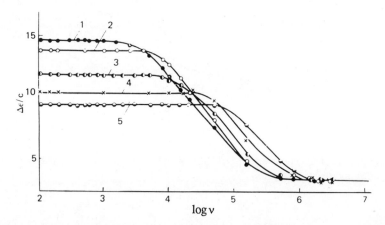

Fig. 7.35. Dielectric increment $\Delta\varepsilon/c \equiv (\varepsilon - \varepsilon_0)/c$ versus frequency v of the sinusoidal field for cellulose carbanilate solutions in dioxane. ε_0, ε are the dielectric permittivities of the solvent and the solution, respectively (at a concentration c). $M \cdot 10^{-3} = 68$ (1), 58 (2), 43 (3), 28 (4), 23 (5) [116].

8.4. Rotational Friction and Geometrical Parameters of Chain Molecules

The orientational mechanism of EB in solutions of rigid-chain polymers and the possibility of determining rotational diffusion coefficients of their molecules from dispersion curves may be utilized for the characterization of equilibrium conformational properties of their chains. The theory of rotational friction of kinetically rigid molecules can be employed for this purpose. The results of this theory for the two limiting cases of molecular conformation referring to the weakly bending rod and the wormlike coil are expressed by Eqs. (2.118) and (2.119) in Chapter 2.

As an example of the use of Eq. (2.119), the data for fractions of cellulose carbanilate, whose molecules can be represented by a partially draining wormlike coil, are plotted in Fig. 7.33. The dependence of the expression on the left side of Eq. (2.119) on $M^{1/2}$ is approximated by a straight line whose slope yields the length of the Kuhn segment A and whose intercept with the ordinate gives the hydrodynamic diameter of the chain d. The curve in Fig. 7.33 gives the values of $A = 160 \cdot 10^{-8}$ cm and $d = 6 \cdot 10^{-8}$ cm for the cellulose carbanilate chain.

For the other limiting case, that of a weakly bending rod, the Hearst equation (2.118) can be represented as [104]

$$\left[\frac{3kT}{\pi\eta_0 D_r}\left(\ln\frac{2\lambda M}{M_0 d}-\gamma\right)\right]^{1/3}\frac{M_0}{M}=\lambda\left\{1-\frac{M}{4M_0 S}\left[1-\left(\ln\frac{2\lambda M}{M_0 d}\right)-\gamma\right)^{-1}\right]\right\}, \quad (7.37)$$

where

$$\gamma = 1.57 - 7\left[\left(\ln\frac{2\lambda M}{M_0 d}\right)^{-1} - 0.28\right]^2,$$

TABLE 7.3. Relaxation Times τ and Coefficients F in Eq. (2.27) and C in Eq. (7.38) according to the Data on the Dispersion of the Kerr Effect in Solutions of Some Polymers

No.	Polymer and solvent	$M \cdot 10^{-5}$	$\tau \cdot 10^5$, s	F	C	Reference
1	Cellulose carbanilate fractions in dioxane	3.4 8.5 15 24 41	2.3 6.9 11.4 14.5 26.5	0.6 1.3 2.3 3.8 4.5	0.83 0.38 0.22 0.13 0.11	[109]
2	Fractions of poly(methacryloyl-phenylic ester) of nonyloxyben-zoic acid (PM-9) in tetrachloromethane	1—17	0.1—1.0	1.5—5	0.3—0.1	[126]
3	Fractions of poly(methacryloyl-phenylic ester) of cetyloxybenzoic acid in tetrachloromethane	3—190	4—6	0.8—7.0	0.6—0.1	[136]

and M_0 is the molecular weight of the monomer unit, S is the number of monomer units in a Kuhn segment, and λ is the length of the monomer unit in the chain direction.

Plotting the left side of Eq. (7.37) (designated by y in Fig. 7.34) versus M, λ is obtained from the initial ordinate of the curve, and S from the initial slope.

This plot (Fig. 7.34) yields the values of $S = 270$ and $\lambda = 2.3 \cdot 10^{-8}$ cm for poly(chlorohexyl isocyanate) [104] and $S = 550$, $\lambda = 2.1 \cdot 10^{-8}$ cm for poly(butyl isocyanate) [119].

The values of A, S, d, and λ obtained from experimental data on the kinetics of the Kerr effect agree qualitatively with those determined for the same polymers by hydrodynamic methods (Chapter 4) and EB (Chapter 6, in particular Fig. 6.28). The agreement between geometrical molecular characteristics obtained from the phenomena of rotational and translational friction indicates that both the hydrodynamic and conformational models on which this theory is based are valid. This is evidence of the kinetic rigidity of the investigated chains manifested in their rotational motion in the electric field.

This conclusion applies equally to both more rigid helical polypeptides and poly(alkyl isocyanate)s, on the one hand, and to less rigid cellulose derivatives on the other.

8.5. Intramolecular Motion in the Electric Field

Although the orientation of macromolecules is the principal mechanism of molecular motion in rigid-chain polymer solutions, solution polarization by the molecular-deformational mechanism can frequently be observed. Figure 7.35 shows, as an example, the dependence of the dielectric increment on the frequency for solutions of cellulose carbanilate fractions in dioxane. The dispersion curves undergo a rapid decline in the frequency range in which the orientational mechanism of polarization of molecules with a given molecular weight are cut off. When v increases further, these curves form a plateau with height independent of molecular weight. This means that for all fractions the residual polarization of the solution in the plateau region is caused by a mechanism which can relax only at higher frequencies than those used in the experiment. Since in the plateau region $\Delta\varepsilon/c$ is independent of M, this mechanism is evidently related to small-scale intramolecular motion, i.e., to the deformational mechanism which is probably determined by the orientation of polar bonds of the molecule in labile side groups of the chain.

However, this mechanism of motion does not provide any noticeable contribution to the Kerr effect since the dispersion curves of EB fall to virtually zero (Figs. 7.22–7.26). This difference may be interpreted by the proportionality of the orientational EB in solution of a rigid-chain polymer to the square of the number of monomer units in a segment S^2 (see Section 8.1), whereas the increment $\Delta\varepsilon/c$ related to the orientational mechanism is proportional to S (see the following sections). Hence, in the case of dielectric polarization the part played by the deformational mechanism as compared to the orientational mechanism can be more important than in the case of EB.

The display of intramolecular motion in rigid-chain polymers can also be observed if the kinetics of the Kerr effect are studied for a series of fractions of relatively high molecular weight. Thus, the study of dispersion curves in Fig. 7.22 and in other works [114], shows that the width of curves usually increases with the molecular weight of the fraction. This also occurs when the molecular-weight polydispersity of fractions (determined by an independent method, such as diffusion–sedimentation analysis) remains virtually invariable in a series of molecular weights. Hence, in these cases the observed increase in the spectrum of relaxation times with increasing M can be either considered as a result of the polydispersity of conformations of kinetically rigid molecules or as an indication of the intramolecular mobility of chain elements. The former interpretation seems more plausible for such kinetically rigid chains as poly(alkyl isocyanate)s, whereas the latter is more probable for less-rigid molecules of high-molecular-weight cellulose esters and ethers.

The consideration of the molecular-weight dependence of the coefficients F in Table 7.2 leads to the same conclusion. Although for all polymers of moderately high molecular weight the experimental values of F are within the range predicted by Eq. (2.27) for rigid particles of various shapes, for higher-molecular-weight fractions (Nos. 13–15) the values of F already exceed these limits. Further increase in F with M is demonstrated by the data given in Table 7.3 for high-molecular-weight cellulose carbanilate fractions.

The increase in F with increasing molecular weight of the polymer virtually implies that the dependence of relaxation time τ on M becomes weaker with increasing M. This weakening may be understood on the basis of general conditions of orientation and deformation of a chain molecule in its motion in the electric field [120]. In fact, as already mentioned, the kinetics of the behavior of a polar chain molecule in the electric field are determined by the relative values of relaxation times of its orientation τ_{or} and deformation τ_{def}. Since τ_{or} increases with molecular weight proportionally to $M[\eta]$, whereas τ_{def} is independent of M, it might be ex-

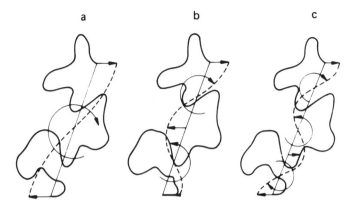

Fig. 7.36. Various modes of intramolecular motion in a Gaussian chain. a) First mode, rotation of a one-segment chain as a whole; b) second mode, rotation of parts of a two-segment chain; c) third mode, rotation of parts of a three-segment chain.

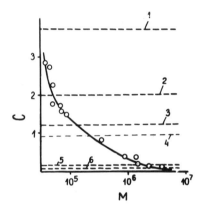

Fig. 7.37. Experimental values of the coefficient C versus molecular weight M for cellulose carbanilate samples in dioxane (points and solid curve). Broken lines correspond to theoretical values of C: 1) rigid rod; 2) rigid coil; 3) rigid sphere; 4) first mode; 5) third mode; 6) fifth mode [109].

pected that at relatively high M the inequality $\tau_{or} > \tau_{def}$ will be fulfilled and hence the polarization and anisotropy of solution in the electric field will follow the deformational mechanism.

The transition from the orientational to the deformational process with increasing M should be manifested in the decrease in the coefficient C in the equation relating the experimental value of τ to $M[\eta]$:

$$\tau = CM\,[\eta]\,\eta_0/RT. \qquad (7.38)$$

In fact, with the orientation mechanism of EB, C is equal to $1/2F$. When M increases and the chain conformation changes from rodlike to spherical, C can change correspondingly from $C = 3.75$ (at $F = 2/15$) to $C = 1.2$ (at $F = 5/12$).

A further decrease in C with increasing M means that the conditions of a purely orientational effect are not observed any longer and intramolecular motion is manifested in EB. With a purely deformational effect, relaxation time τ is independent of molecular weight. Under these conditions the coefficient C in Eq. (7.38) will decrease proportionally to $(M[\eta])^{-1}$ with increasing M. This situation is observed, for example, in solutions of high-molecular-weight DNA [121] and of some mesogenic polymers with comb-shaped molecules (see below, Section 10.2.1).

The transition from the orientational (large-scale) mechanisms of dipole polarization to the deformational (small-scale) mechanism may be demonstrated in terms of normal modes of motion of the polymer chain (Fig. 7.36).

Relaxation times τ_i corresponding to various modes of molecular motion in a shear field are determined by Eq. (5.56), which is equivalent in form to Eq. (7.38).

In the electric field for the same modes relaxation times τ_i' are equal to $2\tau_i$ [122]. In accordance with this, for odd motion modes responsible for the dipole polarization of the chain the coefficients C_i' contained in Eq. (7.38), from Eqs. (5.56) and (5.56'), are equal to $C_1 = 0.844$, $C_3' = 0.142$, and $C_5' = 0.064$.

The points and the curves in Fig. 7.37 show (according to Tables 7.2 and 7.3) the values of the coefficients C obtained experimentally for cellulose carbanilate of various molecular weights M. Broken straight lines (1–6) show the theoretical values of C corresponding to various models for a rigid molecule (1–3) and the motion modes of a flexible Gaussian chain (4–6).

In the range of relatively low molecular weights in which the experimental values of C are higher than $C_1' = 0.844$, corresponding to the first motion modes of a Gaussian chain (broken line 4), the molecule rotates as a whole (kinetically rigid chain). In the range of high M, in which C is less than C_1', the chain motion follows the mechanism with a relaxation time shorter than that of the first mode. This motion is described by higher modes and is to a considerable extent the intramolecular motion occurring by the deformational mechanism.

According to the data shown in Fig. 7.37 for cellulose carbanilate, the transition from the orientational to the deformational EB mechanism takes place in the molecular-weight range $\approx 5 \cdot 10^5 - 5 \cdot 10^6$, which is many times higher than the molecular weight of the Kuhn segment characterizing the equilibrium rigidity of the chains of this polymer.

9. EQUILIBRIUM ELECTRO-OPTICAL PROPERTIES OF RIGID-CHAIN MOLECULES

It has been shown that the primary mechanism responsible for EB in solutions of rigid-chain polymers is the rotation of their polar molecules as a whole, whereas the anisotropy of dielectric polarizability of the macromolecules, $\delta_1 - \delta_2$, does not contribute to the Kerr effect. Hence, the general theory of the Kerr effect for rigid dipole particles with axial symmetry of optical polarizability can be used for the description of equilibrium values of EB in these solutions. Accordingly, assuming that $\delta_1 - \delta_2$ in Eq. (7.8) is zero, the Kerr constant is given by

$$K = B_1 (\gamma_1 - \gamma_2)(\mu^2/M)(3\cos^2\theta - 1), \qquad (7.39)$$

where B_1 is a coefficient equal to

$$B_1 = \pi N_A (n^2 + 2)^2 (\varepsilon + 2)^2 / [1215n\,(kT)^2]. \qquad (7.40)$$

Equation (7.39) represents a general relationship between the Kerr constant K and the dipole and optical properties of a kinetically rigid particle. To establish the quantitative dependence of K on the conformation and structure of a rigid-chain polymer molecule, the molecular model describing its electro-optical properties should be specified. For this purpose, we use a kinetically rigid wormlike chain, just as in the study of the FB problem.

9.1. Dipole Moment

Equation (7.39) shows that the dipole moment of the molecule, μ, is of great importance for EB. For a wormlike model the square of this dipole moment averaged over all chain conformations $\langle\mu^2\rangle$ analogous to Eq. (1.24) is determined from the equation [92]

$$\langle\mu^2\rangle = N\mu_s^2 [1 - (1 - e^{-x})/x], \qquad (7.41)$$

where N is the number of Kuhn segments in the chain and μ_s is the dipole moment of the segment determined as the arithmetical sum of dipole moments of monomer units in the segment (S): $\mu_s = \mu_0 S$. Since $N = M/M_s = M/M_0 S$, Eq. (7.41) is equivalent to the expression

$$\langle\mu^2\rangle/M = (\mu_0^2/M_0)\,S\,[1 - (1 - e^{-x})/x]. \qquad (7.42)$$

Equation (7.42) predicts the dependence of $\langle\mu^2\rangle/M$ on $M = M_0 Sx/2$ coinciding with the dependence of $\langle h^2\rangle/M$ on M shown in Fig. 1.2 (curve 1).

The experimental dependence described by Eq. (7.42) can be checked if $\langle\mu^2\rangle$ is determined for a series of polymer fractions from measurements of dielectric permittivities ε of their dilute solutions using the well-known Debye equation

$$\langle\mu^2\rangle/M = (9kT/4\pi N_A c)\,[(\varepsilon_0 - 1)/(\varepsilon_0 + 2) - (\varepsilon_\infty - 1)(\varepsilon_\infty + 2)], \qquad (7.43)$$

where ε_0 and ε_∞ are limiting values of ε at frequencies $\nu \to 0$ and $\nu \to \infty$, respectively. The experimental dispersion curves which are similar to those in Fig. 7.35 yield the values of $\langle\mu^2\rangle/M$ for polymer fractions. These values for fractions of poly(chlorohexyl isocyanate) and cellulose carbanilate versus molecular weight are presented in Fig. 7.38. The general character of experimental curves corresponds to the theoretical dependence described by Eq. (7.42). This permits the determination of the dipole moment of the monomer unit, μ_0, from the initial slopes $(\mu_0/M_0)^2$ and the calculation of the parameter of chain rigidity S from their limit $(\mu_0^2/M)S$. The values of S obtained in this manner (Table 7.5) and those obtained by other methods (Chapters 4 and 6) are close to each other within experimental error, whereas the values of μ_0 (Table 7.5) and the values that could be expected taking into account the structure of the main chain of these polymers are in reasonable agreement. This means that the equilibrium dielectric properties of rigid-chain polymer solutions can be adequately described in terms of the model of a wormlike chain according to Eqs. (7.42) and (7.43).

It should be noted that at high M, i.e., in the range of curves with saturation (Fig. 7.38), the character of the dependence of $\langle \mu^2 \rangle/M$ on M for rigid-chain polymers and common flexible-chain polymers is the same ($\langle \mu^2 \rangle/M$ is constant). However, the use of the limiting values of $\langle \mu^2 \rangle/M$ for rigid-chain molecules results in S values as high as tens and even several hundreds, whereas S for flexible-chain molecules is invariably less than or equal to 1. This clearly shows the difference in the mechanisms of polarization of rigid- and flexible-chain polymers. For the rigid chain, a Gaussian coil in "frozen" conformation rotates as a whole and in the flexible chain each monomer unit is virtually independent of others (a value of $S < 1$ is usually interpreted as the manifestation of interactions preventing their rotation [123]).

9.2. EB in an Assembly of Wormlike Chains

9.2.1. Dipole Direction in a Wormlike Chain. Equations (7.39) and (7.43) clearly show the difference in the equilibrium dielectric and electro-optical properties of rigid-chain polymer solutions. Dielectric polarization depends on the absolute value of the dipole moment of the molecule, whereas the value and sign of K are also profoundly affected by the angle θ formed by the molecular dipole μ and the optical axis of the molecule γ_1. In an assembly of randomly bent chain molecules, if Eq. (7.39) is used, the optical $(\gamma_1 - \gamma_2)$ and dipole $\mu^2(3\cos^2\theta - 1)$ factors should be averaged over all conformations.

For a wormlike chain, the difference between the main polarizabilities averaged over all conformations $\langle \gamma_1 - \gamma_2 \rangle$ is determined by Eqs. (5.84) and (5.86), where γ_1 corresponds to the direction h of the end-to-end distance. Hence, the angle θ in Eq. (7.39) is the angle formed by the direction of the dipole moment μ of a chain molecule and vector $\mathbf{h}$.

If the dipole moment of the monomer unit μ_0 rigidly attached to the chain forms an angle ϑ with the chain direction, it follows that μ_0 is a geometrical sum of the components parallel ($\mu_{0\parallel} = \mu_0 \cos\vartheta$) and perpendicular ($\mu_{0\perp} = \mu_0 \sin\vartheta$) to the chain direction. Hence

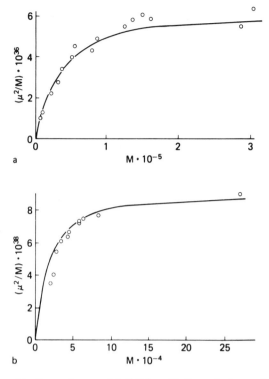

Fig. 7.38. a, b) Dependence of $\langle\mu^2\rangle/M$ on M for solutions of poly(chloro-hexyl isocyanate) fractions in tetrachloromethane (a) and cellulose carbanilate in dioxane (b) [116].

$$\mu_0^2 = \mu_{0\parallel}^2 + \mu_{0\perp}^2, \quad \mu_{s\parallel} = \mu_s \cos\vartheta,$$

$$\mu_{s\perp} = \mu_s \sin\vartheta, \quad \mu_s^2 = \mu_{s\parallel}^2 + \mu_{s\perp}^2, \tag{7.44}$$

where $\mu_{s\parallel}$ and $\mu_{s\perp}$ are the parallel (longitudinal) and perpendicular (normal) components of the dipole moment of the segment μ_s, respectively.

The total dipole moment of a chain molecule μ in any conformation is a geometrical sum of dipoles μ_0 of all its monomer units.

It can be seen (Fig. 7.39) that in this summation the direction of the component of the molecular dipole μ consisting of the sum of longitudinal components of monomer dipoles ($\Sigma\mu_0\cos\vartheta$) coincides with vector **h** and hence in Eq. (7.39) we have $\theta = 0$ and $3\cos^2\theta - 1 = 2$. Consequently, the dipole factor in Eq. (7.39), obtained by summing over all $\mu_{0\parallel}$ of a wormlike chain and taking into account Eq. (7.42), is given by

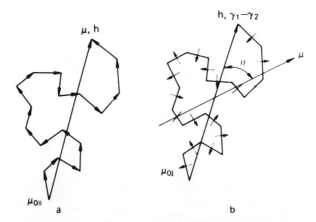

Fig. 7.39. a, b) Direction of the total dipole moment μ in a chain molecule when the dipole moment of the monomer unit $\mu_{0\parallel}$ is parallel to the chain (a) and when this dipole moment $\mu_{0\perp}$ is normal to the chain.

$$\langle (\mu^2/M)(3\cos^2\theta - 1)\rangle_{\mu_{0\parallel}} = (\mu_{0\parallel}^2/M_0)\,S\,[1 - (1 - e^{-x})/x] \cdot 2. \tag{7.45}$$

An essentially different result is obtained if normal components of monomer dipoles $\mu_{0\perp}$ rigidly bonded to the main chain are summed. Kuhn has shown [124] that in this case, for a molecule described by a chain of freely jointed segments, the direction of μ generally does not coincide with $\mathbf{h}$, forming with it angle θ (Fig. 7.39b) determined by the equation

$$\langle\cos^2\theta\rangle = (h/L)/\mathscr{L}^*(h/L), \tag{7.46}$$

where $\mathscr{L}^*$ is the inverse Langevin function represented by Eqs. (5.29) and (5.32).

Equation (7.46) shows that for a relatively long chain, when $H/L \to 0$ and, correspondingly, $\mathscr{L}^* \to 3h/L$, the mean-square cosine ($\langle\cos^2\theta\rangle) \to 1/3$. This means that in the Gaussian range the direction of the dipole moment of a chain molecule μ obtained by summation over all $\mu_{0\perp}$ is not correlated with the h direction. Hence, in the Gaussian range, normal components $\mu_{0\perp}$ do not contribute to the Kerr effect (since for them we have $\langle\cos^2\theta - 1/3\rangle = 0$). In this case, the correlation between the μ and h directions appears only with decreasing L (i.e., with increasing h/L) on passing into the non-Gaussian range and becomes complete for the rodlike conformation when $h/L = 1$, $\mathscr{L}^* \to \infty$, $\langle\cos^2\theta\rangle = 0$ and, correspondingly, $\theta = \pi/2$.

Extending this approach from a chain of freely jointed segments to a wormlike chain [125] and applying Eqs. (7.42) and (7.46), we find, for the dipole factor in Eq. (7.39) obtained by summation over all $\mu_{0\perp}$, instead of Eq. (7.45) the following expression:

$$\langle(\mu^2/M)(3\cos^2\theta-1)\rangle\mu_{0\perp}=(\mu_{0\perp}^2/M_0)S\,[1-(1-e^{-x})/x]\,[(3h/L)/\mathscr{L}^*-1].\qquad(7.47)$$

The total value of the dipole factor in Eq. (7.39) for a wormlike chain is obtained by the addition of Eqs. (7.45) and (7.47). Using Eq. (7.44), we obtain

$$\langle(\mu^2/M)(3\cos^2\theta-1)\rangle=(\mu_0^2/M_0)S\,[1-(1-e^{-x})/x]\,[2\cos^2\vartheta$$
$$-\sin^2\vartheta\,(1-(3h/L)/\mathscr{L}^*)].\qquad(7.48)$$

If Eq. (5.32) is used and $(h/L)^2$ is replaced by $\langle h^2\rangle/L^2$ [just as in Eq. (5.84)] we obtain

$$1-(3h/L)/\mathscr{L}^*=(3/5)(\langle h^2\rangle/L^2)\,[1-(2/5)(\langle h^2\rangle)/L^2].\qquad(7.49)$$

Equation (7.48) shows that the dipole factor is a sum of two terms, the first of which, corresponding to the contribution of longitudinal components of monomer dipoles $\mu_{0\parallel}$, is positive, whereas the second term representing the contribution of their normal components is negative.

9.2.2. The Kerr Constant.
For an assembly of chain molecules of similar length L, the Kerr constant can be obtained by averaging the right side of Eq. (7.39) over all chain conformations

$$K=B_1\langle\gamma_1-\gamma_2\rangle(\mu^2/M)(3\cos^2\theta-1)\rangle.\qquad(7.50)$$

In order to use Eqs. (5.84), (7.48), and (7.49) for this purpose, it should be considered that for a wormlike chain in the Gaussian range where the polydispersity of conformations is most marked, both $\gamma_1-\gamma_2$ [according to Eq. (5.84)] and μ^2 [according to Eq. (7.42)] vary proportionally to h^2 with a changing conformation. Hence, in complete analogy with Eq. (5.104) we obtain

$$\langle(\gamma_1-\gamma_2)\mu^2(3\cos^2\theta-1)\rangle/[\langle\gamma_1-\gamma_2\rangle\langle\mu^2(3\cos^2\theta-1)\rangle]=\langle h^4\rangle/\langle h^2\rangle^2.\qquad(7.51)$$

Using Eq. (7.51), the substitution of Eqs. (5.84), (7.48), and (7.49) into Eq. (7.50) gives for the Kerr constant

$$K=B_1\frac{3}{5}\beta L\frac{\mu_0^2}{M_0}S\left[1-\frac{1}{x}(1-e^{-x})\right]\frac{\langle h^2\rangle/L^2}{1-\frac{2}{5}\langle h^2\rangle/L^2}\frac{\langle h^4\rangle}{\langle h^2\rangle}$$
$$\times\left(2\cos^2\vartheta-\frac{3}{5}\sin^2\vartheta\frac{\langle h^2\rangle/L^2}{1-\frac{2}{5}\langle h^2\rangle/L^2}\right).\qquad(7.52)$$

If $\langle h^2\rangle$ is expressed according to Eq. (1.24) and if it is taken into account that $\beta L=\Delta aSx/2$, Eq. (7.52) becomes

$$K=B_1\frac{6\Delta a\mu_0^2S^2}{5M_0}\frac{(x-1+e^{-x})^2}{x^2-0.8(x-1+e^{-x})}\frac{\langle h^4\rangle}{\langle h^2\rangle^2}\left[\cos^2\vartheta-0.6\sin^2\vartheta\frac{x-1+e^{-x}}{x^2-0.8(x-1+e^{-x})}\right],\qquad(7.53)$$

where B_1 and $\langle h^4\rangle/\langle h^2\rangle^2$ are expressed by Eqs. (7.40) and (1.24), (1.28), respectively.

The Kerr constant in Eq. (7.53) is expressed as a function of the structural parameters Δa, M_0, μ_0, S, ϑ, and the reduced length x of the wormlike chain.

It should be noted that the factor B_1 contained in Eq. (7.53) is expressed by Eq. (7.40) in accordance with the fact that in Eq. (7.8) and all subsequent equations the internal field is expressed according to Lorentz, which is logical if nonpolar liquids are investigated. In other cases it might be more suitable to use Onsager's equation for this purpose. Then, instead of Eq. (7.40), the following expression is used for the factor B_1:

$$B_1 = 4\pi N_A \, (n^2 + 2)^2 \, \varepsilon^2 / [135n \, (2\varepsilon + 1)^2 \, (kT)^2]. \tag{7.54}$$

This form of Eq. (7.53) is also determined by the fact that in this expression Eq. (5.84) is used for the optical anisotropy $\gamma_1 - \gamma_2$ of a wormlike chain. If Eq. (7.97) is used for $\gamma_1 - \gamma_2$, then the expression for K will differ in form from Eq. (7.53). However, this will not introduce any important quantitative changes into the result because the differences in Eqs. (5.84) and (5.97) are very slight and concern only the character of the dependence of anisotropy on chain length but not its limiting values, as is shown in Fig. 5.19. Hence, in consideration of the dependence of K on the structural properties of molecules, Eq. (7.53) will always be used in subsequent discussion.

At low M $(x \to 0)$ it follows from Eq. (7.53) that

$$K_{x \to 0} = B_1 P^2 \Delta a \mu_0^2 (3 \cos^2 \vartheta - 1)/M_0. \tag{7.55}$$

This means that the Kerr constant increases proportionally to the square of the degree of polymerization $P^2 = (M/M_0)^2$, as is to be expected for the rodlike molecule in which both the dipole moment and the difference between the main polarizabilities are proportional to the chain length.

In this case, the sign of EB can either coincide with that of the anisotropy of the monomer Δa or be opposite to it, depending on the value of the angle ϑ in Eq. (7.55). Hence, in principle, the signs of EB and FB for a polymer in the rodlike conformation can either coincide or be opposite to each other.

In the Gaussian range, when $x \to \infty$, the Kerr constant attains the limiting value K_∞ which, according to Eq. (7.53), is given by

$$K_\infty = 2B_1 \Delta a \, (\mu_0^2 \cos^2 \vartheta) \, S^z / M_0. \tag{7.56}$$

Equation (7.56) shows that the sign of K coincides with that of Δa since in this case the Kerr effect is determined by the square of the longitudinal component of the monomer dipole $(\mu_0 \cos \vartheta)^2$. In the Gaussian range the sign of FB always corresponds to that of Δa [Eqs. (5.60) and (5.111)]. Hence, the signs of EB and FB in a solution of kinetically rigid chain molecules of relatively high molecular weight should coincide.

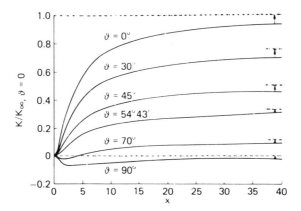

Fig. 7.40. Relative Kerr constant $K/K_{\infty,\vartheta=0}$ versus $x = 2L/A$ for a kinetically rigid wormlike chain. Numbers on the curves show the angle ϑ formed by the dipole of the monomer unit and the chain direction.

TABLE 7.4. Magnitude and Sign of Intrinsic Values of Dynamic $\Delta n/\Delta\tau$ and Electric (K) Birefringence for Some High-Molecular-Weight Polymers in Solution

No.	Polymer	Solvent	$(\Delta n/\Delta\tau) \cdot 10^{10}$, g^{-1} cm s^2	$K \cdot 10^{10}$, cm^5 g^{-1} statvolt^{-2}	Reference
1	Poly(γ-benzyl-L-glutamate)	Dichloroethane	+1100	+30000	[92, 93, 95]
2	Poly(butyl isocyanate)	Tetrachloromethane	+300	+25000	[119, 127]
3	Poly(chlorohexyl isocyanate)	Tetrachloromethane	+250	+6400	[119, 128]
4	Ladder poly(phenyl siloxane)	Benzene	−160	−12.5	[129]
5	Ladder poly(m-chlorophenyl isobutyl siloxane)	Benzene	−300	−15	[130]
6	Ladder poly(phenyl isobutyl siloxane) 1:1	Benzene	−69	−4.9	[129]
7	Ladder poly(phenyl isohexyl siloxane)	Benzene	−81	−8.5	[129]
8	Ethyl cellulose	Dioxane	+30	+25	[110]
9	Benzyl cellulose	Dioxane	+25	+6	[110]
10	Cellulose diphenylphosphono-carbamate	Dioxane	+50	+25	[110]
11	Cellulose butyrate	Dioxane	+4.7	+0.4	[110]
12	Cellulose diphenylacetate	Dioxane	+82	+0.9	[110]
13	Cellulose benzoate	Dioxane	−35	−10	[110]
14	Cellulose carbanilate	Dioxane	−144	−110	[109]
15	Cellulose nitrate, N = 10.7%	Dioxane	+10	+6	[99, 111]
16	Cellulose nitrate, N = 13.2%	Cyclohexanone	−44	−900	[100, 131]
17	Cellulose nitrate, N = 12.1%	Cyclohexanone	−27	−550	[132, 133]
18	Cyanoethyl cellulose	Cyclohexanone	+72	+250	[118, 134]
19	Poly(amide hydrazide)	Dimethylsulfoxide	+280	+35300	[101, 135]

The dependence of the Kerr constant on molecular weight (reduced chain length x) during the corresponding change in the conformation of the molecule from the straight rod to the Gaussian coil is described by Eqs. (7.53) or (7.57):

$$\frac{K}{K_\infty} = \frac{3}{5}\frac{\langle h^4\rangle}{\langle h^2\rangle^2}\frac{(x-1+e^{-x})^2}{x^2-0.8(x-1+e^{-x})}\left[1-0.6\tan^2\vartheta\,\frac{x-1+e^{-x}}{x^2-0.8(x-1+e^{-x})}\right]. \quad (7.57)$$

The character of the dependence of K on x is profoundly affected by the angle ϑ formed by the monomer dipole μ_0 (rigidly bonded to the chain) and the chain direction. The parallel component of this dipole provides a positive (corresponding to the sign of Δa) contribution to the Kerr effect. It is determined by the first term in Eqs. (7.53) or (7.57). The normal component of the monomer dipole $\mu_0 \cos \vartheta$ yields a negative (opposite to the sign of Δa) contribution to EB [second term in Eqs. (7.53) and (7.57)].

The plots of $K/K_{\infty, \vartheta = 0}$ versus x at various ϑ are shown in Fig. 7.40. Here we have

$$K_{\infty, \vartheta = 0} = K_{\infty} / \cos^2 \vartheta = 2B_1 \Delta a \mu_0^2 S^2 / M_0. \tag{7.58}$$

The curves in Fig. 7.40 are distinguished not only by the limiting values (proportional to $\cos^2 \vartheta$) but also by their shape. If the dipole μ_0 is normal to the chain direction ($\vartheta = \pi/2$), the sign of the Kerr effect is opposite to that of Δa at all values of x, and when x increases, its absolute value decreases to zero. At any value of ϑ the contribution of the normal component of the dipole $\mu_0 \sin \vartheta$ rises with decreasing x and hence at $90° > \vartheta > 54.74°$ – according to the theory – a reversal of sign of the curves of K versus x is possible. However, this conclusion is related to a certain extent to the assumption of the axial symmetry of the optical properties of the monomer unit, the segment, and the whole molecule and of the rigidity of attachment of the dipole μ_0 to the polymer chain.

9.3. Some Experimental Data

9.3.1. Value and Sign of the Kerr Constant.

Table 7.4 lists the equilibrium values of the Kerr constants for rigid-chain polymers obtained for high-molecular-weight samples in the range of M in which the chain conformation is far from rodlike. The shear optical coefficients $\Delta n / \Delta \tau$ for the same solutions are given for comparison. The comparison of data in Tables 7.4 and 7.1 is also instructive because it distinctly reveals the differences in the electro-optical properties of flexible-chain and rigid-chain polymers.

In absolute values the K constants for different rigid-chain molecules may differ by several orders of magnitude, depending on the rigidity and the dipole structure of the polymer chain. Nevertheless, even for samples with the lowest Kerr constants listed in Table 7.4, the K values are still higher by two orders of magnitude than those for flexible-chain polymers.

However, the most important difference between the data in Tables 7.1 and 7.4 consists in the fact that for flexible-chain polymers the K values do not correspond either in magnitude or in sign to those of $\Delta n / \Delta \tau$, whereas for all rigid-chain polymers the rule of coincidence in the signs of K and $\Delta n / \Delta \tau$ is obeyed if the measurements of EB and FB are carried out in the same polymer solution. This rule is valid regardless of whether the observed EB and FB are the manifestation of intrinsic chain anisotropy or the effect of its microform (compare data for cellulose nitrate

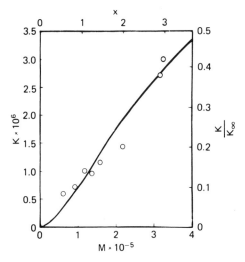

Fig. 7.41. Dependence of K on molecular weight M for poly(γ-benzyl-L-glutamate). The points represent experimental data for polymer fractions in dichloroethane [92]. The curve represents the theoretical dependence of K/K_∞ on x at the values of S, Δa, ϑ, and μ_0 given in Table 7.5.

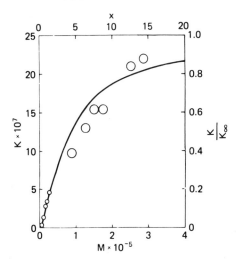

Fig. 7.42. Dependence of K on M for poly(butyl isocyanate). The points represent experimental data for polymer fractions in tetrachloromethane. The curve represents the theoretical dependence of K/K_∞ on x at the values of S, Δa, ϑ, and μ_0 given in Table 7.5.

in cyclohexanone and dioxane). This is in agreement with the conclusions drawn on the basis of Eq. (7.56).

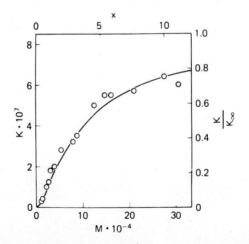

Fig. 7.43. Dependence of K on M for poly(chlorohexyl isocyanate). The points represent experimental data for polymer fractions in tetrachloromethane [104]. The curve represents the theoretical dependence of K/K_∞ on x at the values of S, Δa, ϑ, and μ_0 given in Table 7.5.

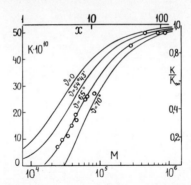

Fig. 7.44. Dependence of K on M for cellulose carbanilate. The points represent experimental data for polymer fractions in dioxane [115]. The curves represent the theoretical dependences at the values of ϑ shown on the curves and the values of S, Δa, and μ_0 given in Table 7.5.

This characteristic property of all rigid-chain polymers demonstrates the similarity between the orientation mechanisms of their molecules in shear and electric fields. In both cases orientation occurs by the mechanism of large-scale motion, i.e., by the rotation of a dipole, aspheric, and optically anisotropic kinetically rigid macromolecule as a whole.

The coincidence of the signs of K and $\Delta n/\Delta\tau$ also implies that, on the average, the three main directions in the molecule coincide: the direction of the greatest geometrical length of the molecule (vector **h**), of the orientational–axial order (respon-

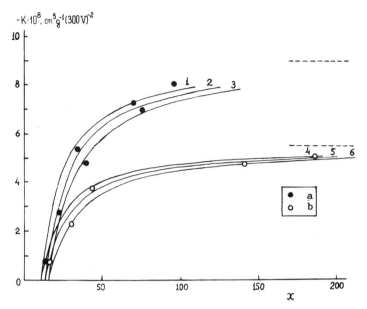

Fig. 7.45. Dependence of K on $x = 2L/A$ for two cellulose nitrate samples in cyclohexanone. a) Experimental points for fractions of the sample with a nitrogen content of 13.4% [100]; b) experimental points for fractions of the sample with nitrogen content of 12.1% [133]. Curves 1–6 represent the theoretical dependences at $\vartheta = 77°$ (1, 4), $78°$ (2, 5), and $79°$ (3, 6), and at the values of S, Δa, and μ_0 given in Table 7.5.

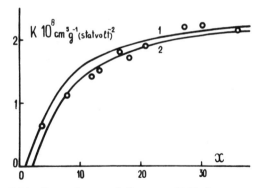

Fig. 7.46. Dependence of K on $x = 2L/A$ for cyanoethyl cellulose. The points represent experimental data for polymer fractions in cyclohexanone [118]. Curves 1 and 2 represent theoretical dependences at the angles $\vartheta = 60°$ (1) and $65°$ (2), and at the values of S, Δa, and μ_0 given in Table 7.5.

sible for the anisotropy $\gamma_1 - \gamma_2$), and of the orientational polar order (determining the total dipole moment μ of the molecule).

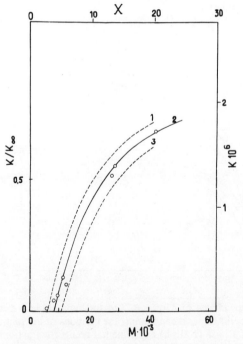

Fig. 7.47. Dependence of K/K_∞ on M for poly(amide hydrazide) in dimethylsulfoxide. The points represent experimental data for samples of various molecular weights [101]. The curves represent theoretical dependences at the values of ϑ = 68° (1), 70° (2), and 72° (3), and at the values of S, Δa, and μ_0 given in Table 7.5.

Fig. 7.48. *Trans*- and *cis*-Structures of the poly(alkyl isocyanate) chain. R is the alkyl side radical.

Hence, in contrast to flexible-chain molecules (see Section 5.2.2), in the interpretation of the EB data in rigid-chain polymer solutions it is quite permissible to compare them with the results of FB investigations in the same solutions for their combined consideration and use.

9.3.2. Molecular-Weight Dependence of *K*. As already repeatedly indicated, a characteristic feature of rigid-chain polymers is the dependence of their Kerr constant on molecular weight (chain length).

This dependence is shown in Figs. 7.41–7.47, in which the points represent the experimental values of *K* as a function of molecular weight or the parameter $x = 2L/A$ for some rigid-chain polymers. For all polymers the Kerr constant increases drastically with increasing chain length. Moreover, at high molecular weights a tendency toward the saturation of this effect is observed. This trend is less pronounced for molecules with high chain rigidity [poly(benzyl glutamate) and polyisocyanates], whereas for more flexible molecules of cellulose esters and ethers in the molecular-weight range of several hundred thousand the *K* value attains virtually the limiting value.

It is possible to use the values of the chain rigidity parameters *S* (or *A*) and chain anisotropy Δa (or β) obtained from FB investigations in the same solutions for the quantitative comparison of experimental values of *K* and of their dependence on *M* with theoretical equations (7.53)–(7.58).

In this case, if the experimental data on EB yield only the limiting value K_∞, Eq. (7.56) makes it possible to estimate directly the longitudinal component $\mu_0 \cos \vartheta = \mu_{0\parallel}$ of the dipole moment of the monomer unit.

If the experimental data on the Kerr effect are more complete and contain a series of points representing the dependence of *K* on *M* (which is shown in Figs. 7.41–7.47), then the comparison of these points with the curve of theoretical dependence (7.57) and the application of Eq. (7.56) allow the determination of both the value μ_0 and the direction, angle ϑ, of the dipole moment of the monomer unit. This comparison is shown in Figs. 7.41–7.47, where the curves represent the theoretical dependence of K/K_∞ on *M* (or on *x*) according to Eqs. (7.56) and (7.57) at the values of the parameters Δa, *S*, μ_0, and ϑ given in Table 7.5 and in the figure captions.

To calculate the angle ϑ from experimental data on EB, it is sufficient to obtain additional information only about the equilibrium rigidity (segment length *A*) of the molecules and the molecular weights *M* of the samples since, according to Eq. (7.57), the value of ϑ is determined only by the character of the dependence of relative values of K/K_∞ on *x*. In this case, both *A* and *M* can be obtained from the data of hydrodynamic investigations of the same polymers.

The experimental points in Figs. 7.41–7.47 agree with the general shape of the theoretical curves at the values of the angle ϑ which are in reasonable agreement with the stereochemical structure of the polymer chain.

Thus, for PBLG molecules in the helical conformation it follows from the dependence of K/K_∞ on *x* that $\vartheta = 0$. It is this result that can be expected for a regular Pauling helix [138] in which all normal components $\mu_{0\perp}$ of dipoles of monomer units should be mutually compensated as a result of the model.

The result $\vartheta = 0$ obtained for poly(alkyl isocyanates) makes it possible to decide unequivocally between the *trans-* and *cis-*structures of the amide groups in the polymer chain (Fig. 7.48). With the *trans-*conformation of all amide groups the longitudinal components of the dipole moment of the monomer unit are mutually compensated and $\mu_{0\parallel} = 0$, whereas normal components should give a high value of $\mu_{0\perp}$, contradicting experimental data. In contrast, for a chain structure with regularly alternating *trans-* and *cis-*conformations of the amide groups (usually called the *cis-*structure of the chain) the normal components of dipoles of valence bonds are mutually compensated, whereas the longitudinal components give the nonzero longitudinal components $\mu_{0\parallel}$ in the chain direction. This condition is obeyed both when the valence angles α and β at the carbon and nitrogen atoms of the amide group are equal and when they are not equal. Hence, the data on the dependence of K/K_∞ on x for poly(alkyl isocyanate)s show that of the two structures shown in Fig. 7.48, only the *cis-*structure of the polyamide chain is possible. This means that for these polymers the signs of EB and FB should coincide over the entire molecular-weight range in which the chain conformation changes from the straight rod to the Gaussian coil, as is confirmed by available experimental data.

For all cellulose esters and ethers investigated, the angle ϑ determined according to the dependence of K/K_∞ on x is high. This means that in contrast to polypeptides and polyisocyanates, for low-molecular-weight samples of cellulose derivatives the normal component of the dipole moment of the monomer unit $\mu_{0\perp}$ plays a considerable part in the observed EB.

The absolute value of μ_0 can be determined for each polymer only from EB measurements in nonpolar solvents using the experimental values of K_∞ and ϑ and Eq. (7.56).

In those cases in which this is possible, the μ_0 values obtained by the EB method are in reasonable agreement with the data of dielectric permittivity measurements for the same solutions (Table 7.5).

The values of μ_0 determined from measurements in polar solvents can correspond to the true values only qualitatively (in order of magnitude) because, in this case, the coefficient B_1 in Eq. (7.56) differs greatly depending on whether it is calculated according to Eq. (7.40) or according to Eq. (7.54). The values of μ_0 given in Table 7.5 have been calculated using Eq. (7.40). If Eq. (7.54) is used for samples 6–9, unreasonably high μ_0 values are obtained.

According to the data listed in Table 7.5, for all cellulose derivatives the values of the longitudinal component $\mu_{0\parallel} = \mu_0 \cos \vartheta$ of the monomer dipole determined from the values of μ_0 and ϑ are close to 1 D. The same result has been obtained for a number of other cellulose esters and ethers from the EB measurements of solutions of their high-molecular-weight samples in nonpolar solvents [110].

The fact that for cellulose esters and ethers with very different structures of substituting side groups the values of $\mu_{0\parallel}$ are similar means that in these polymers

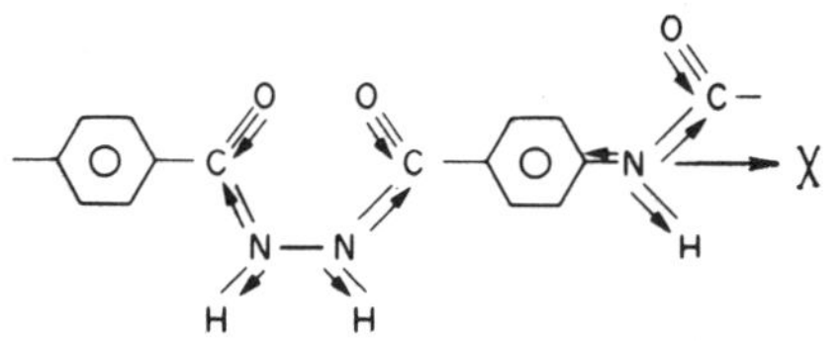

Fig. 7.49. Structure of PAA repeat unit. Arrows show the directions of dipole moments of bonds. Longitudinal (along the x axis) component of dipole of the repeat unit $\mu_{0\parallel} = 1.75$ D.

TABLE 7.5. Dipole Moment of the Monomer Unit μ_0 and Angle ϑ Formed by the Dipole Moment μ_0 and the Chain Direction according to EB and FB Data for Polymer Solutions

No.	Polymer	Solvent	Hydrodynamic and FB data (Chapters 4 and 6)		EB data		Dielectric measurements	
			S	$\Delta a \cdot 10^{25}$, cm³	$\vartheta°$	μ_0, D	S	μ_0, D
1	Poly(γ-benzyl-L-glutamate)	Dichloroethane	1000	+25	0	0.4 [92]	—	—
2	Poly(butyl iso-cyanate)	Tetrachloromethane	400	+15	0	1.6 [119]	500	1.8 [137]
3	Poly(chlorohexyl isocyanate	Tetrachloromethane	250	+15	0	1.3 [104]	250	1.9 [106]
4	Cellulose carb-anilate	Dioxane	40	−50	65	2.3 [115]	50	0.9 [116]
5	Cellulose nitrate, N = 10.7%	Dioxane	30	+4.2	60	2 [111]	—	—
6	Cellulose nitrate, N = 13.4%	Cyclohexanone	60	−9	78	4.5 [100]	—	—
7	Cellulose nitrate, N = 12.1%	Cyclohexanone	35	−8.4	78	6.4 [133]	—	—
8	Cyanoethyl cellulose	Cyclohexanone	50	+17.8	63	1 [118]	—	—
9	Poly(amide hydrazide)	Dimethylsulfoxide	16.8	+200	70	6 [101]	—	—

the longitudinal component of the monomer dipole is determined by the dipole structure of the main glucoside chain containing polar CO bonds rigidly attached to this chain rather than by the side groups.

The differences in sign and the great differences in the value of the limiting Kerr constant for various cellulose esters and ethers are not due to the differences in the dipole properties of their molecules, but are determined (just as in the FB phenomenon) by the anisotropy Δa of the monomer unit profoundly affected by the structure of the substituting side groups.

The same side groups may provide an important contribution to the normal component $\mu_{0\perp}$ of the dipole of the monomer unit, and hence can play a considerable part in EB only in the investigations of low-molecular-weight samples of cellulose derivatives.

The dependence of K/K_∞ on x found experimentally for poly(amide hydrazide) (PAA, Fig. 7.47) also leads to a high value of the angle ϑ, showing that for this polymer at low M a considerable role in the EB phenomenon is played by the normal component of the dipole moment $\mu_{0\perp}$. According to the experimental data given in Table 7.5, the value of $\mu_{0\perp} = \mu_0 \sin \vartheta = 5.6$ D greatly exceeds the longitudinal component of the dipole of the monomer unit $\mu_{0\parallel} = \mu_0 \cos \vartheta = 2$ D.

These values may be compared with those of $\mu_{0\parallel}$ and $\mu_{0\perp}$ that might be expected on the basis of the structure of the repeat ("monomer") unit of the PAH chain shown in Fig. 7.49. The dipole moment of this unit is the vector sum of dipoles of three amide groups O=C–N–H and one C_{ar}–N bond. The chain structure in Fig. 7.49 is represented by an ideal "crankshaft" with equal valence angles $\alpha = \beta = 120°$ between all chain bonds and para-aromatic rotation bonds directed along the X axis. This axis is assumed to be the longitudinal direction of the polymer chain. It can be seen that for this simplified model the longitudinal component (projection on the x axis) of the dipole $\mu_{0\parallel}$ is independent of the azimuthal position of planes of the amide groups rotating about the X axis and is always given by

$$\mu_{0\parallel} = (\mu_{O=C} + \mu_{N-H} + \mu_{N-C}) \cos 60° - \mu_{N-C}.$$

Assuming the following values for the dipoles of valence bonds: $\mu_{O=C} = 2.8$ D, $\mu_{N-H} = 1.3$ D, and $\mu_{N-C} = 0.6$ D [138], we obtain $\mu_{0\parallel} = 1.75$ D, in good agreement with the experimental value 2 D. The normal component (perpendicular to the X axis) of the dipole moment introduced by each of the three amide groups of the unit (Fig. 7.49) is given by $(\mu_{O=C} + \mu_{N-H} - \mu_{N-C}) \sin 60° \approx 3$ D. The total contribution of all three amide groups, equal to $\mu_{0\perp}$, depends on the azimuthal position of the planes of the amide groups in their rotation about the x axis. In the chain conformation shown in Fig. 7.49 the normal components of dipoles of all amide groups are parallel, and hence $\mu_{0\perp}$ has a maximum value equal to 9 D. If the distribution of planes of the amide groups in azimuthal position is symmetrical, the normal component of their common dipole and of the monomer unit of the chain $\mu_{0\perp}$ is equal to zero. Hence, the experimental value of $\mu_{0\perp} = 5.6$ D is within the range of values which might be expected on the basis of the molecular structure of PAH.

Similar calculations for poly(alkyl isocyanate)s in the *cis*-conformation (Fig. 7.48) lead to the value of $\mu_{0\parallel} = (1/2)\mu_{O=C} = 1.4$ D, which is in good agreement with experimental data given in Table 7.5 for these polymers.

These equilibrium properties of the polymers investigated clearly demonstrate the validity of Kuhn's idea [124], according to which the longitudinal component of the dipole moment of chain units is "accumulated" along the chain in contrast to the normal component whose correlation along the chain is much weaker.

However, experiments show that for the manifestation of these properties in the EB phenomenon a sufficiently high kinetic rigidity of the chain is required, at which its orientation in the electric field occurs by the mechanism of large-scale motion.

Stockmayer has pointed out that, in principle, large-scale orientation in the electric field is possible for the molecules of flexible-chain polymers with uniformly directed longitudinal components of dipoles of monomer units [140, 141].

Some indications of this possibility have also been obtained in the experimental investigation of dielectric relaxation in oligomeric samples of flexible-chain polymers [142–144]. However, the manifestations of large-scale molecular motion observed in this case are incomparably weaker than for rigid-chain polymers and rapidly disappear with increasing molecular weights.

10. MESOGENIC POLYMERS

10.1. Lyotropic Mesogenic Molecules

All rigid-chain polymers whose hydrodynamic, dynamo-optical, and electro-optical properties have been considered in the preceding chapters and sections can form a liquid-crystalline phase in concentrated solutions (lyotropic mesomorphism).

A classical example of these polymers is polypeptides in the helical conformation and, primarily, the molecules of PBLG whose lyotropic mesophase has been studied in detail by several researchers [145–147]. Liquid-crystalline solutions of rigid-chain aromatic polyamides and polyesters extensively used in the manufacture of man-made fibers are also well known [148–152]. A stable lyotropic mesomorphic phase can be obtained in concentrated poly(alkyl isocyanate) solutions if alkyl side radicals of appropriate length are chosen [153–156]. Various cellulose derivatives also belong to the class of polymers forming lyotropic mesogenic structures in concentrated solutions [157–163]. Orientational order of the mesomorphic type has been observed in the investigations of films of ladder poly (phenyl siloxane)s with a polarizing microscope [164] and in concentrated solutions of stepladder polyacenaphthylene [165].observed in the investigations of films of ladder poly(phenyl siloxane)s with a polarizing microscope [164] and in concentrated solutions of stepladder polyacenaphthylene [165].

The possibility of the formation of lyotropic mesomorphism in concentrated solutions of all these polymers is due to the high equilibrium rigidity of their chains favoring the orientational order in the arrangement of neighboring molecules in a concentrated solution. Early theories of lyotropic mesomorphism have dealt with the concepts of "rodlike" molecules with a degree of asymmetry, $p = L/d$, proportional to molecular weight $M = LM_L$ considered to be their main characteristic determining the conditions for the formation of the orientationally ordered mesophase [166, 167].

However, subsequent accumulation of quantitative experimental data obtained in dilute solutions (many of these data have been considered in previous sections) has conclusively shown that all the so-called "rodlike" chain molecules actually exhibit finite flexibility and, correspondingly, a curvature determined quantitatively by the persistent length or the length of the Kuhn segment [168]. As we have seen, the ratio of segment length to chain diameter, A/d, plays an important part in the hydrodynamic interaction of polymer molecules.

Accordingly, in modern theories of lyotropic mesomorphism in rigid-polymer solutions the flexibility of polymer chains is taken into account, and instead of a rodlike molecular model a chain of freely jointed segments or a wormlike chain is used [169–172]. In these more realistic theories the main characteristic determining the conditions of the formation of orientational order in the polymer–solvent system is assumed to be the degree of asymmetry of the segment, A/d, instead of the value of L/d.

Hence, the equilibrium rigidity of the chain (its persistent length or the length of the Kuhn segment), determined according to the conformational properties of the molecule in dilute solutions, plays the major part in the processes of formation of the orientationally ordered phase in a concentrated polymer solution.

Consequently, quantitative comparison of modern theories of lyotropic mesomorphism with the results of the corresponding experimental investigations is possible only in those cases in which the values of M, L, A, and d for the polymer are previously determined in dilute solutions by the above methods [163].

10.2. Thermotropic Mesogenic Molecules

The thermotropic mesomorphic state in bulk may be observed for polymers composed of a combination of rigid and flexible chain fragments. The former ensure the mesogenic properties of the molecule, and the latter serve as "plasticizers" facilitating the orientationally ordered mutual arrangement of mesogenic parts. Rigid mesogenic parts can be located both in the main chain and in the side chains. The hydrodynamic and dynamo-optical properties of molecules with mesogenic side groups have already been considered, and their electro-optical characteristics will be dealt with in the next section.

10.2.1. **Molecules Containing Mesogenic Side Chains.** Many comb-shaped molecules are typical of flexible-chain polymers in their hydrodynamic properties, although the equilibrium rigidity of their main chains is slightly higher as a result of the interaction between the side groups. This holds true regardless of whether the side chains of comb-shaped molecules contain or do not contain mesogenic groups (see Chapter 4, Sections 3.1 and 3.2).

The specific features of the comb-shaped structure are manifested to a much greater extent in the FB phenomenon than in the hydrodynamic properties of the molecules. Thus, the structure and length of the side chain profoundly affect the value and sign of FB, and this effect may be described quantitatively in the framework of the wormlike chain model (Chapter 6, Sections 2.9.1 and 2.9.2).

Moreover, the difference in the dynamo-optical properties of comb-shaped molecules containing mesogenic groups and without them is very pronounced: the negative anisotropy of the former is higher by more than one order of magnitude (Chapter 6, Section 2.9.3).

10.2.1.1. **Value and sign of EB.** The presence of mesogenic groups in comb-shaped molecules is manifested most markedly in their electro-optical properties. As already shown (Section 5.2.2 of this chapter and Table 7.1), the behavior in the electric field of comb-shaped molecules without mesogenic groups is typical of all flexible-chain polymers: the orientation of their polar groups follows the mechanism of local motions in the molecule and does not result in the orientation of any considerable parts of the main chain or the side chains. Accordingly, the Kerr constant for these polymers is of the order of magnitude of 10^{-12} cm^5 g^{-1} statvolt^{-2} and is independent of molecular weight, and its sign does not agree with that of FB.

In contrast, the electro-optical properties of comb-shaped molecules with mesogenic side groups are even more unique than the FB of their solutions.

Figure 7.50 can be taken as an example demonstrating these properties. It shows the dependence of EB, Δn, on the square of the field strength, E^2, for tetrachloromethane solutions of poly(methacryloyl phenyl ester) of nonyloxybenzoic acid (PM-9) with the following structure of the monomer unit:

$$H_3C-\underset{\underset{|}{H_2\overset{|}{C}}}{\overset{}{C}}-\underset{\overset{\|}{O}}{C}-O-R, \quad \text{where R is} \quad -\left\langle\bigcirc\right\rangle-\underset{\overset{\|}{O}}{C}-O-\left\langle\bigcirc\right\rangle-O-C_9H_{19}.$$

The solutions of this polymer exhibit high and well-measurable EB coinciding in sign (negative) with FB in the same solutions (Table 6.11, No. 8) and proportional to E^2 (the Kerr law). The extrapolation of the values of $\Delta n/cE^2$ according to the data shown in Fig. 7.50 allows a reliable determination of the Kerr constant, whose value is given in Table 7.6. Similar relationships are observed for all other investigated polymers with mesogenic side groups whose K values are given in Table 7.6. These values are close in order of magnitude to those for such rigid-chain polymers as some cellulose derivatives (Table 7.4) and are higher by two orders of magnitude than those for flexible-chain polymers (Table 7.1).

It is essential that for all the polymers listed in Table 7.6 the Kerr constants are negative, although in solutions of the corresponding monomers they are usually positive. It is also important that for all these polymers the signs of K and $\Delta n/\Delta \tau$ coincide. These relationships can be interpreted only by the fact that the side groups of these comb-shaped molecules are oriented in the electric field with their longitudinal axis normal to the field (in contrast to the molecules of the corresponding monomers) and hence this orientation is not free (as for a flexible-chain polymer), but is correlated with that at the main chain.

This conclusion implies that although for polymer molecules with mesogenic side radicals the equilibrium rigidity of the main chain is very moderate ($A \approx 50 \cdot 10^{-8}$ cm), the orientation of their polar groups in the electric field follows the mechanism of large-scale intramolecular motion characteristic of rigid-chain polymers rather than that of local motion.

10.2.1.2. Nonequilibrium properties. The above conclusion is confirmed by the existence of low-frequency dispersion of EB in solutions of these polymers. This can be seen in Fig. 7.51, which shows the dependence of the relative Kerr constant on the frequency of the sinusoidal field for solutions of PM-9 fractions. For all fractions ($M \approx 1.7 \cdot 10^6 – 0.12 \cdot 10^6$) in the frequency range $10^4 – 10^6$ Hz, the Kerr constant drops sharply with increasing frequency and the dispersion

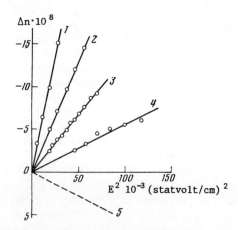

Fig. 7.50. Dependence of Δn on E^2 for solutions of a fraction of PM-9 polymer ($M = 3.6 \cdot 10^5$) in tetrachloromethane at concentrations of: 1) 8.64; 2) 4.92; 3) 2.73; 4) $1.62 \cdot 10^{-3}$ g/cm^3. Broken line – solvent [126].

TABLE 7.6. Kerr Constants K and Shear Optical Coefficients $\Delta n/\Delta \tau$ for Some Comb-Shaped Polymers with Mesogenic Groups in the Side Chains (the first line gives the numbers of those polymers in Table 6.11 or 6.12)*

Polymer No. in Table 6.11	Solvent	$K \cdot 10^{10}$ g^{-1} cm^5 statvolt^{-2}	$(\Delta n/\Delta \tau) \cdot 10^{10}$, cm s^2 g^{-1}	$\mu_{0\parallel}$, D	References
1	Tetrachloromethane	−0.5	−35	0.27	[176]
Monomer	The same	+0.15	—	—	
2	Dichloroacetic acid	−4.9	−22	0.79	[198]
5	Benzene	−2	−30	—	[178]
6	Dioxane	−7	−95	0.67	[177, 198]
7	Tetrachloromethane	From −3 to −7	−50	0.80	[136]
8	The same	From −0.8 to −8.0	−220	0.43	[126]
9	The same	−14	−190	0.61	
10	The same	From −4 to −20	−220	0.70	[136]
	Benzene	−2.2	−160	0.28	[136]
	Benzene + heptane (66:34)	−18	−350	0.55	[136]
	Benzene + heptane (52:48)	−40	−350	0.82	[136]
Monomer	Tetrachloromethane	+0.21	—	—	[179]
11	The same	−6	−85	0.60	[174]
12	The same	−3.4	−55	0.53	[174]
13	Chloroform	−360	−300	1.0	
15	Dioxane	−0.9	−6.5	0.4	[126]
9*	Benzene	−140	−210	—	[175]
(Table 6.12)					

*No. 9 in Table 6.12.

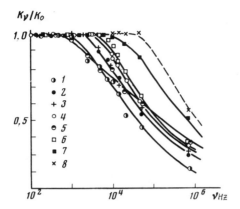

Fig. 7.51. Relative value of the Kerr constant K_ν/K_0 versus frequency ν of the sinusoidal electric field for solutions of PM-9 fractions in tetrachloromethane. Molecular weights of fractions $M \cdot 10^{-5}$ equal to: 1) 17; 2) 9.1; 3) 8.5; 4) 8.4; 5) 4.6; 6) 3.6; 7) 1.8; 8) 1.2 [126].

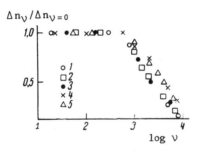

Fig. 7.52. Relative EB, $\Delta n_\nu/\Delta n_{\nu=0}$, versus frequency ν of the sinusoidal electric field for fractions of the poly(methacryloylphenyl ester) of cetyloxybenzoic acid in tetrachloromethane. Molecular weights of fractions $M \cdot 10^6$ equal to: 1) 19.1; 2) 9.2; 3) 6.3; 4) 2.98; 5) 2.3 [136].

curves are displaced toward higher frequencies with decreasing molecular weight. This relationship is characteristic of kinetically rigid polar molecules oriented in the electric field by the mechanism of rotation of the molecule as a whole.

At the same time, the comparison of relaxation times τ (determined from the dispersion curves in Fig. 7.51) with intrinsic viscosities [η] and molecular weights M of fractions according to Eqs. (2.27) and (7.38) leads to the values of the coefficients F and C, with F drastically increasing and C drastically decreasing with increasing M (Table 7.3, sample 2). This relationship indicates (see Section 8.5 of this chapter) that the EB phenomenon is a manifestation not only of the rotation of the molecule as a whole, but also of the deformational mechanism of molecular motion; the significance of this mechanism increases with the molecular weight of the fraction.

For another comb-shaped polymer with mesogenic side chains, poly(meth-acryloylphenyl ester) of cetyloxybenzoic acid (No. 10 in Tables 6.11 and 7.6), the EB of fractions has been investigated in the range of higher molecular weights (up to $M = 19 \cdot 10^6$). In this case, a considerable dispersion of the Kerr effect was also found (Fig. 7.52). However, the relaxation times τ for all fractions virtually coincided, which corresponds to a very sharp increase in the coefficient F [in Eq. (2.27)] and a decrease in the coefficient C [in Eq. (7.38)] with increasing M (Table 7.3, No. 3).

The fact that τ is independent of molecular weight means that for the samples of this comb-shaped polymer the condition $\tau_{or} > \tau_{def}$ is obeyed in this range of M (see Section 8.5), and the chain motion follows the deformational mechanism for which the "kinetic unit" oriented in the electric field is only a part of the molecule rather than the molecule as a whole. In this case, however, the rotational mobility of this kinetic unit (characterized by the value $\tau \approx 10^{-4}$–10^{-5} s) is lower by 5–6 orders of magnitude than in common flexible-chain polymers. Hence, although this motion is intramolecular, it refers to the type of large-scale nonlocal motions.

Hence, the specific orientational interaction between the side groups of a comb-shaped molecule leads to the formation of "kinetic segments," each of which covers a considerable part of the main chain and the corresponding number of monomer units. The correlation between the motions of the side groups and the main chain in the electric field extends to this part of the molecule, just as for kinetically rigid molecules. On the whole the molecule of a high-molecular-weight polymer exhibits the properties of a kinetically flexible chain with local rigidity determined by the dimensions of kinetic segments responsible for intramolecular large-scale motion. In the low-molecular-weight range in which the chain length is less than the "kinetic segment," the molecule as a whole behaves in the electric field as a kinetically rigid chain. As the molecular weight increases, the dynamic properties of the molecule vary from those of a kinetically rigid to those of a kinetically flexible coil.

10.2.1.3. Molecular-weight dependence of K. In the low-molecular-weight range where the dynamics of the molecule correspond to the properties of a kinetically rigid chain, the dependence of the equilibrium Kerr constant on molecular weight should be expected. This dependence is actually observed, as can be seen in Fig. 7.53 where the points represent the experimental values of K obtained for fractions of the PM-9 polymer (Table 7.6, No. 8). The abscissa gives the values $x = 2L/A = 2M/M_0S$ calculated from the experimental values of M, $M_0 = 424$, and $S = 26$ obtained for this polymer by hydrodynamic methods [173]. The general shape of the experimental dependence of K on x is similar to analogous dependences for other rigid-chain polymers (Figs. 7.41–7.47). Curves 1 and 2 in Fig. 7.53 show the theoretical dependence of K on x according to Eqs. (7.56) and (7.57) for a kinetically rigid wormlike chain. The experimental points fit the theoretical curve if the angle ϑ formed by the dipole of the monomer unit and the chain direction is assumed to range from 75° to 80°. At $\vartheta = 77°$, Eq. (7.56) gives $\mu_0 = 1.6$ D.

Similar dependences of K on the molecular weight of fractions have also been observed for other polymers with mesogenic side chains (Table 7.6, Nos. 7 and 10).

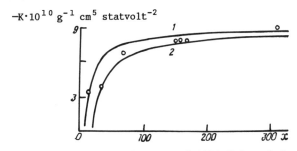

Fig. 7.53. Dependence of K on $x = 2M/(M_0S)$ for solutions of PM-9 fractions in tetrachloromethane. The points represent experimental data [126] at the value of $S = 26$ [173]. Curves 1 and 2 represent theoretical dependences according to Eqs. (7.56) and (7.57) at the values of $\Delta a = -110 \cdot 10^{-25}$ cm^{-3} (Table 6.11, No. 8), $K_\infty = -9 \cdot 10^{-10}$ g^{-1} cm^5 statvolt^{-2}, and angles $\vartheta = 75°$ (1) and 80° (2).

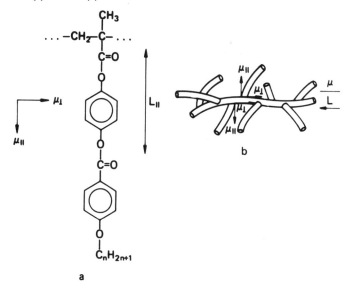

Fig. 7.54. a, b) Dipole structure of the mesogenic side chain (a) and its contributions $\mu_{||}$ and $\mu_\perp$ to the normal and longitudinal components of the dipole moment μ of the main chain L (b).

A considerable deviation of the angle ϑ from 0° corresponds to the fact that the polar side groups of the chain provide a considerable contribution to the dipole moment of the PM-9 monomer unit. At the same time, just as for other rigid-chain polymers, the "accumulation" of the longitudinal component of dipoles of monomer units $\mu_{0||} = \mu_0 \cos \vartheta = 1.6 \cos 77° \approx 0.4$ D occurs with increasing length of the main chain, whereas the contribution of the normal component to the Kerr effect becomes less and less marked.

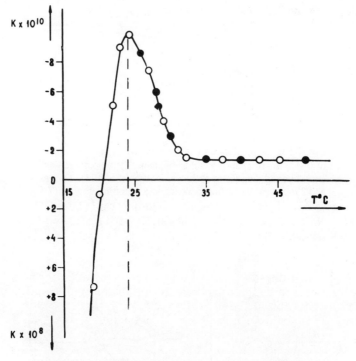

Fig. 7.55. Constant $K = \Delta n/cE^2$ versus temperature for benzene solutions of poly(methacryloylphenyl ester) of hexyl-oxybenzoic acid in benzene. Open circles – at a concentration $c = 0.77\cdot10^{-2}$ g/cm³; filled circles – at a concentration $c = 1.4\cdot10^{-2}$ g/cm³. Broken line corresponds to a change in K when the solution is thermostated at 24°C for 8 h [178].

In the comb-shaped molecules considered here, both the optical anisotropy and the dipole moment of the monomer unit are determined by those of the meso-genic side chain. The positive sign of EB in monomer solutions (Table 7.6), just as in those of other low-molecular-weight mesogenic molecules with similar structures [14–19], indicates that the dipole moment of the side chain is inclined to its longitu-dinal direction $L_\|$ (Fig. 7.54a) by an angle smaller than 54° and hence has a considerable component $\mu_\|$ in the direction $L_\|$. However, in this case, in molecules of the alkoxybenzoic acid type, the normal dipole component $\mu_\perp$ is also not equal to zero (Fig. 7.54a) [14–19].

In the molecule of a comb-shaped polymer the longitudinal axes $L_\|$ of rigid mesogenic side chains are oriented normally to the direction L of the main chain (Fig. 7.54b) and hence the longitudinal components $\mu_\|$ of their dipoles provide a contribution to the normal component of the dipole of the monomer unit of the main chain. As we have seen, this component is not accumulated along the chain. In contrast, the normal components $\mu_\perp$ of dipoles of the side groups in a comb-shaped molecule can contribute to the dipole component parallel to the main chain (Fig.

7.54b). It is this component accumulating along the main chain that determines the electro-optical properties of comb-shaped mesogenic molecules at medium and high molecular weights. Alternatively, the molecules of these polymers exhibit a dipole-orientational order, i.e., a correlation in the orientations of $\mu_\perp$, the components of dipoles of the side groups, as a result of which the macromolecule as a whole or its part can exhibit a high moment μ in the direction of the main chain (Fig. 7.54b). The existence of this dipole is the reason for the orientation of the main chain in the field direction, which leads to a negative EB.

If the rigid mesogenic part of the side chain is separated from the main chain by a flexible sequence, the correlation in the orientations of the main chain and the mesogenic part of the side chain becomes weaker, and this leads to a decrease in EB. This may be demonstrated by a comparison of K for polymer Nos. 8 and 15 in Table 7.6, differing in the presence in the latter of a flexible "spacer" $(CH_2)_{10}$ between the mesogenic core and the main chain. At equal molecular weights ($M \approx 2 \cdot 10^5$) of the two polymers being compared the value of K is lower by one order of magnitude for sample No. 15 than for sample No. 8.

10.2.1.4. Supermolecular mesomorphic order in a polymer solution. The specific dynamo- and electro-optical properties of comb-shaped mesogenic polymers considered in Section 2.9.3 of Chapter 6 and in the previous sections of this chapter are due to the orientational interaction between their side chains. This interaction, leading to the orientational axial and polar intramolecular order of the structural elements of the chain, also depends on the thermodynamic strength of the solvent. Thus, in a thermodynamically poor solvent in which the competing polymer–solvent interactions are weakened, the formation of in-tramolecular ordered structure is facilitated by the polymer–polymer interactions. Hence, as the thermodynamic strength of the solvent becomes lower, the decrease in the size of the coil is accompanied by an increase in FB and EB in solutions of comb-shaped mesogenic polymers [136]. This can be demonstrated by the data for the poly(methacryloylphenyl ester) of cetyloxybenzoic acid in a mixed benzene–heptane solvent (Table 7.6, No. 10), where heptane is the component decreasing polymer solubility.

The interaction between mesogenic side groups is also responsible for the formation of liquid-crystalline order in the polymer bulk or in its concentrated so-lution. However, when a supermolecular ordered structure is formed in the poly-mer–solvent system, the electro-optical properties of this system can differ greatly from those of a dilute solution.

This can be seen in Fig. 7.55, which shows the temperature dependence of the quantity $K = \Delta n/cE^2$ for solutions of the poly(methacryloylphenyl ester) of hexyloxybenzoic acid in benzene. The last one is a poor solvent for this polymer; it is precipitated from solution at 10–12°C at the concentrations used.

As can be seen from Fig. 7.55, at higher temperatures (25–50°C), EB is negative and sharply increases in absolute value with decreasing temperature (i.e., with decreasing thermodynamic strength of the solvent). In this case the value of K is virtually independent of concentration in accordance with the molecular nature of

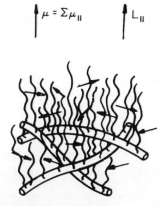

Fig. 7.56. Illustration explaining the value and sign of EB in an aggregated solution of the poly(methacryloylphenyl ester) of alkoxybenzoic acids.

the measured EB. The observed temperature dependence of K reflects the increase in intramolecular orientational order in comb-shaped molecules.

Essentially different phenomena occur in the solutions of the same polymer with a further decrease in temperature: the negative values of K decrease to zero and positive EB appears, exceeding by two orders of magnitude the maximum negative values of EB. Thermostating of the solution at 24°C for long periods of time also leads to these high positive K values. This high positive Kerr effect is of a supermolecular nature and is caused by the formation of molecular aggregates with an ordered internal structure. The higher the solution concentration, the greater is the intensity of the formation of this ordered structure, and at relatively high concentrations the solution acquires the structure of a lyotropic liquid crystal.

In a solution with an ordered structure the orientational interaction between the side chains of comb-shaped molecules occurs not only within a single molecule but also between the side groups of different molecules, forming aggregates (Fig. 7.56). As a result, the axis of the orientational order $L_{\parallel}$ inside the aggregate is determined by the direction of the chains of the side groups rather than by that of the main chains which in the aggregate may remain disordered. In this aggregate (Fig. 7.56) the normal components $\mu_{\perp}$ of dipoles of the side groups do not form an orientationally ordered system (unlike the nonaggregated molecules of comb-shaped polymer; see Fig. 7.54). In contrast, the longitudinal components $\mu_{\parallel}$ of dipoles of the side groups may add up and form the dipole $\mu = \Sigma\mu_{\parallel}$ (Fig. 7.56) of the aggregate along the axis $L_{\parallel}$ of its greatest polarizability (axis of orientational order). This leads to the positive supermolecular Kerr effect coinciding in sign with that in a solution of monomer molecules of the same polymer.

The electro-optical properties of the thermotropic mesophase generated in polymer bulk as a result of the uniaxial orientation of mesogenic side groups belonging to different comb-shaped molecules are determined by a similar molecular mechanism.

10.2.2.   Molecules with Mesogenic Groups in the Main Chain.
Thermotropic polymers with mesogenic groups in the main chain are synthesized
using the molecules of low-molecular-weight compounds capable of forming thermotropic liquid crystals. In the synthesized polymer chain these mesogenic groups
are separated by sequences of a chain molecule of a typical flexible-chain polymer.

Fragments of aromatic polyesters are often used as the mesogenic segments,
and polymethylene chains sequences are used as flexible "spacer" groups. Although the synthesis of these polymers was developed later [180] than that of
polymers with mesogenic side groups [181], many papers dealing with the investigation of their thermotropic phase have already been published [182–187].

However, the study of molecular characteristics of this type of polymers in
dilute solutions is only beginning. Thus, one of the published papers [188] was
concerned with the hydrodynamic and dynamo-optical properties of dilute solutions
of a series of samples and fractions of a mesogenic aromatic polyester (PE–10) with
the following structure:

and of the corresponding "monomer":

Hence, the repeat unit of PE–10 contains the mesogenic core consisting of
three phenyl rings and four ester groups and a polymethylene chain, $C_{10}H_{20}$, as a
flexible spacer.

10.2.2.1.   Data of hydrodynamic investigations and FB.   The
sedimentation-diffusion measurements and the viscometry of dilute solutions of a
series of PE–10 fractions in dichloroacetic acid (DCAA) have permitted the
determination of the equilibrium rigidity of the molecules of this polymer. The
length of the Kuhn segment obtained in this case, $A = 50 \cdot 10^{-8}$ cm, exceeds more
than three times the value of $A = 15 \cdot 10^{-8}$ cm for poly(ethylene terephthalate)
molecules in the same solvent [78].

The comparison of data on FB for PE–10 and poly(ethylene terephthalate)
PETPh) solutions in DCAA gives similar results. These data (Table 7.7) show that
for high-molecular-weight samples of PE–10 and PETPh the values of $\Delta n/\Delta \tau$ differ
by a factor of four. This corresponds to the fourfold difference in the rigidities of

these molecules if the equation $(\Delta n/\Delta\tau)_\infty = B\beta A$ (see Chapter 5) is applied and if it is taken into account that the values of β for the two polymers should be similar.

If the molecular structure of PETPh

$$-\mathrm{CH_2-CH_2-O-\underset{\underset{O}{\|}}{C}-\hspace{-2pt}\langle\!\!\bigcirc\!\!\rangle\hspace{-2pt}-\underset{\underset{O}{\|}}{C}-O-}$$

is compared with that of PE–10, it can be seen that the percentage (of the mass or the chain length) of the rigid component (ester–aromatic groups) in both polymers is approximately equal. However, the redistribution of chain units leading to the formation of longer rigid parts favors an increase in the average equilibrium rigidity of the chain. It should be borne in mind that, according to FB data, the equilibrium rigidity of the chain of a para-aromatic polyester (Table 6.9, No. 14) exceeds by more than one order of magnitude that of PE–10.

The equilibrium rigidity of PE–10 may also be determined from the dependence of $\Delta n/\Delta\tau$ on M according to Eq. (5.112) and the data in Table 7.7. The value of A obtained in this manner is similar to that obtained from hydrodynamic data. The same data make it possible to calculate the anisotropy per unit length of the PE–10 chain, which was found to be $\beta = 4\cdot10^{-17}$ cm^2. This value agrees with that of $\beta = 8.5\cdot10^{-17}$ cm^2 for a para-aromatic polyester (Table 6.9, No. 14) if it is taken into account that the anisotropic ester–aromatic moiety represents only half of the entire PE–10 chain. It follows from the structural formula of PE–10 that the molecular weight and length λ of the repeat unit are $M_0 = 544$ and $\lambda = 33\cdot10^{-8}$ cm. Correspondingly, the number of "monomer units" in a segment is $S = A/\lambda = 50/33 = 1.5$. Hence, the length of the Kuhn segment is greater only by a factor of 1.5 than that of the repeat unit of the chain.

10.2.2.2. Electro-optical properties. The investigations of the EB of PE–10 fractions and samples, and of the corresponding monomer, have been carried out using trifluoroacetic acid (TFAA) as solvent [189]. The conformational and optical characteristics of PE–10 molecules determined from the data of hydrodynamic investigations and FB in two solvents, DCAA and TFAA, are virtually identical.

Although the intrinsic EB of a polar solvent, TFAA, is high ($K = 36.5\cdot10^{-12}$ g^{-1}·cm^5 statvolt^{-2}), the Kerr effect introduced by the dissolved polymer was found to be sufficiently high for all its fractions to allow reliable measurements in a dilute solution (Fig. 7.57).

It can be seen from the linear dependences of Δn on E^2 shown in Fig. 7.57 that for all solutions the Kerr law is obeyed, but the sign of the effect is positive for a high-molecular-weight sample and negative for a low-molecular-weight sample.

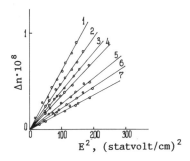

Fig. 7.57. Dependence of Δn on E^2 for solutions of two PM–10 samples in trifluoroacetic acid. Sample 6 in Table 7.7: $c \cdot 10^2 = 6.3$ (1), 3.1 (2), and 1.17 g/cm^3 (3). Sample 8 in Table 7.7: $c \cdot 10^2 = 3.63$ (5), 5.41 (6), and 6.37 g/cm^3 (7); 4) solvent.

TABLE 7.7. Shear Optical Coefficients $\Delta n/\Delta\tau$, Segmental Anisotropy $\alpha_1 - \alpha_2$, and Kerr Constants for Samples and Fractions of Mesogenic Aromatic Polyester PE–10 and Poly(ethylene terephthalate) (PETPh) in Trifluoroacetic Acid [188, 189]

No.	Sample	$M \cdot 10^{-3}$	$(\Delta n/\Delta\tau) \cdot 10^{10}$, cm $s^2 g^{-1}$	$(\alpha_1 - \alpha_2) \cdot 10^{25}$, cm^3	$K \cdot 10^{10}$, $g^{-1} cm^5$ statvolt^{-2}
1	PE-10	64	—	—	16.5
2	PE-10	42	—	—	15.2
3	PE-10	60	16.4	200	17.8
4	PE-10	68	—	—	16.5
5	PE-10	7.1	16	200	5.4
6	PE-10	5	—	—	3.5
7	PE-10	2.7	14.7	200	−0.9
8	PE-10	1.6	—	—	−2.6
9	Monomer PE-10	0.532	8.5	—	−3.2
10	PETPh	50	3.9	49	—

The Kerr constants K obtained for fractions and samples of different molecular weights (including the PE–10 monomer) are given in Table 7.7. At high M the values of K are positive, decrease with decreasing molecular weight, and become negative in the range of low M in which they coincide in sign with the Kerr constant for the monomer. Hence, the signs of EB and FB in PE–10 solutions coincide for high-molecular-weight samples but are opposite for low-molecular-weight fractions. This result in no way implies that the types of molecular motions (local and large-scale motions) of PE–10 in the electric and shear fields are different, as occurs for typical flexible-chain polymers (see Section 5.2.2 in this chapter). Actually, the possibility of opposite signs of EB and FB at low M, i.e., low x, is predicted by Eq. (7.55) for kinetically rigid molecules, the monomer unit of which has a dipole μ_0 forming a large angle ϑ with the chain direction.

Hence, the dependence of the value and sign of K on molecular weight may be discussed using the theory of wormlike chains [Eqs. (7.53)–(7.57)] in which the EB of the solution is assumed to be the result of large-scale orientation of kinetically rigid polar molecules in the electric field.

The points in Fig. 7.58 represent the experimental values of K (according to Table 7.7) as a function of the molecular weight M of PE–10 samples and fractions. Curves 1 and 2 show the theoretical dependences of K/K_∞ on x according to Eq. (7.57) at two similar values of the angle ϑ (see figure caption). The theoretical curves were plotted using the data of hydrodynamic investigations and FB, according to which we have for PE–10 $x = 2M/(M_0 S) = 2.45 \cdot 10^{-3}$ M and $\Delta a = \beta \lambda = 132 \cdot 10^{-25}$ cm^3. The limiting value of the Kerr constant is $K_\infty = 18 \cdot 10^{-10}$ g^{-1} cm^3 statvolt^{-2}.

The comparison of experimental points and theoretical curves gives the angle ϑ, the value of which was found to be close to $\vartheta = 75°$. This fact implies that the normal component of the dipole of the monomer unit $\mu_{0\perp}$ is greater by a factor of 3.7 than its longitudinal component $\mu_{0\parallel}$, which accounts for the negative value of K for the monomer and the low-molecular-weight fractions of PE–10.

However, although the longitudinal component, $\mu_{0\parallel}$, represents only a small part of the dipole moment of the monomer, its significance in the observed EB increases with molecular weight and completely determines the Kerr effect in the range of saturation of the dependence $K = K(M)$, clearly seen in Fig. 7.58. These experimental data again clearly show the fruitfulness of the concept [124] of the different roles played by the longitudinal and normal components of the monomer dipole of the chain molecule in the EB phenomenon of the polymer solution.

The high value of the angle ϑ obtained from the experimental data on EB is in agreement with the values that might be expected on the basis of the above stereochemical structure of the PE–10 chain. The absolute value of the dipole moment μ_0 of the monomer unit is too high ($\mu_0 \approx 13$ D) to be of quantitative importance. This result does not seem unexpected if we take into account the fact that EB measurements have been carried out in such a highly polar solvent as TPhAA. This point has already been emphasized (Section 9.3.2 and Table 7.5).

The above results show that although the PE–10 chains exhibit very moderate equilibrium rigidity ($A = 50 \cdot 10^{-8}$ cm), not only the hydrodynamic and dynamo-optical, but also the electro-optical, properties of the molecules of this polymer may be interpreted in terms of the theory of kinetically rigid wormlike chains.

This feature can be regarded as one of the manifestations of the specific structure and properties of the molecules of the polymer capable of forming the thermotropic mesophase.

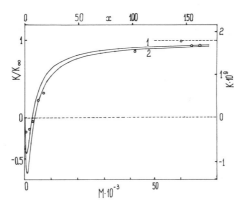

Fig. 7.58. Kerr constant K versus molecular weight K for fractions and samples of PE–10 mesogenic aromatic polyester in trifluoroacetic acid. Points represent experimental data. Curves represent the theoretical dependences of K/K_∞ on x according to Eq. (7.57), plotted at the values of $S = 1.5$, $\Delta a = 132 \cdot 10^{-25}$ cm^3, $K_\infty = 18 \cdot 10^{-10}$ g^{-1} cm^5 statvolt^{-2}, $\vartheta = 74°$ (1) and 76° (2) [189].

11. ELECTRODYNAMIC BIREFRINGENCE

The above data show the possibility and advantage of investigating rigid-chain polymers by combining FB and EB methods since the comparison of the results of these two methods provides complete information on the structural and electro-optical characteristics of the molecules under study.

When this combination of methods is used, it is very important to carry out FB and EB studies of the same polymer at the same concentration and temperature, with the same optical apparatus, etc. For this purpose the method of electrodynamic birefringence (EDB) offers some advantages. In this method the solution under study passes into the state of laminar flow and is simultaneously exposed to the electric field E parallel to the rate gradient g [190–194].

In all cases of practical interest the anisotropy of the solution resulting from the combined action of dynamic and electric fields is not high; the phase difference of two interfering polarized beams usually does not exceed $2\pi \cdot 10^{-2}$. Under these conditions the total birefringence Δn of the solution and the corresponding orientation angle φ_m can be calculated according to the equations [193]

$$\Delta n^2 = \Delta n_g^2 + \Delta n_E^2 - 2\Delta n_g \Delta n_E \cos 2\varphi_{mg}, \qquad (7.59)$$

$$\cot 2\varphi_m = \cot 2\varphi_{mg} - (\Delta n_E/\Delta n_g) \sin 2\varphi_{mg}. \qquad (7.60)$$

where Δn_g and φ_{mg} are the FB and the orientation angle of the optical axis in the absence of the electric field, respectively, and Δn_E is the EB in the absence of the flow.

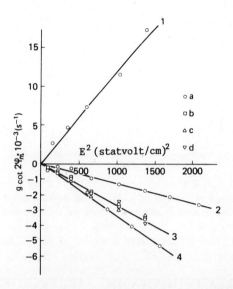

Fig. 7.59. Orientation of electrodynamic birefringence (EDB) of liquids versus applied electric field [193]. 1) Bromoform, $g = 3660$ s^{-1}; 2) α-methylnaphthalene, $g = 1380$ s^{-1}; 3) α-bromonaphthalene: a) $g = 1380$ s^{-1}, b) $g = 1180$ s^{-1}; c) $g = 1035$ s^{-1}; d) $g = 690$ s^{-1}; 4) mixture of α-bromonaphthalene and decalin, $g = 1180$ s^{-1}.

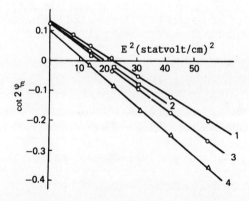

Fig. 7.60. EDB orientation versus E^2 for solutions of PBG ($M = 1.36 \cdot 10^5$) in chloroform [194]. c (in g cm$^{-3} \cdot 10^{-4}$): 1) 9.1; 2) 7.2; 3) 6.7; 4) 2.3; $g = 862$ s^{-1}.

Substitution of the expressions for Δn_g and φ_{mg} from the FB theory [195] of solutions of rigid-chain molecules at low values of g/D_r and for Δn_g from Eqs. (7.2) and (7.8) into Eqs. (7.59) and (7.60) gives

$$\frac{\Delta n}{g\eta_0 c} = \frac{2\pi N_A}{135 M \eta_0 kT} \frac{(n^2+2)^2}{n} b_0 W (\gamma_1 - \gamma_2) \left[1 + \left(\frac{z\mathscr{E}}{kT}\right)^4 \left(\frac{D_r}{b_0 g}\right)^2\right]^{1/2}, \qquad (7.61)$$

$$\cot 2\varphi_m = \frac{g}{D_r} - \left(\frac{z\mathscr{E}}{kT}\right)^2 \frac{D_r}{b_0 g}, \qquad (7.62)$$

with

$$z^2 = (\delta_1 - \delta_2)kT + (\mu^2/2)(3\cos^2\theta - 1),$$

where $\delta_1 - \delta_2$ is the difference between the two main dielectric polarizabilities of the molecule and $\mathscr{E} = [(\varepsilon + 2)/3]E$ is the field strength to which the molecule is exposed [for other symbols see Eqs. (5.2), (5.14), and (7.8)].

Equations (7.61) and (7.62) differ from Eqs. (5.14) and (5.2), respectively, by the presence of the term $(z\mathscr{E}/kT)^2 D_r/(b_0 g)$, which takes into account the changes in the value and orientation of birefringence caused by the effect of the electric field $\mathscr{E}$.

The validity of Eqs. (7.59)–(7.62) for both low-molecular-weight liquids [193] and rigid-chain polymer solutions [194] has been confirmed by appropriate experiments. Examples are given in Figs. 7.59 and 7.60 which reveal the dependence of $\cot 2\varphi_m$ on $E^2 = [3\mathscr{E}/(\varepsilon + 2)]^2 \cdot \mathscr{E}$ at constant g. In accordance with Eqs. (7.60) and (7.62), the points fit straight lines whose intercepts with the ordinate yield the value of $\cot 2\varphi_m$, and the slope can be used for the determination of z. The EDB method provides information on the molecular hydrodynamic, electric, and optical characteristics of the polymer investigated. In principle, these results coincide with those of separate investigations of FB and EB, but the fact that these data are obtained in a single experiment offers advantages.

12. SOME METHODOLOGICAL INFORMATION

The instrumentation used in the investigations of the Kerr effect in liquids or solutions has been described in detail in a number of monographs and papers [11, 30, 61, 62, 28, 197].

Hence, only some methodological approaches used in the determinations of the above experimental data will be described here.

If the Kerr effect is investigated in a solution exhibiting the properties of a good insulator (nonpolar solvent and nonelectrolytic polymer), both the steady-state and sinusoidal field may be used. In this case the optical component of the instrument does not differ from that used in the FB investigations described in Chapter 5, Section 4.2. The measurements may be carried out either by the visual method, by using a half-shadow Brace compensator (Fig. 5.28), or by the photoelectric method in which the half-shadow registering plate of the compensator is replaced with a harmonic modulator of the ellipticity of the polarized light (Fig. 5.28).

Under these conditions, for investigating the kinetics of the Kerr effect in a polymer solution it is convenient to use the sinusoidal electric field in the range of frequency dispersion of EB because the dispersion curves provide the most direct information on the spectrum of relaxation times and the rotational mobility of the macromolecules.

This procedure has usually been employed in the investigations of all the polymers considered above, which are readily soluble in nonpolar organic solvents (tetrachloromethane, dioxane, benzene, etc.).

However, in the investigation of solutions in polar solvent, as a result of their high electric conductivity, inhomogeneities appear in the sinusoidal electric field in the sonic and radiofrequency ranges (and even more inhomogeneities appear in the steady-state field). These inhomogeneities scatter and depolarize the light, masking the original Kerr effect. In these cases measurements are possible only if the field is used in the form of a relatively infrequent sequence of pulses of the steady-state or sinusoidal voltage.

Thus, in the study of EB in solutions of poly(amide hydrazide) in dimethyl sulfoxide [101], a rectangular pulsed field was used and the pulse duration was $3 \cdot 10^{-4}$ s at a frequency of 1–3 pulses per second. This procedure has made it possible to avoid the disturbance due to electric conductivity. In this case the kinetics of the Kerr effect were investigated by studying the rise and decay curves of EB when the field is switched on and off.

The measurement of the EB value in a pulsed field can be carried out with an elliptical compensator without the half-shadow device and the elliptical modulator. A thin mica plate (0.01 wavelength) is used as the compensator, and the intensity of light passing through the system is converted to the photocurrent with the aid of a photomultiplier and recorded with an oscillograph. The value and sign of the observed EB are determined from the angle and direction of rotation of the compensator required for the compensation of the optical pulse fed into the photomultiplier and recorded on the oscillograph screen [101].

The sensitivity of this compensation method used in pulsed fields may be greatly increased if a modulator of the ellipticity of the polarized light is additionally introduced into the optical system [196]. This modulator does not differ from the ellipticity modulator used in FB measurements (Chapter 5, Section 4.2.2) or in EB measurements in the sinusoidal field.

The principal scheme of the instrument used for the measurements of EB in pulsed fields and containing an elliptical compensator and an elliptical modulator is shown in Fig. 7.61.

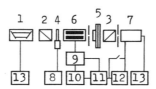

Fig. 7.61. Block diagram of an apparatus for EDB measurements by the compensation method in the pulsed electric field at steady-state and sinusoidal voltage. 1) Light source (laser); 2, 3) polarizer and analyzer; 4) elliptical modulator; 5) elliptical compensator; 6) Kerr .cell; 7) photomultiplier; 8) generator exciting modulator oscillations; 9) pulse former; 10) pulse delay generator; 11) oscillograph; 12) selective system; 13) power supply block [196].

In this scheme, as in all compensation methods, the value and sign of EB are determined by the rotation angle of the compensator and the direction of its rotation to the compensating position. The procedure of measurements in the investigation of the kinetics of the Kerr effect is as follows: the rise and decay curves of the EB when the field is switched on and off are obtained by measuring the values of birefringence at different moments of time.

The application of this procedure has made it possible to carry out a detailed investigation of the Kerr effect in conducting solutions of cellulose nitrates in cyclohexanone and to obtain data on the electro-optical properties of the molecules of these polymers [133].

The advantage of compensation methods over those based on the recording of the value of the light flux is evident. The attendant optical effects appearing in conducting solutions can induce light-flux modulations independent of the changes in the optical anisotropy of the medium, and thus can lead to errors in photoelectrical measurements. In compensation methods, with the application of elliptical modulators and reliable selective devices, the optical system is very sensitive to the changes in the anisotropy of the solution, but is insensitive to attendant light modulations.

REFERENCES

1. J. Kerr, Philos. Mag., (4), **50**, 337, 446 (1875).
2. P. Langevin, Radium, 7, 249 (1910); C. R. Acad. Sci. (Paris), **151**, 475 (1910).
3. M. Born, Ann. Phys., **55**, 177 (1918).
4. P. Debye, Polar Molecules, Dover, New York (1929).
5. P. Debye and H. Sack, Handbuch der Radiologie, Vol. 612, Leipzig (1934), p. 69.
6. J. W. Beams, Rev. Mod. Phys., **4**, 133 (1932).
7. H. A. Stuart, Hand- und Jahrbuch der chemischen Physik, Vol. 10, Part 3, A. Eucken and K. Wolf (eds.), Leipzig (1939).

8. H. A. Stuart, Die Struktur des freien Molekuls, Chapter 7, Springer-Verlag, Berlin (1952), p. 415.
9. J. R. Partington, An Advanced Treatise on Physical Chemistry, Vols. 4 and 5, Longmans and Green, London (1953, 1954).
10. C. G. Le Fevre and R. J. W. Le Fevre, Rev. Pure Appl. Chem., **25**, 261 (1955).
11. C. T. O'Konski (ed.), Molecular Electro-optics, Parts I and II, Marcel Dekker, New York (1976, 1978).
12. J. Kerr, Philos. Mag., (5), **9**, 157 (1880).
13. N. Tolstoi and V. Tsvetkov, Dokl. Akad. Nauk SSSR, **31**, 230 (1941).
14. V. N. Tsvetkov and V. A. Marinin, Zh. Eksp. Teor. Fiz., **18**, 641 (1948).
15. E. I. Rjumtsev and V. N. Tsvetkov, Opt. Spektrosk., **26**, 607 (1969).
16. E. I. Rjumtsev, T. A. Rotinjan, A. P. Kovshik, I. I. Daugvila, G. I. Denis, and V. N. Tsvetkov, Opt. Spektrosk., **41**, 65 (1976).
17. T. A. Rotinjan, H. K. Rout, A. P. Kovshik, and P. V. Adomenas, Kristallografiya, **23**, 578 (1978).
18. E. I. Rjumtsev, S. G. Polushin, A. P. Kovshik, T. A. Rotinjan, and V. N. Tsvetkov, Dokl. Akad. Nauk SSSR, **224**, 1343 (1979).
19. E. I. Rjumtsev, T. A. Rotinjan, and A. P. Kovshik, Dokl. Akad. Nauk SSSR, **257**, 1121; **260**, 77 (1981); **266**, 89 (1982).
20. P. Onsager, J. Am. Chem. Soc., **58**, 1986 (1936).
21. P. Langevin, J. Phys., **4**, 678 (1905).
22. R. Gans, Ann. Phys., **64**, 481 (1921).
23. W. Kuhn, N. Duhrkop, and H. Martin, Z. Phys. Chem., **B45**, 121 (1939).
24. A. Peterlin and H. A. Stuart, Z. Phys., **112**, 129 (1939).
25. C. T. O'Konski, K. Yoshioka, and W. H. Orttung, J. Phys. Chem., **63**, 1558 (1959).
26. M. J. Shah, J. Phys. Chem., **67**, 2215 (1963).
27. D. N. Holcomb and I. Tinoco, J. Phys. Chem., **67**, 2691 (1963).
28. K. Yoshioka and H. Watanabe, Physical Principles and Techniques of Protein Chemistry, Part A, S. J. Leach (ed.), Academic Press, New York (1969), p. 335.
29. C. T. O'Konski, "Theory of the Kerr constant," in: Molecular Electrooptics, S. Krause (ed.), Plenum Press, New York (1981), p. 119.
30. E. Fredericq and C. Houssier, Electric Dichroism and Electric Birefringence, Clarendon Press, Oxford (1973).
31. K. Kikuchi and K. Yoshioka, J. Phys. Chem., **77**, 2101 (1973); Biopolymers, **15**, 583 (1976).
32. M. Tricot and C. Houssier, Macromolecules, **15**, 854 (1982).
33. S. Krause and C. T. O'Konski, "Electric birefringence dynamics," in: Molecular Electro-optics, S. Krause (ed.), Plenum Press, New York (1981), p. 147.
34. H. Benoit, Ann. Phys., **6**, 561 (1951).
35. H. Benoit, C. R. Acad. Sci. (Paris), **228**, 1716; **229**, 30 (1949).
36. I. Tinoco and K. Yamaoka, J. Phys. Chem., **63**, 423 (1959).
37. M. Matsumoto, H. Watanabe, and K. Yoshioka, J. Phys. Chem., **74**, 2182 (1970).
38. A. Peterlin and H. A. Stuart, in: Hand- und Jahrbuch der chemischen Physik, A. Eucken and K. L. Wolf (eds.), Vol. 8, Part IB, Leipzig (1943), p. 1.

39. C. Clement and P. Bothorel, J. Chim. Phys., **61**, 1262 (1964).
40. M. F. Vuks, Light Scattering in Gases, Liquids, and Solutions [in Russian], Leningrad Univ. Press (1977).
41. V. Tsvetkov and V. Marinin, Dokl. Akad. Nauk SSSR, **62**, 67 (1948).
42. G. Khanarian and A. E. Tonelli, Macromolecules, **14**, 5031 (1981).
43. H. A. Stuart and A. Peterlin, J. Polym. Sci., **5**, 551 (1950).
44. H. A. Stuart (ed.), Das Makromolekül in Lösungen, Springer-Verlag, Berlin (1953).
45. M. V. Volkenstein, Configurational Statistics of Polymeric Chains, Wiley–Interscience, New York (1963).
46. D. Dous, J. Chem. Phys., **41**, 2656 (1964).
47. K. Nagai and I. Ishikawa, J. Chem. Phys., **43**, 4508 (1965).
48. P. J. Flory, Statistical Mechanics of Chain Molecules, Wiley–Interscience, New York (1969).
49. K. Nagai, J. Chem. Phys., **51**, 1091 (1972).
50. R. L. Jernigan and D. S. Thompson, "Flexible polymers," in: Molecular Electro-optics, Part I, C. T. O'Konski (ed.), Marcel Dekker, New York (1976).
51. R. L. Jerngan and S. Miyazawa, "Kerr effects of flexible macromolecules," in: Molecular Electro-optics, S. Krause (ed.), Plenum Press, New York (1981).
52. V. N. Tsvetkov, in: Newer Methods of Polymer Characterization, B. Ke (ed.), Wiley–Interscience, New York (1964), p. 563.
53. L. Silberstein, Philos Mag. (6), **33**, 92, 215, 521 (1917).
54. R. L. Rowell and R. S. Stein, J. Chem. Phys., **47**, 2985 (1967).
55. E. M. Mortensen, J. Chem. Phys., **49**, 3732 (1968).
56. G. D. Patterson and P. J. Flory, J. Chem. Soc., Faraday Trans. II, **68**, 1098 (1972).
57. R. T. Ingwall and P. J. Flory, Biopolymers, **11**, 1527 (1972).
58. V. A. Marinin, L. V. Polyakova, and Z. S. Korolkova, Vestn. Leningr. Gos. Univ., Ser. Fiz. Khim., **16**, 73 (1958).
59. J. V. Champion, G. H. Meeten, and G. W. Southwell, Polymer, **17**, 651 (1976).
60. C. G. Le Fevre, R. J. W. Le Fevre, and G. M. Parkins, J. Chem. Soc., 1468 (1968).
61. C. G. Le Fevre and R. J. Le Fevre, in: Techniques of Organic Chemistry, Vol. 1, A. Weissberger (ed.), Wiley–Interscience, New York (1960).
62. R. J. W. Le Fevre, in: Advances in Physical Organic Chemistry, Vol. 3, V. Gold (ed.), Academic Press, London (1965).
63. K. M. Kelly, G. D. Patterson, and A. E. Tonelli, Macromolecules, **10**, 859 (1977).
64. G. Khanarian and A. E. Tonelli, Macromolecules, **15**, 145 (1982).
65. R. J. W. Le Fevre and K. M. S. Sundaram, J. Chem. Soc., 1494 (1962).
66. R. J. W. Le Fevre and K. M. Sundaram, J. Chem. Soc., 4003 (1962).
67. P. Doty, H. Wagner, and S. Singer, J. Phys. Colloid Chem., **51**, 32 (1947).
68. J. Hengstenberg and E. Schuch, Makromol. Chem., **75**, 55 (1964).
69. A. Grugnola and F. Danusso, J. Polym. Sci., **B6**, 535 (1968).
70. G. Khanarian, F. C. Schilling, R. E. Cais, and A. E. Tonelli, Macromolecules, **16**, 287 (1983).

71. V. N. Tsvetkov and N. N. Boitsova, Vysokomol. Soedin., **2**, 1176 (1960).
72. M. J. Aroney, R. J. W. Le Fevre, and G. M. Parkis, J. Chem. Soc., 2890 (1960).
73. G. Khanarian, R. E. Cais, J. M. Kometani, and A. E. Tonelli, Macro-molecules, **15**, 866 (1982).
74. E. V. Frisman, A. M. Martsinovski, and N. A. Domnicheva, Vysokomol. Soedin., **2**, 1148 (1960).
75. W. Zvetkov and E. Frisman, Acta Physicochim. URSS, **21**, 978 (1946).
76. S. P. Mitsengendler, K. I. Sokolova, G. A. Andreeva, A. A. Korotkov, T. Kadyrov, S. I. Klenin, and S. Ya. Magarik, Vysokomol. Soedin., **A9**, 1133 (1967).
77. V. N. Tsvetkov, T. I. Garmonova, and R. P. Stankevich, Vysokomol. Soedin., **8**, 980 (1966).
78. M. L. Wallach, Makromol. Chem., **103**, 19 (1967).
79. V. N. Tsvetkov, D. Hardy, I. N. Shtennikova, E. V. Korneeva, G. F. Pirogova, and K. Nitrai, Vysokomol. Soedin., **A11**, 349 (1969).
80. V. N. Tsvetkov, M. G. Vitovskaya, and S. Ya. Lyubina, Vysokomol. Soedin., **4**, 577 (1962).
81. I. N. Shtennikova, G. F. Kolbina, E. V. Korneeva, M. Kovach, and T. I. Smirnova, Vysokomol. Soedin., **A17**, 404 (1975).
82. V. N. Tsvetkov and I. N. Shtennikova, Zh. Tekh. Fiz., **29**, 885 (1959).
83. V. N. Tsvetkov, L. N. Andreeva, E. V. Korneeva, and P. N. Lavrenko, Dokl. Akad. Nauk SSSR, **205**, 895 (1972).
84. H. Nakayama and K. Yoshioka, J. Polym. Sci., **A3**, 813 (1965).
85. K. Yoshioka and C. T. O'Konski, J. Polym. Sci., **A26**, 421 (1968).
86. V. N. Tsvetkov, S. Ya. Ljubina, V. E. Bychkova, and I. A. Strelina, Vysokomol. Soedin., **8**, 846 (1966).
87. M. Tricot, C. Houssier, V. Desreux, Eur. Polym. J., **12**, 575 (1976); **14**, 307 (1978).
88. A. R. Foweraker and B. R. Jennings, Polymer, **16**, 720 (1975).
89. M. Tricot, C. Houssier, and V. Desreux, Biophys. Chem., **8**, 221 (1978).
90. V. N. Tsvetkov, Yu. V. Mitin, V. R. Glushenkova, A. E. Grischenko, N. N. Boitzova, and S. Ya. Ljubina, Vysokomol. Soedin., **5**, 453 (1963).
91. H. Watanabe, K. Yoshioka, and A. Wada, Biopolymers, **2**, 91 (1964).
92. V. N. Tsvetkov, I. N. Shtennikova, E. I. Rjumtsev, and V. S. Skazka, Vysokomol. Soedin., **7**, 1111 (1965).
93. V. N. Tsvetkov, E. I. Rjumtsev, I. N. Shtennikova, and G. I. Okrhimenko, Vysokomol. Soedin., **8**, 1466 (1966).
94. V. N. Tsvetkov, I. N. Shtennikova, E. I. Rjumtsev, and G. F. Pirogova, Vysokomol. Soedin., **A9**, 1575, 1583 (1967).
95. V. N. Tsvetkov, I. N. Shtennikova, V. S. Skazka, and E. I. Rjumtsev, J. Polym. Sci., **C16**, 3205 (1968).
96. H. Ohe, H. Watanabe, and K. Yoshioka, Colloid Polym. Sci., **252**, 26 (1974).
97. M. Nishioka, K. Kikuchi, and K. Yoshioka, Polymer, **16**, 791 (1975).
98. E. I. Rjumtsev, L. N. Andreeva, N. V. Pogodina, E. U. Urinov, L. I. Kutzenko, and P. N. Lavrenko, Vysokomol. Soedin., **A17**, 61 (1975).
99. N. V. Pogodina, Yu. B. Tarabukina, L. V. Starchenko, G. N. Marchenko, and V. N. Tsvetkov, Vysokomol. Soedin., **A22**, 2219 (1980).

100. V. N. Tsvetkov, I. P. Kolomiets, A. V. Lezov, and G. N. Marchenko, Dokl. Akad. Nauk SSSR, **265**, 1202 (1982).
101. V. N. Tsvetkov, I. P. Kolomiets, and A. V. Lezov, Eur. Polym. J., **18**, 372 (1982).
102. V. N. Tsvetkov, I. N. Shtennikova, E. I. Rjumtsev, L. N. Andreeva, Y. P. Getmanchuk, Y. L. Spirin, and R. I. Dryagileva, Vysokomol. Soedin., **A10**, 2132 (1968).
103. V. N. Tsvetkov, Vysokomol. Soedin., **A16**, 944 (1974).
104. V. N. Tsvetkov, E. I. Rjumtsev, N. V. Pogodina, and I. N. Shtennikova, Eur. Polym. J., **11**, 37 (1975).
105. E. I. Rjumtsev, N. V. Pogodina, and Y. P. Getmanchuk, Vysokomol. Soedin., **A17**, 1719 (1975).
106. E. I. Rjumtsev, F. M. Aliev, and V. N. Tsvetkov, Vysokomol. Soedin., **A17**, 1712 (1975).
107. V. N. Tsvetkov, E. I. Rjumtsev, L. N. Andreeva, N. V. Pogodina, P. N. Lavrenko, and L. I. Kutsenko, Eur. Polym. J., **10**, 563 (1974).
108. P. N. Lavrenko, E. I. Rjumtsev, I. N. Shtennikova, L. N. Andreeva, N. V. Pogodina, and V. N. Tsvetkov, J. Polym. Sci. Symp., **44**, 217 (1974).
109. N. V. Pogodina, K. S. Pojivilko, H. Dautzenberg, K. I. Linow, B. Philipp, E. I. Rjumtsev, and V. N. Tsvetkov, Vysokomol. Soedin., **B19**, 851 (1976).
110. V. N. Tsvetkov, E. I. Rjumtsev, I. N. Shtennikova, T. V. Peker, and N. V. Tsvetkova, Dokl. Akad. Nauk SSSR, **207**, 1173 (1972); Eur. Polym. J., **9**, 1 (1973).
111. N. V. Pogodina, K. S. Pojivilko, N. P. Evlampieva, A. B. Melnikov, S. V. Bushin, S. A. Didenko, G. N. Marchenko, and V. N. Tsvetkov, Vysokomol. Soedin., **A23**, 1252 (1981).
112. E. I. Rjumtsev, I. N. Shtennikova, N. V. Pogodina, and T. V. Peker, Vysokomol. Soedin., **A18**, 743 (1976).
113. V. N. Tsvetkov, Vysokomol. Soedin., **A25**, 1571 (1983).
114. E. I. Rjumtsev, Thesis, Leningrad University (1975).
115. N. V. Pogodina, Thesis, Leningrad University (1975).
116. E. I. Rjumtsev, L. N. Andreeva, F. M. Aliev, L. I. Kutzenko, and V. N. Tsvetkov, Vysokomol. Soedin., **A17**, 1368 (1975).
117. E. I. Rjumtsev, F. M. Aliev, M. G. Vitovskaya, E. U. Urinov, and V. N. Tsvetkov, Vysokomol. Soedin., **A17**, 1368 (1975).
118. I. P. Kolomiets, A. V. Lezov, L. N. Andreeva, and V. N. Tsvetkov, Vysokomol. Soedin., **A26**, (1984).
119. V. N. Tsvetkov, I. N. Shtennikova, E. I. Rjumtsev, and Y. P. Getmanchuk, Eur. Polym. J., **7**, 767 (1971).
120. A. M. North, Chem. Soc. Rev., **1**, 49 (1972).
121. R. S. Wilkinson and G. B. Thurston, Biopolymers, **15**, 1555 (1976).
122. B. Zimm, J. Chem. Phys., **24**, 269 (1956).
123. W. R. Krigbaum and J. V. Dawkins, in: Polymer Handbook, J. Branrup and E. H. Immergut (eds.), Wiley–Interscience, New York (1975).
124. K. Kuhn, Helv. Chim. Acta, **31**, 1092 (1948).
125. V. N. Tsvetkov, Dokl. Akad. Nauk SSSR, **205**, 328 (1972).
126. N. V. Pogodina and V. N. Tsvetkov, Vysokomol. Soedin., **A24**, 2275 (1982).

127. I. N. Shtennikova, V. N. Tsvetkov, T. V. Peker, E. I. Rjumtsev, and Y. P. Getmanchuk, Vysokomol. Soedin., **A16**, 1086 (1974).
128. I. N. Shtennikova, T. V. Peker, Y. P. Getmanchuk, and V. A. Kudrenko, Vysokomol. Soedin., **A20**, 1246 (1978).
129. V. N. Tsvetkov, Makromol. Chem., **160**, 1 (1972).
130. V. N. Tsvetkov, K. A. Andrianov, E. I. Rjumtsev, I. N. Shtennikova, N. V. Pogodina, G. F. Kolbina, and N. N. Makarova, Vysokomol. Soedin., **A17**, 2493 (1975).
131. N. V. Pogodina, K. S. Pojivilko, A. B. Melnikov, S. A. Didenko, G. N. Marchenko, and V. N. Tsvetkov, Vysokomol. Soedin., **A23**, 2454 (1981).
132. N. V. Pogodina, P. V. Lavrenko, K. S. Pojivilko, A. B. Melnikov, T. A. Kolobova, G. N. Marchenko, and V. N. Tsvetkov, Vysokomol. Soedin., **A24**, 332 (1982).
133. I. P. Kolomiets, A. V. Lezov, G. N. Marchenko, and V. N. Tsvetkov, Vysokomol. Soedin., **A27**, 2415 (1985).
134. V. N. Tsvetkov, P. N. Lavrenko, L. N. Andreeva, A. I. Mashoshin, O. V. Okatova, O. I. Mikrjukova, and L. I. Kutzenko, Eur. Polym. J., **20**, 823 (1984).
135. V. N. Tsvetkov, N. V. Pogodina, and L. V. Starchenko, Eur. Polym. J., **19**, 837 (1983).
136. V. N. Tsvetkov, E. I. Rjumtsev, I. N. Shtennikova, I. I. Konstantinov, Y. B. Amerik, and B. A. Krentsel, Vysokomol. Soedin., **A15**, 2270 (1973).
137. V. N. Tsvetkov, E. I. Rjumtsev, F. M. Aliev, and I. N. Shtennikova, Eur. Polym. J., **10**, 55 (1974).
138. L. Pauling, R. V. Corey, and H. R. Branson, Proc. Natl. Acad. Sci. USA, **37**, 205, 235, 241 (1951).
139. V. I. Minkin, D. A. Osipov, and Yu. A. Zhdanov, Dipole Moments in Organic Chemistry, Plenum Press, New York (1970).
140. W. H. Stockmayer and M. E. Baur, J. Am. Chem. Soc., **86**, 3485 (1964).
141. W. H. Stockmayer, Pure Appl. Chem., **15**, 539 (1967).
142. A. M. Narth and P. J. Phillips, Trans. Faraday Soc., **64**, 3235 (1968).
143. W. H. Stockmayer and K. Matsuo, Macromolecules, **5**, 766 (1972).
144. M. E. Baur and W. H. Stockmayer, J. Chem. Phys., **43**, 4319 (1965).
145. C. Robinson, Trans. Faraday Soc., **52**, 571 (1956).
146. E. Iizuka, Adv. Polym. Sci., **20**, 79 (1976).
147. E. T. Samulsi, in: Liquid Crystalline Order in Polymers, A. Blumstein (ed.), Chapter 5, Academic Press, New York (1978).
148. J. Preston, *ibid.*, Chapter 4.
149. P. W. Morgan, Macromolecules, **10**, 1381 (1977).
150. Polymer Preprints, Am. Chem. Soc. Div. Polym. Chem., **17**, No. 1 (1976).
151. J. Polym. Sci., Polym. Symp., **65** (1978).
152. Macromolecules, **14**, No. 4, 900–953 (1981).
153. S. M. Aharoni, Macromolecules, **12**, 95 (1979).
154. S. M. Aharoni, Polymer, **21**, 21 (1980).
155. S. M. Aharoni, J. Polym. Sci., Polym. Phys. Ed., **18**, 1303, 1439 (1980); **19**, 281 (1981).
156. J. K. Moscicki, G. Williams, and S. M. Aharoni, Polymer, **22**, 1361 (1981).

157. R. S. Werbowyj and D. G. Gray, Mol. Cryst. Liq. Cryst., **34**, 1 (1976); Macromolecules, **13**, 69 (1980).
158. S. M. Aharoni, Mol. Cryst. Liq. Cryst., **56**, 237 (1980).
159. P. Navard, J. M. Haudin, S. Dayan, and P. Sixou, Polym. Lett., **19**, 379 (1981).
160. S. N. Bhadani and D. G. Gray, Makromol. Chem. Rapid Commun., **3**, 449 (1982).
161. P. Zugenmaier and U. Vogt, Makromol. Chem., **184**, 1749 (1983).
162. D. G. Gray, J. Appl. Polym. Sci. (Cellulose Symposia) (1983).
163. S. N. Bhadani, So-Lang Tseng, and D. G. Gray, Makromol. Chem., **184**, 1727 (1983).
164. V. N. Tsvetkov, K. A. Andrianov, M. G. Vitovskaya, and E. P. Astapenko, Vysokomol. Soedin., **A14**, 2603 (1972).
165. S. M. Aharoni, J. Macromol. Sci. Phys., **21**, 105 (1982).
166. L. Onsager, Ann. N. Y. Acad. Sci., **51**, 629 (1949).
167. P. J. Flory, Proc. R. Soc. London, **234**, 73 (1956).
168. V. N. Tsvetkov, E. I. Rjumtsev, and I. N. Shtennikova, in: Liquid Crystalline Order in Polymers, A. Blumstein (ed.), Chapter 2, Academic Press, New York (1978).
169. P. J. Flory, Macromolecules, **11**, 1119 (1978).
170. G. Ronca and D. Y. Yoon, J. Chem. Phys., **76**, 3295 (1982).
171. R. R. Matheson and P. J. Flory, Macromolecules, **14**, 954 (1981).
172. A. R. Khokhlov and A. N. Semenov, Physica, **108A**, 546 (1981); **112A**, 605 (1982).
173. S. V. Bushin, Ye. V. Korneeva, I. I. Konstantinov, Y. B. Amerik, S. A. Didenko, I. N. Shtennikova, and V. N. Tsvetkov, Vysokomol. Soedin., **A24**, 1469 (1982).
174. V. N. Tsvetkov, E. I. Rjumtsev, I. N. Shtennikova, E. V. Korneeva, G. I. Okhrimenko, N. A. Mikhailova, A. A. Baturin, Y. B. Amerik, and B. A. Krenzel, Vysokomol. Soedin., **A15**, 2570 (1973).
175. E. I. Rjumtsev, I. N. Shtennikova, N. V. Pogodina, G. F. Kolbina, I. I. Konstantinov, and Y. B. Amerik, Vysokomol. Soedin., **A18**, 439 (1976).
176. V. N. Tsvetkov, I. N. Shtennikova, E. I. Rjumtsev, N. V. Pogodina, G. F. Kolbina, E. V. Korneeva, P. N. Lavrenko, O. V. Okatova, Y. B. Amerik, and A. A. Baturin, Vysokomol. Soedin., **A18**, 2016 (1976).
177. G. F. Kolbina, Thesis, Institute for Macromolecular Compounds, Academy of Sciences of the USSR, Leningrad (1981).
178. V. N. Tsvetkov, E. I. Rjumtsev, I. I. Konstantinov, Y. B. Amerik, and B. A. Krenzel, Vysokomol. Soedin., **A14**, 67 (1972).
179. V. N. Tsvetkov, I. N. Shtennikova, E. I. Rjumtsev, G. F. Kolbina, I. I. Konstantinov, Y. B. Amerik, and B. A. Krentsel, Vysokomol. Soedin., **A11**, 2528 (1969).
180. A. Roviello and A. Sirigu, J. Polym. Sci., Polym. Lett. Ed., **13**, 455 (1975).
181. A. Blumstein (ed.), Liquid Crystalline Order in Polymers, Academic Press, New York (1978).
182. A. C. Griffin and S. J. Havens, J. Polym. Sci., Polym. Phys. Ed., **19**, 951 (1981).
183. G. Galli, E. Chiellini, C. K. Ober, and R. W. Lenz, Makromol. Chem., **183**, 2693 (1982).

184. A. Blumstein, K. N. Sivaramakrishnan, R. B. Blumstein, and S. B. Clough, Polymer, **23**, 47 (1982).
185. V. Frosini and A. Marchetti, Makromol. Chem. Rapid Commun., **3**, 795 (1982).
186. C. K. Ober, J. Jin, and R. W. Lenz, Makromol. Chem. Rapid Commun., **4**, 49 (1983).
187. Polymer Preprints, Am. Chem. Soc., **24**, No. 2 (1983), pp. 245–311.
188. V. N. Tsvetkov, L. N. Andreeva, P. N. Lavrenko, E. V. Beliaeva, O. V. Okatova, A. Y. Bilibin, and S. S. Skorokhodov, Eur. Polym. J., **20**, 817 (1984).
189. V. N. Tsvetkov, I. P. Kolomiets, A. V. Lezov, L. N. Andreeva, A. Y. Bilibin, and S. S. Skorokhodov, Dokl. Akad. Nauk SSSR, **282**, 147 (1984).
190. Y. Mukohata, S. Ikeda, and T. Isemura, J. Mol. Biol., **5**, 570 (1962).
191. S. Ikeda, J. Chem. Phys., **38**, 1062 (1963).
192. Y. Mukohata, J. Mol. Biol., **7**, 442 (1963).
193. V. N. Tsvetkov and E. L. Vinogradov, Opt. Spektrosk., **21**, 603 (1966).
194. V. N. Tsvetkov and E. L. Vinogradov, Vysokomol. Soedin., **8**, 662 (1966).
195. A. Peterlin and H. A. Stuart, Z. Phys., **112**, 1 (1939).
196. V. N. Tsvetkov, I. P. Kolomiets, A. V. Lesov, and A. S. Stepchenkov, Vysokomol. Soedin., **A25**, 1327 (1983).
197. C. Houssier and C. T. O'Konski, in: Molecular Electro-optics, S. Krause (ed.), Plenum Press, New York (1981), p. 309.